EFFLUENTS FROM LIVESTOCK

Proceedings of a Seminar to discuss work carried out within the EEC under the programme Effluents from Intensive Livestock, organised by Professor Dr H. Vetter, and held at Bad Zwischenahn, 2–5 October, 1979.

The Seminar was sponsored by the Commission of the European Communities, Directorate-General for Agriculture, Co-ordination of Agricultural Research.

WITHDRAWN
S.F.B.I.U.
LIBRARY

EFFLUENTS FROM LIVESTOCK

Edited by

J. K. R. GASSER

Agricultural Research Council, London, UK

assisted by

J. C. Hawkins, J. R. O'Callaghan, B. F. Pain

APPLIED SCIENCE PUBLISHERS LTD

LONDON

APPLIED SCIENCE PUBLISHERS LTD
RIPPLE ROAD, BARKING, ESSEX, ENGLAND

British Library Cataloguing in Publication Data

Effluents from livestock.
1. Animal waste—Congresses
2. Farm manure—Congresses
I. Gasser, John Kenneth Russell
628′.7 TD930

ISBN 0-85334-895-2

WITH 206 TABLES AND 127 ILLUSTRATIONS

Publication arrangements by: Commission of the European Communities, Directorate-General for Scientific and Technical Information and Information Management, Luxembourg

EUR 6633 EN

Printed in Great Britain by Galliard (Printers) Ltd, Great Yarmouth

TABLE OF CONTENTS

Page

Preface 1

Opening Remarks 3

SESSION I: SLURRY CHARACTERISATION AND ITS USE FOR ARABLE CROPS

THE INFLUENCE OF THE CONTENT OF TRACE ELEMENTS IN THE FEED ON THE COMPOSITION OF LIQUID MANURE OF PIGS
R. Priem and A. Maton 9

DISCUSSION 21

EFFECTS OF LANDSPREADING OF LARGE AMOUNTS OF LIVESTOCK EXCRETA ON CROP YIELD AND CROP AND WATER QUALITY
K.W. Smilde 23

DISCUSSION 30

LANDSPREADING OF LIQUID PIG MANURE:
I. EFFECTS ON YIELD AND QUALITY OF CROPS
C. Duthion 32

LANDSPREADING OF LIQUID PIG MANURE:
II. NUTRIENT BALANCES AND EFFECTS ON DRAINAGE WATER
C. Duthion, G. Catroux and J.C. Germon .. 59

LANDSPREADING OF LIQUID PIG MANURE:
III.1. SURVEY OF THE PIG FARMS IN THE 'BRESSE'
J.C. Germon, C. Duthion, Y. Couton, R. Grosman, L. Guenot and J. Mortier 80

LANDSPREADING OF PIG MANURE:
III.2. PIG MANURE COMPOSITION - CORRELATIONS BETWEEN COMPONENTS
C. Duthion and J.C. Germon 96

LANDSPREADING OF PIG MANURES:
IV. EFFECT ON SOIL NEMATODES
G. de Guiran, L. Bonnel and Mona Abirached .. 109

LANDSPREADING OF LIQUID PIG MANURE:
V. EFFECT OF SOIL MICROFLORA
G. Delpui 120

DISCUSSION 138

THE INFLUENCE OF AGRONOMIC APPLICATION OF SLURRY ON THE YIELD AND COMPOSITION OF ARABLE CROPS AND GRASSLAND AND ON CHANGES IN SOIL PROPERTIES
R. Lecomte 139

DISCUSSION 181

EXPERIMENTS ON HEAVY APPLICATIONS OF ANIMAL MANURE TO LAND
A. Dam Kofoed and O. Nemming 184

DISCUSSION 218

INFLUENCE OF DIFFERENT SLURRY DRESSINGS ON THE YIELD AND QUALITY OF PLANTS, AND THE NUTRIENT CONTENTS OF THE SHALLOW GROUNDWATER AND OF THE SOIL
H. Vetter and G. Steffens 219

SPREADING OF PIG AND CATTLE SLURRIES ON ARABLE LAND: LYSIMETER AND FIELD EXPERIMENTS
P. Spallacci and V. Boschi 241

DISCUSSION 275

SUMMARY OF SESSION I
H. Vetter 276

GENERAL DISCUSSION 284

SESSION II: THE USE OF SLURRY FOR GRASSLAND AND FORAGE CROPS

COMPARISON OF THE LEACHING PATTERNS OF NUTRIENT ELEMENTS FROM MINERAL FERTILISERS AND LIQUID MANURE
F. Van de Maele and A. Cottenie 287

DISCUSSION 298

EFFLUENTS FROM INTENSIVE LIVESTOCK UNITS FERTILISER EQUIVALENTS OF CATTLE SLURRY FOR GRASS AND FORAGE MAIZE
B.F. Pain and Lesley T. Sanders 300

DISCUSSION 312

EFFECTS OF LANDSPREAD ANIMAL MANURES ON THE FAUNA OF GRASSLAND
J.P. Curry, D.C.F. Cotton, T. Bolger and V. O'Brien 314

DISCUSSION 326

EFFECTS OF CATTLE SLURRY, PIG SLURRY AND FERTILISER ON YIELD AND QUALITY OF GRASS SILAGE
H. Tunney, S. Molloy and F. Codd .. 327

THE USE OF ANIMAL MANURES ON PASTURE FOR GRAZING
D.P. Collins 344

DISCUSSION 360

TIME AND RATE OF APPLICATION OF ANIMAL MANURES
P.V. Kiely 361

DISCUSSION 378

THE EFFECTS OF LANDSPREADING OF ANIMAL MANURES ON WATER QUALITY
Marie T. Sherwood 379

DISCUSSION 391

SUMMARY OF SESSION II
H. Tunney 393

GENERAL DISCUSSION 396

SESSION III: PROBLEMS OF COPPER IN SLURRY; MODELLING; TREATMENT OF SLURRY

REAL OR POTENTIAL RISK OF POLLUTION OF SOILS, CROPS, SURFACE AND GROUNDWATER DUE TO LANDSPREADING OF LIQUID MANURE
K. Meeus-Verdinne, G. Neirinckx, X. Monseur and R. de Borger 399

DISCUSSION 409

TOXICITY OF COPPER
Th. M. Lexmond and F.A.M. de Haan .. 410

DISCUSSION 419

HAZARDS ARISING FROM APPLICATION TO GRASSLAND OF COPPER-RICH PIG FAECAL SLURRY
D. McGrath, D.B.R. Poole and G.A. Fleming .. 420

DISCUSSION 432

SIMULATION OF ENVIRONMENTAL POLLUTION BY LANDSPREADING OF MANURE
H. Laudelout and R. Lambert 433

THE USE OF MATHEMATICAL MODELS IN STUDYING THE LANDSPREADING OF ANIMAL MANURES
P.D. Herlihy 446

A MODEL FOR THE SIMULATION OF THE FATE OF NITROGEN IN FARM WASTES ON LAND APPLICATION
K.K.S. Bhat, T.H. Flowers and J.R. O'Callaghan 459

DISCUSSION 478

TRANSFER OF OXYGEN IN PIGGERY WASTES
A. Heduit 481

BALANCE AND EVOLUTION OF NITROGEN COMPOUNDS DURING THE TREATMENT OF SLURRY
C. Besnard 496

DISCUSSION 507

SUMMARY OF SESSION III
J.K.R. Gasser 508

GENERAL DISCUSSION 510

SESSION IV: ODOUR CHARACTERISATION AND MEASUREMENT

DEVELOPMENT OF INSTRUMENTAL METHODS FOR MEASURING ODOUR LEVELS IN INTENSIVE LIVESTOCK BUILDINGS
J. Schaefer 513

DISCUSSION 535

TENTATIVE MEASUREMENTS OF ANIMAL MANURE ODOUR
J.L. Roustan, A. Aumaitre and C. Bernard .. 536

DISCUSSION 560

ODOUR CHARACTERISATION IN ANIMAL HOUSES BY GAS CHROMATOGRAPHIC ANALYSIS ON THE BASIS OF LOW TEMPERATURE SORPTION
J. Hartung and H.G. Hilliger 561

DISCUSSION 579

INVESTIGATIONS INTO THE TREATMENT OF ANIMAL EXCREMENTS BY AERATION TO REDUCE SMELL, AID DISINFECTION AND REDUCE VOLUME
R. Vetter and W. Rüprich 580

DISCUSSION 608

MEASUREMENT OF ODOUR EMISSIONS AND IMISSIONS
H.H. Kowalewsky, R. Scheu and H. Vetter .. 609

DISCUSSION 626

ODOUR MEASUREMENT RESEARCH IN INTENSIVE LIVESTOCK PRODUCTION UNITS
A.A. Jongebreur and J.V. Klarenbeek .. 627

DISCUSSION 632

SUMMARY OF SESSION IV
J.H. Voorburg 633

SESSION V: TRANSMISSION OF DISEASES AND PARASITES THROUGH SLURRY

A STUDY OF THE EFFECTS OF ANIMAL EFFLUENTS UTILISATION FOR GRASSLAND PRODUCTION ON THE LEVELS OF CERTAIN PATHOGENIC BACTERIA IN FARM ANIMALS AND THEIR CARCASES
W.R. Kelly 637

THE POSSIBLE ROLE OF ANIMAL MANURES IN THE DISSEMINATION OF LIVESTOCK PARASITES
N.E. Downey and J.F. Moore 653

DISCUSSION 672

SUMMARY OF SESSION V
W.R. Kelly 673

GENERAL DISCUSSION 674

SESSION VI: REVIEW OF THE 'EFFLUENTS FROM INTENSIVE LIVESTOCK' PROGRAMME

1975-1978: RESEARCH ON EFFLUENTS FROM INTENSIVE LIVESTOCK IN THE EUROPEAN COMMUNITIES

P. L'Hermite and J. Dehandtschutter .. 677

DISCUSSION 703

Closing Remarks 705

List of Participants 707

PREFACE

The initial programme on Effluents from Intensive Livestock funded by the Commission of the European Communities placed much emphasis on the utilisation of slurry by spreading on agricultural land. This topic received the major part of money available for Common Action (projects partly funded by CEC) and was also the subject of a Seminar at Modena in 1976. The objective measurement of odour has proved an intractable problem in spite of its great importance and the effort devoted to it, including part of the present programme. A Seminar was held at Gent in 1976 to discuss odour characterisation and control. The handling and treatment of slurry was considered to be a topic more appropriate for each country to undertake so that it could match the experimental work to its own particular needs. Engineering, however, has to provide the necessary equipment for controlling odours, storing slurry and spreading it on the land. The considerable amount of work being done was reviewed and discussed during the Seminar at Cambridge in 1978.

For the farmer to be able to use slurry to his best advantage and with minimum risk to the environment, he requires a framework within which to fit his particular conditions. This can be provided by fairly simple models. More complex models are needed by investigators in order to follow the processes involved in slurry use and indicate important areas for further work. Both aspects have received some attention in the Common programme and were discussed at a Seminar at Chimay in 1978.

Finally, the subject matter of Effluents from Intensive Livestock included aspects of human and livestock health hazards associated with the utilisation of animal effluents. This topic was the subject of a Workshop in Dublin in 1977.

These proceedings record the work done and progress made in the areas funded by the CEC following the advice of the Expert Working Group on Effluents from Intensive Livestock. Some topics are complete; others need to be continued in order to assess long term effects. The areas of common interest with other investigations, such as the recycling of animal wastes in relation to work on secondary raw materials, will require greater co-ordination in the future if the best use is to be made of the scientific skills and resources available.

OPENING REMARKS

G. Wansink
TNO, Adelheidstraat 84, Postbus 257,
The Netherlands.

In opening this meeting my pleasure is to welcome all of you here to these discussions which are intended to continue previous successful meetings and provide a review of work done under the programme on Effluents from Livestock. As you may recall, the original co-ordinator on the Standing Committee for Agricultural Research for this programme was Mr. Bruggemans. Unfortunately, for personal reasons he was no longer able to give the necessary attention to this programme, and therefore resigned as Chairman. The Belgian delegation has asked me, as a mark of the Benelux relationship, to take his place in this programme and I am pleased to do so. I am from the Netherlands and I am Secretary to the Agricultural Research Council there.

The success of the previous seminars in this programme have already been mentioned and the results have been disseminated all over the world. Directorate General XIII in Luxembourg houses The Office of Offical Publications of the European Community, and from there I have been told that the Proceedings of the Modena Conference on 'Utilisation of Manure by Landspreading' is the best seller of the EEC.

We are now at the border of the old and the new programmes. Our old programme ended in December 1978 and regrettably a lot of participating research workers in various laboratories faced difficulties due to administrative troubles in Brussels. We can only deplore that in a situation where money is available, the procedures and regulations do not make it possible for us to proceed as we should. We cannot blame individual members of the EEC staff for this because they are not responsible for the situation. On the contrary, Mr. Dehandtschutter has always been most anxious to do whatever he could to help this programme.

At the end of this year, the Agricultural Research Programme will have quite a lot of money to spend which could have been used for this programme had we been in a position to settle programme details last year. In due course the new published programme, as agreed by the Council in October last year, will be distributed to you, containing the specifications of the new programme and the budget. The budget is larger than before and is divided between research and co-ordination. This meeting is financed by the co-ordination budget which is designed for the exchange of information on scientific matters and discussion of results.

You will see also that a new element has been introduced into the programme under the title 'Waste'. We are moving more from manure, slurry, or whatever other animal excreta might be called, to farm wastes generally. In doing so we have discovered that the scientific division of the European Community, Directorate General XII, under its programme for the environment, also has proposed a programme related to agricultural waste. The committee for this programme, together with the Agricultural Research Committee, have made an attempt to integrate both programmes. The Director of the DG XII programme, Mr. Bourdeau, will attend the meeting of our Expert Working Group on 5 October 1979 to see how we can co-operate.

The second point you will see in the new programme is that animal pathology is no longer included as it has been taken over by the veterinary sector of the Agricultural Research Programme.

We have a very full programme for this meeting, which has been divided into five sessions, with each session under a different Chairman. Our host, Professor H. Vetter, will be Chairman of the first session.

H. Vetter *(West Germany)*

Ladies and gentlemen, I welcome you here to this meeting on behalf of myself and also on behalf of the President and the Director of our Landwirtschaftkammer at Oldenburg. I also welcome you on behalf of my co-workers, Mr. Steffens, Dr. Klasink and Mr. Kowalesky.

I hope that we will be able to learn a lot from each other at this meeting, in these pleasant and comfortable surroundings.

SESSION I

SLURRY CHARACTERISATION AND ITS USE FOR ARABLE CROPS

Chairman: H. Vetter

THE INFLUENCE OF THE CONTENT OF TRACE ELEMENTS IN THE FEED ON THE COMPOSITION OF LIQUID MANURE OF PIGS

R. Priem and A. Maton
National Institute of Agricultural Engineering,
Merelbeke, Belgium.

ABSTRACT

The purpose of the research (EEC-project 253) was to establish a relationship between the content of trace elements which are taken up by fattening pigs via the feed and the quantity which is excreted via the liquid manure. It seems that for copper 72 to 80 percent is excreted, and for zinc 92 to 96 percent. Final results for arsenic and selenium are not very clearly understood, although it seems that both these elements are also for the greater part found in the liquid manure. Furthermore, during this research practically no influence was detected of the different contents of trace elements on the growth and the feed conversion of fattening pigs, nor on the environmental-technical parameters of the liquid manure.

1. PURPOSE OF THE RESEARCH

In an EEC-project concerning animal wastes, research was set up, in close collaboration with the Institute of Chemical Research (ISO) in Tervuren, to obtain information about the relationship between uptake and excretion of trace elements by fattening pigs.

2. CONDITIONS OF THE RESEARCH

In consultation with the staffs of the ISO, Tervuren, the National Institute of Cattle Feeding, Melle, and the National Institute of Agricultural Engineering, Merelbeke, it was decided that, in this research, three different concentrations of trace elements would be added to the feed. The composition of the feed was changed in respect of the content of copper, zinc, arsenic and selenium, also an adaptation was carried out with mercury, but later on it was found that the content in the feed was lower than the analytically detectable limit. It was also found that the amount of As, which was fixed at 2 ppm for all feeds, showed a large deviation of value. Probably this was due to the problem of mixing small quantities of arsenic-salts with a large quantity of basic feed.

The research was carried out with three groups of fattening pigs, each group receiving the same feed but containing different amounts of trace elements. In order to formulate the feed, special permission was granted by the Service for the Inspection of Raw Materials of the Ministry of Agriculture. In order to be certain that the same feed was used over the entire period of investigation, the complete quantity of feed for each group of animals was produced in one batch and stored at the farm. The respective amounts of trace elements are shown in Table 1.

TABLE 1

CONTENT OF TRACE ELEMENTS IN THE DIFFERENT FEEDS

	A	B	C
Cu in ppm in the DM	116	233	189
Zn "	194	300	260
Hg	<0.01	<0.01	<0.01
As	1.8	1.4	1.6
Se	0.16	0.28	0.24

The research was carried out at an existing farm where the fattening house was equipped with 2 x 20 pens. The pens are each 2 m x 4 m and separated from one another by a brick wall. The floor of the pens is of the fully slatted type and consists for two-thirds of normal concrete gratings and for one-third of concrete vaults, between which a slit of 3 cm is left open. The aeration of the piggery is done by natural ventilation, via an open ridge. The feeding is carried out by a stock-feeding system and the drinking water is provided via drinking-bowls. During the experiment, the automatic feeding of the animals involved in the test, was interrupted and feeding was done manually. The drinking water supply was also detached and a meter was placed in the circuit in order to measure consumption of drinking water, provision was also made to collect spilled water separately.

In order to collect the liquid manure from the animals, large metal containers were hung under the slatted floor, from which the liquid manure could be removed regularly. Every three weeks, or month, all liquid manure was removed and weighed. By means of the specific gravity, the volume could easily be calculated. From each test object, after thoroughly mixing it, 4 samples of 2 litres were taken. Two of these samples were used by the National Institute of Agricultural Engineering to carry out the following determinations: BOD, COD, DM, ash,

P, NH_4-N and total-N. The other two samples were collected by ISO, Tervuren and were used for the determination of trace elements.

Each pen was occupied by 11 pigs, viz. 6 hogs and 5 sows. The average initial weight of these animals was 25 kg. All animals were checked every other day and the consumption of feed and water was noted, while both were replenished.

A few animals needed treatment for pneumonia and diarrhoea by the supervising veterinary surgeon. None of the animals died during the experiment, while in both groups B and C one animal was slow in growing.

3. RESULTS OF THE RESEARCH

a) Performances of the animals

All data of the fattening period of 119 days are given in Table 2. In groups B and C one animal was excluded for the calculation of daily growth because of its poor growth (final weight 35.8 kg and 53.2 kg respectively).

From these results it seems that growth and feed conversion of the test animals were not at all, or very slightly, influenced by the varying content of trace elements in the feed.

b) Manure production

The manure produced was removed five times and the manure production was calculated for each respective period. These data are collected in Table 3. The dry-matter content was abnormally high in this liquid manure. This was caused by the fact that evaporation was possible since the metal collecting containers for liquid manure were only 24 cm deep. For this reason manure production is also given in this table after conversion to a normal dry matter content of 8 percent.

TABLE 2

AVERAGE VALUES CONCERNING GROWTH AND FEED CONVERSION OF THE ANIMALS DURING THE PERIOD OF INVESTIGATION

	Weight of the animals (kg)		Daily growth (g)	Feed conversion kg kg^{-1}	Water consumption l day^{-1} $animal^{-1}$
	Initial	Final			
Group A					
Sows	22.32	87.80	550		
Hogs	24.68	94.98	591		
Total $\bar{x}$	23.61	91.72	572	3.198	3.42
$\sigma/\bar{x}$ in %	18.79	12.99	12.76		
Group B					
Sows	21.97	83.68	518		
Hogs	23.70	99.98	641		
Total $\bar{x}$	22.66	90.20	567	3.168	3.24
$\sigma/\bar{x}$ in %	11.68	13.69	18.34		
Group C					
Sows	31.30	81.78	424		
Hogs	24.28	94.07	586		
Total $\bar{x}$	27.09	89.15	521	3.346	3.31
$\sigma x/$ in %	18.13	16.02	25.34		

Little or no influence could be detected on the liquid manure production of the test animals from the content of trace elements in their feed, as shown in Table 3. Groups A, B, C produced on average respectively 4.20 l, 3.93 l and 3.90 l $animal^{-1}$ day^{-1}. After conversion to a DM content of 8 percent, the general average production of liquid manure amounted to 4.1 l $animal^{-1}$ day^{-1}. The rate of liquid manure production (calculated to a DM content of 8 percent) as an average for all test animals is given in Table 4.

TABLE 3

MANURE PRODUCTION OF THE ANIMALS DURING THE PERIOD OF INVESTIGATION

Period	Waste kg	Specific weight kg l^{-1}	Waste l	Production l $animal^{-1}$ day^{-1}	Dry matter %	Dry matter kg	Manure production at a DM content of 8% l	Manure production at a DM content of 8% l $animal^{-1}$ day^{-1}
Group A								
27 d.	384	1.043	368	1.24	15.64	60	751	2.53
24 d.	440	0.973	452	1.71	14.18	62	780	2.95
28 d.	690	1.022	675	2.19	17.26	119	1 489	4.83
24 d.	770	1.036	743	2.82	16.89	130	1 626	6.16
16 d.	400	0.961	416	2.36	17.02	68	851	4.84
Group B								
27 d.	210	1.023	205	0.69	14.84	31	400	1.31
24 d.	562	0.960	585	2.22	15.84	89	1 113	4.22
28 d.	676	1.007	671	2.18	17.15	116	1 449	4.71
24 d.	700	1.015	690	2.61	15.89	111	1 390	5.27
16 d.	440	0.956	460	2.62	16.16	71	889	5.05
Group C								
27 d.	356	1.007	354	1.19	12.26	44	546	1.84
24 d.	504	0.986	511	1.94	14.81	75	933	3.53
28 d.	740	1.047	707	2.29	16.95	125	1 568	5.09
24 d.	672	1.019	659	2.50	15.42	104	1 295	4.91
16 d.	370	1.016	364	2.07	16.01	59	740	4.21

TABLE 4

AVERAGE LIQUID MANURE PRODUCTION (CONVERTED TO 8 PERCENT DM) OF ALL ANIMALS DURING THE PERIOD OF INVESTIGATION

Period in days	Manure production l animal^{-1} day^{-1}
0 - 27	1.89
27 - 51	3.57
51 - 79	4.88
79 - 103	5.45
103 - 119	4.70
During 119 days	4.10

For the average production of liquid manure of fattening pigs (with dry feeding) 5 l animal^{-1} day^{-1} is often accepted as a rule of thumb. The amount which was obtained viz. 4.1 l animal^{-1} day^{-1} is close to the above mentioned figure, certainly if one knows that the spilled water during the experiment is not added to the liquid manure but was collected in separate containers which were placed under the drinking bowls.

c) The composition of the liquid manure

In order to evaluate the influence of the varying content of trace elements in the feed on the composition of the liquid manure of pigs it was necessary to determine the content. Other environmental-technical parameters of the samples were also determined. Per test-object a total of 11 samples were analysed and the average results are collected in Table 5. Because of the high dry matter content, all environmental-technical parameters were converted to a liquid manure with a dry matter content of 8 percent. Table 5 shows clearly that the amount of trace elements in the feed does not produce any influence on the environmental-technical parameters (BOD & COD) nor on the fertilising parameters (N & P) of the liquid manure obtained. For the content of trace elements in the liquid manure it is completely different. Especially for copper and

TABLE 5

AVERAGE COMPOSITION OF THE LIQUID MANURE OF PIGS. THE RESPECTIVE AVERAGE VALUES, CONVERTED TO A DRY MATTER CONTENT OF 8 PERCENT ARE FOUND UNDER A', B' AND C'.

	DM %	ash in DM %	BOD mg l^{-1}	COD mg l^{-1}	P mg l^{-1}	N-NH_4 mg l^{-1}	NK_j mg l^{-1}	Cu* ppm in DM	Zn* ppm in DM	As* ppm in DM	Se* ppm in DM
Group A											
$\bar{x}$	16.15	23.87	41 807	163 539	3 548	6 606	10 345	416	851	8.50	0.63
σ	1.19	2.20	1 293	26 051	546	1 343	1 272	44.12	74.57	0.83	0.05
$\sigma/\bar{x}$ in %	7.37	9.22	3.09	15.93	15.37	20.32	12.29	10.62	8.76	9.75	8.40
Group B											
$\bar{x}$	15.92	24.56	41 957	154 026	3 491	6 351	9 998	859	1 385	10.93	1.57
σ	0.75	1.79	2 504	15 867	360	1 097	1 000	68.20	107.44	1.76	0.17
$\sigma/\bar{x}$ in %	4.71	7.29	5.97	10.30	10.30	17.27	10.01	7.94	7.76	16.07	10.53
Group C											
$\bar{x}$	14.63	25.58	35 546	143 178	3 294	5 774	9 083	754	1 180	11.13	1.08
σ	2.14	2.48	5 253	19 977	838	1 209	1 367	66.26	31.27	0.87	0.08
$\sigma/\bar{x}$ in %	14.63	9.70	14.78	13.95	25.43	20.94	15.05	8.78	2.65	7.85	7.40
A'	8	23.87	20 709	81 010	1 758	3 272	5 124	-	-	-	-
B'	8	24.58	21 084	77 400	1 754	3 191	5 019	-	-	-	-
C'	8	25.58	19 437	78 293	1 801	3 157	4 967				

* The determinations of trace elements were carried out by the Institute for Chemical Research, ISO, in Tervuren.

zinc there is a clear parallel in the concentration ratio between the feeds, and the concentration ratio of the respective manure, i.e. a double dose of copper or zinc in the feed results in a double amount in the liquid manure. This is clearly illustrated in Table 6. The picture is however, less clear for arsenic and selenium. The low amount of these components in the feed has probably been the reason for a less homogeneous distribution of these elements in the total feed mass and the results in Table 1 are, for this reason, probably not very reliable.

TABLE 6

RATIO OF THE CONCENTRATION OF TRACE ELEMENTS BETWEEN THE FEEDS ON ONE HAND AND THE OBTAINED LIQUID MANURE, ON THE OTHER HAND

	B/A		C/A		B/C	
	Feed	Manure	Feed	Manure	Feed	Manure
Cu	2.01	2.08	1.63	1.81	1.23	1.14
Zn	1.55	1.63	1.34	1.39	1.15	1.17
As	0.78	1.29	0.89	1.31	0.88	0.98
Se	1.75	2.49	1.50	1.71	1.17	1.45

d) Relation between the amount of trace elements administered via the feed and the amount excreted via the liquid manure

For the undermentioned results, the applied dose of trace elements via the feed was calculated by means of the average values of Table 1. For the amount excreted via the liquid manure, the content of trace elements was calculated by means of the dry matter content and the respective concentration of each evacuated quantity of liquid manure separately.

For the calculation of the supplemented quantity of trace elements, the addition via the drinking water is not taken into account because of the very low concentration in it. Each group received only 14 mg copper and 480 mg zinc via the drinking water during the complete duration of the fattening period.

TABLE 7

THE QUANTITY OF TRACE ELEMENTS ADMINISTERED VIA THE FEED AND EXCRETED VIA THE LIQUID MANURE

	A				B				C			
	Cu	Zn	As	Se	Cu	Zn	As	Se	Cu	Zn	As	Se
Amount administered via the feed, g.	249	416	3.86	0.343	451	581	2.71	0.542	369	507	3.12	0.468
Excreted amount via the manure, g.	180	382	3.88	0.274	343	561	4.22	0.629	292	489	4.45	0.438
Amount excreted compared to administered dose in %	72.18	91.72	100.52	79.82	76.14	96.58	155.72	115.92	79.17	96.48	142.68	93.44

Table 7 clearly shows that the largest part of the trace elements administered via the feed is excreted via the liquid manure: for copper 72 to 80 percent of the administered dose is excreted, this value rises to 92 - 96 percent for zinc. No conclusion is possible for arsenic nor for selenium when the obtained results are taken into account. As already mentioned, this is probably due to problems concerning the homogeneity of the feeds during the production.

4. CONCLUSIONS

From the experiments which were carried out, it is clear that an increased amount of trace elements in the feed of fattening pigs has no influence or practically none, on the growth or the feed conversion of these animals. Also no influence could be detected on the environmental-technical or fertilising parameters of the liquid manure. Since copper and zinc are respectively excreted at the rate of 72 to 80 percent and 92 to 96 percent, it is ecologically advisable to limit the addition of these elements to the feed of fattening pigs to a minimum, otherwise this might lead to contamination of the soil. From the results for arsenic and selenium it is impossible to come to a conclusion, although it is likely that both elements are also practically completely excreted.

5. ACKNOWLEDGEMENTS

This research was possible thanks to the support of the Directorate-General of Agriculture (Dir. E4) of the EEC (programme no. 253).

We are also grateful to the collaborators of the Institute of Chemical Research (ISO) in Tervuren for the determinations of trace elements.

We also wish to thank Mr. F. Lunn and Mr. R. Janssen, collaborators at the National Institute for Agricultural

Engineering in Merelbeke for the other analyses and Mr. L. Van Der Beken and Mr. R. Van Den Meersschaut for the collection of the samples. Last but not least we wish to express our thanks to Mr. Moerman in Kruishoutem on whose farm this research was carried out.

DISCUSSION

J.K. Grundey *(UK)*

Dr. Maton, how much drinking water was spilled, and what effect would including this have on the dry matter content of the slurry?

A. Maton *(Belgium)*

About 10 percent of drinking water and this would have given a dry matter content of perhaps 12 percent, but the main factor was the evaporation. The flat containers used allowed much evaporation of water, which explains the values of 14 to 15 percent which we found.

J.H. Voorburg *(The Netherlands)*

What was the copper content of the pigs because 20 percent of the copper was retained, and the liver might contain 9 g of copper, which is very toxic.

A. Maton

As I said, copper is mainly stored in the liver, and about 15 to 20 percent of the copper which is administered is retained but not zinc and other elements, so that the liver may be dangerous to eat. That is why I said that it is better to limit the amount of copper. In our experiment, Group B received 233 ppm, which is almost double the maximum allowed. About 110 - 120 ppm is used in commercial practice similar to that given to the animals in Group A.

D.B.R. Poole *(Ireland)*

Dr. Maton, I would like to question your finding in relation to liveweight gain. First of all the large response to copper, which has been published many times in the literature, between an unsupplemented feed with maybe 20 ppm copper and a supplemented feed such as all of yours is missing.

Secondly, by the design of the experiment you are unable to separate the effects of different minerals as added. Looking at the feed components, the level of zinc is increasing in parallel with the level of copper to very high levels. Certainly in Ireland we would not consider adding 300 ppm zinc to a pig feed; 150 ppm would be the maximum. There appears to be some danger in extrapolating from this useful work in relation to the liveweight gain of the animals.

H. Vetter *(West Germany)*

The copper content in your slurry is extremely high compared with ours, or with slurry from other countries. Is it really necessary or profitable for the growth of the animals?

A. Maton

First of all,we did not use any unsupplemented feed, because the purpose of the research was only to establish the relationship between trace elements in the feed and trace elements in the manure. We also used high levels of zinc, because large amounts of zinc are necessary if much copper is given in order to prevent copper intoxication of the animal.

H. Vetter

Thank you. I think we must now close the discussion on this very interesting report.

EFFECTS OF LANDSPREADING OF LARGE AMOUNTS OF LIVESTOCK EXCRETA ON CROP YIELD AND CROP AND WATER QUALITY

K.W. Smilde+
Institute for Soil Fertility
Oosterweg 92, Haren (Gr.),
The Netherlands.

INTRODUCTION

In some areas of the Netherlands the supply of minerals in livestock excreta exceeds crop requirements, with enrichment of the soil as a result. Mobile constituents (NO_3,Cl) are easily transported to a considerable depth down the soil profile and may reach the groundwater and the surface waters. Ions of lower mobility (PO_4, Cu, Zn) accumulate in the top soil and may pollute surface waters under conditions conducive to runoff.

In the long term, enrichment of the soil with minerals may have adverse effects on crop production and crop quality, such as lodging of cereals, decreases in starch (potatoes) and sugar content (sugar beets); other effects may adversely affect animal health, for example, increases in nitrate and potassium and decreases in calcium and magnesium concentrations in pasture herbage, increased copper concentrations, toxic to sheep, in the upper part of the sod following application of large amounts of pig slurry.

The aim of the research, in which five institutions*were participating, was to study short and long term effects of very large dressings of manure on crop production and crop quality, accumulation of minerals in the soil and possible pollution of soil and surface waters. Some pertinent data and conclusions are reported here.

* EEC Contract No.290:
Institute for Soil Fertility, Haren (Gr.); Research Institute for Forestry and Landscape Planning, Wageningen; Research Station for Arable Farming and Field Production of Vegetables, Lelystad; Research and Advisory Institute for Cattle Husbandry, Lelystad; Institute for Land and Water Management Research, Wageningen.

+Project Leader

FORESTRY

Long term effects of livestock manure disposal in conifer stands are considered harmful because of the more rapid mineralisation of the raw humus and litter layer (containing a large part of the roots) and the resulting stimulation of root fungi like *Fomus annosus*. Hardwoods, like poplar, tolerate up to 400 - 500 kg N ha^{-1} for some years (100 tonnes pig slurry ha^{-1}). Much less is known about the tolerance of other hardwoods, but spacing also plays a part.

GRASSLAND

Research included disposal of slurry of housed cattle after dilution with ten volumes of water by sprinkler irrigation. Generally, undiluted slurry cannot be applied to pasture in summer because it scorches the grass and reduces intake of herbage by cattle. So disposal occurs in winter and early spring resulting in considerable nutrient losses (mainly nitrogen). Irrigation with diluted slurry in summer presents few problems, and serves the dual purpose of supplying water and nutrients.

ARABLE CROPS

Sugar beet

The effect of large amounts of pig slurry, containing about 3 kg (effective) N, 2 kg P_2O_5 and 4 kg K_2O $tonne^{-1}$ fresh material, on sugar content and juice quality is shown in Table 1.

Sugar content decreases and concentrations of monovalent cations (Na + K) and α-amino N increase, following increasing applications of pig slurry. The overall effect of these constituents is shown by the alkalinity coefficient (AC), a value of 1.8 being optimal. With ACs higher than 1.8 monovalent cations interfere with sugar crystallisation, so that more sugar remains in the molasses. With ACs lower than 1.8 extractable sugar mainly depends on α-amino N content, which decreases juice pH and stimulates sugar decomposition. If present in excessive amounts, sodium carbonate has to be added, resulting in a lower extractability.

TABLE 1.

EFFECT OF PIG SLURRY ON SUGAR CONTENT AND JUICE QUALITY OF SUGAR BEET.

	Sugar %	Sugar yield kg ha^{-1}	m val 100 g^{-1} sugar Na + K	m val 100 g^{-1} sugar α-amino N	AC*
0 tonne pig slurry/ha	17.4	9 740	24.4	13.4	1.8
40 tonnes " " "	17.4	9 670	25.5	12.6	2.0
60 tonnes " " "	17.4	9 960	25.8	11.8	2.2
80 tonnes " " "	16.7	10 620	31.2	19.1	1.6
100 tonnes " " "	16.5	10 580	31.9	22.9	1.4
120 tonnes " " "	16.1	10 690	33.8	22.6	1.5

$$* \text{ AC = alkalinity coefficient} = \frac{\text{m val (Na + K) 100 g}^{-1}\text{ sugar}}{\text{m val }\alpha\text{-amino N 100 g}^{-1}\text{ sugar}}$$

Silage maize

A long term field experiment tested annual dressings of 50, 100, 150, 200, 250 or 300 tonnes ha^{-1} cattle slurry applied to a sandy soil (2.8% organic matter; pH-KCl 4.9) in the period December - April, annually since 1973 (Table 2).

Apart from mineral-N in 50 tonnes cattle slurry ha^{-1} nutrient supply always exceeded crop uptake. Results in Figure 1 demonstrate this clearly for the mineral-N (ammonium + nitrate N) concentrations in soil determined in 20 cm layers to a depth of 1 metre in spring (after the last slurry application) and in autumn (after harvesting). A considerable portion of

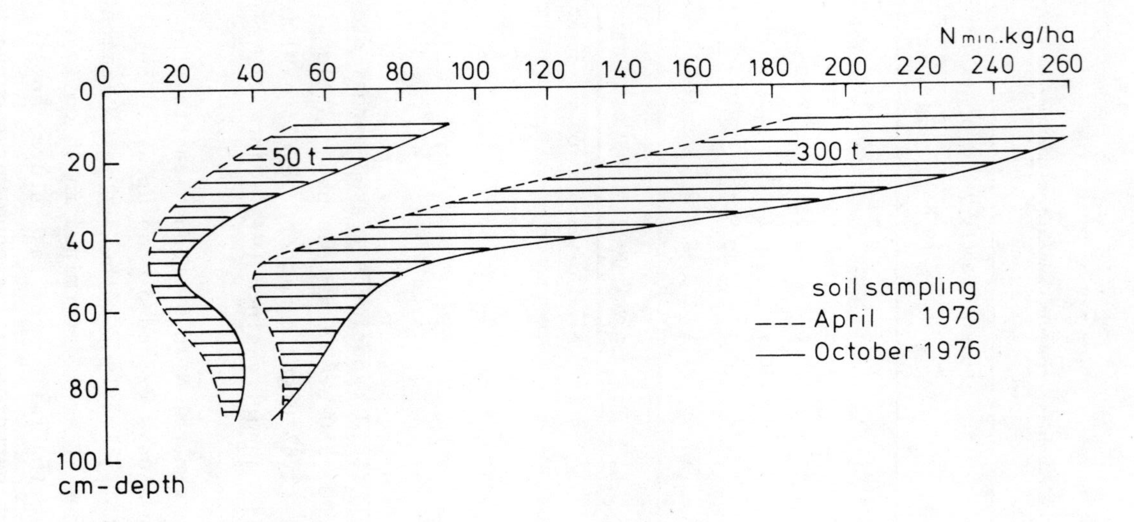

Fig. 1. Soil enrichment (0 - 100 cm) with mineral N following 3 years' application with 50 or 300 tonnes ha^{-1} cattle slurry, in kg ha^{-1} per layer of 20 cm.

nitrate-N left in the soil after harvesting is likely to be lost by leaching during the winter. Moreover, losses of mineral-N contained in the cattle slurry applied in winter will occur as a result of volatilisation as ammonia. Losses by denitrification to gaseous nitrogen products are thought to be small because of low winter temperatures. The average amounts of nitrogen not accounted for in the balance sheet for soil mineral-N during the winter periods (1975 - 1978) vary from 120 to 630 kg ha^{-1} with slurry applications of 50 and 300 tonnes ha^{-1} respectively. Part of it may not be lost but temporarily immobilised in a plant-unavailable form. However, considering the substantial rise in nitrate concentration of the ground water at 1.5 metre depth, shown in Figure 2, with the increasing supply of mineral-N, originating from the soil reserve in autumn plus the slurry application in winter, a substantial part is probably lost by leaching. The nitrate concentrations in ground water following the various applications of cattle slurry far exceed the EEC limit for potable water (11 mg NO_3-N $litre^{-1}$).

Following three annual applications of large quantities of cattle slurry, water-soluble P had penetrated to a depth of 40 cm and exchangeable K to 80 cm down the soil profile.

TABLE 2.

AVERAGE ANNUAL SUPPLY OF NUTRIENTS IN CATTLE SLURRY AND UPTAKE OF NUTRIENTS BY SILAGE MAIZE (1974 - 1977).

	Mineral N* kg ha^{-1}	P_2O_5 kg ha^{-1}	K_2O kg ha^{-1}	CaO kg ha^{-1}	MgO kg ha^{-1}	Yield t ha^{-1}
Nutrient supply						
50 t ha^{-1} slurry	120	100	270	120	60	12
300 t ha^{-1} slurry	730	685	1 500	615	365	14
Crop uptake						
50 t ha^{-1} slurry	175	50	215	25	25	
300 t ha^{-1} slurry	210	65	325	50	35	

* 50 % of total N

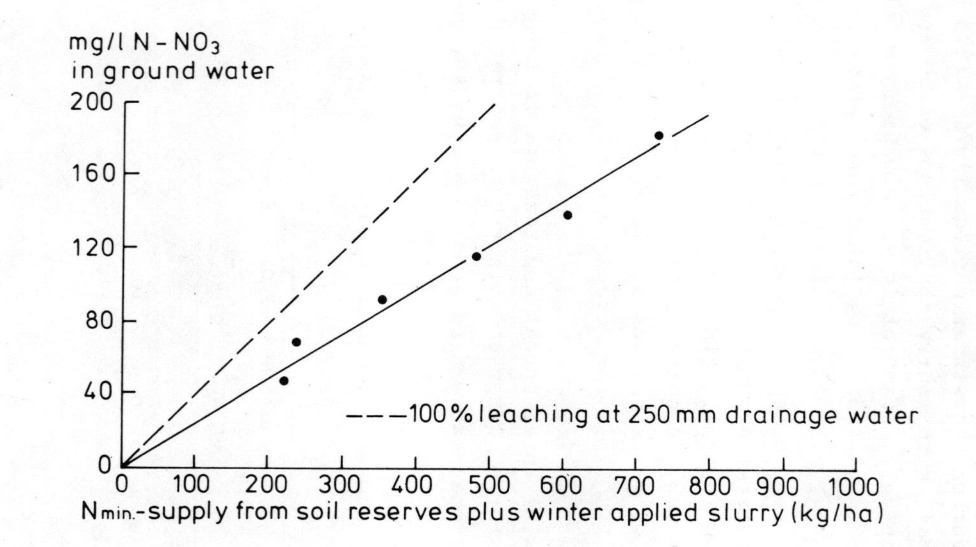

Fig. 2. Mineral N supply from the soil (0 - 100 cm; autumn) plus (winter) applied cattle slurry as influencing NO_3-concentration of groundwater (1.5 - 2 m depth).

CONCLUSIONS

Stands of hardwoods, like poplar, may be used for disposal of livestock excreta, but this practice is not recommended for conifers.

An effective way of applying livestock slurries to grassland is by sprinkler irrigation of the diluted material during the growing season.

Care should be taken not to overdose arable crops such as sugar beet with livestock excreta because of deterioration in crop quality and sugar extractability. Crops such as silage maize tolerate large doses of livestock excreta, the excess of nutrients, however, pollute the ground water (if soluble such as nitrate) or accumulate in the soil profile, if immobile such as phosphate.

DISCUSSION

H. Vetter *(West Germany)*

Thank you for your report Dr. Smilde. I noticed that the soil contains more mineral-N in October than in April. You say that this is due to the rapid decomposition of the organic matter but in our experiments we have not found an increase. Can you explain this effect, please?

K.W. Smilde *(The Netherlands)*

Figure 1 shows the relevant results. These experiments were started in 1974 and every year, from 50 t up to 300 t of slurry were applied giving an accumulation of organic matter and of organic nitrogen in the soil. In 1976, negligible amounts of mineral-N were lost by denitrification or leaching, so during the hot summer the difference between April and October is the amount of N which was mineralised.

Marie Sherwood *(Ireland)*

Do these figures refer to arable land? You mentioned in your paper that disposal through an irrigation system on grassland does not present many problems: did you separate the solids before irrigating with the sprinkler system?

K.W. Smilde

The crops were all silage maize. There was not any separation of solids, but there was a dilution of 1 : 10, which may have solved any problems.

A. Maton *(Belgium)*

On the soil on which you were doing your experiments, what would have been the maximum amount of liquid pig manure you could use and still avoid the groundwater becoming undrinkable?

K.W. Smilde

On sandy soils, even with a normal application of manure, the groundwater already contains too much nitrate to be used as potable water. On clay soils the situation is better. But even without applying large quantities of pig slurry you already have too high a content of nitrogen in the groundwater.

H. Vetter

I think it is necessary to limit the amounts of slurry to the amounts of nitrogen that are put on as mineral fertiliser. The amount of nitrogen leached in your experiment was 80 kg ha^{-1} from 360 kg nitrogen. Such levels have not been recorded by us for mineral fertiliser.

A. Dam Kofoed *(Denmark)*

From the Danish point of view, when nitrogen is present in the drainage water, then drinking water is polluted. In Denmark 90 to 93 percent of all drinking water is totally free of nitrogen, so one should be careful to avoid equating drainage water containing nitrate with drinking water.

K.W. Smilde

Yes, that is a good point. Of course, before the drainage water reaches the groundwater it may be diluted and denitrification may occur.

H. Vetter

We must finish now. Thank you for your interesting report.

LANDSPREADING OF LIQUID PIG MANURE:
1. EFFECTS ON YIELD AND QUALITY OF CROPS

C. Duthion
(with the technical collaboration of J. Mortier and L. Guenot)
INRA, BV 1540,
21034 Dijon Cedex, France.

ABSTRACT

The effects of liquid pig manure on yields and quality of crops were studied in lysimeter and field experiments. The influence of soil type, amount supplied and time of application were studied on Fescue, ryegrass, maize and wheat. Quality of crops **was** *determined. Total-N in pig manure was 40 to 57 percent as efficient as fertiliser-N.*

1. INTRODUCTION

France produces more than 13 million pigs year^{-1}, using mainly liquid manure systems for effluent handling. Land-spreading is the most used procedure for disposal of pig manures. Their use is more or less on an empirical basis and generally environmental protection is not considered.

Due to increasing water pollution problems and to increasing fertiliser costs, manure recycling needs to be optimised from both environmental and agricultural points of view. This work tried to relate integrated studies on the effects of liquid pig manure on crops, soils and water.

2. MATERIALS AND METHODS

2.1 Experimental design

The programme included field and lysimeter trials to permit the study of possible effects of pig manure as shown in Table 1.

TABLE 1

PROGRAMME SUMMARY AND PARAMETERS STUDIED

	Lysimeters	Field experiments	
Purpose	Nutrient balances	Effects on plants	Effects on Environment
Parameters	- crops, leachates, soils - Influence of: amount spread time of application types of soils	- plant: - grass - maize - barley - wheat	- waters and soils - soil microflora - soil nematodes - earthworms

Lysimeter experiments

Twenty lysimeters, made from concrete pipes of 1 m diameter and 1 m deep, with the interior painted with epoxy paint, were filled with 0.15 m gravel at the bottom for drainage and 0.85 m of soil. Lysimeters were built in two groups of 10 each and put in a concrete tank filled with soil as insulating material. Four different soils were used.

Nine lysimeters were started in 1976 using three soil types (see Table 2) clay, calcareous loam and loam and with 3 treatments: fertilisers, liquid pig manure supplying total-N equivalent to fertiliser-N and pig manure supplying five times as much total-N. They were cropped with Fescue (Festuca elatior, cv Manade).

In 1977, eleven more lysimeters were started with two soil types: loam (the same as 1976) and a sand, in order to test the influence of amount applied and time of application. The seven treatments applied were: fertilisers, pig manure supplying total-N equivalent to fertiliser-N, threefold and fivefold pig manure; threefold pig manure applied in spring, threefold applied in autumn; sixfold pig manure applied in alternate autumns. Table 3 gives a summary of the treatments.

Field experiments

Four field experiments were made.

Experiment 1 was on a clay soil (which was also used in the lysimeters) on three plots each with an area of 210 m^2 planted in 1976 with Fescue (Festuca elatior, cv Manade) and receiving fertilisers, pig manure supplying total-N equivalent to fertiliser-N and a fivefold N equivalent in pig manure. In each plot, a control lysimeter of 1.5 m x 1.5 m width and 1 m deep was used.

TABLE 2

SOIL CHARACTERISTICS

	Clay		Calcareous loam	Loam		Sand	
	0 - 30 cm	30 - 80 cm	0 - 80 cm	0 - 30 cm	30 - 80 cm	0 - 35 cm	< 35 cm
Mechanical analysis %							
2 - 0.2 mm	1.5	2.1	24.4	5.1	4.4	50.0	43.5
0.2 - 0.05 mm	3.5	4.9	14.7	10.2	6.1	11.0	11.5
0.05 - 0.02 mm	20.5	20.2	15.6	29.0	14.6	8.5	10.0
0.02 - 0.002 mm	31.5	31.4	27.5	35.3	26.0	17.0	20.5
< 0.002 mm	43.0	41.4	17.8	20.4	48.9	13.5	14.5
pH	6.6	7.2		6.9	5.3	5.1[1]	4.8[1]
$CaCO_3$ total %	0	0.6	80.1	0	0	0	0
Organic-C %	1.79	1.05	1.80	1.11	0.28	1.50	0.28
Total-N %	0.2	0.13	0.22	0.09	0.05	0.10	0.03
P_2O_5 oxalate ppm	30	40	150				
Dyer ppm				170	10	12	4
Cation exchange Capacity meg 100 g^{-1}	26.4	23.8	8.8	9.3	21.5	7.5	7.0
K exchangeable meg 100 g^{-1}	0.30	0.36	0.38	0.29	0.39	0.11	0.077

(1) Before supply of CaO to increase soil pH to 6.5

TABLE 3

SUMMARY OF LYSIMETER STUDIES

Study	Treatments		Soils
1976 started: 3 treatments Lysimeters	(Fertilisers (Pig manure supplying total-N equivalent (Pig manure x 5	x 3 soils	(clay (calcareous loam (loam
1977 started: 7 treatments Lysimeters	(Fertilisers) (Pig manure supplying total-N equivalent) (Pig manure x 3) (Pig manure x 3 - spring) (Pig manure x 3 - autumn) (Pig manure x 6 alternate years) (Pig manure x 5)	x 2 soils	(loam ((sand

Experiments 2 and 3 were done on a loamy soil. Experiment 2, started in 1977 on three plots each with an area of 210 m^2 with three treatments - fertilisers, manure supplying total-N equivalent to fertiliser-N, manure at threefold total-N equivalent. Lysimeters, with the same dimensions as in Experiment 1, were set in each plot. The area was sown with Fescue (Festuca elatior, cv Manade). Experiment 3 was similar to Experiment 2 except that the plots were planted in 1977 with maize, in 1978 with winter wheat and in 1979 with winter barley.

Experiment 4 was designed to study the efficiency of N in pig manure applied to Italian ryegrass. Five treatments were applied to a loamy soil: fertiliser-N, pig slurry supplying total-N equivalent to fertiliser-N x 1, x 1.5 x 1, x 3 with five replicates on plots of 18 m^2. There were also plots without added N. Plots without N, with fertiliser-N and with manure supplying equivalent N received the same amounts of P and K.

2.2 Weather

Experiments started in March 1976 and the weather during the experimental periods were:

March 1976 - March 1977 was very dry in spring and summer, with 700 mm rain mainly in autumn, and 1 182 mm potential evapotranspiration calculated with Piche evaporation, using the Bouchet formula with a coefficient $\alpha = 0.37$. Irrigation was necessary to establish the Fescue. 337 mm were applied to the lysimeters and 68 mm to Field Experiment 1.

March 1977 - March 1978 The weather was wet for the whole year with 1 110 mm rain and 626 mm potential evapotranspiration for lysimeters and Field Experiment 1, and 1 260 mm rain for Field Experiments 2, 3 and 4.

March 1978 - March 1979 was very dry in the autumn with 695 mm rain and 749 mm potential evapotranspiration for lysimeters and Field Experiment 1; 45 mm of irrigation water was applied to lysimeters and 70 mm to Field Experiment 1. There was 972 mm rainfall in Field Experiments 2 and 3.

2.3 Treatments

Fertilisers were supplied to compensate for plant removal as ammonium nitrate (34.5 percent N), triple superphosphate and potassium chloride to supply

- for Fescue : 300 kg N; 50 kg P; 350 kg k ha^{-1}
- for maize : 300 kg N; 65 kg P; 125 kg K ha^{-1}
- for winter wheat : 160 kg N; 45 kg P; 85 kg K ha^{-1}

Manures were applied following the fertiliser dressings on the basis of equivalent total-N (as described in 2.1). Additional K was supplied for treatments with small amounts of manure to grass. Several times during dry weather in summer, all the plots with Fescue received, 50 kg fertiliser-N after cutting, to avoid damage to the sward from manure.

Time of application

Fertilisers and manure were normally applied to grass in spring before growth and after each cut, at rates indicated in 2.1. For the 'time of application study', manure was applied in spring before start of growth (3 x total-N equivalent) or in autumn (3 x total-N equivalent and 6 x total-N equivalent in alternate years).

For maize, fertilisers and manure were spread one month before sowing.

For winter wheat, fertilisers and manure were spread at three times (sowing, tillering and shooting).

Manures applied came from different pig fattening units, were transported by tank and sampled for analysis just before, during and at the end of spreading.

An approximate estimation of N content was made on the field by using a densimeter (Tunney and Molloy, 1975). Manures were spread by hand on lysimeters and fields. In the field, each plot was divided into areas of 2 m^2 or 5 m^2 depending on the amount to be applied; thus allowing the volume of manure to be accurately measured.

Manures varied extremely in their composition, depending on farms, sampling difficulties and dates. In the period 1976 - 1978, the composition ranged between the limits given in Table 4. Consequently, the desired amounts could not be obtained and amounts actually applied only approached them.

TABLE 4

MEAN CHARACTERISTICS OF MANURES APPLIED

		Mean	Range
pH		7.4	6.9 - 7.6
Specific gravity		1.023	1.011 - 1.052
Dry matter	kg m^{-3}	45.3	8.5 - 128
Ash	kg m^{-3}	14.1	5.8 - 37
COD	kg m^{-3}	46.8	9 - 127
Total Nitrogen	kg m^{-3}	4.48	2.52 - 9.38
NH_4 - Nitrogen	kg m^{-3}	2.86	1.41 - 4.84
P	kg m^{-3}	1.31	0.14 - 4.30
K	kg m^{-3}	1.73	0.75 - 3.20
Ca	kg m^{-3}	1.39	0.06 - 5.20
Mg	kg m^{-3}	0.60	0.05 - 1.83
Zn	g m^{-3}	50	4 - 115
Cu	g m^{-3}	37	3 - 99

Amounts of elements applied

Table 5 for lysimeter studies and Table 6 for field trials give NPK supplied for each year. To allow balances to be calculated for the elements, data are given for one year periods from March to March corresponding to the active growth of plants plus winter drainage. Consequently in Experiment 3 (Table 6) nutrients added in 1977 were the sum of dressings for maize and a part of those for winter wheat. It was the same in 1978 for wheat and winter barley.

In Experiment 4 with ryegrass, fertilisers and manure were applied as three dressings supplying totals of:

- for mineral fertiliser: 179 kg N as ammonium nitrate,
- for manure x 1: 178 kg total-N including 111 kg as mineral-N
- other manure treatments were x 1.5, x 2 and x 3 dressings of manure x 1.

2.4 Analytical determinations

For all the experiments, samples of each cut were taken after measuring yields of dry matter. Chemical determinations were made on samples of dry matter for ash, organic matter, nitrogen (total-N, NO_3^--N), total P, Ca, K, Mg, Cu, Zn.

Experiment 3 was completed by measuring sward composition in the 2nd year.

3. RESULTS

3.1 Yields and quality

3.1.1 Temporary grassland

A. Lysimeter studies

In 1976, yields of Fescue from the lysimeters were not influenced by soil type, so that Figure 1 shows mean annual

TABLE 5

AMOUNTS OF NPK APPLIED TO LYSIMETERS (kg hectare^{-1})

	Fertiliser		Pig manure							
	F		M		3 M	3 MS	3 MA	6 M/2	5 M	
Total Nitrogen										
1976	265[a]		265[a]						1123[a]	
1977	269[a]	286	409[a]	283	740	1267	912	1718	1828[a]	957
1978	319[a]	319	273[a]	274	797	565	573	-	1340[a]	1340
Total	953[a]	605	948[a]	557	1537	1832	1485	1718	4291[a]	2296
Mineral Nitrogen										
1976	265[a]		203[a]						815	
1977	369[a]	286	237[a]	186	402	692	641	1176	978	556
1978	319[a]	319	155[a]	155	465	274	393	-	775	775
Total	953[a]	605	595[a]	341	867	966	1034	1176	2568	1331
Total Phosphorus										
1976	45[a]		37[a]						186[a]	
1977	127[a]	127	107[a]	191	280	621	335	670	532[a]	448
1978	50[a]	50	108[a]	108	321	230	300	-	540[a]	540
Total	222[a]	177	252[a]	299	601	851	635	670	1258[a]	998
Potassium										
1976	126[a]		68[a]						335[a]	
1977	347[a]	124	428[a]	179	282	273	389	779	450[a]	290
1978	600[a]	400	293[a]	293	230	146	210	-	862[a]	463
Total	1073[a]	524	789[a]	472	512	419	599	779	1647[a]	753

a = 1976 lysimeters started; F: fertiliser, M: pig manure supplying total-N equivalent to 3 M = M x 3, 3 MS = 3 M once a year in spring, 3 MA = 3 M once a year in autumn, 6 M/2 = M x 6 in alternate years, 5 M = M x 5.

TABLE 6

AMOUNTS OF NPK APPLIED TO FIELD EXPERIMENTS (kg $hectare^{-1}$)

	Experiment 1 Clay soil Fescue			Experiment 2 Loamy soil Fescue			Experiment 3 Loamy soil Maize, wheat		
	F	M	5 M	F	M	3 M	F	M	3 M
Total Nitrogen									
1976	265	331	1195						
1977	367	378	1791	263	310	830	347	406	1220
1978	248	226	1130	320	339	886	140	207	371
Total	880	955	4116	583	649	1716	487	613	1591
Mineral Nitrogen									
1976	265	252	852						
1977	367	234	970	263	229	588	347	288	860
1978	248	165	828	320	252	664	140	151	240
Total	880	651	2650	583	481	1252	487	439	1100
Total Phosphorus									
1976	45	44	195						
1977	127	107	439	52	69	208	109	105	315
1978	124	89	352	101	59	142	35	33	113
Total	296	240	986	153	128	350	144	138	428
Potassium									
1976	126	88	355						
1977	347	428	477	125	102	305	208	185	555
1978	572	343	818	399	312	564	67	135	200
Total	1045	859	1650	524	414	869	275	320	755

F = Fertilisers, M = Pig manure at total-N equivalent to F, 3 M = 3 x M, 5 M = M x 5.

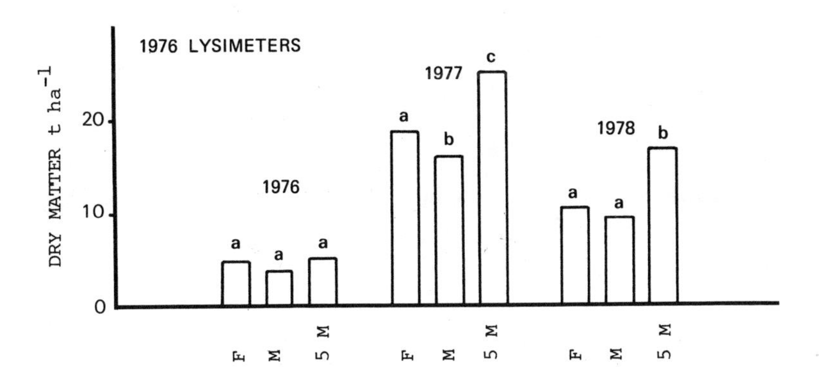

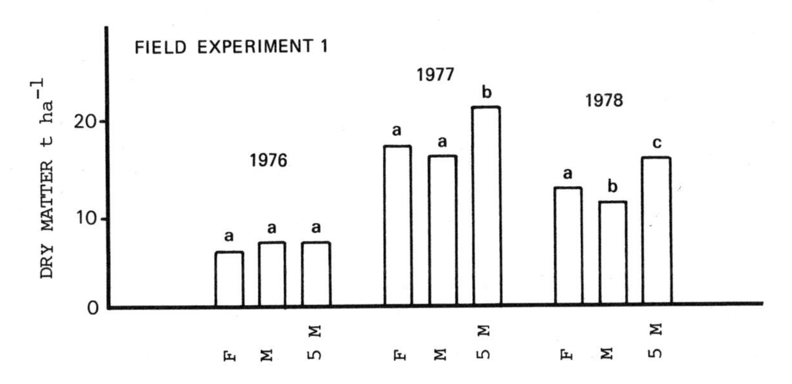

Fig. 1. Fescue dry matter yields for 1976 lysimeters (mean for 3 soils) and Field Experiment 1.

F = fertiliser

M = manure supplying equivalent N

5 M = manure x 5

Yields with the same letter above do not differ at 5 percent level.

yields for the three soils. Considering all 11 cuts during the 3 years of the experiment, the following points may be noted:

1) Pig manure (M) supplying total-N equivalent to fertiliser-N (F) increased yields significantly less, (on average 15 percent lower).
2) Large amounts of pig manure (5 M) greatly increased the yields of dry matter (on average, 37.5 percent increase).

The 1977 lysimeters did not give usable results in 1977 because of the need to start some treatments in autumn.

In 1978 usable results were obtained from all the lysimeters. The following points may be noted:

1) Yields on the sandy soil were smaller than on other soils but treatment effects were similar.
2) Dry matter yields increased with the amount of manure applied. However, treatment 5 M did not increase yield significantly more than treatment 3 M. The same was true for treatment M and 3 M which did not differ statistically from treatment F (fertiliser).
3) Time and frequency of speading did not influence yield of Fescue in 1978.

The mean composition of yields of dry matter are presented in Table 7 for lysimeters started in 1976. Statistical analysis of all available results from these lysimeters (11 cuts) clearly showed an effect of soil type on the contents of most elements. Treatment effects are indicated in Table 8.

Nitrogen contents are especially affected by treatments because total-N content was significantly less for treatment M than for treatment F. Treatment 5 M greatly increased total-N and simultaneously increased nitrate-N content considerably

TABLE 7

AVERAGE COMPOSITION OF DRY MATTER OF FESCUE FOR LYSIMETER EXPERIMENTS STARTED IN 1976 AND FIELD EXPERIMENT 1 ON AVERAGE OF 11 CUTS FROM 1976 TO 1978

	1976 Lysimeters									Field Experiment 1		
	Clay soil			Calcareous loam			Loamy soil					
	F	M	5 M	F	M	5 M	F	M	5 M	F	M	5 M
Total-N %	2.38	2.23	3.00	2.28	1.99	2.90	2.27	1.97	2.94	2.33	2.12	3.01
NO_3-N %	0.03	0.03	0.30	0.03	0.01	0.23	0.05	0.02	0.30	0.05	0.05	0.25
P %	0.26	0.31	0.28	0.23	0.28	0.26	0.28	0.32	0.31	0.31	0.35	0.34
K %	2.35	2.44	2.40	2.88	2.55	2.88	2.71	2.52	2.76	2.45	2.45	2.67
Ca %	0.46	0.48	0.52	0.48	0.46	0.48	0.40	0.39	0.46	0.40	0.38	0.42
Mg %	0.23	0.25	0.27	0.14	0.18	0.19	0.19	0.21	0.24	0.22	0.21	0.26
Cu ppm	13	20	27	11	13	15	12	16	22	12	14	17
Zn ppm	35	52	60	31	39	41	31	39	47	36	37	50

F = fertiliser, M = manure supplying total-N equivalent to F, 5 M = M x 5.

(average 0.27 percent in dry matter and higher values up to 0.50 percent).

TABLE 8

TREATMENT EFFECT ON FESCUE COMPOSITION - 1976 LYSIMETERS

Element	Treatment effect	Element	Treatment effect
Total-N	M F 5 M	Ca	M F 5 M
NO_3-N	M F 5 M	Mg	F M 5 M
P	F 5 M M	Cu	F M 5 M
K	M F 5 M	Zn	F M 5 M

F = fertiliser, M = manure supplying total-N equivalent to F., 5 M = M x 5.
Treatments connected by line do not differ at the 5 percent level; concentrations increase from left to right.

Copper and zinc contents were significantly increased over the whole experiment but not for all the cuttings, suggesting external grass contamination more than plant absorption.

Table 9 gives the mean composition of dry matter yields from 1978, for 1977 lysimeters for sandy and loamy soils only (the corresponding results from 1976 lysimeters are included). Amount of manure applied influenced the nitrogen content. Comparing treatments supplying the same volume of manure (supplying similar amounts of total-N) and differing in time and frequency of application showed that:

1) Mean total-N content after spreading once a year in autumn (3 M/A) was significantly less than for other treatments,
2) spreading once a year in spring (3 M/S) increased total-N content only at the first cut,

TABLE 9

AVERAGE COMPOSITION OF THE DRY MATTER OF FESCUE FOR LYSIMETER EXPERIMENTS IN 1977 AND FIELD EXPERIMENT 2, ON AVERAGE OF 4 CUTS IN 1978

	1977 Lysimeters Dijon														Field Experiment 2		
	Sandy soil							Loamy soil									
	F	M	3 M	5 M	3 M/S	3 M/A	6 M/2	F*	M*	3 M	5 M*	3 M/S	3 M/A	6 M/2	F	M	3 M
Total-N %	2.28	1.77	2.73	2.80	2.78	2.28	2.55	2.22	1.87	2.49	2.71	2.58	2.09	2.81	2.20	1.96	2.31
NO_3-N %	0.02	0.01	0.11	0.20	0.13	0.08	0.20	0.01	0.01	0.05	0.16	0.12	0.05	0.26	0.07	0.04	0.11
P %	0.19	0.22	0.23	0.26	0.25	0.17	0.21	0.25	0.33	0.27	0.30	0.29	0.22	0.29	0.33	0.30	0.33
K %	2.34	2.24	2.19	2.21	1.83	1.90	2.05	2.48	2.16	2.47	2.36	2.04	2.34	2.67	2.01	1.97	2.37
Ca %	0.47	0.44	0.48	0.43	0.56	0.49	0.51	0.45	0.45	0.39	0.57	0.42	0.44	0.54	0.45	0.42	0.41
Mg %	0.16	0.21	0.29	0.33	0.33	0.20	0.25	0.22	0.24	0.30	0.32	0.37	0.33	0.31	0.20	0.19	0.19
Cu ppm	10	16	19	18	16	11	27	9	14	16	21	18	12	14	13	12	13
Zn ppm	32	48	55	54	46	50	74	40	47	47	59	60	52	52	39	38	42

F = fertiliser, M = manure supplying total-N equivalent to F., 3 M = M x 3

3 M/S = M x 3 once a year in spring, 3 M/A = M x 3 once a year in autumn

6 M/2 = M x 6 in alternate years, 5 M = M x 5.

* 1976 Lysimeters.

3) spreading a double amount in alternate years (6 M/2) increased total-N content at the first and second cuts in the following year. Nitrogen content decreased at the third and fourth cuts to values less than other treatments,

4) nitrate-N contents were higher at the start of plant growth with treatments 3 M/S and 6 M/2 and the annual mean NO_3-N concentration was significantly higher with treatment 6 M/2,

5) phosphorus content of grass was also significantly affected by treatments.

B. Field experiments

Yields of Fescue for Experiments 1 and 2 are given in Figures 1 and 2. Yields of Fescue in Experiment 1, with fertiliser were comparable with those from lysimeters on the same clay soil with same rate supplied. Fescue with manure yielded less in the field than from lysimeters.

Fescue with treatment M (manure supplying total-N equivalent to fertiliser) yielded an average 7 percent less than with fertiliser (F); the difference was not statistically significant. In the second and third years of the experiments yields with treatment M were 9.5 percent less than with treatment F, and the difference was significant.

Treatment 5 M gave the best yields (22.2 percent greater) over the whole period, which were significantly better.

The mean composition of yields of dry matter given in Table 7 are comparable with those of the 1976 lysimeter study.

For Experiment 2, yields of Fescue for the years 1977 and 1978 did not differ significantly between treatment F and M. Treatment 3 M gave yields 15 percent more on average than treatment F. The mean compositions of yields of dry matter

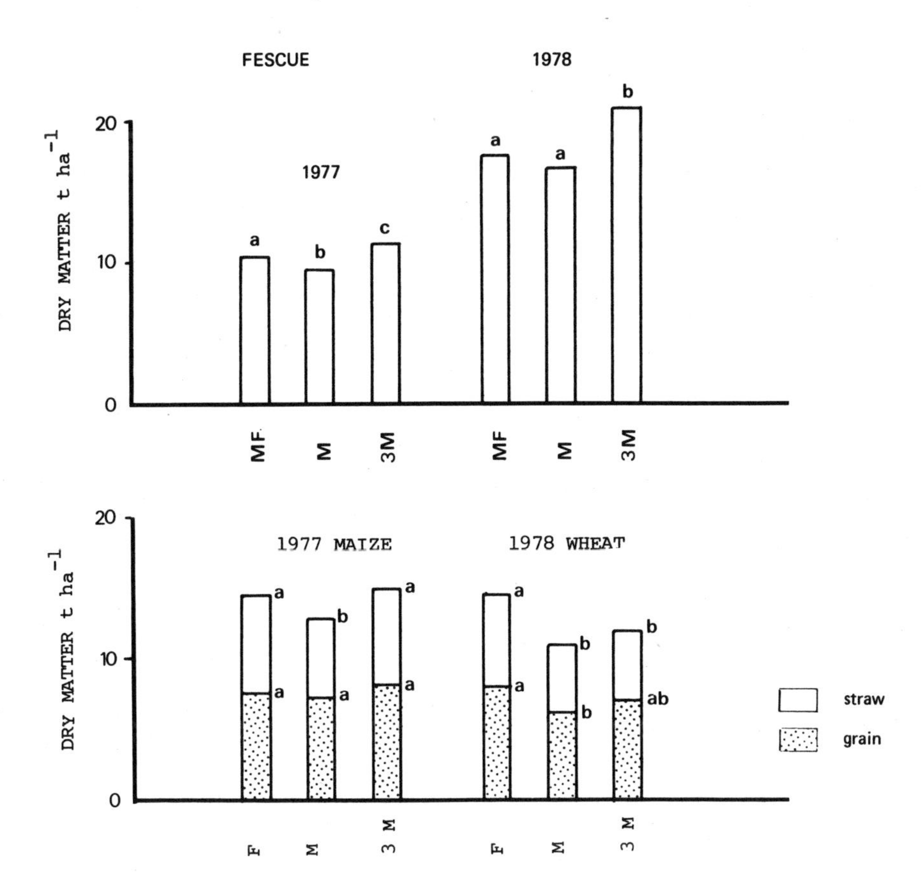

Fig. 2. Crop yields for Field Experiment 2

F = fertiliser

M = manure supplying equivalent N

3 M = manure x 3

Treatments with same letter above yield do not differ at the 5 percent level.

for the 7 cuts are given in Table 9, and indicate significantly smaller total-N in grass with treatment M.

The botanical composition of the plots of Experiment 2 was determined in May 1979, two years after sowing on ploughed out permanent grassland. Results in Table 10 show that Fescue was still dominant with most on the plot with the large dressing of manure. There were no clearly defined trends in the percentages of other species.

TABLE 10

BOTANICAL COMPOSITION IN EXPERIMENT 2 EXPRESSED AS PERCENT OF TOTAL DRY MATTER

	Fertiliser	Manure	Manure x 3
Festuca elatior L.	71.2	72.7	87.7
Dactylis glomerata L.	7.1	3.6	4.5
Poa trivialis L.	15.5	13.6	6.8
Trifolium pratense L.	1.6	8.1	1.0
Taraxacum dens. leonis Desp.	2.5	0.3	x
Rumex crispus L.	2.1	1.7	xx
Other weeds: *Poa annua L. Stellaria media Cry, Cerastium coespitosum gilib. etc.*	xx	xx	x

x = present, xx = numerous.

3.1.2 Annual crops

Yields of grain and straw from Experiment 3 are given in Figure 2. Grain yields of maize are large and do not differ significantly between treatments; treatment M produced significantly less stover than the other treatments. Plant composition during growth and grain composition at harvest were affected by treatments as shown in Table 11, the main effect being that the total-N content of crops with treatment M was smaller than with the other treatments.

TABLE 11

COMPOSITION OF DRY MATTER OF CROPS IN EXPERIMENT 3

	Maize			Winter wheat					
	Grain			Grain			Straw		
	F	M	3 M	F	M	3 M	F	M	3 M
Total-N %	1.55	1.30	1.51	1.80	1.58	1.74	0.49	0.47	0.52
P %	0.34	0.34	0.36	0.35	0.35	0.37	0.17	0.13	0.13
K %	0.24	0.24	0.24	0.28	0.29	0.29	0.76	0.70	0.88
Ca %	0.01	0.01	0.01	0.04	0.03	0.03	0.26	0.16	0.16
Mg %	0.11	0.11	0.12	0.10	0.10	0.10	0.05	0.15	0.10

F = fertiliser, M = manure supplying total-N equivalent to F., 3 M = M x 3.

Winter wheat harvested in 1978 gave statistically different yields of grain and straw with treatments M and F, and the total-N content also differed (Table 11). The full amount of manure was not applied for treatment 3 M, because of excess nitrogen from 1977. Manure was applied twice, in autumn at sowing and in spring at shooting instead of three times. However, wet weather during the winter of 1977 - 1978 prevented excess nitrogen, probably by leaching. Yield was slightly less and composition did not differ statistically from those with treatment F.

3.2 Efficiency of nitrogen in pig manure

Experiment 4 was done during 1977 under wet field conditions (Campardon, 1977) and gave results on the immediate value of manure nitrogen for ryegrass. Figure 3 shows the mean yields obtained for each treatment and Table 12 compares the amount of nitrogen as manure needed to give the same dry matter yield as given by fertiliser-N. Expressing the manure nitrogen as a proportion of the fertiliser-N, twice as much total-N as manure is required to give yields equal to those with fertiliser-N and about 1.3 times as much mineral-N.

TABLE 12

COMPARISON BETWEEN MANURE NITROGEN NEEDED TO OBTAIN SAME YIELD AS WITH FERTILISER

Cutting	Fertiliser nitrogen	Manure nitrogen		Manure-N / Fertiliser-N	
		Total-N	Mineral-N	Total-N	Mineral-N
1	78	138.8	103.1	1.78	1.32
2	50	110.5	60.8	2.21	1.22
1 + 2	128	262.0	173.9	2.05	1.36
1 + 2 + 3	178	384.0	239.5	2.15	1.35

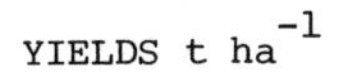

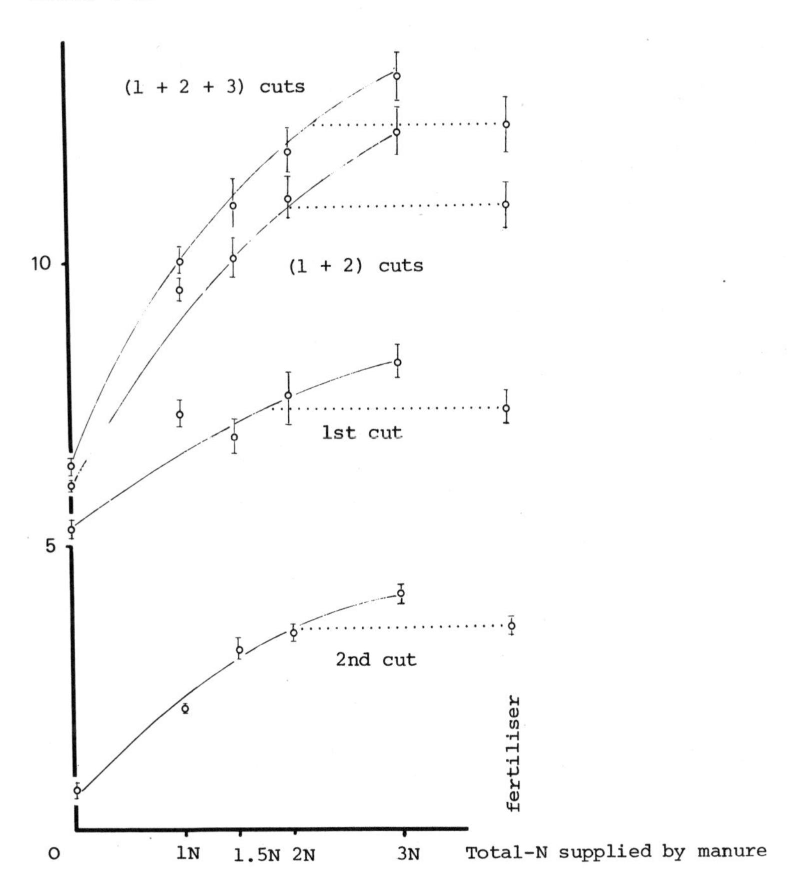

Fig. 3. Dry matter yields of ryegrass in Field Trial 4.

(1 N = Pig manure supply at nitrogen equivalent of mineral fertiliser, etc.)

Nitrogen utilisation coefficients calculated from plant removals and expressed as:

$$\frac{\text{nitrogen in crop with treatment - nitrogen in crop without added N}}{\text{amount of N applied}}$$

are indicated in Table 13.

TABLE 13

COMPARISON OF MANURE AND FERTILISER NITROGEN UTILISATION COEFFICIENTS. FIELD EXPERIMENT 4

Treatment	Nitrogen utilisation coefficient		Manure nitrogen utilisation coeff. / Fertiliser nitrogen utilisation coeff.	
	Total nitrogen	Mineral nitrogen	Total nitrogen	Mineral Nitrogen
F	0.82	0.82	1.00	1.00
M	0.47	0.70	0.57	0.87
1.5 M	0.38	0.58	0.47	0.72
2 M	0.39	0.59	0.48	0.73
3 M	0.32	0.49	0.40	0.60

Manure nitrogen utilisation coefficients are always lower than for ammonium nitrate fertiliser. Between 30 and 50 percent of mineral nitrogen from manure was not effective in the first year or lost compared to only 18 percent for fertiliser-N.

4. DISCUSSION AND CONCLUSION

Mansat (1969) suggested that normal yields of dry matter of Fescue were 12 - 16 tonne ha^{-1} year^{-1} with a maximum yield of possibly 20 tonne. Cooper (1969) proposed a potential yield for ryegrass of 25 tonne dry matter ha^{-1} year^{-1} on small plots. Our results in lysimeters and field experiments were in this range. Therefore we conclude that pig manure, even with very large amounts supplied, did not inhibit the growth of grass (the yield in 1977 was 25 tonne hectare^{-1} with treatment 5 M).

This study considered essentially the nitrogen value of pig manure, independently of other fertiliser elements. Mineral nitrogen in manure is less effective than fertiliser-N with a ratio between manure utilisation and fertiliser utilisation varying from 0.6 to 0.87 depending on the amount of manure applied (Table 13).

These results differ slightly from those obtained in this laboratory by Mohaemen (1978) in pot trials with ryegrass (Figure 4), when yields were proportional to mineral nitrogen supplied as manure or fertiliser in agreement with those of MacAllister (1977). However, the nitrogen utilisation coefficient found by Mohaemen (1978) is slightly smaller for manure than for fertiliser varying from 0.44 to 0.48 for total-N and from 0.54 to 0.76 for mineral-N. The explanation for this difference from Field Experiment 4 may be the probable absence of ammonia volatilisation from soil in the pots where manure was mixed with soil before sowing.

In our experiment, with the same amount of total nitrogen spread on the soil surface, plots with manure yielded less than with fertiliser, but not always significantly so. Nitrogen efficiency defined as the fertiliser-N equivalent of manure can be estimated from the ratio of total nitrogen utilisation coefficents of manure and fertiliser (Table 13). For the first year an efficiency of 0.40 - 0.57 percent was found depending on the amount applied which agrees with the efficiency coefficient of 0.55 noted in a Dutch report (CCE, 1978) for the first year with pig manure applied in spring.

However, these authors noted a mineral-N of 50 percent total-N in pig manure while we have found a mean value near 60 percent. In addition, they considered that 27 percent of the organic-N was mineralised in the first year, whereas

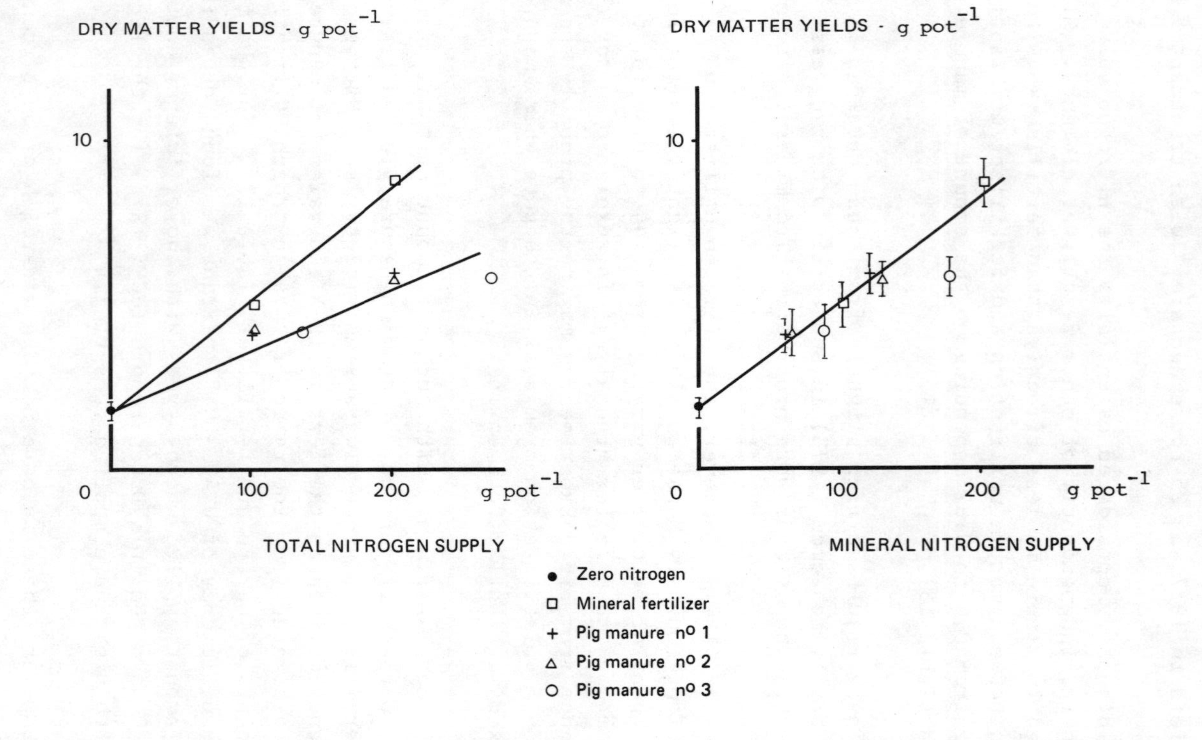

Fig. 4. Dry matter yields of ryegrass (g pot^{-1}) total for three cuts; comparison between three different pig manures and fertiliser, each at two rates.

incubations of pig manure with soils done in this laboratory by Mohaemen (1978) and by Germon et al. (1978) did not clearly demonstrate mineralisation to this extent.

Therefore, from the agricultural point of view in the short term, optimal yield and quality of grass can be obtained by supplying manure containing 2 to 2.5 total-N as would normally be applied as fertiliser. Experiments are now being done to determine the effect of the application of manure for a longer period. In addition, no detrimental effects on the botanical composition of Fescue grassland was observed after two years.

Finally pig manure, used in good weather conditions at convenient times (shortly before sowing and avoiding dry weather) may be of interest as a fertiliser for annual crops as indicated by Experiment 3 which has continued for 1979 with winter barley producing a good yield.

REFERENCES

Campardon, P. 1977. Fertilisation à base de lisier de porc. Expérimentation sur une culture de Ray-grass d'Italie en limon de Bresse. Memorie ENSSAA 53 p.

CCE, 1978. L'épandage des effluents d'élevage sur les sols agricoles dans le CE. I. Bases scientifiques pour une limitation des épandages et critères pour des dispositions réglementaires. CB. NA. 78-047-FR-C. 180 p.

Cooper, J.P., 1969. Potentialités des productions fourragères. Fourrages, 38, 3-19.

Germon, J.C., Giraud, J.J., Chaussod, R. and Duthion, C., 1978. Nitrogen mineralisation-nitrification of pig slurry added to soil in laboratory conditions. CEC Seminar, 10 - 11 October 1978. Chimay (Belgium).

MacAllister, J.S.V., 1977. Efficient recycling of nutrients In: Voorburg, J.H. Utilisation of manure by landspreading. 87-103. Luxembourg.

Mansat, P., 1969. Les potentiels de productions fourragères d'après des références acquises à l'échelle expérimentale. Fourrages, 38, 75-88.

Mohaemen, B., 1978. Evolution dans le sol de l'azote des lisiers., Sa disponibilité à l'égard des végétaux. DEA Biochimie appliquée. Université de Dijon. 70 p.

LANDSPREADING OF LIQUID PIG MANURE: II. NUTRIENT BALANCES AND EFFECTS ON DRAINAGE WATER

C. Duthion, G. Catroux and J.C. Germon
(with the technical collaboration of Y. Couton and L. Guenot)
INRA, BV 1540,
21034 Dijon Cedex, France.

ABSTRACT

Lysimeter studies on the landspreading of liquid pig manure were conducted on 4 soils supplied with different amounts of manure at several application times. The effects of soils and treatments on the composition of the drainage water are discussed. Tentative nutrient balances were made, particularly for nitrogen. The environmental consequences are discussed.

1. INTRODUCTION

Pig manure may have good fertilising value for grasses and annual crops. However, some of the components of the manure may be lost by leaching into underground waters and may cause undesirable eutrophication.

Pig manure, like all organic materials, mineralises at a rate and to an extent depending on soil conditions and may give a temporary excess of mineral nitrogen. Moreover, time of application (during dry or wet periods in the year) application pattern (once a year or several times) may play an important role in possible nutrient leaching. Furthermore, there is often too little land on the farm to allow pig slurry to be safely spread, leading to an overloading of soils with manure which must be followed by leaching of nutrients.

This paper describes results obtained in lysimeter studies of landspreading pig manure on four soils at different levels and times of application.

2. MATERIAL AND METHODS

2.1. Experimental design

The experimental design has been described in detail elsewhere (Duthion, 1979) and comprised field and lysimeter experiments. Field experiments were:

- Experiment 1 on clay soil planted with Fescue
- Experiment 2 on loamy soil with Fescue
- Experiment 3 on loamy soil with annual crops

For all these field experiments a control lysimeter was set up in the same field. Simultaneously, 20 lysimeters were built as described previously (Duthion, 1979) and filled with four different soils.

2.2 Drainage water sampling and analysis

Drainage water was sampled twice each week and after immediate determination of volume, pH and conductivity, an aliquot proportional to the collected volume was frozen. Aliquots from 2 weeks were mixed and analyses were performed on the mixture, giving average determinations for a 2-week period. A sampling frequency of twice a week does not affect the nitrogen and COD composition of samples even in summer with warm conditions (20°C).

The analytical procedures were:

Kjeldahl-N by autoanalyser (perchloric digestion and indophenol blue colorimetry)

NH_4^+-N by indophenol blue colorimetry (autoanalyser)

NO_2^--N by diazotisation and colorimetry (autoanalyser)

NO_3^--N by reduction with Cd-Cu catalyst and colorimetry as nitrate (autoanalyser)

Cations: calcium and potassium - flame photometry

magnesium, zinc and copper - atomic absorption spectrometry

Phosphorus as phosphate by vanado-molybdate colorimetry (autoanalyser) and hydrolysable phosphorus by H_2SO_4 hydrolysis at 120°C and colorimetry as phosphate

Chemical Oxygen Demand by $K_2Cr_2O_7$, H_2SO_4, $AgSO_4$, oxidation and colorimetry (autoanalyser).

2.3 Pig manure analysis (see Duthion, 1979)

2.4 Treatments studied

The purpose of the work is to study the influence of soil type, amount of manure, spreading time and application pattern, and the crops on the composition of the drainage water below a metre in depth. Details of treatments and crops were given previously (Duthion, 1979).

The treatments applied were increasing manure amounts, from treatment M at a total N equivalent to the fertiliser N need up to 5 M which was equivalent to five times treatment M. In fact, they correspond to levels of total nitrogen from 955 to 4 116 kg ha^{-1} for a 3 year period with Fescue in comparison with 880 kg mineral nitrogen from fertiliser.

Treatments related to time and type of application varied from treatment M (total-N equivalent to the fertiliser N need of crops) up to 3 M (equivalent to three times treatment M), supplied in several applications during the year, or once in spring before plant growth or sowing, or once in autumn, or once every 2 years as treatment 6 M (equivalent to six times treatment M). The amounts of total nitrogen spread varied from 557 to 1 832 kg N ha^{-1} for a 2 year period with Fescue, in comparison with 605 kg N from fertiliser.

3. RESULTS

3.1 Drainage water and leaching data

3.1.1. Volumes of drainage water

Drainage water collected from the 1976 to 1977 lysimeters is indicated in Figures 1 and 2. Every year, lysimeters gave drainage water over a 3 to 4 month period: November 1976 to March 1977, November 1977 to April 1978 and January 1979 to April 1979.

From an hydraulic point of view, the lysimeters functioned satisfactorily: several verifications made at periods when evapotranspiration was low and soils at field capacity proved that rain was followed by a corresponding amount of drainage (more than 90 percent recovery).

Lysimeters from Field Experiment 1 yielded 30 to 40 mm more drainage water than a 1976 lysimeter filled with the same clay soil.

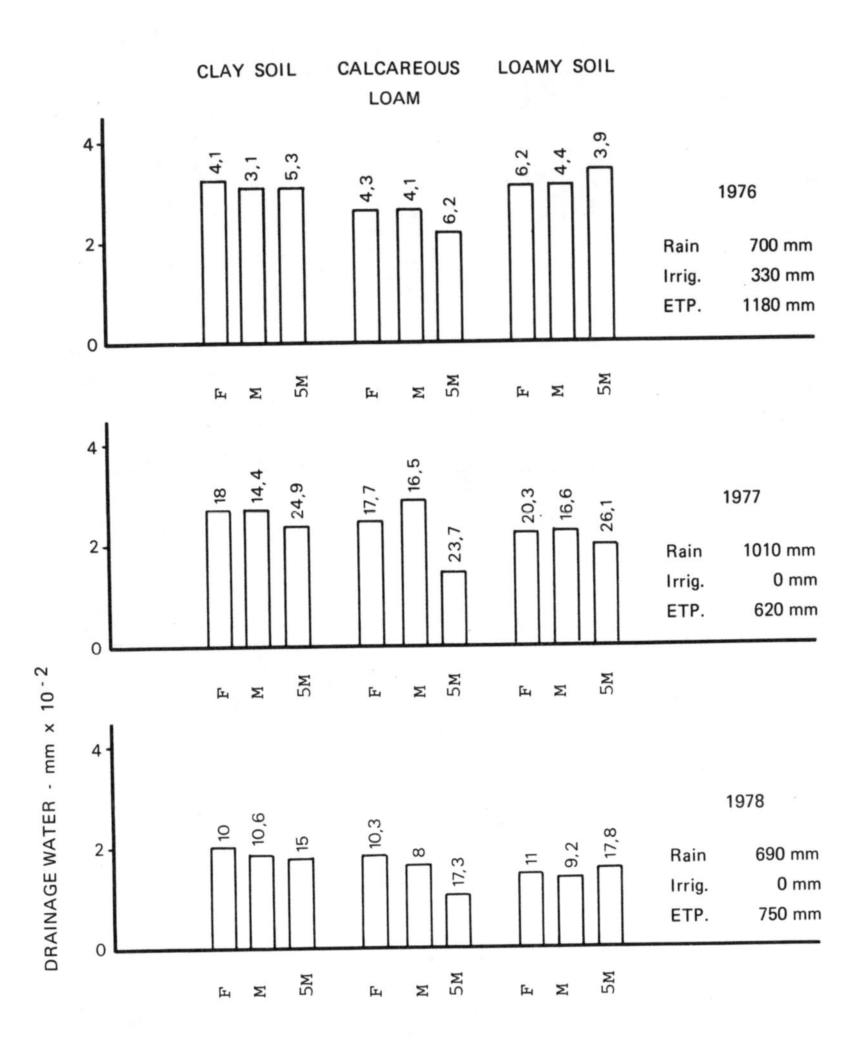

Fig. 1. Drainage water volumes for 1976 lysimeters.
Fescue dry matter yield is indicated above each volume in t ha^{-1}

F = fertiliser

M = manure at N equivalent

5M = manure x 5.

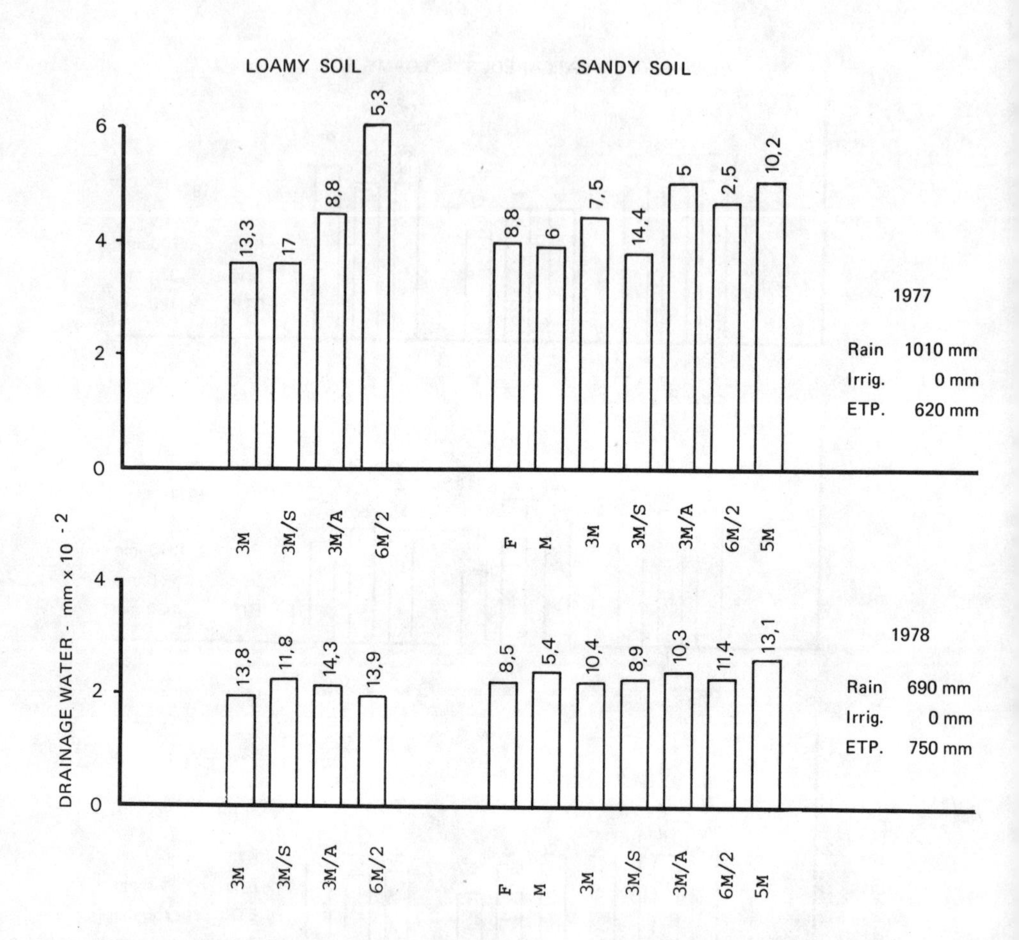

Fig. 2. Drainage water volumes for 1977 lysimeters.
Fescue dry matter yield is indicated above each volume, in t ha^{-1}.

F = fertiliser

M = manure at N equivalent

3 M = manure x 3

3 M/S = manure x 3, spring

3 M/A = manure x 3, autumn

6 M/2 = manure x 6 once, 2 years

5 M = manure x 5.

For lysimeters from Field Experiments 2 and 3, drainage was in the range of 500 mm in 1977 (March 1977 - March 1978) and 430 mm in 1978 (March 1978 - April 1979) for Fescue, corresponding to 535 mm in 1977 and 485 mm in 1978 for annual crops.

3.1.2. Amounts of leached elements

- Treatment effects

Average amounts leached each year for the 1976 lysimeters are given in Table 1. It appears that there was a significant treatment effect. Treatment 5 M is different from treatments F and M although they are not different from each other.

Amounts of total-N, NO_3-N, calcium and magnesium, differ with treatments. In contrast, chemical oxygen demand (COD) does not differ and is very low.

Amounts leached varied from year to year, particularly of nitrate nitrogen, with a smaller quantity in 1977 - a year with favourable conditions for grass growth in the autumn - than in 1978 with very dry conditions in the autumn which stopped plant growth.

Results from the 1977 lysimeters are available for only 2 years and so do not allow any conclusions on the treatment effects to be drawn. This part of the study is still in progress. Partial results are indicated in Table 2.

- Soil effects

With the 1976 lysimeters, some differences are observed between soils for the cumulative amounts leached during a 3 year period, as indicated in Table 3. However, the experimental design does not allow any statistical difference to be proved.

TABLE 1

AVERAGE AMOUNTS OF LEACHED ELEMENTS FOR THE 3 SOILS 1976 LYSIMETERS IN kg ha^{-1}

	Year 1976 (19/3/76 - 22/3/77)			Year 1977 (23/3/77 - 13/3/78)			Year 1978 (14/3/78 - 20/4/79)		
	F	M	5 M	F	M	5 M	F	M	5 M
N Kjeldahl	< 1	3	9	2	2	4	1	< 1	2
N NO_3	15	11	122	6	< 1	60	48	11	149
P PO_4	1	2	2	< 1	< 1	1	< 1	< 1	3
K	69	50	60	28	39	30	37	31	35
Ca	126	111	197	95	113	130	145	102	250
Mg	4	5	11	4	4	5	8	3	9
Cu	0	0	0	0	0	0^+	0	0	0
Zn	0	0	0	0^+	0^+	0^+	0	0	0
COD	104	83	141	117	114	86	81	55	70

F = fertilise; M = pig manure at total-N equivalent to F, 5 M = M x 5

$^+$ traces in one or several lysimeters $treatment^{-1}$.

TABLE 2

AVERAGE AMOUNTS OF PERCOLATED ELEMENTS FOR 2 SOILS, 1977 LYSIMETERS IN kg ha^{-1}

	Year 1977 (23/3/77 - 13/3/78)				Year 1978 (14/3/78 - 20/4/79)			
	3 M	3 M/S	3/A	6 M/2	3 M	3 M/S	3 M/A	6 M/2
N Kjeldahl	11.5	29.5	24.5	61	3	3.5	1	7
N NO_3	30.5	78	55	55.5	55	10	105.5	63
P PO_4	3	1	2.5	5	5.5	2	13	3
K	79	151	115	82	51.5	55.5	65	48.5
Ca	114.5	130	152	200	66.5	50	92.5	78.5
Mg	10	23.5	20	25.5	10	10	19	9.5
Cu	0	0	0	0^+	0	0	0	0
Zn	0^+	0^+	0^+	0^+	0	0	0^+	0^+
COD	128	370	286.5	461	207	40	80	213

3 M = pig manure at 3 x total-N equivalent to F, 2 M/S = 3 M once in spring, 3 M/F = 3 M once in autumn, 6 M/2 = 6 M once every 2 years.

$^+$traces in one or several lysimeters treatment^{-1}

TABLE 3

AMOUNTS OF PERCOLATED ELEMENTS FOR 1976 LYSIMETERS IN kg ha^{-1} DURING A 3 YEAR PERIOD (MARCH 1976 - APRIL 1979).

	N Kjeldahl			N NO_3			K			Ca			Mg		
Treatment	F	M	5 M	F	M	5 M	F	M	5 M	F	M	5 M	F	M	5 M
Soil:															
Clay soil	4	7	18	71	36	401	134	76	101	297	302	603	23	11	26
Calcareous loam	2	4	5	108	13	270	151	112	92	583	459	548	13	8	11
Loamy soil	4	5	24	26	15	312	116	173	180	217	218	577	16	14	41

F = fertiliser, M = pig manure at total-N equivalent to M, 5 M = M x 5.

With the lysimeters in Field Experiment 1 on clay soil, with the same treatments as the 1976 lysimeter, nitrogen leaching was slightly higher but with correspondingly greater quantities of drainage water. Moreover, calcium leaching was higher and potassium leaching lower than for the 1976 lysimeters.

- Crop effects

There are insufficient comparisons between grass and annual crops, in Field Experiments 2 and 3 after only 2 years; they must be continued and not analysed until after at least one crop rotation. Nevertheless, it is possible to identify, for 1977, a higher nitrogen retention by grass than by corn.

3.2 Tentative nutrient balances

Results presented here and previously (Duthion, 1979) make it possible to calculate an apparent element balance between the amount of an element supplied and the amounts removed by plants or by drainage water.

3.2.1. Nitrogen balances

The apparent nitrogen balances for the 1976 and 1977 lysimeters are indicated in Figure 3. Corresponding nitrogen balances for lysimeters from Field Experiments 1, 2 and 3 are indicated in Table 4 for the period 1977 - 1979.

3.2.2. Nitrate concentrations

Average nitrate nitrogen concentrations in the drainage from lysimeters were calculated by dividing the total annual amount of nitrate nitrogen leached by the volume of the drainage water. Results are indicated in Table 5 for the 1976 lysimeters for each year and for the entire 3 year period. There are no evident differences between soils. Even if treatment M was usually lower than treatment F, differences between treatments are not significant. The drainage from treatment 5 M has the higher nitrate nitrogen concentration - more than five times the 10 mg NO_3-N l^{-1} limit.

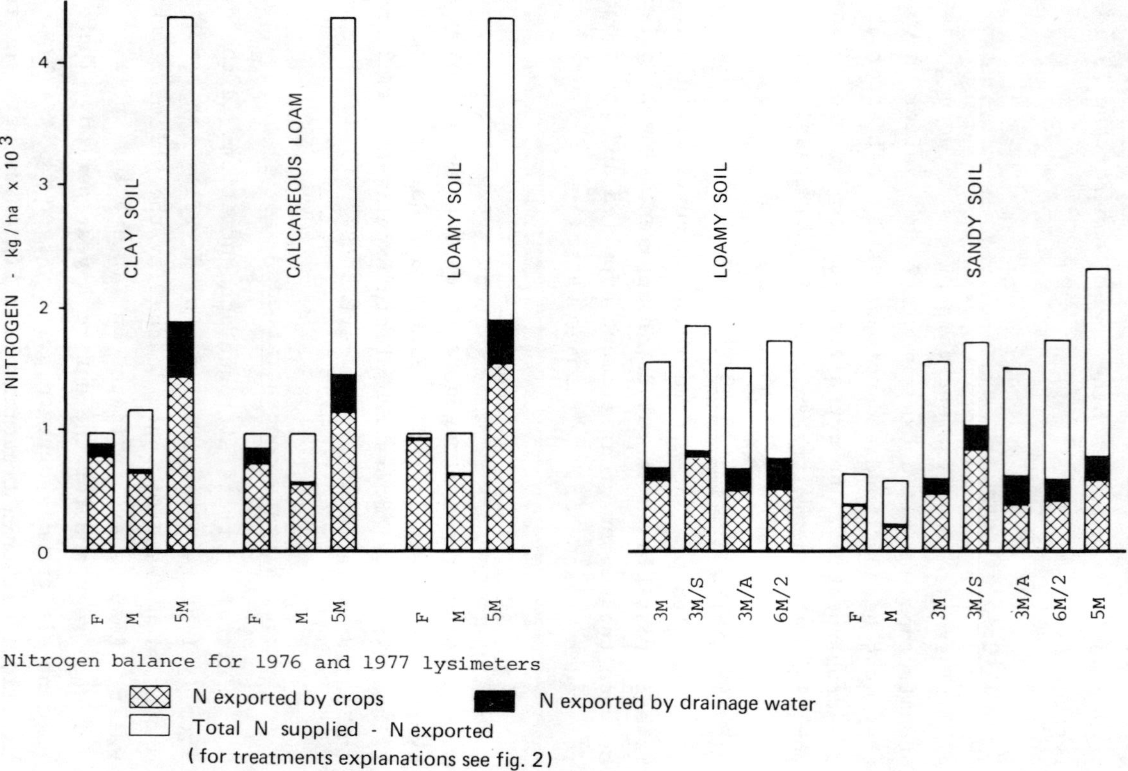

Fig. 3. Nitrogen balance for 1976 and 1977 lysimeters

N exported by crops
N exported by drainage water
Total N supplied - N exported
(for treatments explanations see fig. 2)

TABLE 4

NITROGEN BALANCE FOR FIELD EXPERIMENTS 1, 2 AND 3 IN kg ha^{-1}

	Experiment 1			Experiment 2			Experiment 3		
Treatment	F	M	5 M	F	M	3 M	F	M	3 M
Total N supplied	615	624	2 921	583	649	1 716	487	613	1 591
N removed by crops	602	511	1 134	637	515	810	291	214	269
N removed by drainage	34	6	393	68	70	327	390	201	564
(Total-N supplied - N removed)	- 21	+ 107	+ 1 394	- 172	+ 64	+ 579	- 194	+ 198	+ 758

F = fertiliser, M = pig manure at total N equivalent to F, 3 M = M x 3, 5 M = M x 5.

TABLE 5

AVERAGE N NO_3 CONCENTRATIONS IN DRAINAGE WATERS FROM 1976 LYSIMETERS IN mg N l^{-1}

Soil / Treatment		F	M	5 M
Clay soil	1976	3.7	2.3	28.6
	1977	0.2	0.2	47.2
	1978	29.0	15.8	112.7
1976 - 1978		8.9	4.7	55.8
Calcareous loam	1976	3.4	3.8	43.8
	1977	6.5	0.2	23.0
	1978	44.1	1.8	137.9
1976 - 1978		15.4	1.8	58.3
Loamy soil	1976	8.0	4.8	54.1
	1977	0.2	0.2	18.8
	1978	0.7	0.4	58.3
1976 - 1978		3.8	2.2	45.0

F = fertiliser, M = pig manure at total N equivalent as F, 5 M = M x 5

TABLE 6

AVERAGE N NO_3 CONCENTRATIONS IN DRAINAGE WATERS FROM 1977 LYSIMETERS IN mg l^{-1} FOR 2 YEARS (MARCH 1977 - APRIL 1979). FOR TREATMENTS EXPLANATIONS SEE TABLES 1 AND 2

Soil / Treatment	F	M	3 M	5 M	3 M/S	3 M/A	6 M/2
Sandy soil	2.9	1.1	16.5	21.0	23.1	25.3	17.4
Loamy soil			11.0		6.0	20.2	14.3

Frequencies of nitrate nitrogen concentrations are indicated in Figure 4, by dividing the concentration range into 4 steps. These results confirm those observed with average concentration in that treatment 5 M produced drainage water heavily loaded with nitrate much more frequently.

Average nitrate concentrations in the 1977 lysimeter drainage water over the 2-year period, are given in Table 6.

3.2.3. Other elements

Phosphorus was always supplied in excess by manure, even with treatment M which was not sufficient for optimal plant growth, from the nitrogen point of view. Due to the virtual immobility of phosphorus, a build-up of 1 100 kg of P ha^{-1} occurred in 3 years for treatment 5 M in the 1977 lysimeters. Potassium was not supplied in sufficient quantity by manure on grassland except by treatment 5 M, while for annual crops, potassium needs might have been satisfied.

Calcium balances depended on soils and situations. Generally, calcium supplied by manure was adequate and sometimes excessive with a build-up of around 1 000 kg Ca ha^{-1} for 3 years with treatment 5 M. The situation was quite different in Field Experiments 2 and 3 on loamy soil where calcium leaching was higher.

Magnesium, zinc and copper were supplied in excess with a build-up in soil after 3 years varying from 9 to 45 kg Zn ha^{-1} and from 7 to 42 kg Ca ha^{-1} for treatment M and 5 M in the 1976 lysimeters.

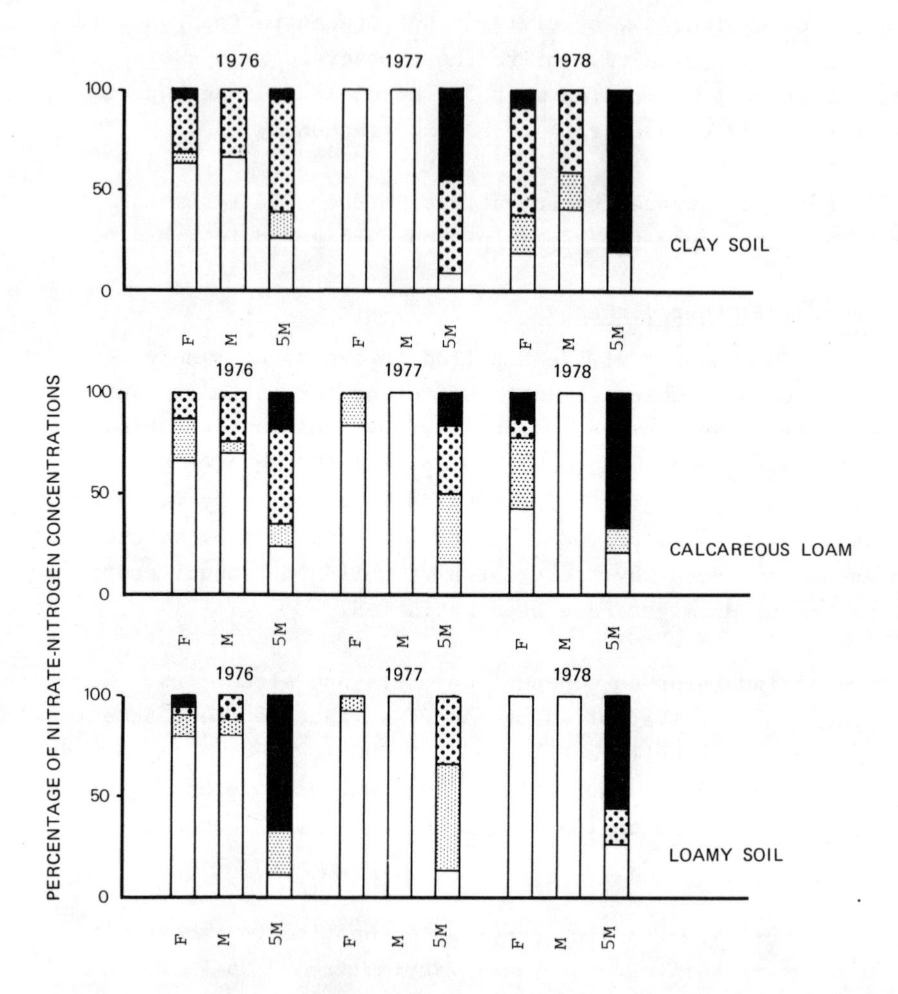

Fig. 4. Percentage of nitrate nitrogen concentrations in drainage waters 1976 lysimeters

F = fertiliser

M = manure at N equivalent

5 M = manure x 5

< 10 mg l^{-1} N

10.20 mg l^{-1} N

20 - 50 mg l^{-1} N

> 50 mg l^{-1} N

4. DISCUSSION

Volumes of drainage water during a one year period varied within lysimeters containing the same soil, due probably to differences in filling as well as in treatment. For example, in 1977 - a year favourable to plant growth - treatment 5 M which gave the highest grass yield gave also the lowest drainage volume. This is probably due to the dry matter yield increase of 5 000 to 6 000 kg DM ha^{-1} over treatment F and also to an increasing influence of edge effects.

Volumes of drainage water also varied with the soil according to its water-holding capacity. In the case of calcareous loam, the difference between treatment 5 M and treatments F and M, were higher than for the other soils. We suggest that this was an effect of carbon dioxide evolved from the action of manure on limestone introducing modifications of soil properties.

COD reduction was sufficient with all manure treatments in all soils since for an application of 42 000 kg COD ha^{-1} over 3 years, the total COD leached was around 300 kg for treatment 5 M. This was the same as for treatment F.

Mineral phosphate leaching was insignificant and determinations of hydrolysable phosphorus in 1978 - 1979 demonstrate very little phosphorus leaching.

Copper and zinc were sometimes detected as traces at levels less than 0.2 mg l^{-1}. Practically speaking nitrogen leaching involved only nitrate. Treatment M, supplying the same total nitrogen equivalent, gave lower amounts of leached nitrate than treatment F due to the lower amount of potentially mineralisable nitrogen than total nitrogen in manure. However, treatments M and F are not statistically different. Nitrate leaching increased with the amount of manure spread as follows: treatment M < treatment 3 M < treatment 5 M.

However, for treatment 5 M on the 1976 lysimeters, nitrate leaching did not exceed 6 to 9 percent of the total nitrogen supplied to the 3 soils over 3 years. Hewgill and Le Grice (1976) give, for a total application of 600 m^3 ha^{-1} of pig manure, a value of 3 - 4 percent but this was for applications made only in summer and with drainage very inferior to that which we obtained. Moreover we noted rather big differences from one year to another in climatic conditions favourable to the growth of grass and/or the evolution of nitrogen applied to soil.

Leaching increased with the amount of total nitrogen applied although there was no very precise relationship between these two quantities, even in the same year. The result was the same if instead of total applied nitrogen we considered excess of nitrogen over that removed by crops, or only the last application in the autumn.

The results obtained in the comparison of split applications with single applications of the same total volume must be interpreted with some caution. However, some practical facts are evident. A single application in autumn seems potentially more polluting than other patterns of application. If too late, it will result in rapid leaching of Kjeldahl nitrogen because it coincides with the resumption of drainage; if earlier in a relatively dry year which favours nitrification and not the growth of grass, it will give high nitrate levels in drainage water. On the other hand single application in spring, coinciding with the beginning of plant growth, seems to limit nitrogen leaching as long as the moisture deficit of the soil is adequate for the volume of manure applied. With heavy applications of cattle manure in March, Burford et al. (1976) recorded in water reaching the drainage network, less than 1 percent of the applied nitrogen.

The crop has an evident influence on the behaviour of nitrogen as shown in a comparison between Field Trials 2 and 3. More important, the drainage on these two trials shows higher percolation rates even on grassland than in the lysimeters.

After manure applications containing 5 times the nitrogen needs of the crop, nitrate nitrogen concentrations in the drainage water are rarely within the limits of 10 mg NO_3^--N l^{-1} fixed by health authorities. The average concentrations varied from 20 to 130 mg N l^{-1} according to the year and the soil. In contrast, with dose M, they are generally lower than this limit, and always below those from the F treatment. At dose 3, concentrations are intermediate.

Mineral nitrogen removed by plants and drainage represented on average, for the 3 soils of the 1976 lysimeters over the 3 years, 97 percent of the mineral nitrogen supplied by pig manure at the M level and 66 percent of the mineral nitrogen at the 5 M level. Apparent gains, given in the nitrogen balances in Table 4 and Figure 3 are equivalent to the organic nitrogen supplied by dose M: these gains reach 60 percent of the total supplied with dose 5 M. However, these apparent gains include losses by NH_3 volatilisation and possible denitrification. Our results suggest that volatilisation, estimated at 20 percent of the ammoniacal nitrogen applied by Dutch Document CCE (1978), could vary very much according to the amounts supplied. It is also important to evaluate more precisely the volatilisation which takes place.

Our trials do not permit a precise evaluation of the maximum dose of pig manure which will not result in nitrogen leaching higher than is observed after a mineral N dressing. Without taking into account any evolution of nitrogen from exsiting sources in the soil (for example, Importance of denitrification, Erickson et al., 1974) the dose 5 M - included in our trials because it was considered to be an extreme dose - is very excessive. However at one metre depth of soil, the quantities leached represent only a small percentage of that supplied. The dose 3 M can, depending on the year, constitute a risk to drainage water. Until more precise results are available, it seems reasonable to assume that an application of manure of 2 to 2.5 M will give the same yield as mineral fertilisation, providing only for the requirements for plant growth.

REFERENCES

Burford, J.R., Greenland, D.J. and Pain, B.F., 1976. Effects heavy dressings of slurry and inorganic fertilisers applied to grassland on the composition of drainage waters and the soil atmosphere. Minist. Agric. Fish. Bull. (London), 32, 432-443.

CCE, 1978. L'épandage des effluents d'élevage sur les sols agricoles dans la CE. I. Bases scientifiques pour une limitation des épandages et critères pour des dispositions réglementaires. CB. NA. 78-047--FR-C, 180 p.

Duthion, C., 1979. Landspreading of pig manure. I. Effects on yield and quality of crops. CEC Seminar, Oldenburg, 2 - 5 October 1979.

Erickson, A.E., et al., 1974. Soil modification for denitrification and phosphate reduction of feedlot waste Environ. Prot. Ag. Document 660, 2-74-0057, 120 p.

Hewgill, D. and Le Grice, S., 1976. Lysimeters study with pig slurry. Minist. Agric. Fish. Bull. (London), 32, 444-460.

LANDSPREADING OF LIQUID PIG MANURE:
III.1. SURVEY OF THE PIG FARMS IN THE 'BRESSE'

J.C. Germon, C. Duthion, Y. Couton, R. Grosman
L. Guenot and J. Mortier
INRA, BV 1540,
21034 Dijon Cedex, France.

ABSTRACT

Forty eight samples of liquid pig manure were collected from pig fattening units and breeding units in two areas in the East of France. Analyses were made of the homogenised samples and of the liquid supernatant fraction after centrifuging: pH, density, ash, COD, Kjeldahl and NH_4^+ *nitrogen, P, K, Ca, Mg, Na, Cu and Zn were determined. Results and their variations are discussed.*

1. INTRODUCTION

During the last 20 years, the development of pig breeding in France has been characterised by an increase in the size of the production units and their concentration in some well-defined areas. This was followed by two consequences for the disposal of manure:

- The high density of pigs on limited land areas has involved massive, and generally polluting, landspreading or the search for solutions involving treatment. In both cases the plant nutrients are very badly utilised.
- The problems of waste disposal are of a marked regional character which must be taken into account in the studies undertaken to resolve them.

Data on the composition of pig manure are abundant, but the results can show very important fluctuations according to the authors and areas or countries (Heduit et al., 1977). Therefore, we decided to define more accurately the pig manure composition in Bresse within the framework of a more general study on the consequences of landspreading manure. This work had two aims: on the one hand, to bring a direct answer to a request from the local professionals and on the other hand, to try to display simple statistical relationships between the concentrations of different elements in order to define the best fertilising value of manure from a limited number of observations.

The study was made in two departments of Bresse: Ain and Saône-et-Loire. In this first part, after the description of the methodology used, we present the analytical results and the different comparisons made on the composition of the manure according to its geographical origins (Ain or Saône-et-Loire) according to the enterprise (fattening or breeding), and to the dates of the different samplings. The analysis of the relationships between the different components of the manure is given in the second part.

2. MATERIAL AND METHODS

Origin of the samples

In Saône-et-Loire, 20 manure tanks were sampled (samples 1 to 20) among which there were 8 from breeding units, (samples 2, 9, 10, 11, 12, 13, 14, 20) in the cantons of Cuiseaux, Buxy, Sennecey-le-Grand, Verdun-sur-le-Doubs, Pierre-de-Bresse, Montret and Seurre, in Côte d'Or).

In Ain, 28 samples (numbers 21 to 49) were taken from fattening units only in the cantons of Pont-de-Vaux, Saint-Trivier de Courtes, Montrevel-en-Bresse, Treffort, Coligny, Bage-le-Chatel.

The size of the piggeries studied is appreciably more important in Ain (350 pigs on average with extremes of 70 and 1 200) than in Saône-et-Loire (250 pigs on average in fattening units, with extremes of 115 and 530).

The samples were taken in February 1978. In 7 piggeries in Ain, samples were also taken in June and September 1978 and January 1979. In these piggeries, the food consists mainly of pellets, and water.

Method of sampling

The samples were taken over the whole depth of the tank with a circular pipe (diameter of 8 cm) the bottom end of which can be closed by a stopper operated by a nylon thread which goes through the pipe. The samples were made in triplicate at least taking into account the layout of the tank. Then, the samples were cooled to 4°C before the treatment preceding the analyses.

Samples treatment

In the laboratory the samples were divided into two parts: the first part was analysed in the raw state; the second part was centrifuged at 16 000 g for 20 minutes and the soluble elements in the supernatant were determined. The density and pH were measured before this treatment, the other elements after.

Analysis

The pH was measured of the raw and centrifuged manure, after dilution at 1 : 2 with demineralised water. The density was determined by weighing a volume of 50 ml of manure. The dry matter was determined after drying 50 ml at 105°C to uniform weight. These dry matter samples were heated twice for six hours at 550°C with a treatment of 10 percent NH_4NO_3 in between the combustions to achieve oxidation of the organic matter. Ash was weighed, dissolved in 1 ml of concentrated HC1, then made to 100 ml. In this solution different cations were estimated: K^+, Na^+, Ca^{++} by flame-emission, and Mg^{++}, Cu^{++}, Zn^{++} by atomic absorption spectrometry.

Ammoniacal nitrogen was obtained by distillation of 25 ml of manure with 5 g of magnesium oxide. Kjeldahl nitrogen was obtained after digestion of 25 ml of manure with 3 g of selenium catalyst and 25 ml of concentrated H_2SO_4, by steam distillation of one aliquot and titration of ammonia trapped in boric acid.

Total phosphorus was measured in the ash after combustion at 450°C, precipitating silica and dissolving in concentrated HNO_3 and titration with nitrovanado - molybdic reagent. Chemical Oxygen Demand was determined according to AFNOR NFT 90-101 norm.

3. RESULTS

3.1. Composition of manures

Tables 1 and 3 give the composition of the manure in the 2 departments, Tables 2 and 4 give the concentrations of the soluble elements expressed on the basis of the initial volume of raw manure. Table 5 gives the mean values of the two departments together.

First of all, we can observe that these manures are fairly concentrated: 60 to 70 g l^{-1} of dry matter with a difference between two types of manures: some with a dry matter concentration above 100 g l^{-1} - one-third of samples in Saône-et-Loire and only one-tenth in Ain - and the others with a concentration often below 50 g l^{-1}. This difference is very clear for the manure in Saône-et-Loire, but less clear in Ain. The concentrations of main fertilising elements are within the limits of values observed in the literature: appreciably lower than those of Heduit et al. (1977) but also with less dry matter and comparable with those of Tunney (1975), whose manures contain more dry matter than ours.

In Table 5 we observe that the solid fractions contain 80 percent of dry matter, 67 percent of ash, 75 percent of COD and calcium and almost all the phosphorus, magnesium, copper and zinc. This last point can explain, in part, the small movement of these elements observed in the landspreading of manure (Duthion et al., 1979). At least 60 percent of total nitrogen is in the soluble ammoniacal form.

3.2. Comparison of the manures coming from fattening units of the two departments

The comparison between the results from the two departments shows that the manures from the fattening units of Saône-et-Loire are significantly higher in dry matter than those of Ain as indicated by variance analyses in Table 6. Such a difference can be explained, in part, by a smaller depth

TABLE 1

COMPOSITION OF MANURES (TOTAL). DEPARTMENT OF 'SAONE-ET-LOIRE', MEANS, MINIMA, MAXIMA AND STANDARD DEVIATIONS

	pH	Density kg m^{-3}	Dry matter kg m^{-3}	Ash kg m^{-3}	COD kg m^{-3}	Nitrogen kg m^{-3}	
						Total Kjeldahl	Ammoniacal
$\bar{x}$	7.3	1031.6	69.4	21.0	67.15	5.4	3.3
Minimum	6.7	1012	25	7.7	28	2.4	1.0
Maximum	7.9	1046	138	43	152	7.9	5.1
S	0.3	11.8	37.3	9.7	34.9	1.9	1.1

	P kg m^{-3}	K kg m^{-3}	Ca kg m^{-3}	Mg kg m^{-3}	Na kg m^{-3}	Cu g m^{-3}	Zn g m^{-3}
$\bar{x}$	2.55	2.3	5.1	0.72	0.43	52.8	91.7
Minimum	0.8	0.6	2.3	0.27	0.15	6.4	23
Maximum	5.2	4.2	9.1	1.49	0.86	114	162
S	1.25	0.9	2.1	0.32	0.19	38.6	41.8

TABLE 2

SOLUBLE ELEMENT CONCENTRATIONS IN MANURE - DEPARTMENT OF 'SAONE-ET-LOIRE'. MEANS, MINIMA AND MAXIMA, STANDARD DEVIATIONS

	Dry matter kg m^{-3}	Ash kg m^{-3}	COD kg m^{-3}	Nitrogen kg m^{-3}	
				Total Kjeldahl	Ammoniacal
$\bar{x}$	9.9	5.9	16.3	3.35	3.0
Minimum	3.5	1.7	4.6	1.0	0.8
Maximum	19.1	9.6	32.2	5.0	4.6
S	4.7	2.3	7.9	1.15	1.1

	P kg m^{-3}	K kg m^{-3}	Ca kg m^{-3}	Mg kg m^{-3}	Na kg m^{-3}	Cu g m^{-3}	Zn g m^{-3}
$\bar{x}$	0.21	1.85	1.24	0.027	0.38	1.35	1.97
Minimum	0.10	0.3	1	0.013	0.15	0.15	0.7
Maximum	0.39	3.6	1.70	0.045	0.84	15	14.8
S	0.07	0.80	0.20	0.010	0.20	3.29	3.11

TABLE 3

COMPOSITION OF MANURES (TOTAL). DEPARTMENT OF 'AIN'. MEANS, MINIMA, MAXIMA, STANDARD DEVIATIONS

	pH	Density kg m^{-3}	Dry matter kg m^{-3}	Ash kg m^{-3}	COD kg m^{-3}	Nitrogen kg m^{-3}	
						Total Kjeldahl	Ammoniacal
$\bar{x}$	7.3	1027.7	57.0	18.8	56.5	4.95	2.85
Minimum	6.9	1010	13.5	6.4	24	2.3	1.4
Maximum	7.6	1046	114	63	124	7.6	4.2
S	0.22	9.2	24.3	10.9	21.9	1.3	0.8

	P kg m^{-3}	K kg m^{-3}	Ca kg m^{-3}	Mg kg m^{-3}	Na kg m^{-3}	Cu g m^{-3}	Zn g m^{-3}
$\bar{x}$	1.8	2.1	3.3	0.85	0.55	53.0	54.0
Minimum	0.25	0.9	1.6	0.11	0.30	2.4	6.0
Maximum	3.3	3.4	6.8	1.86	0.77	102	197
S	0.9	0.6	1.25	0.39	0.13	26.2	40.1

TABLE 4

SOLUBLE ELEMENT CONCENTRATIONS IN MANURE - DEPARTMENT OF 'AIN'. MEANS, MINIMA AND MAXIMA, STANDARD DEVIATIONS

	Dry matter kg m^{-3}	Ash kg m^{-3}	COD kg m^{-3}	Nitrogen kg m^{-3}	
				Total Kjeldhal	Ammoniacal
$\bar{x}$	9.2	6.65	14.2	3.0	2.7
Minimum	4.6	3.1	7.5	1.4	1.3
Maximum	14.9	11.3	20.4	4.5	4.2
S	2.65	2.10	3.6	0.9	0.75

	P kg m^{-3}	K kg m^{-3}	Ca kg m^{-3}	Mg kg m^{-3}	Na kg m^{-3}	Cu g m^{-3}	Zn g m^{-3}
$\bar{x}$	0.11	1.9	1.13	0.037	0.48	0.70	0.62
Minimum	0.03	0.9	0.9	0.015	0.25	0.1	0.29
Maximum	0.20	3.3	2.3	0.085	0.79	2.2	1.1
S	0.05	0.6	0.26	0.018	0.14	0.43	0.20

TABLE 5

AVERAGE COMPOSITION FOR THE TWO DEPARTMENTS

		pH	Density kg m^{-3}	Dry matter kg m^{-3}	Ash kg m^{-3}	COD kg m^{-3}	Nitrogen kg m^{-3}	
							Total Kjeldahl	Ammoniacal
1	$\bar{x}$	7.28	1029.4	62.2	19.7	60.9	5.1	3.0
	s	0.25	10.4	30.3	10.4	28.0	1.6	0.95
2	$\bar{x}$	-	-	9.5	6.3	15.05	3.2	2.8
	s	-	-	3.6	2.2	5.8	1.0	0.9

		P kg m^{-3}	K kg m^{-3}	Ca kg m^{-3}	Mg kg m^{-3}	Na kg m^{-3}	Cu g m^{-3}	Zn g m^{-3}
1	$\bar{x}$	2.1	2.2	4.0	0.79	0.50	52.9	69.7
	s	1.05	0.8	1.7	0.36	0.16	31.9	40.8
2	$\bar{x}$	0.15	1.9	1.2	0.04	0.44	0.97	1.2
	s	0.06	0.7	0.2	0.02	0.17	2.14	2.0

1 = Total elements in bulk manure; 2 = Soluble elements in bulk manure.

TABLE 6

COMPARISON OF MANURES FROM PIG FATTENING UNITS FOR THE TWO DEPARTMENTS. MEAN CONCENTRATIONS AND VARIANCE ANALYSIS

		pH	Density kg m^{-3}	Dry matter kg m^{-3}	Ash kg m^{-3}	COD kg m^{-3}	Nitrogen kg m^{-3}	
							Total Kjeldahl	Ammoniacal
Total elements	$\bar{x}$ (Ain)	7.28	1027.75	57.0	18.8	56.5	4.95	2.85
	$\bar{x}$ (Saône-et-Loire)	7.19	1034.9	85.1	24.0	83.25	6.20	3.5
	F	1.3	4.4	7.5**	1.9	8.2**	5.8*	4.25*
Soluble elements	$\bar{x}$ (Ain)	-	-	9.2	6.65	14.2	3.0	2.7
	$\bar{x}$ (Saône-et-Loire)	-	-	11.2	6.35	19.5	3.6	3.2
	F	-	-	2.4	0.14	7.8**	2.8	3.2

		P kg m^{-3}	K kg m^{-3}	Ca kg m^{-3}	Mg kg m^{-3}	Na kg m^{-3}	Cu g m^{-3}	Zn g m^{-3}
Total elements	$\bar{x}$ (Ain)	1.8	2.1	3.3	0.84	0.55	53.0	54
	$\bar{x}$ (Saône-et-Loire)	2.7	2.6	5.1	0.79	0.50	72.6	92.8
	F	5.5*	3.4	14.8**	0.15	1.8	3.7	6.95*
Soluble elements	$\bar{x}$ (Ain)	0.1	1.9	1.1	0.04	0.5	0.7	0.6
	$\bar{x}$ (Saône-et-Loire	0.25	2.0	1.25	0.04	0.4	2.0	2.5
	F	48.9**	0.2	2.0	0.2	1.2	2.8	6.5*

* Significant $P < 0.05$ ** Highly significant $P < 0.01$

of the manure tanks: on average 1.90 m in Saône-et-Loire and 2.40 m in Ain. Generally the tanks are not fully emptied and the sediment present in all tanks is proportionately more important in the shallower tanks.

Moreover, we have observed a highly significant negative correlation between the depth and the dry matter content in the department of Saône-et-Loire ($r = 0.91$). However, such a correlation was not observed for the manures of Ain.

3.3. Comparison between the manures coming from fattening units or from breeding units

This comparison was made with the manures from Saône-et-Loire; the data are given in Table 7. The manures from breeding units contain significantly less dry matter than the manures from fattening units: they are also lower in COD and nitrogen. Copper from the rations is found in effluents from fattening units.

3.4. Comparison between the manures taken at different periods of the year

Table 8 gives the mean composition of raw manures taken in February, June, September 1978 and January 1979 in some piggeries of Ain, and the values for F (variance ratio of 2 factors - date and sample origin - without repetition).

Such is the heterogeneity, that we can osberve a significant influence of the sampling date on the manure composition only for nitrogen, sodium and zinc. However, total nitrogen excepted, the samples taken in June show the lowest concentrations in all elements; these lower values can be related to the animals' weight at the same period: in June the mean weight for the different piggeries was 50 kg pig^{-1} whereas it was 70 kg in September, and 75 - 80 kg in January and February. However, the correlation between the estimated weight and the values for different parameters proved to be without significance.

TABLE 7

COMPARISON BETWEEN MANURE OF PIG FATTENING UNITS AND NURSERIES. MEAN CONCENTRATIONS AND VARIANCE ANALYSIS FOR EACH DETERMINED ELEMENT

		pH	Density kg m^{-3}	Dry matter kg m^{-3}	Ash kg m^{-3}	COD kg m^{-3}	Nitrogen kg m^{-3}	
							Total Kjeldahl	Ammoniacal
Total elements	$\bar{x}$ fattening units	7.19	1034.9	85.1	24.0	83.25	6.2	3.5
	$\bar{x}$ nurseries	7.4	1026.9	46.0	16.4	43.0	4.2	2.9
	F fattening units	2.5	2.5	6.9*	3.3	9.1**	6.6*	1.6
Soluble elements	$\bar{x}$ nurseries	-	-	11.2	6.4	19.5	3.6	3.2
	$\bar{x}$	-	-	8.0	5.2	11.5	3.0	2.6
	F	-	-	2.3	1.2	6.3*	1.5	1.6

		P kg m^{-3}	K kg m^{-3}	Ca kg m^{-3}	Mg kg m^{-3}	Na kg m^{-3}	Cu g m^{-3}	Zn g m^{-3}
Total elements	$\bar{x}$ fattening units	2.7	2.6	5.55	0.79	0.48	72.6	92.8
	$\bar{x}$ nurseries	2.4	2	4.35	0.60	0.36	23.2	90.1
	F fattening units	0.2	1.8	1.6	1.6	2.0	12.6**	0.02
Soluble elements	$\bar{x}$ nurseries	0.25	2.0	1.25	0.04	0.42	2.0	2.5
	$\bar{x}$	0.15	1.65	1.2	0.02	0.32	0.4	1.2
	F	8.7**	0.8	0.2	1.5	1.0	1.2	0.9

* Significant $P < 0.05$ ** Highly significant $P < 0.01$

TABLE 8

MEAN VALUES FOR THE 7 STUDIED FARMS AT DIFFERENT DATES AND VARIANCE ANALYSIS

	Dry matter kg m^{-3}	Ash kg m^{-3}	COD kg m^{-3}	Nitrogen kg m^{-3}	
				Total Kjeldahl	Ammoniacal
$\bar{x}$ February 1978	60.3	18.7	59.7	5.2	3.2
$\bar{x}$ June 1978	50	14.8	43.7	4.6	2.4
$\bar{x}$ September 1978	57.7	19.8	64	3.7	2.7
$\bar{x}$ January 1979	80.7	26	63.9	5.1	3.1
Calculated F	1.91	2.43	1.04	3.47*	3.78*

	P kg m^{-3}	K kg m^{-3}	Ca kg m^{-3}	Mg kg m^{-3}	Na kg m^{-3}	Cu g m^{-3}	Zn g m^{-3}
$\bar{x}$ February 1978	2.1	2.25	3.18	0.94	0.51	68.8	88.8
$\bar{x}$ June 1978	1.1	2.1	1.65	0.78	0.46	66.3	21.5
$\bar{x}$ September 1978	1.5	2.15	2.15	0.92	0.60	70.8	80.7
$\bar{x}$ January 1979	1.9	2.5	2.16	0.93	0.58	70.8	73.1
Calculated F	2.45	2.0	0.79	0.26	4.2*	0.03	4.65*

* Significant $P < 0.05$

The analysis of variance allowed us to test the influence of the piggery on the composition of manure. We were able to show significant differences between the piggeries for potassium and sodium. For the other elements, we have not observed such an 'origin' effect.

CONCLUSION

The concentration of the fertilising elements NPK are near to 5; 2; 2 g l^{-1} for the analysed manures in the two departments; 60 percent of the nitrogen is in the soluble ammoniacal form; potassium is almost fully soluble, whereas we find in the suspended solids 80 percent of the dry matter and almost all the phosphorus, copper and zinc. Therefore, the separation between the suspended solids and the aqueous phase, recommended to reduce the polluting load of liquid manure, will involve the production, on the one hand, of a solution containing the greater part of the available nitrogen and potassium, and on the other hand, of a product with much less available nitrogen and all the phosphorus.

Besides the leakage of water into tanks, the great fluctuations in dry matter around the mean can have different explanations: a difference between nurseries and fattening units and different tank depths affecting the relative importance of deposits. Moreover, though the differences are not always significant, the manure in summer is less concentrated than in winter.

REFERENCES

Duthion, C. et coll., 1979. Landspreading of liquid pig manure. II. Nutrients balance and effects on drainage water. To be published.

Heduit, M. et coll., 1977. Composition du lisier de porc: influence du mode d'exploitation, journées 'Recherche porcine en France', pp. 305-310. Ed. Inst. Tech. Porc. Paris.

Tunney, H. and Molloy, S. 1975. Variation between farms in N, P, K, Mg and dry matter composition of cattle, pig and poultry manures, In Journ. Agric. Res., 14, pp. 71-79.

LANDSPREADING OF PIG MANURE:
III. 2. PIG MANURE COMPOSITION. CORRELATIONS BETWEEN COMPONENTS

C. Duthion and J.C. Germon
INRA, BP 1540,
21034 Dijon Cedex, France.

ABSTRACT

Results of the analyses of manures sampled in forty fattening piggeries in two French departments, Ain and Saône-et-Loire, show that there are correlations between some of their different components. The most interesting of these relationships are between percentages of components, such as nitrogen or phosphorus, and the dry matter content or specific gravity of manures. They permit in particular rapid estimation of the total nitrogen content of manures and therefore facilitate a rational utilisation of manures by landspreading.

Many factors, such as piggery types or storage conditions, influence liquid pig manure composition. The resulting great variability in composition, therefore, makes the management of reasonable crop fertilisation difficult, to permit the best use of pig manure without creating risks to the environment. But, if there are connections between the different components of manures, we can devise a simple and quick method for estimating the fertilising value of manures. We have previously observed, for example, that levels of phosphorus, copper or zinc and to a lesser extent nitrogen, are connected to solid matter levels and ammoniacal nitrogen, which is easily assimilable, represents on average 50 percent of total nitrogen.

Heduit et al. (1977) presented linear regression equations between the dry matter content of manures and the main elements of their composition; moreover, Tunney and Molloy (1975b) suggested a specific gravity method for estimating their phosphorus and nitrogen contents, based on other equations.

To assess the value of this approach, we have extracted from data collected during the survey, possible relationships between the mean variables measured, then we have tried to apply them to pig manures spread on our experimental plots.

RESULTS

From analytical data collected by means of the survey made during February 1978, we have calculated the results given in Tables 1 to 3, utilising the characteristics of manures from fattening pigs (40 piggeries). Some of these calculations are also made from all the available data and some from comparisons between populations (fattening piggeries, breeding units, from Ain department or Saône-et-Loire department).

TABLE 1

VALUES OF SOME CORRELATION COEFFICIENTS BETWEEN COMPONENTS OF FATTENING PIGGERY MANURES (AT 1 PERCENT LEVEL r = 0.403)

	Specific gravity	Dry matter	COD	N total	P
Dry matter	0.904				
COD	0.847	0.926			
N total	0.873	0.764	0.660		
N NH_4	0.557	0.425	0.312	0.780	
P	0.868	0.906	0.877	0.785	
K	0.476	0.296	0.169	0.731	
Ca	0.793	0.914	0.920	0.632	0.838
Mg	0.790	0.740	0.674	0.700	0.824
Cu	0.855	0.886	0.896	0.672	0.877
Zn	0.720	0.769	0.790	0.563	0.748

Table 1 gives some of the calculated correlation coefficients. The close connections between dry matter amounts in manures and COD value or phosphorus, calcium, copper, zinc and even nitrogen or magnesium, contents are well illustrated. High correlations between these different variables and the specific gravity of manures are also emphasised. There are, too, good correlations between contents of several elements; some are certainly of causal value (connections between phosphorus and calcium, copper, zinc).

We indicate in Tables 2 and 3 linear regression equations which connect contents of manure components and fertilising and pollutant elements to dry matter amounts or specific gravity; indeed we have considered these two variables as independent variables, for the purpose of this work. Other functions (logarithmic, parabolic, exponential or power functions) were also tested when diagrams suggested them; we indicate only those which improve the interpretation of the experimental points, in comparison with linear regression.

TABLE 2

RELATIONSHIPS BETWEEN DRY MATTER AMOUNT AND SPECIFIC GRAVITY OF LIQUID PIG MANURE AND COD, NITROGEN AND PHOSPHORUS CONTENTS. FOR COMPARISON RELATIONSHIPS BY HEDUIT ET AL. (1977) AND TUNNEY AND MOLLOY (1975a) ARE INCLUDED

Dependent variable	Independent variable	Correlation coefficient	Regression equation	Equation coefficients standard errors (relation $y = a + bx$ or $y = a + bx + cx^2$) s_a	s_b	s_c
Specific gravity g l^{-1}	Dry matter g l^{-1}	0.904	$d = 1010.9 + 21.9\ 10^{-2}$ DM	1.62	$2.22\ 10^{-2}$	
		0.926	$d = 956.3 + 41.8 \log$ DM	4.91	2.77	
		0.934	$d = 999.9 + 65.10^{-1}$ DM $- 2.4\ 10^{-3}$ (DM)^2	3.07	$0.93\ 10^{-1}$	$0.60\ 10^{-3}$
COD g l^{-1}	Dry matter g l^{-1}	0.926	COD $= 9.06 + 8.48\ 10^{-1}$ DM	4.08	$0.56\ 10^{-1}$	
	Specific gravity g l^{-1}	0.847	COD $= -2420 + 2412\ 10^{-3}$ d	253	$246\ 10^{-3}$	
		0.902	log COD $= -15.83 + 17.09\ 10^{-3}$ d	1.37	$1.33\ 10^{-3}$	
N Kjeldahl g l^{-1}	Dry matter g l^{-1}	0.764	$N = 2.88 + 3.74\ 10^{-2}$ DM	0.37	$0.51\ 10^{-2}$	
		0.788	$N = 1.50 + 8.30\ 10^{-2}$ DM $- 3.02\ 10^{-4}$ (DM)^2	0.81	$2.44\ 10^{-2}$	$1.59\ 10^{-4}$
		0.811	$\log N = -0.14 + 4.81\ 10^{-1} \log$ DM	0.10	$0.56\ 10^{-1}$	
		0.80	$N = 3.16 + 5.3\ 10^{-2}$ DM (Heduit et al.)			
		0.95	$N = 0.76 + 6.18\ 10^{-2}$ DM $- 1.61\ 10^{-4}$ (DM)^2 (Tunney and Molloy)			

(Table 2 cont.)

TABLE 2 (Cont.)

Dependent variable	Independent variable	Correlation coefficient	Regression equation	Equation coefficients standard errors (relation $y = a + bx$ or $y = a + bx + cx^2$)		
				s_a	s_b	s_c
N Kjel-dahl $g\ l^{-1}$	Specific gravity $g\ l^{-1}$	0.873	$N = -134.84 + 133.18\ 10^{-3}\ d$	12.40	$12.04\ 10^{-3}$	
$P\ g\ l^{-1}$	Dry matter $g\ l^{-1}$	0.906	$P = 0.03 + 3.12\ 10^{-2}\ DM$	0.17	$0.24\ 10^{-2}$	
		0.84	$P = 0.54 + 0.03\ DM$ (Heduit et al.)			
		0.86	$P = -0.34 + 3.77\ 10^{-2}\ MS - 1.15\ 10^{-4}\ (DM)^2$ (Tunney and Molloy)			
	Specific gravity $g\ l^{-1}$	0.868	$P = -93.77 + 93.06\ 10^{-3}\ d$	8.88	$8.62\ 10^{-3}$	
		0.895	$\log P = -22.17 + 21.77\ 10^{-3}\ d$	1.81	$1.76\ 10^{-3}$	

TABLE 3

RELATIONSHIPS BETWEEN DRY MATTER AMOUNT OR SPECIFIC GRAVITY OF LIQUID PIG MANURE AND CALCIUM, MAGNESIUM, COPPER AND ZINC CONTENTS. SOME OTHER RELATIONSHIPS BETWEEN COMPONENTS ARE ALSO INCLUDED FOR COMPARISON, FROM HEDUIT ET AL. (1977) AND TUNNEY AND MOLLOY

Dependent variable	Independent variable	Correlation coefficient	Regression equation	Equation coefficients standard errors (Relation $y = a + bx$ or $y = a + bx + cx^2$) s_a	s_b	s_c
Ca $g\ l^{-1}$	Dry matter $g\ l^{-1}$	0.912	$Ca = 0.31 + 0.56\ 10^{-2}\ DM$	0.30	$0.04\ 10^{-2}$	
	Specific gravity $g\ l^{-1}$	0.793	$Ca = -153.04 + 152.45\ 10^{-3}\ d$	19.56	$19.00\ 10^{-3}$	
		0.815	$\log Ca = -15.39 + 15.48\ 10^{-3}\ d$	1.84	$1.82\ 10^{-3}$	
Mg $g\ l^{-1}$	Dry matter $g\ l^{-1}$	0.740	$Mg = 0.26 + 0.87\ 10^{-2}\ DM$	0.09	$0.13\ 10^{-2}$	
		0.88	$Mg = -0.13 + 1.53\ 10^{-2}\ DM - 0.48\ 10^{-4}\ (DM)^2$ (Tunney and Molloy)			
	Specific gravity $g\ l^{-1}$	0.790	$Mg = -29.13 + 29.09\ 10^{-3}\ d$	3.78	$3.66\ 10^{-3}$	
Cu $mg\ l^{-1}$	Dry matter $g\ l^{-1}$	0.886	$Cu = 6.35 + 0.81\ DM$	5.15	0.07	
	Specific gravity $g\ l^{-1}$	0.855	$Cu = -2537 + 2520\ 10^{-3}\ d$	255	248.10^{-3}	

(Table 3 Cont.)

TABLE 3 (Cont.)

Dependent variable	Independent variable	Correlation coefficient	Regression equation	Equation coefficients standard errors (relation $y = a + bx$ or $y = a + bx + cx^2$)		
				s_a	s_b	s_c
Zn mg l^{-1}	Dry matter g l^{-1}	0.768	$Zn = -5.80 + 1.09\ DM$	0.55	0.15	
	Specific gravity g l^{-1}	0.720	$Zn = -3217 + 3187\ 10^{-3}\ d$	513	$498\ 10^{-3}$	
		0.807	$\log Zn = -25.60 + 26.52\ 10^{-3}\ d$	5.46	$5.31\ 10^{-3}$	
N Kjeldahl g l^{-1}	$N\ NH_4$ g l^{-1}	0.880	$N = 0.28 + 1.60\ N\ NH_4$	0.47	0.14	
		0.86	$N = 0.63 + 1.33\ N\ NH_4$ (Heduit et al.)			
K g l^{-1}	N Kjeldahl g l^{-1}	0.731	$K = 0.27 + 0.37\ N$	0.31	0.05	
Ca g l^{-1}	P g l^{-1}	0.835	$Ca = 0.82 + 1.51\ P$	0.38	0.16	

The standard errors of the coefficients of the different equations are given. Some regressions between components, which have particular interest, are also given in Table 3. To permit comparison with our results, we give as well in Table 3 equations produced by Tunney and Molloy (1975a) and Heduit et al. (1977) from their data; if necessary, we transposed them into the units that we have used.

DISCUSSION

Spreading piggery effluents on soils is certainly the simplest and most inexpensive manner to dispose of them. A planned utilisation, which permits yields and quality in crops without creating risks to water and soil, however, needs knowledge of the fertilising value and pollutant load of manures.

Nitrogen (total and ammoniacal), phosphorus, and, to a lesser extent, copper and zinc contents, are sufficient to characterise the pollutant load of a manure which is spread on land. After spreading on soil, indeed, COD is easily removed (Duthion et al., 1979). On the other hand risks are high, with nitrogen, which is transformed into nitrates and is easily leached to surface or deep waters, and with phosphorus which is washed away by runoff. Copper and zinc may become phytotoxic from accumulations over a long period of time.

The fertilising value of manures depends on nitrogen and phosphorus contents, but also on potassium and possibly calcium and magnesium contents. Knowledge, even approximate, of these amounts, therefore is necessary; but is also sufficent to make possible landspreading in the most effective way. The relationships in Tables 2 and 3 provide indirect, simple and fast ways of reaching approximations for most of them (and also for COD values, useful for dealing with manures other than by landspreading) after determination of dry matter amounts and/or specific gravity. Only potassium content is insufficiently correlated with these two variables.

Tables 1 and 3 indicate also a high correlation between total nitrogen and ammoniacal nitrogen; the ratio NH_4-N/total-N is nearly 0.6. Relationships between phosphorus and calcium and copper and zinc are obvious. Gerritse (1977) has shown that mineral compounds which are slightly soluble, or insoluble, correspond to about 80 percent of the total phosphorus; from the value of the regression for the Ca-P coefficient, we may estimate that 65 - 70 percent of mineral phosphorus is di-calcium phosphate.

There is a lack of connection between liquid manure pH and other variables, nitrogen in particular ($r = 0.253$ with total nitrogen, $r = 0.521$ with ammoniacal nitrogen).

Tables 2 and 3 equations are valid only between the limits of observed values for every variable.

When we distinguish piggeries by their location (department) or their type of enterprise, regressions obtained for the same pair of variables are not significantly different; indeed types of piggeries are generally similar and the feed which is used varies little in spite of several origins.

Composition of a manure spread on a soil is always different from the average composition of a pit from which it is taken, because of lack of, or deficiency in, mixing before filling tankers. Do quantitative relationships between components then continue to apply? On two of our field experiments we supplied pig manure over three years in 3 or 4 dressings $year^{-1}$. For one of these, manure came from a fattening piggery with 250 animals (piggery 1) located in the survey zone, for the other from a fattening piggery with 500 animals (piggery 2) outside this zone by 50 km. Comparison between estimated values and actual values for the nitrogen and phosphorus contents of manures spread in succession (Figure 1) confirm the practical value of Table 2 relationships. Total nitrogen content is within less than 20 percent

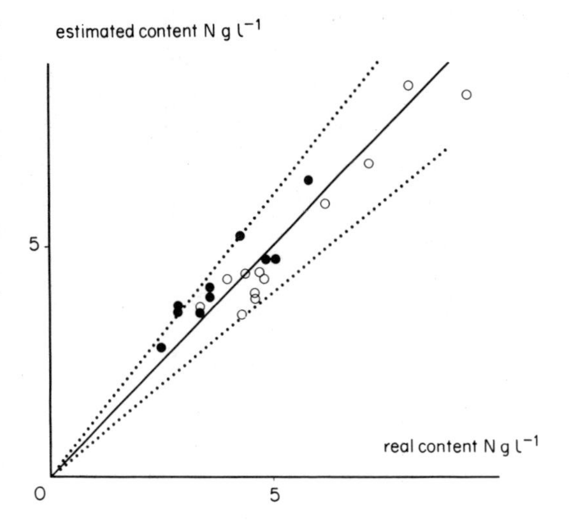

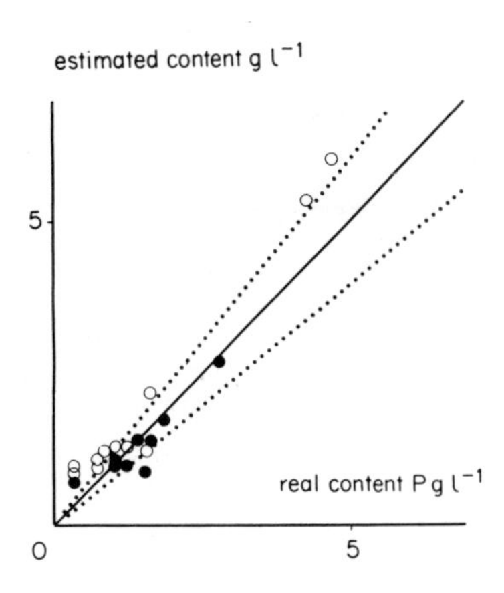

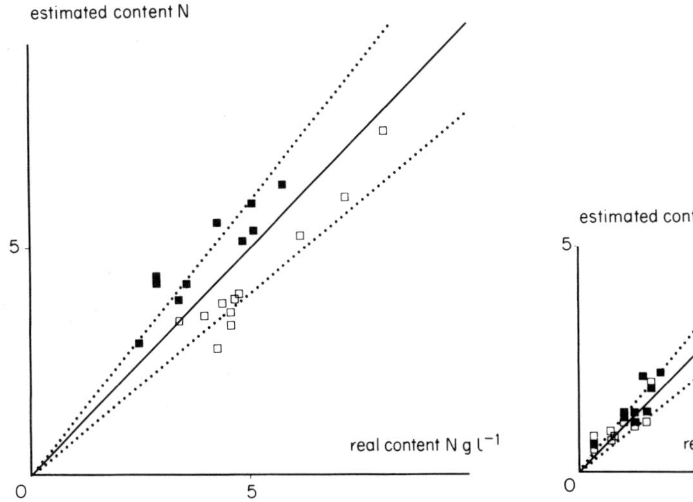

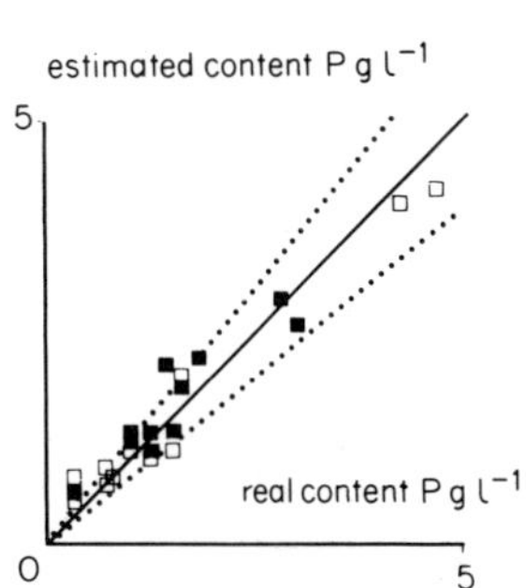

Fig. 1. Comparison of values esimated, from dry matter amount or specific gravity, and real values for nitrogen and phosphorus contents of spread liquid manures.

○ estimation from specific gravity ○□ piggery 1

□ estimation from dry matter amount ●■ piggery 2

Two broken straight lines on each side of solid line, represent real value ± 20 percent.

by estimation from the specific gravity of manure with a slightly higher error from the dry matter amount. The latter, on the other hand permits the better estimate of phosphorus content. Applications of the Heduit et al. (1977) and Tunney and Molloy (1975) formulae to the same manures give N and P content estimations of which the precision varies very much with the piggery or date of sampling.

This last observation suggests that we cannot utilisise relationships like those in Tables 2 and 3, out of the area for which they are established, without modification. Rearing conditions (rearing types, feeding...) indeed influence very much composition of effluents (Salmon-Legagneur et al., 1973; Salmon-Legagneur and Bernard, 1975).

Deciding manure application rates from their nitrogen contents is an optimal manner of making planned fertilisation. Knowledge of dry matter and/or specific gravity will permit it. Determination of dry matter level requires preliminary drying of a known volume or weight of manure, so it is not instantaneous. Specific gravity of a manure which is to be spread may on the contrary be determined easily in two ways:

- With an hydrometer, as Tunney and Molloy (1975b) propose. This method is very easy but necessitates some precautions in reading. Also, variation between values determined by an hydrometer and by weighing ($r = 0.97$ between these two values) is enough to require a readjustment of the relationships which we have presented.
- By weighing, provided that we can estimate to 1.0 g or better to 0.5 g. To minimise mistakes from balance defects, the weighing of equal volumes (exactly known or not) of manure and distilled water is essential.

CONCLUSION

From the results of a survey made in the two departments of Ain and Saône-et-Loire (France), we have calculated relationships of a very high probability, between characteristics of pig manures. Indeed we understand that in similar enterprises, with small or medium sized piggeries which produce animals of a well-defined weight from almost the same feed, ratios between effluent components vary little about average values. Further physico-chemical conditions (pH, anaerobiosis) limit the kinds of chemical compounds which are formed.

High correlations between the dry matter content or specific gravity of manures and their nitrogen, phosphorus, calcium, magnesium, copper and zinc contents give an indirect way of determining these quantities.

The prominent part of nitrogen in yield and the potential risk of water pollution after excessive applications suggest controlling the spreading of pig manure on land according to its total nitrogen content. Rapid estimation of this is possible from a measurement of manure specific gravity before spreading. This gives a good approximation (relative error lower than 20 percent) and is best obtained by weighing.

REFERENCES

Duthion, C., Catroux, G., Germon, J.C., 1979. Landspreading of pig manure. II. Nutrient balances and effects on drainage water. CEE Seminar, Oldenburg, 2 - 5 October 1979.

Gerritse, K.J., 1977. Phosphorus compounds in pig slurry and their retention in the soil. In: J.H. Voorburg, 1977. Utilisation of manure by landspreading 257-266, Commission of the European Communities, Luxembourg.

Heduit, M., Roustan, J.L., Aumaitre, A., Seguin, M., 1977. Composition du lisier de porc: influence du mode d'exploitation. Journées Rech. porcine en France, 305-320.

Salmon-Legagneur, E., Gayral, J.P., Leveau, J.M. and Rettagliati, J., 1973. Etude de quelques paramètres de variation de la composition des effluents de porcherie. Journées Rech. porcine en France, 285-291.

Salmon-Legagneur, E. and Bernard, C.R., 1975. Composition minérale des lisiers de porcs. Valeur agronomique. Journées Rech. porcine en France, 323-330.

Tunney, H. and Molloy, S., 1975

a) Variations between farms in N, P, K, Mg and dry matter composition of cattle, pig and poultry manures. Ir. J. agric. Res, 14, 71-79.

b) Field test for estimating dry matter and fertiliser value of slurry. Preliminary report. Ir. J. agric. Res. 14, 84-86.

LANDSPREADING OF PIG MANURES
IV. EFFECT ON SOIL NEMATODES

G. de Guiran*, L. Bonnel and Mona Abirached

*INRA, Station de Recherches sur les Nematodes, BP 78, 06602 Antibes, France.

INRA, Station de Recherches sur la Faune du Sol, BP 1540 Dijon, France.

ABSTRACT

Spreading a large amount of pig manure (equivalent to 5 x mineral fertiliser N) on a Festuca elatior *meadow increased the number of microphagous but decreased phytophagous nematodes within 24 hours. These effects were less marked within a few weeks of spreading but successive manure applications led to a small but stable increase in total nematode population. This was especially noticeable when numbers were at peak levels in spring and autumn. Lower rates of manure application (equivalent to mineral fertiliser N application) did not affect numbers or population composition.*

Every arable soil, whether cultivated or not, contains a large number of nematodes, each species having its own trophic requirements. For example, some feed on plants, generally on roots, others on bacteria or fungi. Some species are predators on microscopic animals or parasites on larger ones (insects, earthworms, etc.).

In an undisturbed environment, such as natural meadow, the most important trophic groups are the phytophagous and the microphagous (bacteria + mycophagous), each representing 40 to 50 percent of the total nematofauna. There is also a small number (5 to 10 percent) of predatory species feeding particularly on nematodes. Cultivation tends to break this equilibrium to the benefit of adapted species. In this instance, these are the phytophagous nematodes which can reach 90 percent of the total number.

Organic manures also tend to modify the proportions of the nematode trophic groups, the literature on the subject having been reviewed by Sayre (1971) and Marshall (1977). The principal effect of such manures is to increase the number of microphagous and to decrease the phytophagous. Increase in microphagous is most probably related to increase in bacteria and fungi. The decrease of phytophagous is not always clear. Linford et al. (1938) have suggested that the increase of microphagous nematodes induces a multiplication of the natural enemies of all kinds of nematodes. Phytophagous species, destroyed by these enemies, then decrease in the soil. This possibility has not been confirmed by in situ counting of populations.

Direct toxic effects of degradation by-products (butyric acid or ammonia) on plant nematodes has been demonstrated. In water saturated soils, soluble sulphides also have this toxic effect (Fortuner and Jacq, 1976).

The effects of landspreading pig manures on different characteristics of the soil has been studied and the present paper gives the results concerning the effects on soil nematodes. Two series of observations were conducted: i) the populations were followed on treated plots for 29 months by monthly or bimonthly sampling, ii) the same populations were measured just before spreading and at increasing intervals from one to seven days. These observations showed the short and long term changes in nematofauna.

MATERIAL AND METHODS

Field trial

Three adjacent plots of 210 m^2 each (21 x 10 m) at the Epoisse Farm (21110 Bretenières, France) were planted with Fescue (*Festuca elatior*) cv. Manade. One plot (control) received 60 kg N ha^{-1} as mineral fertiliser at planting and 50 kg N ha^{-1} at each cut. The two other plots were spread with pig manures on the same dates: one with an equivalent amount of N as the control (plot M), the other with five times this amount (plot 5M). General characteristics of the trial are given in this publication by Duthion et al. (1979).

Sampling

In the absence of replications, the global method of sampling was used (Merny and Luc, 1969). On each plot, 40 sub-samples were taken in the root zone (0 - 3 cm) and pooled.

Nematode extraction

Nematodes were extracted from 250 cc of each sample using the elutriation method (Seinhorst, 1962) with a 100 ml mn^{-1} flow. Nematode suspensions thus obtained were poured on six superposed sieves: 50 µm (4 sieves), 28 µm and 10 µm aperture. The extracts were washed and centrifuged using the method of Gooris and D'Herde (1972) before counting the nematodes. Results were expressed as number of nematodes 100 ml^{-1} of soil.

RESULTS

Composition of the nematofauna

The trophic requirements of all soil nematode species are not well known. Most of the tylenchids are phytophagous and most of the rhabditids are thought to be bacteriophagous. Among the species extracted from the field, only *Aphelenchus avenae* is known to be mycophagous; *Tylenchus* and dorylaimids could possibly be included in this group also. Only mononchids are regarded as predators of nematodes. With the above restrictions, the nematofauna can be classified as in Table 1. It seems to be of the meadow type but with a rather low number of phytophagous species.

TABLE 1

COMPOSITION OF THE SOIL NEMATOFAUNA AT THE BEGINNING OF THE TRIAL MADE IN BRETENIERES (CONTROL PLOT) (Numbers of nematodes 100 m^{-1} soil).

		Number	Percent
- Phytophagous		972	36
- *Pratylenchus thornei*	220		
- *Merlinius brevidens*	540		
- *Helicotylenchus vulgaris*	210		
- *Macrotrophurus arbusticola*	traces		
- Bacteriophagous (Rhabditids)		370	51
- *Aphelenchus avenae*		40	2
- *Tylenchus* sp.		200	8
- Dorylaimids		80	3
- Predators (Mononchids)		traces	-
Total		2 660	100

Long term changes

The three plots were sampled each month (sometimes each fortnight). Figure 1 shows the changes in the total nematofauna from June 16th, 1976 to November 6th, 1978 on the control plot and on plot 5M. No difference was found between plot 5M and the control plot, either in the total number of nematodes, or in the proportion of the different trophic groups. A seasonal fluctuation was observed with two peaks, one in spring and the other in autumn, separated by a wide valley in winter and a narrow one in summer. This is a classical pattern (Berge et al., 1973) where the fauna are reduced in winter by low temperatures and in summer by soil dryness; it was particularly pronounced in 1976.

The treated plot 5M contained a more abundant fauna on the spring and autumn peaks from the beginning of the observations. This difference was noticeable and throughout the year from autumn 1977 on, could be related to differences between the respective proportions of phytophagous and microphagous nematodes shown in Figure 2. Here the numbers of these two trophic groups on plot 5M are expressed as a percentage of those on the control plot at the same dates. In the treated plot, an increase in phytophagous nematodes was recorded on the first sampling date and numbers reached a maximum on October 10th, 1979. This can be related to a rapid growth of the Fescue. After spreading manure on October 10th, 1979, this situation was reversed and there were less phytophagous for the rest of the period, apart from two occasions. This was in spite of stronger plant growth on the treated plot.

The microphagous nematodes showed the inverse tendency: the treated plot contained more than the control, particularly after each spreading.

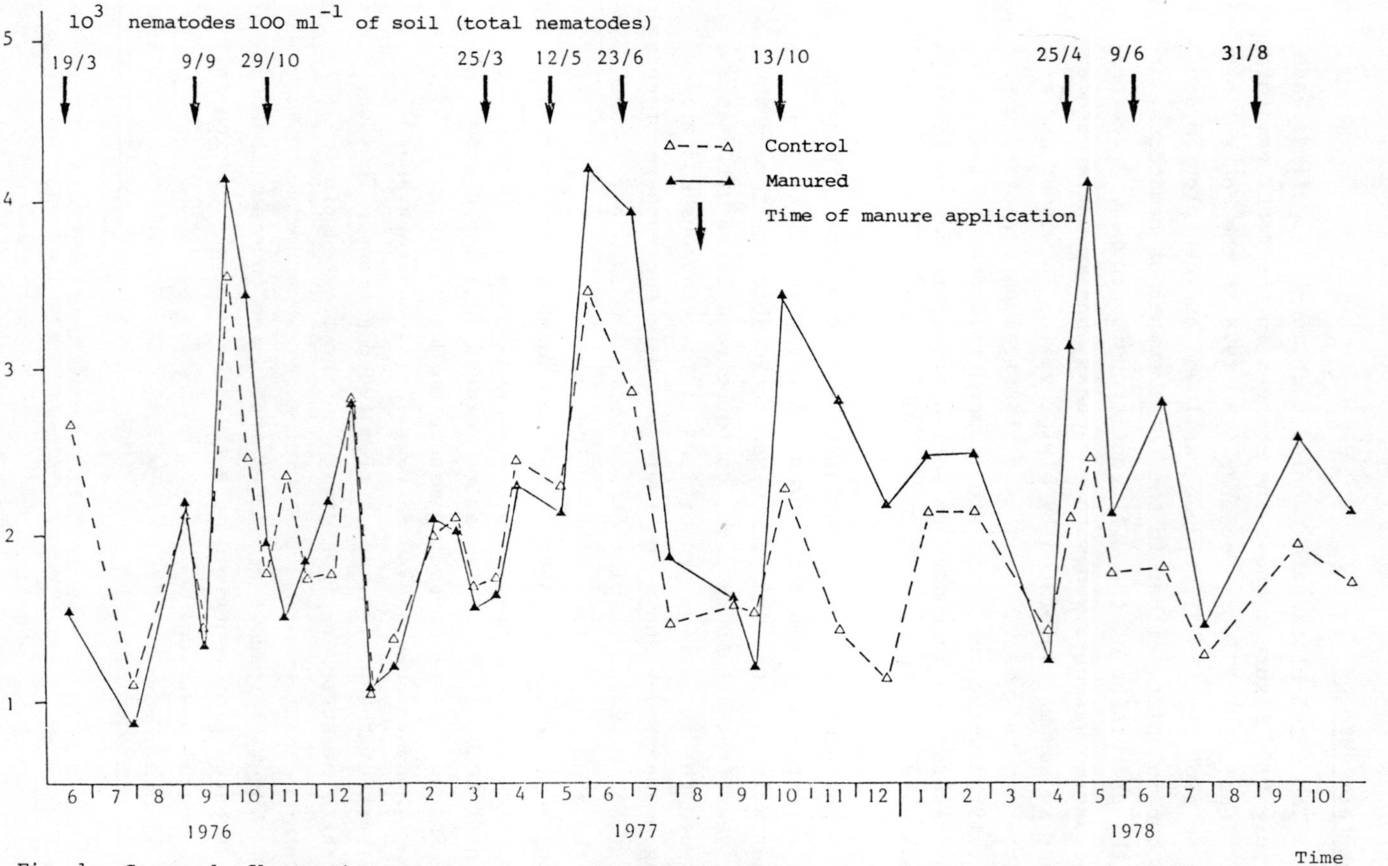

Fig. 1. Seasonal fluctuations of the total nematofauna in Fescue plots receiving pig manure or mineral fertiliser.

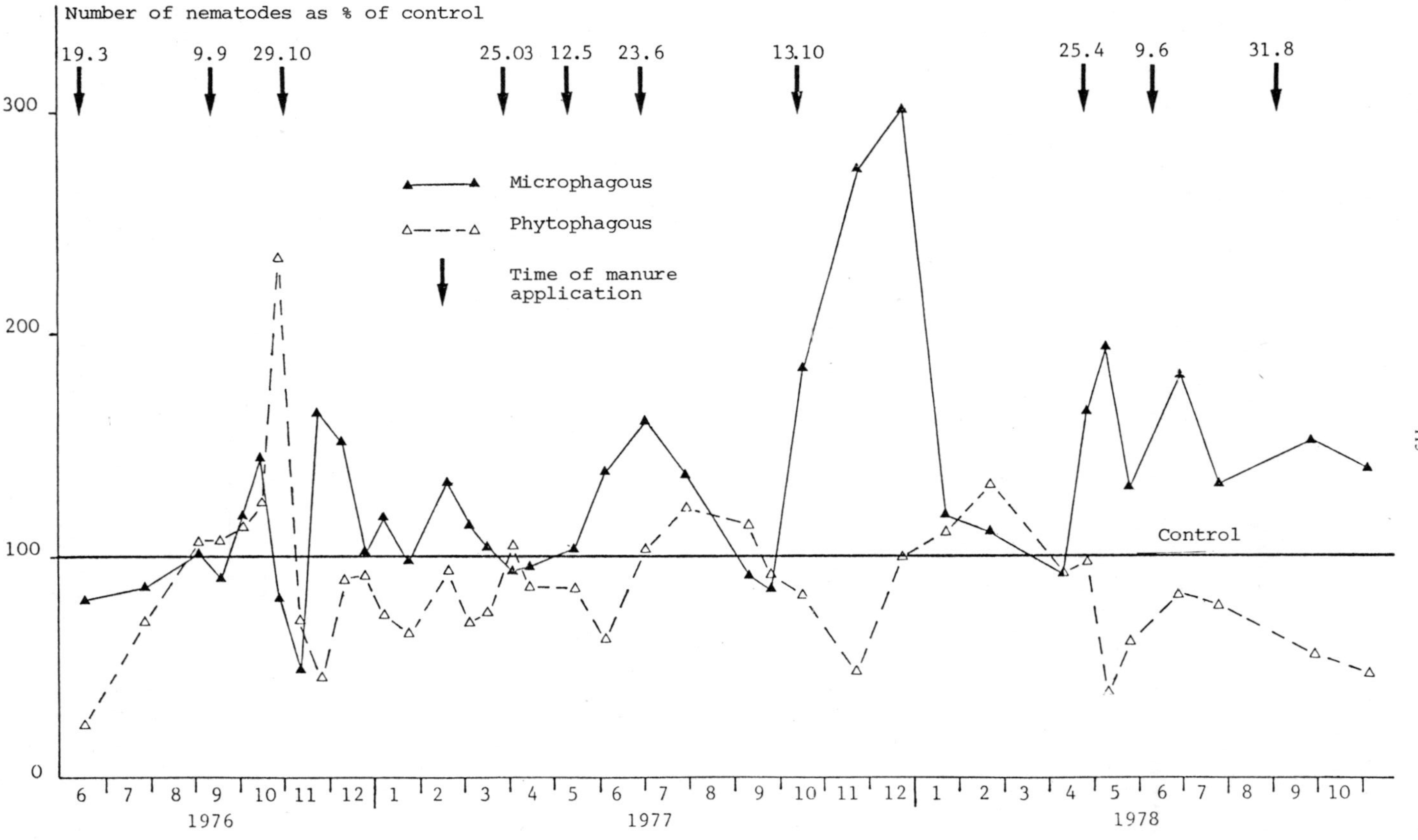

Fig. 2. Changes in numbers of microphagous and phytophagous nematodes in a manured plot compared to control

Short term changes

On the 5M plot and the control plot, soil was sampled just before spreading manure on April 25th, 1978 and immediately after, each day for a week, every two days for the following week, then on the 17th, 21st, 28th and 35th days after spreading. Figures 3 and 4 show the changes in numbers of microphagous and phytophagous nematodes on the treated and control plots.

Spreading of pig manures resulted in a rapid increase of microphagous nematodes within 24 hours (Figure 3) and their number more than doubled after three days. This could not be accounted for by the number of nematodes in pig manure applied to the plots. The number of bacteria feeding nematodes were probably stimulated by manure application. After three days, this effect was less obvious and after three weeks the number of microphagous nematodes was back to the level measured before spreading.

The converse was true for the phytophagous nematodes (Figure 4): numbers decreased 24 hours after treatment but then stabilised and after about two weeks showed the same pattern as the control.

A difference persisted between the treated plot and the control plot for both trophic groups. This difference was recorded in the long term study and can be explained by the cumulative effect of successive treatments.

CONCLUSION

Spreading a large amount of pig manure (5 x N equivalent or a normal fertiliser application) resulted in a rapid and important change in the composition of the nematofauna: microphagous were greatly increased and phytophagous decreased. This effect was rapidly attenuated and, after some weeks, the numbers of the different trophic groups followed a normal

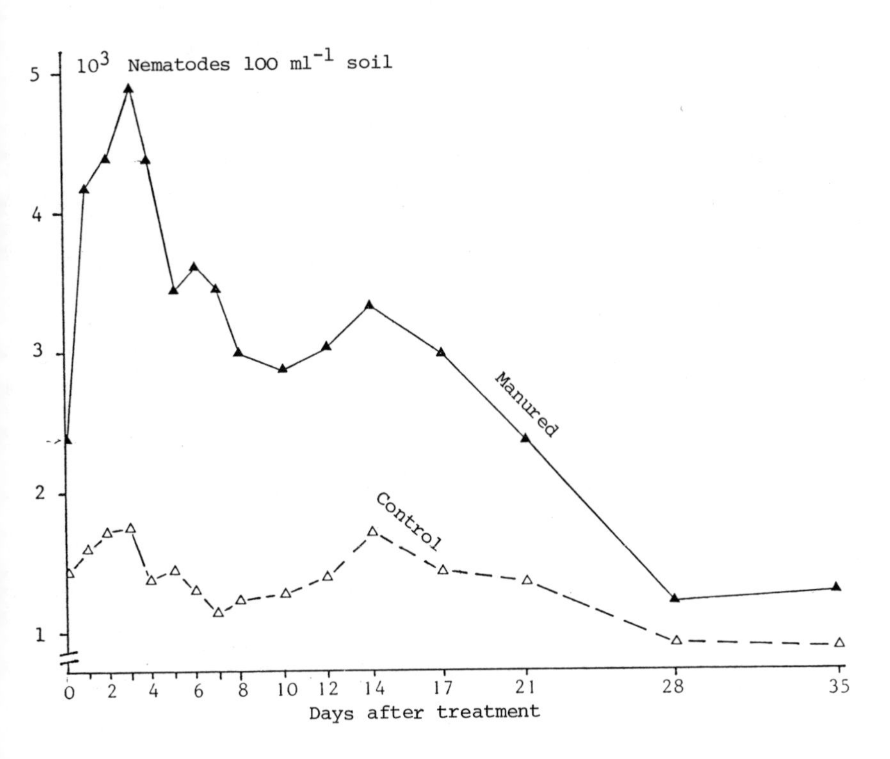

Fig. 3. Numbers of microphagous nematodes after spreading.

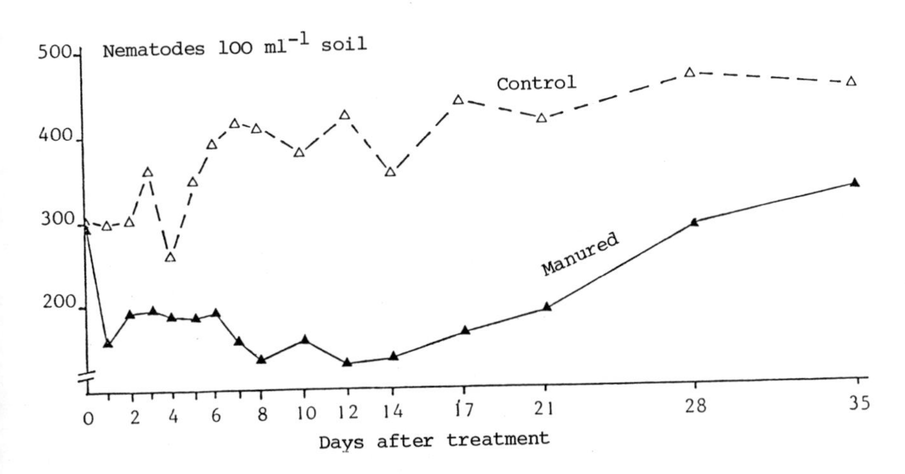

Fig. 4. Numbers of phytophagous nematodes after spreading.

pattern. However, successive applications had a cumulative effect and the treated soil contained higher numbers of nematodes particularly in spring and autumn, due primarily to increase in the number of microphagous species. This modification did not significantly affect the equilibrium of the nematofauna. Spreading pig manures in amounts equivalent to a normal fertiliser application did not affect the nematofauna.

The influence of pig manures on each species of nematodes was beyond the aim of this study and can only be determined by specific observations or experiments.

REFERENCES

Berge, J.B., Dalmasso, A. and Kermarrec, A., 1973. Etude des fluctuations des populations d'une nématofaune prairiale. Rev. Ecol. Biol. Sol. 10, 271-285.

Duthion, C. et al., 1979. Landspreading of liquid pig manure. This publication.

Fortuner, R. and Jacq, V.A., 1976. In vitro study of toxicity of soluble sulfides to three nematodes parasitic on rice in Senegal. Nematologica, 22, 343-351.

Gooris, J. and D'Herde, J.C., 1972. A method for the quantitative extraction of eggs and second stage juveniles of *Meloidogyne* spp. from soil. Publ. Gov. Res. Sta. Nematol. Entomol. Merelbeke (Belg.): 1-36.

Linford, M.B., Yap, F. and Oliveira, J.M., 1938. Reduction of soil population of the root-knot nematode during decomposition of organic matter. Soil Sci., 45, 127-140.

Marshall, V.G., 1977. Effects of manure and fertilisers on soil fauna: A review. Commonw. agric. Bur. Spec. Publ. No.3: 79 pp.

Merny, G. and Luc, M., 1969. Les techniques d'évaluation des populations de nématodes dans le sol et les tissus végétaux. In: Problèmes d'Ecologie. L'échantillonnage des peuplements animaux dans les milieux terrestres, M. Lamotte and F. Bourlière, Masson et Compagnie, Paris: 257-292.

Sayre, R.M., 1971. Biotic influnces in soil environment. In: Zuckerman, B.M., Mai, W.F. and Rhode, R.A. Plant parasitic Nematodes. Academic Press, N.Y. and London, Edit.: 235-256.

Seinhorst, J.W., 1962. Modification of the elutriation method for extracting nematodes from soil. Nematologica, 8, 117-128.

LANDSPREADING OF LIQUID PIG MANURE
V. EFFECT OF SOIL MICROFLORA

G. Delpui
(with the technical collaboration of Regine Pedergnana)
Laboratoire de recherches de la chaire de Microbiologie,
INA - PG, 9 rue de l'Arbalete,
75231 Paris, Cedex 05, France.

ABSTRACT

The influence of manure on soil microflora was studied by comparison of control and heavily manured grass plots. The main physiological microbial groups were enumerated after spreading but did not reveal any detrimental effect. The survival of intestinal bacteria was also studied.

1. INTRODUCTION

Spreading of pig manure on arable lands is one of the simplest and cheapest methods of disposal for wastes produced by pig husbandry. Furthermore, it has the advantage of recycling plant nutrients in crop production (Ballay and Catroux, 1974). However, pig manure also supplies phosphorus and heavy metals which can constitute hazards to soil microflora equilibrium. A second potential problem concerns the hygienic aspects of landspreading: manures contain an especially rich microflora, possibly with some strains pathogenic to humans and animals.

This two year study comprised two parts: the study of pig intestinal bacteria in soil, and the behaviour of soil microflora samples from plots which received pig manure or mineral fertiliser.

2. MATERIAL AND METHODS

Microbial analyses were made immediately after receiving soil and manure samples. Soil analyses were made after dilution, following the Pochon and Tardieux method (1962). Manure analyses were made after dilution in Tryptone 0.1 percent, 1 percent, NaCl 0.85 percent solution.

Decimal dilutions inoculated semi-solid media (3 tubes or 3 Petri dishes dilution^{-1}) or liquid media (3 tubes dilution^{-1}).

2.1. Pig manure analysis

Three different samples were analysed by the following methods:

- Total microflora (aerobic): Tryptone - Glucose Yeast Extract medium (Bio-Mérieux). Five days incubation at 30°C.
- Total microflora (anaerobic): The medium of Mossel et al. (1962); regenerated media was poured on 1 ml dilution sample in 14 x 180 mm tubes. An Agar stopper was poured on the media top. Five days incubation at 30°C.
- Viable enteric bacteria: violet red bile Agar (Difco) with 1 percent mannitol added - 24 hours incubation at 37°C.
- Faecal *Escherichia coli*: Chapman medium (1951), Tergitol 7 Agar (Difco) with 40 mg l^{-1} of TTC, surface inoculated with 0.1 ml dilution. Twenty four hours incubation at 37°C (Buttiaux et al., 1953).
- Faecal streptococci: The medium of Hajna and Perry (1943) with sodium azide. Forty eight hours incubation at 37°C. Identification was confirmed by the medium of Litsky et al. (1955), ethyl - violet and sodium azide broth (Bio-Mérieux).
- Sulphite - reducing clostridia: The medium of Rivière et al. (1974). Three days incubation at 30°C in anaerobic jars. Spore counts were made on samples heated for 10 minutes at 80°C.
- Indol - producing bacteria: 48 hours incubation at 30°C in peptone broth and indol detection with the Kowacs reagent.
- Ureolytic bacteria: The medium of Rivière (1974). Fifteen days incubation at 30°C. Sulphate reducing bacteria: Gatellier's medium (1962). Three weeks incubation at 30°C.
- H_2S producing bacteria: the medium (meat-yeast) from Buttiaux et al., 1969, was used to detect the H_2S production after 7 days incubation at 30°C.

- Salmonella: the enrichment culture was made on selenite broth from Leifson (Bio-Mérieux). After 24 hours at 37°C, strains were isolated on selective media: Gélose SS (Bio-Mérieux) and desoxycholate citrate Agar (Difco). After 48 hours incubation, colonies presumably identified as salmonella, colourless or colourless with central black point, were sampled and tested on different metabolic characteristics from genus salmonella (Le Minor, 1972): urease, indole, beta galactosidase, glucose, lactose, mannitol fermentation, lysine decarboxylase, ornithine, citrate utilisation, phenylalanine. H_2S production.

2.2. Soil analysis

Experimental design was described by Duthion (1979). This included 3 plots on a clay soil cropped with Fescue and receiving mineral fertiliser, pig manure equivalent to mineral fertiliser N and five fold pig manure. First results did not show any differences between plots, so subsequently microbial analyses were carried out only on the mineral fertiliser plot and the heavily manured plot.

- Total microflora: soil extract Agar (Pochon and Tardieux, 1962) and Soumare and Blondeau medium (1972).
- Gram-negative bacilli: Holding's medium (1960).
- Actinomycetes: actidione medium of Porter et al. (1960).
- Fungi: rose bengal and streptomycine medium (Pochon and Tardieux, 1962).
- Coliform bacteria: lactose, bile, brilliant green medium (Institut Pasteur Production).
- Faecal *Escherichia coli* : see above.
- Cellulolytic bacteria: liquid medium of Rivière (1964) with a paper strip and without antibiotic.
- Ammonifying, nitrifying and proteolytic bacteria: Media of Pochon and Tardieux (1962).

- Denitrifying bacteria: Tryptone 1 percent, yeast extract 0.2 percent, meat extract 0.3 percent, glucose 1 percent, KNO_3 1 percent, pH 7.2, 17 ml in 14 x 180 mm tubes fitted with a gas lock. Control tubes without KNO_3 were inoculated. Seven days incubation at 30°C. (Riviere, unpublished results).
- Strains identification: 100 colonies were sampled on the media of Soumare and Blondeau, purified and identified to genus following the 8th edition of Bergey's Manual (Buchanan and Gibbons, 1974).

3. RESULTS AND DISCUSSION

3.1. Pig intestinal microflora

3.1.1. Manure analysis

Manure composition was very variable and the results reflected this variability. This was especially true for odour producing micro-organisms but total microflora and faecal *Escherichia coli* remained stable in the 3 samples of manure which were analysed (Table 1). Results obtained were in the same range as data published by Rivière et al. (1974). Faecal bacteria from manure in decreasing order of size were: streptococci, *Escherichia coli* and sulphite reducing clostridia.

No salmonellae were found in 100 ml from each of the 3 samples. The presence of other pathogens was not investigated.

3.1.2. Enterobacteria

Escherichia coli and coliform bacteria (Table 2) were not detected in the control plot nor in the other plots. One day after spreading, coliform bacteria and *Escherichia coli* were detected but disappeared after 1 to 3 months.

TABLE 1

MICROBIOLOGICAL ANALYSIS OF MANURE SAMPLES (Number ml^{-1})

	June 1976	October 1978	June 1978
Aerobic total microflora	2 10^7	9 10^7	3 10^7
Aerobic total microflora	2.5 10^7	1 10^7	3 10^7
Enterobacteria	1 10^4	7.5 10^4	1.5 10^5
Escherichia coli faecal	1 10^4	6 10^4	3 10^4
Faecal *Streptococci*	3 10^6	2.5 10^7	1.1 10^7
Sulphide-reducing Clostridia			
- vegetative cells	3 10^2	2.5 10^3	2 10^4
- spores	2 10^4	8 10^3	5 10^3
Indole producing bacteria	1.1 10^6	3 10^3	3 10^4
Ureolytic bacteria	2.5 10^4	2.5 10^6	7.5 10^6
Sulphate reducing bacteria	1.1 10^3	1.1 10^5	1.4 10^5
H_2S producing bacteria	9.5 10^5	4.5 10^6	1.1 10^7
Salmonella	not detected in 100 ml	not detected in 100 ml	not detected in 100 ml

In a second set of analyses, enteric bacteria were detected in the manured plot, 30 hours after spreading and remained viable 1 month later. This experiment was not continued for a longer period, but results agreed with previous experiments.

3.1.3. Faecal streptococci and clostridia

Faecal streptococci (Figure 1) and sulphite-reducing clostridia were present and remained viable in control and manured plots. It was not possible to conclude if streptococci and clostridia found in soil were indigenous soil bacteria or were added to soil by manure. These data must be explained by the results of Prevot (1961) who found that numerous soil clostridium strains can reduce sulphite. Considering faecal streptococci, the results obtained indicate that it is not possible to use these organisms as a test of faecal contamination in soils, as published by Kibbey et al. (1978).

TABLE 2

COLIFORM BACTERIA AND FAECAL *Escherichia coli* IN CONTROL SOIL AND AFTER MANURE SPREADING (Number g^{-1})

	Manured soil		Control soil	
	Coliform	*E. coli*	Coliform	*E. coli*
June 1976			0	0
October 1976				
- before spreading	0	0	0	-
- time after spreading				
24 h	$1.1\ 10^3$	$2\ 10^3$	-	-
7 days	$2.5\ 10^2$	$1\ 10^2$	-	-
15 days	$4.5\ 10^3$	$2\ 10^2$	-	-
1 month	$1.5\ 10^2$	0	0	0
3 months	0	0	0	0
8 months	0	0	0	0
January 1978	0	0	0	0
June 1978				
- time after spreading				
30 h	$1.4\ 10^4$	$3\ 10^5$	0	0
10 days	$1.4\ 10^4$	$7\ 10^4$	0	0
1 month	$1.1\ 10^4$	$2\ 10^5$	0	0

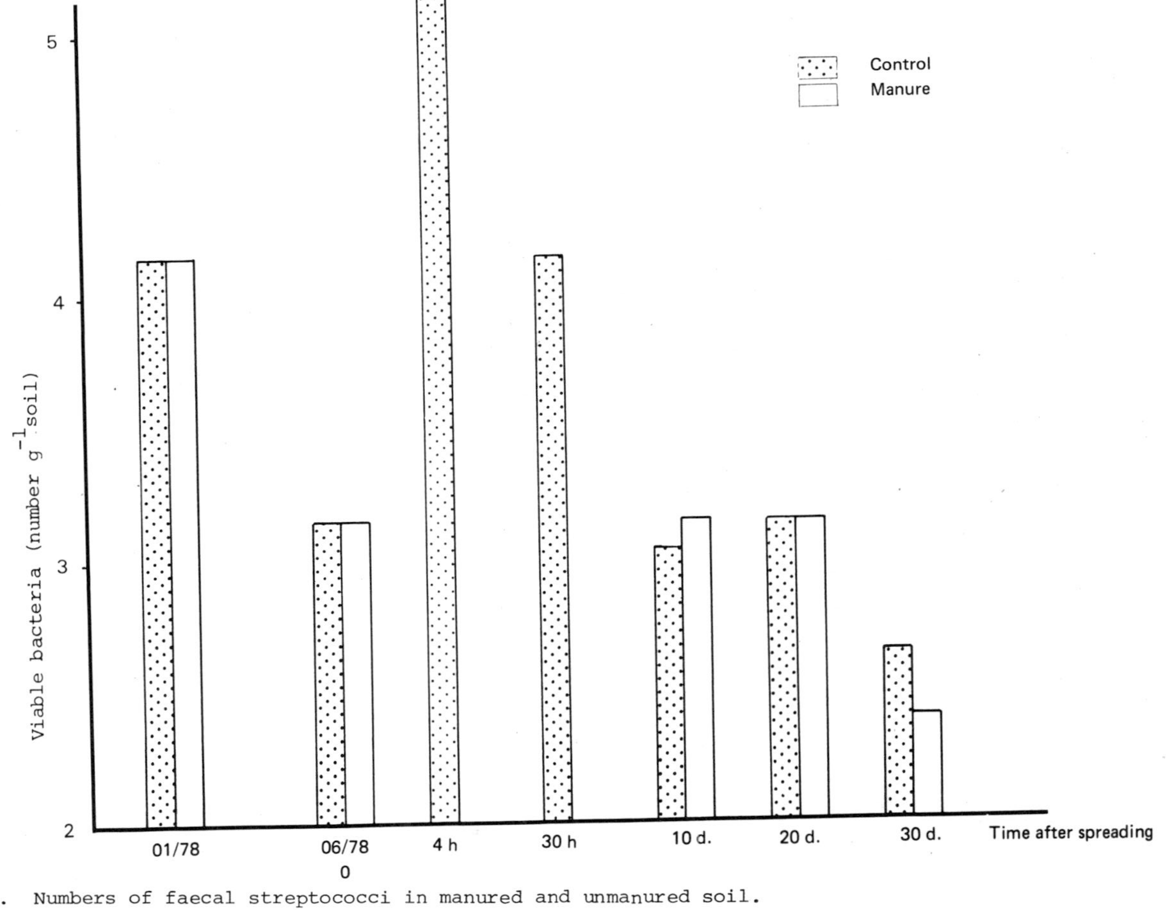

Fig. 1. Numbers of faecal streptococci in manured and unmanured soil.

3.2. Soil microflora

Although populations were very variable, comparison between total microflora (bacterial flora on Soumare and Blondeau medium, gram-negative bacilli, fungi, actinomycetes) in the control and manured plots did not show a significant increase (Table 3). Similar variations, apparently unrelated to treatment are shown in Table 4. A temporary stimulation of aerobic cellulolytic bacteria was detected at the end of 1976. However, the growth rate of cellulolytic bacteria is too slow to account for a large increase within 24 hours of spreading manure.

The identification of isolated strains (on non-selective media) from the control plot and the manured plot 1 and 3 months after spreading is shown in Figures 3 and 4. Observed differences were not due to manure application, because the variation between counts for control plots were too high.

4. CONCLUSION

1. Since faecal streptococci and sulphite-reducing clostridia were detected in the soil from the control plot it was not possible to conclude whether these were added in the pig manure or were already present. Coliform bacteria and *Escherichia coli* were a good test of faecal contamination but could not be detected 1 or 2 months after applying the manure.

2. Spreading pig manure did not significantly affect the soil microflora under the conditions of the experiment.

3 The only remaining problem is hygienic: in this experiment, entero-bacteria and faecal *Escherichia coli* disappeared quite quickly but in other circumstances, these bacteria can survive for longer periods (White and Spencer, 1976), indicating a possible survival of pathogenic micro-organisms under the same soil conditions.

TABLE 3

MICROBIAL TYPES IN MANURED (M) AND CONTROL (C) SOILS (Viable units g^{-1} soil)

Date	Time after spreading	Total microflora Soumare and Blondeau		Gram negative Bacilli		Fungi		Actinomycetes	
		M	C	M	C	M	C	M	C
June 1976	-	-	7 10^6	-	1 10^6	-	5 10^4	-	8 10^6
October 1976	0 before	1 10^8	3 10^7	1 10^7	1 10^7	4 10^5	3 10^5	4 10^5	1 10^4
	24 h	2 10^8	-	1 10^8	-	3 10^5	-	1 10^5	
	1 week	3 10^8	-	4 10^7	-	3 10^5	-	5 10^5	-
	2 weeks	4 10^7	-	1 10^7	-	3 10^4	-	6 10^5	-
	1 month	1.5 10^8	4.5 10^8	7 10^6	4.5 10^6	1.3 10^4	4 10^4	4 10^6	1 10^5
	3 months	6 10^7	2 10^7	2 10^7	3 10^5	3 10^5	3 10^5	7 10^5	1 10^6
June 1977	8 months	1 10^8	4 10^7	1 10^7	1 10^6	4 10^5	4 10^5	4 10^6	5 10^6
January 1978	-	6 10^7	3 10^7	1 10^7	3 10^6	3 10^5	2 10^5	4 10^7	4 10^7
June 1978	0 before	5.5 10^7	1.2 10^8	1.5 10^7	1.2 10^7	3 10^4	2 10^5	4 10^7	7 10^7
	4 h	3.5 10^8	-	4 10^7	-	2 10^5	-	2 10^8	-
	30 h	2.5 10^8	-	1.1 10^8	-	3 10^5	-	2 10^8	-
	10 days	4 10^8	6 10^8	3 10^8	4 10^7	2 10^4	4 10^4	5 10^8	3 10^8
	20 days	1 10^8	1 10^8	5 10^7	5 10^6	3 10^4	1 10^4	8 10^7	3 10^7
July 1978	30 days	2 10^8	1 10^8	1 10^8	4 10^6	1 10^5	1 10^5	8 10^7	1 10^8

TABLE 4

PHYSIOLOGICAL GROUPS IN MANURED (M) AND CONTROL (C) SOILS (Number g^{-1} soil)

Date	Time after spreading	Proteolytic bacteria		Ammonifiers		Nitrifying bacteria	
		M	C	M	C	M	C
June 1976	-	-	$1.1\ 10^7$	-	$4.5\ 10^7$	-	$2.5\ 10^2$
October 1976	0 before	$2.5\ 10^7$	$4.5\ 10^7$	$1.1\ 10^8$	$1.1\ 10^8$	45	$1.1\ 10^2$
	24 h	$4.5\ 10^7$	-	$1.1\ 10^8$	-	$3\ 10^2$	-
	1 week	$1.1\ 10^8$	-	$1.1\ 10^8$	-	4	-
	2 weeks	$1.1\ 10^8$	-	$1.1\ 10^8$	-	$4.5\ 10^2$	-
	1 month	$1.1\ 10^8$	$7.5\ 10^6$	$1.1\ 10^8$	$4.5\ 10^8$	25	25
	3 months	$1.5\ 10^7$	$2\ 10^5$	$4.5\ 10^8$	$2\ 10^7$	45	$1.1\ 10^2$
June 1977	8 months	$2.5\ 10^6$	$2.5\ 10^5$	$4.5\ 10^7$	$7.5\ 10^7$	$3.5\ 10^2$	$2.5\ 10^2$
January 1978	-	$1.5\ 10^7$	$1.1\ 10^7$	$1.4\ 10^8$	$1.4\ 10^7$	$4\ 10^3$	$1.4\ 10^3$
June 1978	0 before	$1.4\ 10^5$	$4.5\ 10^5$	$1.4\ 10^9$	$1.4\ 10^9$	$2.5.10^2$	$1.1\ 10^3$
	4 h	$1.1\ 10^6$	-	$1.4\ 10^9$	-	$2.5\ 10^2$	-
	30 h	$1.4\ 10^5$	-	$1.1\ 10^8$	-	$1.1\ 10^3$	-
	10 days	$1.4\ 10^5$	$3\ 10^6$	$1.4\ 10^9$	$1.1\ 10^8$	25	9
	20 days	$1.4\ 10^6$	$1.1\ 10^6$	$1.1\ 10^9$	$2\ 10^8$	15	25
July 1978	30 days	$4.5\ 10^7$	$1.1\ 10^6$	$1.4\ 10^9$	$1.4\ 10^9$	25	25

(Table 4 cont.)

TABLE 4 (Cont.)

Date	Time after spreading	Dentrifying bacteria		Cellulolytic aerobic bacteria		Sulphate reducing bacteria	
		M	C	M	C	M	C
June 1976	-	-	$1.5\ 10^{5}$	-	$4.5\ 10^{2}$	-	-
October 1976	0 before	$4.5\ 10^{5}$	$1.1\ 10^{7}$	1	1	-	-
	24 h	$1.5\ 10^{7}$	-	$4.5\ 10^{2}$	-	-	-
	1 week	$1.1\ 10^{8}$	-	$2.5\ 10^{3}$	-	-	-
	2 weeks	$7.5\ 10^{6}$	-	$1.1\ 10^{3}$	-	-	-
	1 month	$4.5\ 10^{5}$	$4.5\ 10^{5}$	9.5	2	-	-
	3 months	$2.5\ 10^{6}$	$2.5\ 10^{6}$	2	0	-	-
June 1977	8 months	$1.5\ 10^{5}$	$4.5\ 10^{6}$	0	0	-	-
January 1978	-	$4.5\ 10^{6}$	$3.5\ 10^{5}$	25	0	$2.5\ 10^{2}$	$2.5\ 10^{2}$
June 1978	0 before	$1.4\ 10^{5}$	$1.4\ 10^{6}$	15	$1.1\ 10^{2}$	$7.5\ 10^{2}$	95
	4 h	$1.4\ 10^{6}$	-	15	-	$1.4\ 10^{3}$	-
	30 h	$1.4\ 10^{7}$	-	25	-	$1.4\ 10^{4}$	-
	10 days	$1.1\ 10^{7}$	$2\ 10^{7}$	9	2	$1.1\ 10^{2}$	$4.5\ 10^{2}$
	20 days	$1.1\ 10^{7}$	$1.4\ 10^{6}$	4	2	$4.5\ 10^{3}$	9
July 1978	30 days	$1.4\ 10^{7}$	$7.5\ 10^{5}$	2	2	3	45

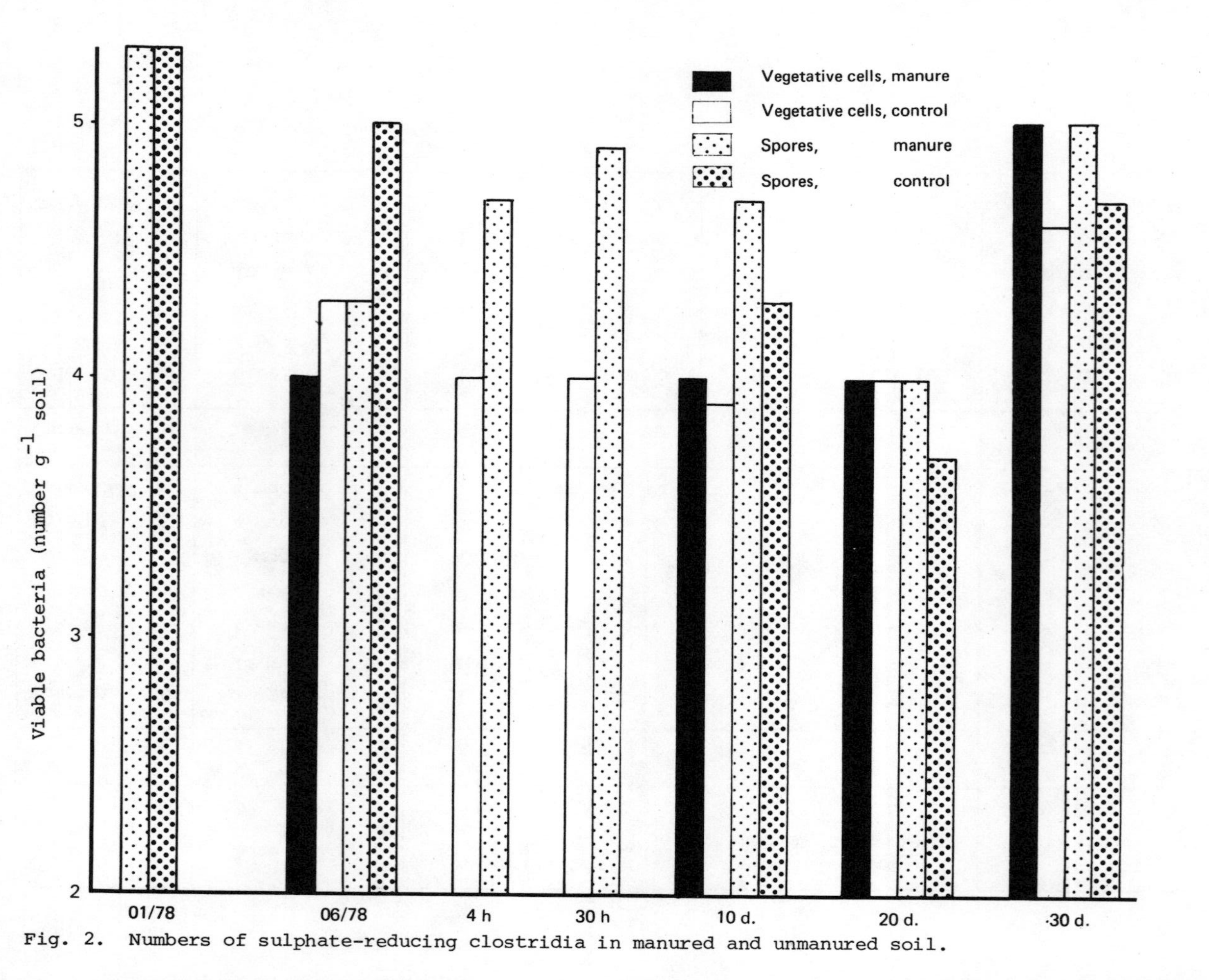

Fig. 2. Numbers of sulphate-reducing clostridia in manured and unmanured soil.

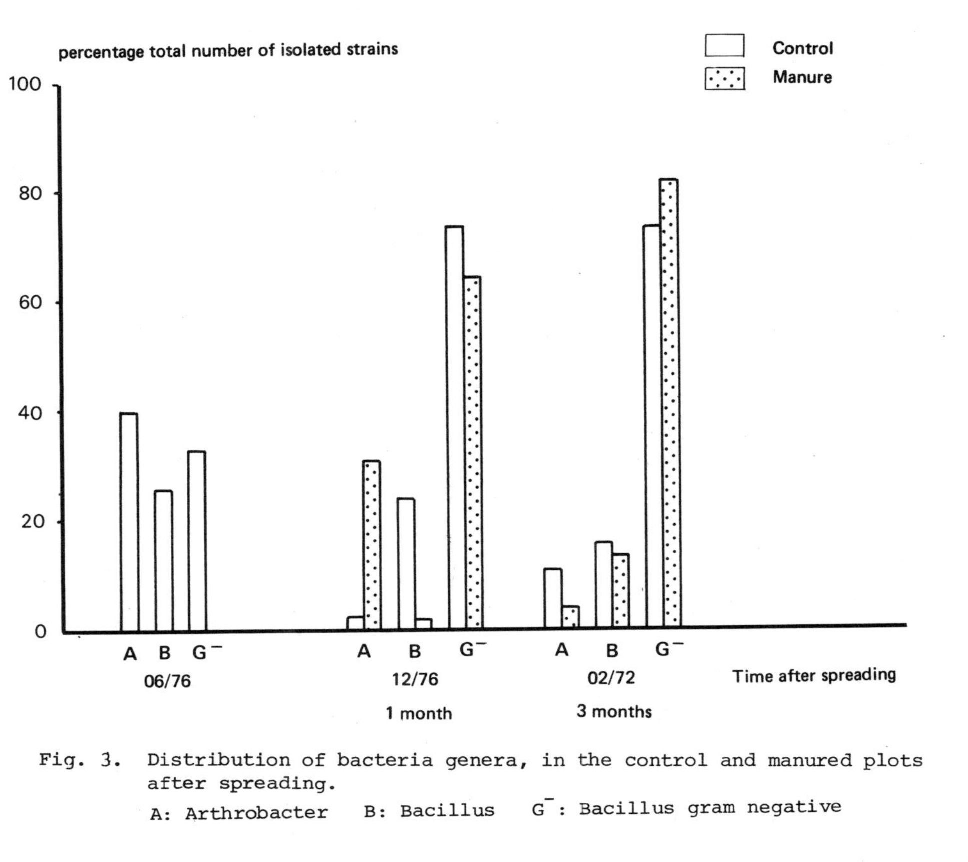

Fig. 3. Distribution of bacteria genera, in the control and manured plots after spreading.
A: Arthrobacter B: Bacillus G^-: Bacillus gram negative

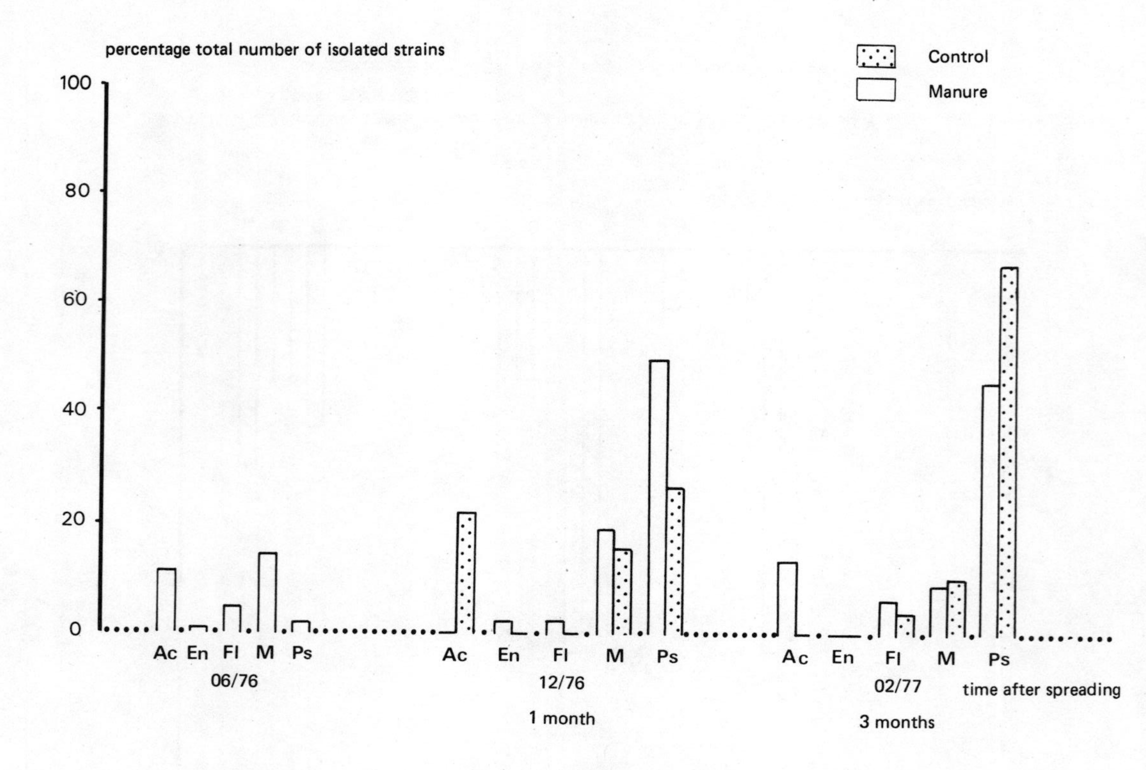

Fig. 4. Distribution of genera of negative gram bacillus in the control and manured plots after spreading.
AC: Acinetobacter EN: Enterobacter FL: Flavobacterium
M: Moraxella PS: Pseudomonas

Consequently, even if faecal micro-organisms decrease during storage (Rivière et al., 1974), hygienic problems remain important.

REFERENCES

Ballay, D. and Catroux, G., 1974. Possibilités de limitation des nuisances et des pollutions dues aux elevages porcins. Ann. Agron. 25, 351-381.

Buchanan, R.E. and Gibbons, N.E., 1974. Bergey's Manual of Determinative Bacteriology, 8th Ed., Baltimore, Williams & Wilkins.

Buttiaux, R., Muchemble, G. and Leurs, T., 1953. La colométrie de l'eau sur membranes filtrantes. Ann. Inst. Pasteur, 84, 1010-1025.

Buttiaux, R., Beerens, H. and Tacqueta, A., 1969. Manuel de technique bactériologique, Paris, Editions Medicales Flammarion.

Chapman, G.H., 1951. A culture medium for detecting and conforming *Escherichia coli* in ten hours, Ann. J. Pub. Health, 41, 1381.

Gatellier, C., 1962. Importance du rôle des bactéries dans la corrosion du matériel. Bull. AFTP. No. 152.

Hajna, A.A. and Perry, C.A., 1943. Comprative study of presumptive and confirming media for bacteria of the coliform group and for faecal streptoccoci. Ann. J. Pub. Health, 33, 550-556.

Holding, A.J., 1960. The properties and classification of the predominant gram negative bacteria occurring in soil. J. Appl. Bacteriol., 23, 515-525.

Kibbey, H.J., Hagedorn, C. and McCoy, E.L., 1978. Use of faecal streptoccoci as indicators of pollution in soil. Appl. Environ. Microbiol. 35, 711-717.

Le Minor, L., 1972. Le diagnostic de laboratoire des bacilles à Gram négatif. Saint-Mandé, Editions de la Tourelle.

Litsky, W., Hallmann, W.L. and Fifield, C.W., 1955. Ann. J. Pub. Health, 45, 1049.

Mossel, D.A.A., Bechet, J. and Lambion, R., 1962. La prévention des infections et des toxi-infections alimentaires. Bruxelles, CEPIA.

Pochon, J. and Tardieux, P., 1962. Techniques d'analyse en microbiologie du sol. Saint-Mandé, Editions de la Tourelle.

Porter, J.N., Wilhem, J.J. and Tresner, H.D., 1960. Method for the preferential isolation of actinomycète. Appl. Microbiol., 8, 174.

Prevot, A.R., 1961. Traité de systématique bactérienne, Paris, Dunod.

Rivière, J., 1961. Isolement et purification des bactéries cellulolytiques du sol. Ann. Inst. Pasteur, 101, 253-258.

Rivière, J., 1974. Evolution de la microflore au cours d'un traitement d'epuration d'eaux urbaines par boues activées. Ann. Agron. 25, 515-533.

Rivière, J., Subtil, J.C. and Catroux, G., 1974. Etude de l'évolution physico-chimique et microbiologique du lisier de porc pendant le stockage anaérobie. Ann. Agron., 25, 383-401.

Soumare, J. and Blondeau, R., 1972. Caractéristiques microbiologiques des sols de la région du Nord de la France. Ann. Inst. Pasteur, 123, 239-249.

White, L.A. and Spence, M.R., 1976. Persistence of human enteric bacteria in the Canadian North. Can. J. Pub. Health. 67, 25-29.

DISCUSSION

H. Tunney *(Ireland)*

Your figures for ryegrass yields show that your pig slurry gave approximately half of the yield of the fertiliser nitrogen, based on the total-N content.

C. Duthion *(France)*

To obtain the same yield as with mineral fertiliser it is necessary to spread twice as much total nitrogen as manure, because it contains 60 percent of the total as mineral-N.

H. Tunney

But you also got a higher maximum yield with pig slurry, and yields were still increasing.

C. Duthion

The yield increased with the amount of manure spread. The details are in the paper.

H. Vetter *(West Germany)*

Thank you very much.

THE INFLUENCE OF AGRONOMIC APPLICATION OF SLURRY ON THE YIELD AND COMPOSITION OF ARABLE CROPS AND GRASSLAND AND ON CHANGES IN SOIL PROPERTIES

Project Leader: R. Lecomte
Collaborators: L. Couvreur, A. Crohain, J.P. Destain, G. Droeven, M. Frankinet, J. Guiot, Y. Raimond, L. Rixhon.

Ministere de L'Agriculture,
Administration de la Recherche Agronomique,
Centre de Recherches agronomiques de Gembloux, Belgium.

I INTRODUCTION

A network of field experiments was carried out at different locations covering the pedological and ecological conditions encountered in various natural regions of Belgium. These experiments were designed to assess the effects of slurry application on the yield, chemical composition and quality of grass and arable crops. The changes in the soil properties were also investigated with particular reference to nitrogen. Topics studied in the project included rate, time, method and frequency of slurry application.

II CHEMICAL CHARACTERISATION OF SLURRIES

The different types of slurries used in the experiments had very different chemical properties (Tables 1 and 2). Pig slurry was rich in N and also in P_2O_5 and K_2O. The ratio of P_2O_5 : K_2O varied considerably between samples. In contrast, cattle slurry contained much K_2O but less N and P_2O_5. Poultry slurry was particularly rich in N and CaO.

The high coefficients of variation calculated for the different parameters (Tables 1 and 2) showed that the slurries were extremely heterogeneous. Dry matter content did not appear to be the main factor responsible for the variability, because expressing the results on a dry matter basis did not reduce the coefficients of variation. Other factors such as degree of mixing of the slurry, type of farming and animal diet were also important.

TABLE 1

CHEMICAL CHARACTERISATION OF SLURRY (WET MATTER BASIS)

Slurry	Parameter	Dry matter %	Mineral matter %	Organic matter %	Total N %	NH_4-H %	P_2O_5 %	K_2O %	CaO %	MgO %
Pig n = 234	Mean	7.43	1.94	5.50	.55	.33	.40	.4C	.33	.11
	Mean - MS	3.60	1.12	2.46	.37	.23	.15	.28	.14	.06
	Mean + MS	11.26	2.76	8.54	.74	.43	.64	.51	.53	.17
	CV %	51.58	42.27	55.19	32.94	30.60	61.22	29.42	57.68	51.32
Cattle n = 82	Mean	8.31	1.82	6.50	.37	.18	.23	.58	.39	.16
	Mean - MS	5.70	1.35	4.17	.30	.14	.10	.38	.17	.08
	Mean + MS	10.93	2.29	8.83	.44	.22	.36	.78	.60	.24
	CV %	31.53	25.70	35.90	19.16	23.00	56.95	33.49	55.10	52.54
Poultry n = 7	Mean	7.51	2.50	5.01	.85	.54	.39	.50	.80	.16
	CV %	32.15	28.91	35.88	5.63	6.70	36.64	12.92	34.13	41.30

n = number of samples
MS = mean square
CV = Coefficient of variation (%)

TABLE 2

CHEMICAL CHARACTERISATION OF SLURRY (DRY MATTER BASIS)

Slurry	Parameter	Mineral matter %	Organic matter %	Total-N %	NH_4-N %	P_2O_5 %	K_2O %	CaO %	MgO %
Pig n = 234	Mean	28.11	71.89	8.89	5.50	5.37	6.97	4.70	1.55
	Mean - MS	21.61	65.56	4.90	2.85	3.27	2.46	2.63	0.94
	Mean + MS	34.61	78.22	12.89	8.15	7.47	11.47	6.78	2.16
	CV %	23.13	8.81	44.91	48.17	39.04	64.65	44.13	39.52
Cattle n = 82	Mean	24.06	75.94	4.90	2.42	2.80	7.44	5.18	1.98
	Mean - MS	13.57	61.64	2.31	1.17	1.50	4.20	1.79	0.73
	Mean + MS	34.55	90.24	7.50	3.66	4.10	10.68	8.57	3.24
	CV %	43.62	18.83	53.00	51.46	46.50	43.52	65.37	63.20
Poultry n = 7	Mean	34.06	65.94	12.10	7.66	5.16	7.27	11.00	2.18
	CV %	14.39	7.44	23.86	24.67	10.53	33.11	17.91	32.86

n = number of samples
MS = mean square
CV = coefficient of variation (%)

TABLE 3

PIG SLURRY: COEFFICIENTS OF CORRELATION (r) BETWEEN CHEMICAL CHARACTERISTICS

I - on wet matter

	DM	MM	OM	Total-N	NH_4-N	P_2O_5	K_2O	CaO	MgO
DM	-	+0.934	+0.966	+0.807	-	+0.734	+0.445	+0.739	+0.738
MM		-	+0.902	+0.540	-	+0.640	+0.520	+0.655	+0.634
OM			-	+0.752	-	+0.742	+0.417	+0.743	+0.736
Total-N					+0.856	+0.704	+0.461	+0.633	+0.677
NH_4-N					-	-	-	-	-
P_2O_5						-	-0.061	+0.657	+0.684
K_2O							-	+0.354	+0.281
CaO								-	+0.777
MgO									-

II - on dry matter

	MM	OM	Total-N	NH_4-N	P_2O_5	K_2O	CaO	MgO
MM	-	-0.992	+0.461	-	-0.013	+0.786	+0.069	+0.111
OM		-	-0.774	-	+0.013	-0.786	-0.069	-0.111
Total-N			-	+0.958	+0.173	+0.764	+0.277	+0.269
NH_4-N				-	-	-	-	-
P_2O_5					-	-0.238	+0.249	+0.482
K_2O						-	+0.223	+0.11
CaO							-	+0.627
MgO								-

TABLE 4

CATTLE SLURRY: COEFFICIENTS OF CORRELATION (r) BETWEEN CHEMICAL CHARACTERISTICS

I - on wet matter

	DM	MM	OM	Total-N	NH_4-N	P_2O_5	K_2O	CaO	MgO
DM	-	+0.683	+0.978	+0.664	-	+0.731	+0.579	+0.151	+0.264
MM		-	+0.522	+0.642	-	+0.535	+0.474	-0.065	+0.008
OM			-	+0.519	-	+0.691	+0.442	+0.167	+0.278
Total-N				-	+0.777	+0.571	+0.520	+0.179	-0.125
NH_4-N					-	-	-	-	-
P_2O_5						-	+0.689	+0.490	+0.606
K_2O							-	+0.242	+0.358
CaO								-	+0.898
MgO									-

II - on dry matter

	MM	OM	Total-N	NH_4-N	P_2O_5	K_2O	CaO	MgO
MM	-	-1.000	+0.876	-	-0.173	+0.712	+0.077	-0.063
OM		-	-0.785	-	+0.241	-0.740	+0.033	+0.164
Total-N			-	+0.968	-0.176	+0.519	+0.409	+0.225
NH_4-N				-	-	-	-	-
P_2O_5					-	+0.086	+0.269	+0.406
K_2O						-	-0.114	-0.222
CaO							-	+0.844
MgO								-

If there were close correlations interlinking various analytical parameters, then 1, or even 2 or 3, constituent elements (preferably ones which were easy to measure - e.g. dry matter) could be taken as a basis on which to determine the value of the other elements. This would avoid the need for complete analysis of the manures. The various coefficients of correlation therefore, are given in Tables 3 and 4. The r values which are underlined were calculated from all the individual samples taken, the remaining values were calculated from the means in each trial.

The values established were not always very high. For example, coefficients linking total-N to other parameters, were:

- the dry matter (r = +0.807 for pig slurry and + 0.664 for cattle slurry)
- the ammoniacal N (r = +0.856 and 0.777)
- the P_2O_5 for pig slurry (r = +0.704)
- the K_2O for liquid cattle manure (r = +0.520)

Finally, a standard relationship appears between CaO and MgO (r = +0.777 for pig slurry and r = +0.898 for cattle slurry).

The considerable heterogeneity of liquid manures, together with the absence of any sufficently close correlations between the various chemical components, suggest that complete analysis is required before land application. This is especially necessary when changing this type of operation (e.g. dairy farming to suckling calves, or from rearing sows to fattening piglets) or when just changing the type of fodder during the course of the year.

Particularly accurate information is needed for the total nitrogen content. Nitrogen is firstly the key component of the manure, and secondly, correlation calculations have shown us that once the nitrogen content is known it is possible to estimate with a good degree of aproximation the not inconsiderable phospho-potassium value of a liquid manure.

III THE EFFECT OF SLURRY ON THE CHIEF CHEMICAL PROPERTIES OF THE SOIL

A) pH, humus, P, K, Ca, Mg

Soil samples were taken from the majority of experimental plots to check on the homogeneity of the field, and assess the effect of the various treatments. As when studying yields, a comparison was made between treatment with nitrogenous mineral fertiliser and treatment with slurry.

1) pH water

In the light of results obtained so far there is nothing to suggest that incorporating animal effluent into the soil affects soil pH. The reasons behind the acidification of our Belgian soils are well known (reduction in the consumption of lime, increased use of acidifying fertilisers and deeper ploughing) and dressings of slurry will thus not fundamentally alter the problem.

2) Humus

Liquid manure had a positive effect on humus level at one location only. Even though slurry supplies 3 to 4 times less organic matter than a normal manure, it is still a considerable source of humus since application of 40 000 l is equivalent to ploughing in the straw. However, increase in the level of humus is a slow process (as moreover is its reduction) which can only be revealed after many years of returning organic matter to the soil.

3) Exchangeable phosphorus

There were few significant effects of treatment on soil phosphorus. However, spreading 40 000 l ha^{-1} could supply some 160 kg ha^{-1} of P_2O_5 which would enrich the stock of exchangeable P in the top soil by some 2 mg 100 g^{-1}. This is only a theoretical calculation though and in actual fact only part of the P supplied by the slurry is in mineral form; the organic component must be mineralised. Also, the PO_4 ions which are supplied or liberated do not all remain in an exchangeable form. They could undergo reversion (due to the effect of Ca, Al, Fe ions) and even be reincorporated into organic matter (micro-organisms).

Intensive use of slurries will therefore have a long-term effect on the overall stock of phosphorus reserves in the soil and could give substantial savings on mineral fertiliser. Tests which have been underway for more than 11 years now relating to the problem of phospho-potassium fertiliser for mass crops, have shown that P - K applications in Belgium were generally too high and, as a result, the soil reserves were more than sufficient. Farmers will thus be recommended only to apply a quantity of P - K mineral fertiliser which is equivalent to the amount that the plants remove, with perhaps a little more to off-set potential loss due to leaching or fixation, for example. Allowances should be made for the P - K returned to the soil in the form of harvest by-products (leaves of beet, straw etc.) or in the form of other organic matter such as manure or liquid manure, carbonate sludge of sugar refining.

Finally, in view of the fact that slurry application was limited to crop requirements in this study, there is no risk of pollution. Indeed, even if the top soil were to be enriched with P the element is not very mobile and the danger of percolation and P being carried towards the underground water would be very slight, even in sandy loamy soils.

4) Exchangeable potassium

A number of results show a significant increase in the exchangeable K in the soil following applications of liquid cattle and pig manure. A dressing of 40 000 l could supply 160 (pig slurry)or 240 (cattle slurry) kg ha^{-1} of K_2O. Here however, unlike phosphorus, the plant can make direct use of the K (K^+ in solution). The K could, however, undergo a fairly intense fixation process (inside clay mineral lattices) particularly in clayey and loamy soils. It would thus be temporarily removed from the soil's stock of exchangeable K. All the K applied is likely to contribute towards feeding the plant sooner or later. If the amount of mineral or slurry applied is greater than the crop requirements then soil K content will increase. Here again the farmer must reconsider the amount of potassium fertiliser he applies when slurries are used. Considerable savings are possible and indeed this is desirable since increased K content of soil is a problem in Belgium (with the risk of bringing the K : Mg ratio out of balance).

5) Exchangeable calcium

The amount of CaO supplied by slurries is relatively high (e.g. poultry slurry) but compared with soil content of exchangeable Ca (from 150 mg 100 g^{-1} for sandy or sandy loamy soils up to 250 mg and more for loamy and clayey soils) it is quite low. Obviously then, no significant results were produced. Dressings of liquid manure do not, of course, help towards resolving the problem of liming (cf pH).

6) Exchangeable magnesium

One sole case of an increase in Mg reserves was found following applications of slurry. The amount of MgO suppplied by slurry is very low (60 kg ha^{-1} per 40 000 l of liquid manure). Furthermore, liquid cattle manure (rich in K) increases the imbalance between K and Mg already apparent in Belgian soils and extra magnesium fertiliser may have to be applied.

Conclusions

Even though most of the tests have only run for a limited period of time so far (one farming year, or at most three farming years) it can be concluded that dressings of slurry should not cause too marked an increase in the soil content of P and K, providing that the amount of mineral fertiliser is adjusted accordingly. In addition to being a valuable source of organic matter, slurry should thus enable the farmer to make substantial savings on mineral fertilisers. Finally, as far as the elements Ca and Mg are concerned, animal effluent could increase the problems of soil reaction and magnesium deficiency frequently observed by the soil analysis departments in Belgium.

B) Dynamics of mineral nitrogen in soils treated with slurry

1) Influence of the rate and time of slurry application on the evolution of mineral nitrogen in the soil

This study was carried out in 1975 - 1976 on bare loamy soil and involved 10 treatments. Five quantities of mineral N were applied in the spring: 0, 40, 80, 120 and 160 kg ha^{-1}. Three quantities of pig slurry were spread before the winter: 20 000 l ha^{-1}, 40 000 l ha^{-1} and 60 000 l ha^{-1}. After the winter 40 000 l ha^{-1} was also applied on the ploughed land. Finally, a quantity of 80 000 l ha^{-1} was applied in two amounts of 40 000 l ha^{-1} before and after the winter.

TABLE 5

EFFECT OF APPLYING SLURRY BEFORE WINTER ON N CONTENT OF SOIL

Treatments	Total-N kg ha^{-1}	Natural soil reserve in November	Slurry-induced increase	Soil mineral-isation	Total
Control	0	60		180	240
20 000 l slurry	132	60	60	180	300
40 000 l slurry	265	60	90	180	330
60 000 l slurry	394	60	160	180	400

1.1. Changes in the control plot without slurry or nitrogenous fertiliser: The test was carried out on very rich land which, at the time the liquid manure was spread, had a mineral N content of some 60 kg ha^{-1} to a depth of 120 cm. During the winter, mineralisation continued and released 60 kg ha^{-1} mineral N. During the growing period the exceptional climatic conditions of 1976 favoured production of 120 kg ha^{-1} mineral N. At the beginning of August, the total content to a depth of 1.20 m was 240 kg ha^{-1}, 180 kg of which was located in the top 30 cm.

1.2. Application of slurry before the winter: Three quantities of slurry were applied at the end of November 1975, i.e. 20, 40 and 60 000 l, which increased the mineral nitrogen content of the soil by 60, 90 and 165 kg respectively.

During the winter and the growing period, mineralisation was comparable with that of the control plot, i.e. 180 kg ha^{-1} N. No penetration of N was observed below a depth of 90 cm during the winter which was abnormally dry.

1.3. Application of slurry after the winter: A quantity of 40 000 l ha^{-1} (136 kg ha^{-1} of total nitrogen) applied at the beginning of March gave an increase of 90 kg ha^{-1} mineral N compared with the control plot. The amount of mineralisation during the growing period (85 kg ha^{-1}) was lower than the amount observed for the control plot. The level over a 1.20 m profile during August was as follows:

Reference sample on 1st March	Increase due to liquid manure	Mineralisation of the soil	Total
120	90	85	295

1.4. Application of slurry before the winter and after the winter: (40 000 l for each application: 410 kg ha^{-1} total nitrogen). The results were similar to those obtained from two separate treatments before and after the winter. The overall level during August was as follows:

Level in soil before winter	Increase from liquid manure before winter	Winter mineral-isation	Increase from liquid manure after winter	Summer mineral-isation	Total
60	90	60	90	85	385

1.5. Slurry values compared with an increasing scale of nitrogenous mineral fertiliser: A comparison of N levels in the soils which received the various slurry treatments with the levels obtained after application of amounts of mineral N at 10, 40, 80, 120 and 160 kg ha^{-1} leads to the following ratings:

20 000 l before the winter was equivalent to 100 kg ha^{-1} of nitrogen

40 000 l before the winter was equivalent to 150 kg ha^{-1} of nitrogen

60 000 l before the winter was equivalent to more than 160 kg ha^{-1} of nitrogen

40 000 l after the winter is equivalent to 100 kg N ha^{-1}

40 000 l before and 40 000 l after the winter is equivalent to more than 160 kg N ha^{-1}.

It should be noted that the amount of liquid manure is not proportional to its equivalent in nitrogen.

2) Influence of the quantities applied, the date of application and delay before working slurry into the soil on the evolution of mineral nitrogen to a depth of 1.20 m

The production of mineral N was studied on sugar beet for different treatments as given in Table 6.

TABLE 6

Treatment	Mineral nitrogen kg ha^{-1}	Total nitrogen kg ha^{-1}	Spreading of pig slurry	Delay before working into soil
1	0	-	-	-
2	150	-	-	-
3	-	370	End of Aug.	0 month
4	-	370	"	1 month
5	-	370	"	3 months
6	-	353	Begin. of Autumn	0 month
7	-	353	"	2 months
8	-	436	Begin. of Spring	0 month

2.1. Changes in the untreated control plot (Treatment 1): At harvest time the soil content was 40 kg ha^{-1} mineral N to a depth of 1.20 m. During the autumn 35 kg ha^{-1} mineral N was released by mineralisation. Over the winter, with a slightly above-average rainfall, enrichment of the lower horizons showed that the nitrogen penetrated deeper than 1.20 m. During the spring there was an increase of 28 kg ha^{-1} in the top 30 cm. From June onwards the total N content of the profile began to fall and mineralisation provided less than the amount removed by the crops. This started at the surface and then gradually worked its way down so that the whole of the profile was exhausted at the end of July. At harvest time approximately 10 kg ha^{-1} N remained.

2.2. Application of slurry after cereal harvest - three digging in dates: Application of slurry increased mineral N content by 140 kg ha^{-1}.

2.2.1. Digging in the slurry directly after spreading (Treatment 3): After spreading at the end of the autumn, 70 kg ha^{-1} mineral N was released compared with 35 kg in the control plot. During the winter, enrichment of the lower horizons showed that the nitrogen penetrated below the profile under study (1.20 m). As in the case of the control plot, mineralisation kept the N level constant. In the spring the balance became positive and released some 40 kg ha^{-1} mineral N. In June, as for the control plot, mineralisation did not provide as much mineral N as the beet crop removed. This first became evident at the surface and then extended over the whole of the profile by the end of July. At harvest time the soil was not fully exhausted - some 30 kg ha^{-1} mineral N remained, with two-thirds of this located at a depth of 1 m to 1.20 m.

<u>2.2.2 Digging in one month after spreading</u> (Treatment 4): A considerable loss of N occurred after the slurry was spread and continued until the manure was dug in. At this point the level over the whole profile was 90 kg ha^{-1} lower than the level observed with Treatment 3. The loss compared with the level recorded just after application of the slurry was 50 kg ha^{-1}. Hence the real loss lies somewhere between these two values. Once the slurry was worked into the soil,mineralisation became positive again and released 60 kg ha^{-1} mineral N. During the winter any nitrogen washed to depths greater than 1.20 m by the rain was off-set by mineralisation. Removal of mineral N by the crop followed the same pattern over the profile as for Treatment 3. However, the soil was exhausted to a greater degree because only 15 kg ha^{-1} remained.

<u>2.2.3. Digging in 3 months after spreading</u> (Treatment 5): Nitrogen development was similar to that observed for Treatment 4. The content in the profile continued to drop, however, until mid-October, one month before digging in. The loss estimated as for Treatment 4 lies between 60 and 130 kg. After the slurry was worked into the soil the mineral N content followed the same pattern as for Treatment 4 but at a level lower by 30 kg ha^{-1}.

The amount remaining at harvest time was identical, i.e. 15 kg ha^{-1}.

2.3. Application of slurry at the beginning of the autumn: The liquid manure increased soil N content to 195 kg ha^{-1}.

2.3.1. Digging in directly after spreading (Treatment 6): The level recorded after application was maintained during the autumn and winter since mineralisation compensated for any N that penetrated below 1.20 m. In the spring mineralisation provided some 15 kg. The crop began to remove nitrogen in June and followed the same development pattern as for Treatment 3. The amount remaining at harvest time was identical.

2.3.2. Digging in 2 months after spreading (Treatment 7): As for Treatments 4 and 5 a rapid drop in the N level of the profile was observed up to the start of November when the loss compared with Treatment 6 was estimated at 80 kg ha^{-1}, i.e. slightly above one-third. From November onwards the pattern was identical to Treatment 4 but at a level some 20 kg higher.

2.4. Application of slurry in the spring with digging in immediately after spreading (Treatment 8): Application of slurry in this way produced the same effects as application of 200 kg of mineral nitrogen (cf curves 2 and 8 on Figure 1), and did not penetrate below 60 cm. Uptake of N by the crop started earlier towards 15th May and reached a peak rate during July. The large quantity of nitrogen located in the top 15 cm - more than two-thirds - was without doubt the main reason behind this. The pattern of N loss in the profile was similar to that of the control plot.

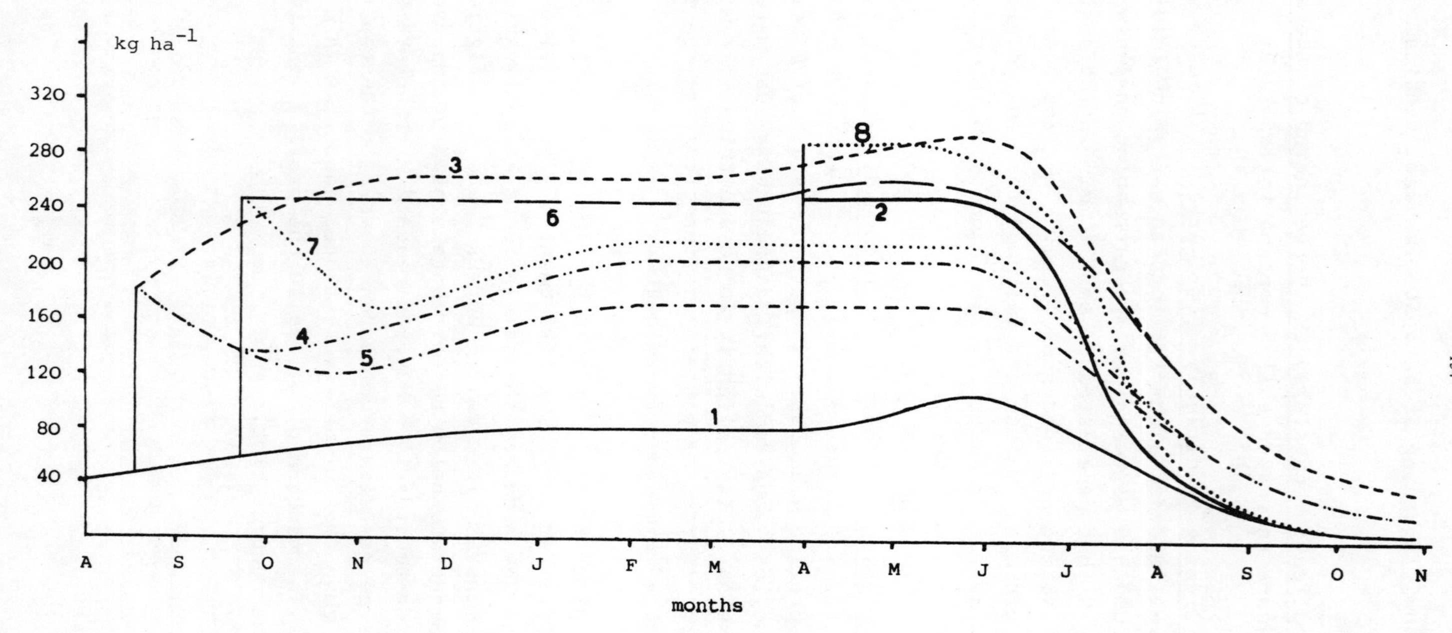

Fig. 1. Evolution of mineral nitrogen in the soil to 1.20 m (see Table 6 for treatments 1 - 8).

2.5. Conclusions

2.5.1. Slurry applied before the winter should be dug in immediately after spreading.

2.5.2. If the slurry is left on the surface then nitrogen loss increases with the amount of time that elapses before it is dug in.

2.5.3. During the winter the nitrogen migrates rapidly down the profile. Part of it penetrates below 1.20 m and may not be recovered by crop roots.

2.5.4. Slurry applied in the spring migrates only a little. Its development pattern in land under cultivation is similar to an application of mineral fertiliser.

2.5.5. Since beet starts to remove nitrogen from the surface downwards, slurry is more effective if located in the upper soil layers.

2.5.6. Application in the spring is preferable to application before the winter. The nitrogen in slurry applied before the winter is subject to considerable leaching, it is not recovered in full and thus this recovery takes place at a later stage.

IV INFLUENCE OF THE APPLICATION OF LIQUID MANURE ON YIELD OF GRASS AND ARABLE CROPS

A) Quantitative effects

The principal aim of this study was to determine the practical fertilising value of slurry viewed essentially in terms of the nitrogen supply for the plants under cultivation. Since the soils on which the trials were conducted had a sufficient level of phosphorus and potassium, the influence of these two elements was not investigated. The fertilising value was established by first of all examining the effect of different quantities of slurry on crop production, then the effect of the time lag before working into the soil and the season in which it is spread. This was followed by the influence of application frequency and finally the interaction between liquid manure and other organic matter was investigated.

1) Quantities of slurry

A so-called 'nitrogen effect' of the slurry was derived by comparing the effects of different rates of slurry and mineral nitrogen application on grass and arable crops. As a rule the slurry was spread in moderate quantities, similar to those applied in agricultural practice. However, in certain tests, massive quantities, up to 180 000 l ha^{-1}, were applied. On arable land the slurry was dug in as soon as possible after spreading. Except where specified otherwise in the tests, the slurry was applied before ploughing on land for arable crops and at the beginning of spring for grassland.

In this summary the amount of slurry applied will be defined in each case in terms of the quantity of total N actually applied. This was calculated on the basis of the amount applied and the N content of a slurry sample taken at the time of application from the spreader cart. The overall 'nitrogen effect' of the slurry was defined by estimating the 'nitrogen effect' on the crop following spreading (direct effect) and, where possible, by measuring changes in subsequent years (after effects).

1.1. Direct effect of slurry: The direct effect on the principal arable crops will be considered, followed by the direct effect on grassland.

1.1.1. Cereals: The effect of slurry was investigated solely for winter wheat and winter barley since these two crops alone account for almost 2/3 of all the cereal crop in Belgium. The slurry was always applied before ploughing, i.e. in September or October, since pedo-climatic conditions were often unsuitable for landspreading in the spring. The overall 'nitrogen effect' of slurry spread before the winter on yield was compared with the effect of quantities of mineral N applied in instalments during the growing period. Tests on wheat were set up in deep loamy soil only, with one test sited on a drainage network. For winter barley, tests were located on deep loamy soil or on shallow soil. The amount of N supplied by the slurry varied between 88 and 341 kg ha^{-1}.

As a rule, for similar quantities of N applied, the effect of the slurry was always the same whatever its origin (cattle, pig, poultry) (Table 7). In the case of winter wheat the mean observed 'nitrogen effect' of the liquid manure was 85 kg ha^{-1} (range 60 - 120 kg ha^{-1}). The efficiency rate of N in the slurry (ratio of the 'nitrogen effect' value to the quantity of nitrogen applied) was on average 50 percent but this percentage tended to drop as the quantity applied increased.

For nitrogen applications of 108 to 142 kg ha^{-1} (mean = 127 kg ha^{-1}) the mean 'nitrogen effect' was about 77 kg ha^{-1} (efficiency rate = 61 percent). For an application of 274 kg ha^{-1} (the mean of applications ranging from 207 to 341) the 'nitrogen effect' recorded was some 92 kg ha^{-1} (efficiency rate = 34 percent). This was only 15 kg ha^{-1} more for twice the amount of N applied in the slurry. The fertilising value of slurry observed in the case of winter barley (Table 8) was influenced to a large extent by the type of soil on which the test was carried out. In deep soil the 'nitrogen effect' was 90 and 135 kg ha^{-1} when the slurry supplied 167 and 261 kg N ha^{-1} respectively. This gave an efficiency slightly above 50 percent, i.e. very close to the rate obtained for winter wheat. On shallower soil, however, the 'nitrogen effect' was less, i.e. in the 10 to 30 kg ha^{-1} range for 88 to 181 kg ha^{-1} applied N. In this case, the efficiency rate for slurry N did not exceed 22 percent.

TABLE 7

'NITROGEN EFFECT' OF SLURRY ON WINTER WHEAT

Type of slurry	Location of trial	Total N applied kg ha^{-1}	'Nitrogen effect' kg ha^{-1}	Efficiency %
Pig	Marcq	121	60	50
	Gembloux	127	75	59
	Florée	139	80	58
	Marcq	247	60	24
	Gembloux	285	110	39
	Cortil	341	70	21
Cattle	Marcq	108	100	90
	Marcq	207	100	48
Poultry	Marcq	142	70	49
	Marcq	288	120	42

TABLE 8

'NITROGEN EFFECT' OF SLURRY ON WINTER BARLEY

Type of slurry	Location of trial	Total N applied kg ha^{-1}	'Nitrogen effect' kg ha^{-1}	Efficiency %
	Shallow soil			
Cattle	Emptinne	88	10	11
	Braibant	91	20	22
	Emptinne	152	10	7
	Braibant	157	20	13
	Emptinne	172	10	6
	Braibant	181	30	17
	Deep loamy soil			
Pig	Piétrebais	167	90	54
	Gembloux	261	135	52

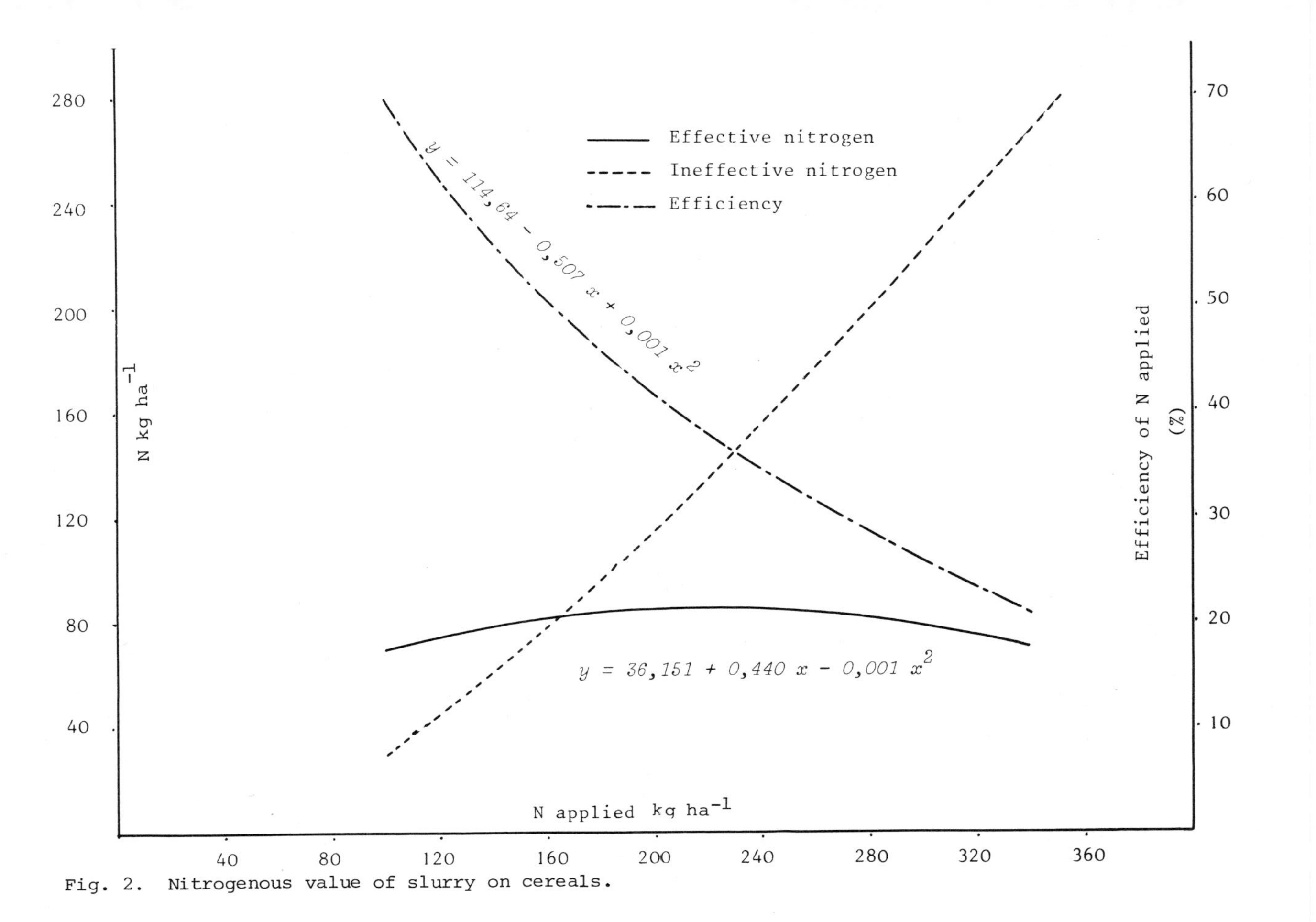

Fig. 2. Nitrogenous value of slurry on cereals.

Implications

On shallow soils (of the Gbap type) it would thus seem unwise to apply slurry before ploughing as a nitrogenous fertiliser for winter barley. The efficiency rate of the slurry N on deeper soils averaged 50 percent but decreased as the amount of slurry applied increased. Hence, there is no benefit to be gained from increasing rates of application before sowing winter cereals. The curvilinear relationship established between the 'nitrogen effect' (y) and the amount of N applied (x) is expressed by the following equation:

$$y = 36.151 + 0.440 x - 0.001 x^2$$

This gives a maximum value for a slurry application containing 220 kg N ha^{-1} and corresponds to a practical effect on the crops of 80 kg ha^{-1}. Since the practical 'nitrogen effect' is virtually constant (80 kg ha^{-1}) for N applications of 120 to 320 kg ha^{-1}, application of 120 kg ha^{-1} slurry N would seen the optimum rate of application for winter cereals, particularly wheat.

Given the mean composition of slurries reported above, this quantity corresponds to ± 20 - 25 000 l ha^{-1} pig slurry, ± 30 - 35 000 l ha^{-1} cattle slurry and ± 10 - 15 000 l ha^{-1} poultry slurry. Larger quantities did not increase the 'nitrogen effect' and only served to increase the proportion of ineffective nitrogen. Application of larger amounts of manure did not, however, have any harmful effect on wheat during the experiments.

1.1.2. Sugar beet: Two tests on loamy soil (Gembloux and Malèves) were set up in order to estimate the 'nitrogen effect' of pig slurry on sugar beet.
The results of these two tests were influenced to a large extent by the mineral N content of the soil (Table 9). Slurry is usually applied in the autumn on loamy soils, since ploughing generally takes place before

TABLE 9

YIELDS OF SUGAR BEET

Treatments	N applied kg ha^{-1}	Root yield kg ha^{-1}	Sugar yield kg ha^{-1}	Sugar content %	Leaf production	
					Wet matter kg ha^{-1}	Dry matter kg ha^{-1}
GEMBLOUX						
mineral	0	55 633	9 800	17.63	12 123	1 568
manure	40	54 784	9 605	17.52	12 376	1 593
	80	60 478	10 248	16.94	16 176	1 916
	120	56 744	9 463	16.70	15 127	1 852
	160	59 896	9 794	16.35	17 303	2 052
20 000 l pig slurry	132	58 550	10 386	17.75	12 658	1 629
40 000 l "	265	62 398	10 561	16.94	14 898	1 837
60 000 l "	394	62 431	9 825	15.75	17 083	1 946
80 000 l "	410	62 208	9 902	15.79	18 244	2 158
MALEVES						
mineral	0	48 931	8 363	17.09	40 084	5 404
fertiliser	50	54 958	9 528	17.35	45 083	5 955
	100	58 667	10 063	17.15	48 167	6 582
	150	59 278	10 288	17.36	46 847	5 779
	200	60 514	10 025	16.58	52 445	6 369
pig slurry	427	55 204	9 021	16.34	58 306	6 778
" + 25 kg N ha^{-1}	452	56 692	9 331	16.46	57 245	6 764

the winter in these regions. Application of slurry on ploughed land in the spring will be covered at a later stage (see 'Influence of the time of year at which the liquid manure is applied').

Very few differences emerged between the various treatments in the first test because of the very high mineral N reserves in the soil. The amounts of mineral N applied (0 to 160 kg ha^{-1}) gave virtually equal sugar yields with the slight increase in root production going hand in hand with a drop in sugar content. Under these conditions the slurry produced root and sugar yields equal to, if not higher than those obtained with mineral fertiliser. As is the case with mineral fertilisers, sugar content showed a decrease as a function of the amount of slurry applied. Leaf production increased as the amount of slurry applied was increased. The very low 'nitrogen effect' of 20 000 l ha^{-1} on sugar content and leaf production, increased as the amount applied was increased (Table 10).

TABLE 10

NITROGEN EFFECT OF SLURRY ON SUGAR BEET

Quantities of slurry	'Nitrogen effect' (kg N ha^{-1}) on		
	Sugar content	Leaf production	
		WM	DM
20 000 l	5	20	25
40 000 l	90	90	90
60 000 l	160	155	130
80 000 l	160	160	160

On this same type of soil, slurry was applied on a larger scale up to 180 000 l ha^{-1}. Amounts of 120 000 l ha^{-1} and 180 000 l ha^{-1} gave good root production but low sugar content, resulting in poorer sugar yield (Table 11).

TABLE 11

EFFECTS OF PIG SLURRY ON SUGAR BEET YIELD

Treatments	Root yield kg ha^{-1}	Sugar yield kg ha^{-1}	Sugar content %
175 kg ha^{-1} mineral nitrogen	60 514	10 333	17.07
60 000 1 (353 kg ha^{-1} total-N)	59 414	9 906	16.65
120 000 1 (753 kg ha^{-1} total-N)	63 559	9 833	15.47
180 000 1 (1 095 kg ha^{-1} total-N)	62 716	9 712	15.48

In the second test (Malèves), where the level of mineral N in the soil was lower, slurry application before the winter produced a lower root yield than that obtained from mineral fertiliser (150 kg ha^{-1}). This was despite the fact that the slurry supplied considerably more total N than the mineral fertilisers (total N supplied by the liquid manure varying from 370 to 491 kg ha^{-1}). The 'nitrogen effect' of the slurry was some 60 kg ha^{-1} on average (efficiency rate ± 15 percent).

In all the cases studied slurry application gave a lower sugar content than the highest rate of mineral fertiliser (200 kg ha^{-1}). Slurry had a very low 'nitrogen effect' on sugar production, with an efficiency rate in the region of 10 percent. By applying an additional 25 kg ha^{-1} of mineral N at the time of sowing, over and above the slurry, it was possible to improve yield but not, however to attain optimum production figures.

Implications

The 'nitrogen effect' of slurry applied before winter for beet takes longer to show in the crop than mineral fertiliser and this consequently delays the ripening of the beet. Nitrogen from the slurry pentrates deep into the soil and recovered at a later stage, giving rise to lower sugar contents and higher leaf production.

On soil with a standard level of N the efficiency rate is some 10 percent. It is more difficult to assess the effects on soil with a high level of mineral nitrogen. The sugar content and leaf production are influenced in the same direction as for soil with a standard level of nitrogen, but to a lesser extent. Application of massive quantities of liquid manure, even in excess of 1 000 kg N ha^{-1}, does not harm the beet except for causing a marked drop in sugar content.

1.1.3. Potatoes: The 'nitrogen effect' of slurry on potatoes was assessed in 1976 and 1977 on sandy loam soil and in 1976 on loamy soil. On each occasion the slurry was applied in spring before ploughing. The climatic conditions recorded in 1976 had a marked influence on the results obtained.

On sandy loamy soil the 'nitrogen effect' was 138 kg ha^{-1} on average, with a maximum value of 160 kg ha^{-1} and a minimum value of 115 kg ha^{-1} for slurry applications of 150 to 565 kg N ha^{-1} (Table 12). The efficiency rate decreased appreciably with increasing rate of slurry application. Hence, for 1977, application of 310 kg N ha^{-1} as pig or poultry slurry gave a 'nitrogen effect' of 128 kg ha^{-1} (efficiency rate: 41 percent). Application of 550 kg ha^{-1} gave a 'nitrogen effect' of some 153 kg ha^{-1} (efficiency rate: 28 percent), i.e. some 25 kg ha^{-1} more. By increasing the amount supplied by 240 kg N ha^{-1}, i.e. an increase of 77 percent, the 'nitrogen effect' increased by 16 percent only which was equivalent to a 6 percent rise in production.

On loamy soil in 1976 slurry application produced a tuber yield greater than that obtained with the highest quantity of mineral nitrogen (200 kg ha^{-1}) (Table 13). In this case yield increased as the quantity of slurry applied increased.

TABLE 12

'NITROGEN EFFECT' OF SLURRY ON POTATOES ON SANDY LOAMY SOIL

Year	Origin of slurry	Quantities l ha^{-1}	Total-N applied kg ha^{-1}	'Nitrogen effect' kg N ha^{-1}	Efficiency %
1976	Pig	33 000	150	156	100
		53 000	245	116	68
1977	Pig	41 000	310	119	38
		82 000	536	159	30
1977	Poultry	34 000	312	137	44
		68 000	565	146	25

TABLE 13

PRODUCTION OF POTATOES ON LOAMY SOIL IN 1976

Treatments		Total nitrogen kg ha^{-1}	Tuber yields (size > 25)	
			kg ha^{-1}	%
Mineral manure		80	18 123	79.16
		120	19 827	86.59
		160	20 112	87.84
		200	21 626	94.78
pig	33 000 l ha^{-1}	122	23 760	103.78
slurry	53 000 l ha^{-1}	252	27 422	119.77
	86 000 l ha^{-1}	495	29 398	128.40

Implications

The results show that the potato derives great value from slurry applied during the spring. However, it would seem appropriate to apply quantities providing no more than 300 kg N ha^{-1}. As in the case of beet, a small amount of mineral N added at the time of planting could be effective and enable optimum production to be attained. Application of large quantities of slurry does not reduce production. However, large quantities with lower rates of efficiency increase the amount of non-effective N for the crop considerably.

1.1.4. Ryegrass and temporary grassland: The effect of liquid manure was considered for Italian ryegrass on sandy soil and for temporary grassland on shallow soil in the Ardennes. The slurry was applied at the start of the spring before vegetation recommenced growing. The climatic conditions which followed slurry application had a marked influence on its fertilising value. For example, at Destelbergen in 1978 a cold north wind after spreading, combined with the effect of the liquid manure, produced scorching of the foliage and stunted the growth of the ryegrass. The results obtained from this test (Table 14) differ appreciably from the others. The dry matter content of the grass at the first cutting date was influenced by the type of fertiliser applied. Slurry produced dry matter contents of the same order of magnitude as those produced by large amounts of mineral fertilisers. However, the yield from slurry was frequently less than that obtained from mineral fertilisers. It would thus seem that, even with favourable climatic conditions, slurry application reduces growth rates and results in a lesser degree of maturity at harvest time. The 'nitrogen effect' derived from comparing dry matter was, therefore, lower than that derived from the comparison of fresh weights. For the two cases, under favourable climatic conditions, the efficiency rate at the time of the first cut was estimated at 45 percent and 35 percent for fresh and dry matter respectively. This efficiency rate tended to drop slightly as the amount of slurry applied was increased. The 'nitrogen effect' observed for grass yield varied from 40 to 100 kg ha^{-1} for nitrogen applications of 74 and 250 kg ha^{-1}. In the case of dry yield this 'nitrogen effect' was between 35 and 70 kg ha^{-1}. Under bad climatic conditions the 'nitrogen effect' was less pronounced and

TABLE 14

FERTILISING VALUE OF SLURRY ON TEMPORARY GRASSLAND

Trial	Quantity of N applied kg ha^{-1}	First cut				Second cut				Total (1 + 2)			
		'Nitrogen effect'		Efficiency %		'Nitrogen effect'		Efficiency %		'Nitrogen effect'		Efficiency %	
		WM	DM	WM	DM	WM	DM	WM	DM	WM	DM	WM	DM
Libramont	74	40	34	54	46	5	3	7	4	45	37	61	50
Destelbergen 76	78	47	35	60	45	-	-	-	-	-	-	-	-
Destelbergen 76	164	62	42	38	26	0	0	0	0	62	42	38	26
Libramont	169	82	66	49	39	10	8	6	5	92	74	54	44
Destelbergen 76	250	100	70	40	28	15	31	6	12	115	101	46	40
Destelbergen 78	289	75	10	26	3	35	55	12	19	110	65	38	22

even negligible as far as dry matter yield was concerned (efficiency rate in the region of 3 percent). For the second cut, the 'nitrogen effect' on dry matter yield was low, attaining a maximum of 30 kg ha^{-1}. For grass yield it was even lower (± 15 kg ha^{-1}). However, at Destelbergen in 1978 the 'nitrogen effect' was higher and thus compensated for the low fertilising value recorded during the first cut. Taking the two cuts together, the efficiency rate varied between 38 and 61 percent for fresh matter yields or between 22 and 50 percent for dry matter yield. For grass, though, dry matter production is most important and this is the sole value that should be taken.

Implications

On temporary grassland and for ryegrass the efficiency rate is in the region of 35 percent. This rate falls as the quantity of slurry is increased, though not to the same extent as for the other crops discussed above. On sandy soil, under dry conditions such as those in 1976, a large application of slurry for the first cut is more effective than a smaller application, on account of its effect on the second cut. In the case where the climatic conditions caused scorching of the foliage, the recovery of production during the second cut was insufficient to compensate for the low yields of the first cut.

1.1.5. Permanent hay field: Results were similar to those obtained for temporary grassland. At Izier in 1977, for instance, the growth of grass on plots treated with slurry was considerably slowed down by the north wind which prevailed after spreading slurry. In June, at the time of the first cut, these plots and in particular those which were treated with the low rate of slurry application (± 25 000 l ha^{-1}), only produced a small amount of grass and were, upon regrowth, taken over almost entirely by clover.

In all these tests the slurry was applied in the spring (Table 15). Where climatic conditions after spreading were favourable the 'nitrogen effect' of the slurry on the first cut was similar to that observed for temporary grassland. However, where small quantities were applied the efficiency rate of the nitrogen in the slurry was minimal or even negligible. The efficiency rate was between 23 and 43 percent or between 9 and 49 percent, depending on whether fresh or dry matter production was used as the basis for comparison of liquid manure with mineral fertilisers. In the specific case of Izier in 1977 the climatic conditions caused blight on the vegetation and the 'nitrogen effect' was zero for all the amounts of slurry applied. For the second cut the efficiency rate was almost 20 percent for both fresh and dry matter yields. This corresponded to a 'nitrogen effect' of 30 to 70 kg ha^{-1} depending on the quantity spread. Taking the two cuts together the mean efficiency rate of nitrogen in the slurry was 61 percent and 35 percent for fresh and dry matter production respectively.

Implications

Similarly, with temporary grassland and ryegrass, adverse weather after slurry application affects the herbage in hay fields and reduces growth rate as compared with mineral fertiliser. This gave an efficiency rate in the region of 61 percent for fresh yields and 35 percent for dry matter yields. As in the case of temporary grassland and pasture land, the latter figure is the more important.

1.1.6. Permanent pasture land: Fertilisation of permanent pasture land with cattle slurry was investigated in the Ardennes (Chenogne in 1976 and Morhet in 1977 and 1978) and in Condroz (Arsimont in 1978), i.e. on shallow soil in both cases. The slurry was applied at the end of the winter and its effect was compared with that of mineral N fertilisers.

TABLE 15

FERTILISING VALUE OF SLURRY ON PERMANENT HAY FIELD

Trial	Slurry	Quantity of N applied kg ha^{-1}	First cut				Second cut				Total (1 + 2)			
			'Nitrogen effect'		Efficiency %		'Nitrogen effect'		Efficiency %		'Nitrogen effect'		Efficiency %	
			WM	DM	WM	DM	WM	DM	WM	DM	WM	DM	WM	DM
Melen	Cattle	76	17	0	23	0	-	-	-	-	-	-	-	-
Izier	Cattle	100	0	0	0	0	-	-	-	-	-	-	-	-
Izier	Cattle	122	52	60	43	49	-	-	-	-	-	-	-	-
Melen	Cattle	137	62	34	45	34	33	22	24	16	95	69	69	41
Melen	Pig	174	58	15	33	9	44	43	25	25	102	58	59	33
Izier	Cattle	196	0	0	0	0	-	-	-	-	-	-	-	-
Izier	Cattle	244	81	51	33	22	-	-	-	-	-	-	-	-
Melen 78	Pig	345	113	39	33	11	75	67	22	19	188	106	54	31

First grazing period: The 'nitrogen effect' of the slurry increased as the amount applied was increased (mean efficiency rate in the region of 22 percent) (Table 16). In contrast to temporary grassland and permanent hay fields, the efficiency rate, as calculated from fresh and dry matter yields, was similar to that obtained with mineral fertilisers.

Second grazing period: In all the tests, slurry applied before the first grazing period had a positive influence on the yield measured during the second grazing period. The amount of grass left uneaten was greater after application of liquid manure than after application of a mineral N fertiliser. In certain cases, such as at Arsimont in 1978, the amount of grass left uneaten on plots treated with slurry was 1/3 of production or 2.5 times higher than the amount left on plots treated with mineral fertiliser. Slurry had a very small influence on subsequent yields and the maximum 'nitrogen effect' was only 10 kg ha^{-1}.

1.2. After effects of liquid manure (Table 17): Slurry had the largest effect on the crop growth after its application when only moderate quantities were applied. With quantities of up to 400 kg N ha^{-1} a very slight after effect was observed for all crops. This was equivalent to 0 - 20 kg ha^{-1} mineral N and became negligible after two years. For higher quantities, above 500 kg N ha^{-1}, this after effect was equivalent to a minimum of 60 kg ha^{-1}, and was still apparent after two years. In the test at Gembloux, for instance, the after effect measured two years after application of 120 000 l ha^{-1} and 180 l ha^{-1} was still some 50 kg ha^{-1}.

TABLE 16

FERTILISING VALUE OF LIQUID MANURE ON PERMANENT PASTURE LAND

Trial	Quantity of N applied (kg ha^{-1})	First cut			
		'Nitrogen effect'		Efficiency %	
		WM	DM	WM	DM
Chenogne	76	20	30	26	39
Chenogne	151	30	45	20	30
Morhet 77	159	25	20	16	13
Arsimont	161	50	50	31	31
Morhet 78	173	60	60	35	35
Morhet 77	314	60	50	19	16
Arsimont	324	40	40	12	12
Morhet 78	346	75	75	22	22

TABLE 17

AFTER EFFECT OF LIQUID MANURE

Type of slurry	Trial	Crop	Quantity of slurry l ha^{-1}	Quantity of N applied kg ha^{-1}	After effect
Cattle	Morhet 78	Grassland	78 000	314	0
Pig	Florée	Winter barley	20 000	138	0
	Gembloux	Winter wheat	20 000	132	10
			40 000	265	10
			60 000	394	20
			80 000	410	15
			120 000	752	60
			180 000	1 095	60
	Otegem	Winter wheat	41 000	310	0
			82 000	536	60
Poultry	Otegem	Winter wheat	34 000	312	0
			68 000	565	60

2) Time lag before digging in

Immediate digging in of the slurry has the advantage of reducing odour after spreading and reducing N loss through volatilisation. The influence of rapid or delayed digging in of slurry on sugar beet yield and on the development of mineral nitrogen in the soil was compared on a loam soil. Slurry was applied on the following dates: 19/8/76 and 23/9/76. Slurry applied on the first date was dug in either immediately, one month after spreading (23/9/76) or three months after spreading (24/11/76). For the second date, slurry was dug in immediately or two months later (21/11/76) (Table 18). These various treatments were then compared with increasing quantities of mineral N, ranging from 0 to 200 kg ha^{-1}. Delay in digging in the slurry did not seriously affect beet yield except in the case of a three month time lag. This observation is confirmed by the virtually identical quantities of mineral N (cf. Chapter III B) found in the profile at the time when the beet started to remove nitrogen.

Implications

To obtain maximum benefit from the fertilising value of slurry it would seem wise, in the light of comments made in Chapter III B, to dig in the slurry as soon as possible after application.

3) Time of year of application

The influence of slurry on crop yield is highly dependent upon the date of application. In particular, the fertiliser value of slurry nitrogen was investigated for a crop of sugar beet by comparing dressings applied before the winter with those applied after. For a period of two years, 1976 and 1977, pig slurry was applied on deep loamy soil. Part of the slurry was spread in November before ploughing and part in March a few days before sowing. The quantities of nitrogen spread were virtually identical for the two dates of application: 132 - 136 kg ha^{-1} for the Gembloux test and 405 - 436 kg ha^{-1} for the Malèves trial (Table 19).

TABLE 18

INFLUENCE OF DELAY IN DIGGING IN ANIMAL MANURE ON THE YIELDS OF SUGAR BEET

Dates		Quantities of nitrogen applied from slurry kg ha^{-1}	Root yield kg ha^{-1}	Sugar yield kg ha^{-1}	Sugar content %
Application of the slurry	Digging in of the slurry				
19/8/76	19/8/76	370	54 292	8 778	16.18
	23/9/76		57 917	9 498	16·41
	24/11/76		52 375	8 646	16.50
23/9/76	23/9/76	353	52 667	8 610	16.35
	24/11/76		58 042	9 410	16.19

TABLE 19

INFLUENCE OF THE DATE OF APPLICATION ON A CROP OF SUGAR BEET

Trials	Dates of application	Quantity of total N applied kg ha^{-1}	Root yield		Sugar yield		Sugar content	Leaf production (kg ha^{-1})	
			kg ha^{-1}	%	kg ha^{-1}	%	%	WM	DM
Gembloux	Nov.	132	56 214	100	9 979	100	17.75	12 658	1 629
Persin	March	136	62 485	111	10 665	107	17.07	15 930	1 943
1976									
Malèves	Nov.	405	55 931	100	9 183	100	16.43	58 847	6 903
1977	March	436	58 917	105	10 009	109	16.99	55 958	7 227

For a given quantity of nitrogen, the slurry spread in the spring gave higher root and sugar yield, 8 percent higher than that applied before the winter. The dry matter yield of leaves also showed evidence of better use of the slurry applied in March. Fresh weight produced by leaves responded differently from one test to the other.

Implications

Hence it would seem that, even on deep soil, better use is made of the slurry when it is spread at a time when the crop has a high requirement for fertilising elements. Nevertheless, in the case of winter cereals planted on heavy, deep soil top dressing is impractical because of the ground conditions in spring (lack of load bearing capacity). In this case, slurry must be applied before ploughing, preferably as close to the sowing date as possible. For spring crops it would seem wise to combine slurry application with ploughing, in the spring. However, the pedoclimatic conditions are such that spring ploughing does not generally guarantee such a successful crop as winter ploughing, thus making application before the winter the more attractive solution. In lighter soil, where spring ploughing is less hazardous, the slurry should be applied as near as possible to planting time to ensure that optimum value is obtained.

4) Frequency of application

In the light of the low after effect levels for moderate applications of slurry it would seem that the largest effect occurs during the year of spreading. The influence of repeated slurry application on yields in a rotation confirms this observation.

5) Conclusions

The various tests conducted over the period 1975 to 1978 show that:

a) Only in exceptional cases does the efficiency rate of slurry N exceed 50 percent. All in all, the efficiency rate tends to fall as the amount of N supplied by the slurry increases.

b) With the exception of a few tests, the 'nitrogen effect' of the liquid manure is generally below 80 kg ha^{-1} and occurs primarily in the year in which the manure is spread, whatever the amount applied. In the case of moderate quantities, the after effect is slight in the year following application, and zero after two years. With larger quantities, in the region of at least 500 kg N ha^{-1}, the after effect is more pronounced one year after application and in some cases is still evident two years afterwards.

c) Maximum value can be obtained from the N supplied by the slurry by digging in immediately after application and by applying it at a time when the plants have the greatest demand for fertilising elements. Slurry applied before the winter is less effective than that applied after the winter, especially on shallow soil.

d) With grassland, slurry has adverse effects on the herbage, particularly when climatic conditions are unfavourable. It is particularly effective for the first cut in hay fields and for the first two cattle grazing periods on pasture land.

B) Qualitative effects

1) Chemical composition

Samples of the crops harvested in the field experiments were analysed for total- N, P, K, Ca and Mg to compare the effects of slurry and mineral fertiliser. The crops or products analysed included cereal grain, potato, tubers, maize, ryegrass, turnips, sugar beet pulp and leaves, grass. The main results are summarised below.

1.1. Total nitrogen: The N content of almost all the crops harvested, though of grasses in particular, increased with increasing rate of mineral N application. It is well known that large amounts of nitrogenous fertiliser, particularly when applied at a late stage of growth, increase the N content of the plants. Liquid manure had a much less marked effect. These results obviously have to be compared with results relating to the yields of the various crops, which showed that the efficiency rate of N in the slurry was less than for the mineral N.

1.2. Phosphorus: As far as phosphorus is concerned, the highest quantities of pig and cattle slurry increased the P content in ryegrass and grass. This is beneficial because a high level of phosphate is important to the health of cattle and thus in breeding profitability. High rates of nitrogenous fertiliser increased the phosphorus level in the plant even higher, particularly in the case of grain.

1.3. Potassium: Slurry application did not affect the K content of grain, but an appreciable and unwelcome increase was observed in beet pulp, turnips, ryegrass and grass after application of cattle or pig slurry. However, it is important to note that this was certainly no higher than with nitrogenous mineral fertiliser.

1.4. Calcium: Only very few significant results were obtained. It should of course be remembered that the quantities supplied by slurries are negligible in comparison with the amount of exchangeable Ca in the soil.

1.5. Magnesium: A number of positive effects of cattle and pig slurry on plant Mg content occurred, but also with nitrogen fertiliser.

Conclusions

Slurry does not seem to alter the composition of grain, tubers and roots in any way; nitrogenous fertiliser, on the other hand, can cause an increase in N content and sometimes in the level of other elements (P, K and Mg). As for fodder plants, interesting modifications in P content were observed at times after applying slurry. Even though increases in K content were observed regularly, this also happened with high rates of nitrogenous fertilisers. Furthermore, if it is considered that the supply of biogenic elements to the plant is dependent upon the level of reserves in the soil, then by readjusting this level with a suitable mineral fertiliser, any excess or imbalance will be avoided.

2) Culinary value of potatoes

There was no evidence of any appreciable modification in the culinary properties of tubers produced with slurry as compared with those obtained using a mineral fertiliser. The slurry did not harm the culinary qualities of the potato and no off-taste was found. However, large scale applications of liquid manure did as a rule delay the maturing process.

GENERAL CONCLUSIONS

- The large number of analyses carried out during the course of this study have shown the wide diversity in the composition of effluent from pig rearing enterprises. To appreciate the fertilising value of slurry the farmer needs to have a complete analysis of the manure carried out, but total nitrogen content is most important. Once the total nitrogen content is known correlations exist for estimating the content of other elements.
- Over the 4 years in which experiments have been carried out slurry, at moderate rates of application rarely exceeding 60 000 l ha^{-1}, did not lead to major modifications in the main chemical properties of the

soil. More specifically, aplied P_2O_5 and K_2O contribute towards the overall stock of soil reserves, which may sooner or later be utilised by the plant.

- The nitrogen in the slurry will be used to a greater degree if it is dug in immediately and applied at a time close to the period when plants have a considerable requirement for N. The liquid manure has its chief effect in the year following spreading. After effects in subsequent years are slight where moderate quantities are applied.
- As a rule, whatever the crop in question, it is inadvisable to apply massive quantities of slurry. The trials have shown that small quantities, with their higher efficiency rate, often have an effect equivalent to that of massive quantities. However, it should be pointed out that these quantities never caused any harm to the crops but serve to increase the amount of non-effective nitrogen.
- Use of liquid manure has not caused any important modifications in the composition of plants, or culinary properties of potatoes.

DISCUSSION

J.C. Hawkins *(UK)*

Mr. Lecomte, on your Table 2 you show a mean of 7.51 percent dry matter for poultry slurry. Could you explain why the slurry contained so little?

R. Lecomte *(Belgium)*

All the samples were taken at spreading time directly from the tank where water had been added to make spreading easier.

G. Catroux *(France)*

Your Figure 1 shows an increase in nitrogen mineralisation, is this from organic manure?

R. Lecomte

Yes, there is no other explanation.

J.R. O'Callaghan *(UK)*

With regard to the application in August, nitrogen moved down the soil profile and the efficiencies were very low. Has that nitrogen not just moved below the plant zone altogether and into the groundwater?

R. Lecomte

Some of that nitrogen was found as deep as 120 cm, but in very low concentrations. The organic nitrogen will enter the nitrogen cycle. The amount lost to the ground water also depends on the type of soil. These studies are in the programme of dynamics of nitrogen in the soil.

J.R. O'Callaghan

The movement of nitrogen is also a reason why it is not good practice to spread the liquid manure in August, it is easier for the farmer but bad for the environment.

L. Rhixon *(Belgium)*

Certainly, spreading in August is not good but a vehicle can be taken onto the field until September; in April that is quite impossible.

H. Vetter *(West Germany)*

We should be looking for a way of spreading the slurry in the spring, perhaps by building special vehicles to allow slurry to be spread as easily as mineral fertiliser.

J.C. Hawkins

The solution is already available for some farmers because if you use a machine to separate slurry you have a liquid that can be handled like water and a solid which can be stacked like farmyard manure. The liquid can be put on with a travelling irrigator, using just a bar which dribbles it on to the surface. The irrigator can go through a crop without causing damage.

K.W. Smilde *(The Netherlands)*

May I ask Mr. Hawkins how you separate the liquid and solid components?

J.C. Hawkins

You buy a machine of which there are now four or five available on the market. There is no other way.

J.H. Voorburg *(The Netherlands)*

That would only be a solution for sandy soils, because clay soils have to be ploughed before the winter. Even on sandy soils, there will be difficulties in the spring because farmers do not have the time.

H. Vetter

I think we must return to this topic again in connection with other reports. Now I must finish this session and say thank you to Dr. Lecomte.

EXPERIMENTS ON HEAVY APPLICATIONS OF ANIMAL MANURE TO LAND

A. Dam Kofoed and O. Nemming
Askov Experimental Station, 6600 Vejen, Denmark.

How to utilise animal manure has become a serious problem following the concentration of animal husbandry in fewer and bigger units. This has been the case in Denmark as well as in other EEC countries. Very often the increase in numbers of animals was not followed by an increase of area for disposal of the effluent, so that under these circumstances very much larger amounts of animal manure might be available for each hectare than in the old system.

Problems of the utilisation of animal manure must be solved by field experiments.

Some of the questions are:

1. How much manure can be applied without damage to plants?
2. How much manure can be economically applied?
3. How much manure can be applied without risk to the environment?

Besides these, other effects of animal manure on soil structure and soil fauna and on the chemical composition of plants will also require attention. Investigations on some of these problems were stated at Askov Experimental Station in 1973. Results of the work to date are reported here.

EXPERIMENT 1 - DESIGN

Field experiments

Treatment no. 1. fertiliser, 1 NPK
2. fertiliser, 2 NPK
3. fertiliser, 4 NPK
4. animal manure, annually 25 t ha^{-1}
5. animal manure, annually 50 t ha^{-1}
6. animal manure, annually 100 t ha^{-1}
7. animal manure, every 2nd year, 50 t ha^{-1}
8. animal manure, every 2nd year, 100 t ha^{-1}
9. animal manure, every 2nd year, 200 t ha^{-1}
10. animal manure, every 4th year, 100 t ha^{-1}
11. animal manure, every 4th year, 200 t ha^{-1}
12. animal manure, every 4th year, 400 t ha^{-1}

Fertilisers, kg ha^{-1}

	For barley	For beet and grass
1 N	40	80
P and K	equal to P and K in manure (treatments 4 - 6)	

Soil types: light sandy soil and light loam

Crops: barley, beet and grass or maize

Crop analyses: DM, total-N, NO_3-N, P, K, Na, Ca, Mg, Cu, Mn and Zn

Soil analyses: Every year (in November - December) soils were sampled to a depth of 2 - 3 m, and were analysed for P, K, Mg, Cl, Cu, Zn, NO_3-N, NH_3-N and total-N

Manure analyses: Total-N, NH_3-N, P, K, Ca, Na, Mg, Cu, Zn and Mn

Types of animal manure: FYM and slurry

Times of application: December - January

Lysimeter experiments: same scheme as for field experiment

EXPERIMENTS WITH SLURRY

Analyses of slurry

Slurry was analysed every year and Table 1 gives the average for the years 1973 - 1977. Table 2a shows the amounts of N, P and K applied over the four year period at Askov; and Table 2b shows the amounts applied at Lundgård.

TABLE 1

COMPOSITION OF SLURRY. AVERAGE FOR YEARS 1973-77

pH	7.30
	%
DM	9.10
Sand	-
Total-N	0.43
NH_3-N	0.25
P	0.08
K	0.35
Na	0.06
Ca	0.12
Mg	0.05
	ppm
Cu	4
Mn	27
Zn	15

Crop yields

Barley: In the period of 1973 - 1978 barley was used as a test crop five times for both sites together. Table 3 shows the yields of grain. Results in Table 3 show that the best yields of grain were obtained when 50 t ha^{-1} of slurry was applied. However, considering the utilisation of plant

TABLE 2

TOTAL AMOUNTS OF DIFFERENT NUTRIENTS, kg ha^{-1}, SUPPLIED AS FERTILISER OR SLURRY

A) Askov for the four years 1973-76

		Total-N	NH_3-N	P	K
Fertiliser applied annually	1 NPK	240	-	83	369
	2 NPK	480	-	166	737
	4 NPK	960	-	331	1 474
Slurry applied annually	25 t ha^{-1}	454	266	80	375
	50 t ha^{-1}	907	533	160	751
	100 t ha^{-1}	1 814	1 065	321	1 502
Slurry applied every 2nd year	50 t ha^{-1}	434	257	74	362
	100 t ha^{-1}	867	513	148	724
	200 t ha^{-1}	1 735	1 027	296	1 447
Slurry applied every 4th year	100 t ha^{-1}	430	256	60	378
	200 t ha^{-1}	860	512	120	756
	400 t ha^{-1}	1 720	1 024	240	1 512

B) Lundgård for the four years 1974-77

		Total-N	NH_3-N	P	K
Fertiliser applied annually	1 NPK	240	-	79	328
	2 NPK	480	-	158	656
	4 NPK	960	-	315	1 312
Slurry applied annually	25 t ha^{-1}	414	242	79	328
	50 t ha^{-1}	828	483	158	656
	100 t ha^{-1}	1 655	966	315	1 312
Slurry applied every 2nd year	50 t ha^{-1}	459	267	85	368
	100 t ha^{-1}	918	534	170	736
	200 t ha^{-1}	1 836	1 068	340	1 472
Slurry applied every 4th year	100 t ha^{-1}	475	273	81	348
	200 t ha^{-1}	950	546	162	696
	400 t ha^{-1}	1 900	1 092	324	1 392

nutrients and their potential leaching from the soil 50 t ha^{-1} of slurry is adequate for barley.

Beet: Table 4 shows the results of the four test crops of beet. On average of the four crops the best yield was obtained with 200 t ha^{-1} of slurry, although the last 100 t ha^{-1} only increased yield by 0.78 t of DM. Taking into account yield, quality and the environment, 75 - 100 t ha^{-1} of slurry for beet appears to be adequate.

Italian ryegrass and maize: Table 5 gives the results from one experiment with ryegrass at Askov and one experiment with maize at Lundgård.

For ryegrass, the slurry was given in three dressings, half just before the winter ploughing in December and one-quarter each after the first and second cuts. Although yields of dry matter were increased with up to 200 t ha^{-1}, the maximum amount of slurry should not exceed 75 - 100 t ha^{-1} in order to minimise the risk to the environment.

For maize, no generalisation could be made from a single experiment but other experiences suggested that 100 t ha^{-1} of slurry is the largest amount that should be applied if the nutrients are to be reasonably used.

Average yields for the rotation

In the previous section the effect of slurry on single crops was discussed. However, the effect of treatments on the average of a four year rotation, has to be considered.

Table 6 gives the effect of fertiliser and slurry given every year, alternate years and every four years, on yields of single crops and averages for the rotation at Askov. P and K have been supplied as equal amounts in fertiliser and slurry whether more N was given in slurry than as fertiliser. Table 7 gives the corresponding results for Lundgård, except that maize was grown instead of ryegrass.

TABLE 3

BARLEY YIELDS OF GRAIN, FOR THE YEARS 1973-78, t ha^{-1}

	Askov 3 crops	Lundgård 2 crops	Average 5 crops
Fertiliser			
1 NPK	3.86	1.91	3.08
2 NPK	4.83	2.74	3.99
4 NPK	4.46	3.10	3.92
Slurry			
25 t ha^{-1}	4.44	1.95	3.44
50 t ha^{-1}	4.94	2.72	4.05
100 t ha^{-1}	4.65	2.68	3.86

TABLE 4

BEET YIELDS OF DRY MATTER, FOR THE YEARS 1973-78, t ha^{-1}

	Askov, 2 crops		Lundgård, 2 crops		Average of 4 crops	
	root	tops	root	tops	root	tops
Fertiliser						
1 NPK	8.96	3.20	8.66	3.08	8.81	3.14
2 NPK	10.47	3.81	9.16	3.28	9.82	3.55
4 NPK	11.59	4.75	9.34	3.87	10.47	4.31
Slurry						
25 t ha^{-1}	7.94	3.07	6.70	1.71	7.32	2.34
50 t ha^{-1}	8.37	3.37	10.14	3.02	9.25	3.20
100 t ha^{-1}	9.65	3.97	12.19	4.18	10.92	4.08
200 t ha^{-1}	10.80	4.43	12.58	4.91	11.70	4.67
400 t ha^{-1}	11.00	4.45	11.90	5.17	11.47	4.81

TABLE 5

YIELDS OF DRY MATTER, t ha^{-1}, OF ITALIAN RYEGRASS AT ASKOV, 1975; AND MAIZE AT LUNDGÅRD, 1976.

	Ryegrass	Maize
Fertiliser		
1 NPK	4.40	6.67
2 NPK	6.21	6.74
4 NPK	7.14	6.46
Slurry		
25 t ha^{-1}	4.31	5.18
50 t ha^{-1}	5.28	6.50
100 t ha^{-1}	6.72	5.98
200 t ha^{-1}	8.20	6.63

TABLE 6

AVERAGE YIELDS OF GRAIN OR DRY MATTER, t ha^{-1} AND cu ha^{-1} AT ASKOV FOR THE YEARS 1973 - 1976 (1 crop unit = 1 cu = 1 t barley)

	Beet		Barley	Italian ryegrass	Barley	Average cu ha^{-1}
	Root	Tops				
Fertiliser, annually						
1 NPK	7.25	2.75	4.05	4.40	3.65	5.40
2 NPK	8.43	3.30	5.31	6.21	43.1	6.66
4 NPK	9.60	4.21	6.26	7.14	3.32	7.43
Slurry, annually						
25 t ha^{-1}	64.1	2.51	4.34	4.31	4.32	5.46
50 t ha^{-1}	7.02	2.79	5.35	5.52	4.22	6.22
100 t ha^{-1}	8.05	3.23	6.07	69.8	3.45	6.93
Slurry, alternate years						
50 t ha^{-1}	5.83	2.55	2.61	5.04	2.71	4.44
100 t ha^{-1}	7.21	3.06	2.90	6.45	3.56	5.49
200 t ha^{-1}	7.97	3.60	3.79	8.20	4.13	6.60
Slurry, every 4th year						
100 t ha^{-1}	7.61	3.09	3.17	2.51	1.83	4.33
200 t ha^{-1}	8.47	3.51	3.78	3.15	2.29	5.08
400 t ha^{-1}	8.57	3.44	5.07	4.24	2.77	5.86

TABLE 7

AVERAGE YIELDS OF GRAIN OR DRY MATTER, t ha^{-1} AND cu ha^{-1} AT LUNDGÅRD FOR THE YEARS 1974-77 (1 cu = 1 crop unit = 1 t barley)

	Beet		Barley	Maize	Barley	Average cu ha^{-1}
	Root	Tops				
Fertiliser, annually						
1 NPK	11.72	4.17	1.60	6.67	2.22	6.03
2 NPK	10.96	3.95	1.82	6.74	3.66	6.34
4 NPK	9.66	4.70	1.78	6.46	4.41	6.47
Slurry, annually						
25 t ha^{-1}	9.16	2.11	1.83	5.18	2.07	4.73
50 t ha^{-1}	12.66	3.71	1.74	5.62	3.70	6.50
100 t ha^{-1}	13.86	4.92	1.33	5.31	4.02	7.00
Slurry, alternate years						
50 t ha^{-1}	11.97	3.56	1.31	7.38	0.93	5.66
100 t ha^{-1}	13.04	43.5	1.54	6.64	1.11	6.02
200 t ha^{-1}	13.62	5.83	1.81	6.63	1.51	6.65
Slurry, every 4th year						
100 t ha^{-1}	13.48	5.18	1.54	5.15	0.84	5.91
200 t ha^{-1}	13.38	5.68	1.70	5.53	0.90	6.14
400 t ha^{-1}	12.86	5.89	2.01	5.75	1.07	6.26

Results in Tables 6 and 7 give crop averages calculated as crop units (cu) equal to 1 tonne of barley.

Yields of crops with 25 t ha^{-1} of slurry every year, 50 t ha^{-1} of slurry every second year and 100 t ha^{-1} of slurry every fourth year, can be compared over four years because all received 100 t ha^{-1} of slurry. Results in Table 6 show yields of 5.46, 4.44 and 4.34 cu ha^{-1} respectively. Moderate manuring every year is better than heavier dressings at less frequent intervals, the same trend is seen for the rest of the treatments.

Table 7 gives similar results for Lundgård.

On sandy soil the same trend is seen as on Askov loamy soil except for crops receiving 25 t ha^{-1} of slurry every year, which yielded less well. Apart from this treatment better yields were obtained from slurry given to the light sandy soil at Lundgård than to the loamy soil at Askov. This result is in agreement with numerous other experiments with slurry, which have shown that slurry increases yields more on sandy soil than on loam soil.

Residual effects of slurry

The residual effects of slurry were calculated and showed that even after the largest amounts of slurry the residual effects in the first and second years were smaller than the increases from the smallest dressing of fertiliser, and in the third year practically no effect was measured.

Crop analyses

Every crop was analysed for its content of different nutrients. Results were consistent for nitrogen and increasing amounts of N as fertiliser or slurry. The figures show increased N-percent. For barley and maize the content of NO_3-N was not affected, but for beet and ryegrass percent NO_3-N in the crops also increased with increasing N supply. The potassium content of beet, ryegrass and maize was also increased by increasing fertiliser or slurry applications. Other nutrients were much less and irregularly affected.

Removal of nutrients in crops

A reasonable utilisation of plant nutrients in the manure requires a balance between supply and uptake of nutrients. Such balances were calculated for these experiments.

In general, for all nutrients more had been supplied in slurry than was removed by the crops, except for potassium in slurry applied annually at 25 t ha^{-1}, when more was removed than supplied. Treatments supplying the equivalent of 100 t ha^{-1} of slurry annually provided a considerable surplus of nutrients.

Soil analyses

In order to follow leaching of plant nutrients, soil samples were taken every year in December at 25 cm intervals down to 200 cm, from 1973 - 1978 at Askov and 1974 - 1978 at Lundgard.

The results for N are shown in Figures 1 and 2, where lines are drawn showing the content of NO_3-N and NH_4-N in the soil after the first, fourth and sixth years of the experiment at Askov for the treatment given slurry annually.

Figure 1 shows the content of NO_3-N in the Askov soil and that after a period of six years with 50 and 100 t ha^{-1} of slurry annually, NO_3-N content was increased down to a depth of 1 m. Figure 2 has values for the NH_4-N contents which show the same trends.

For phosphorus, there was no effect on extractable-P, even from the large surplus of organic P given with slurry.

For potassium, 100 t ha^{-1} of slurry given annually increased extractable-K most in the surface layer and slightly down to 2 m.

For magnesium, chloride, copper and zinc, no effect of slurry was found on the soil indices.

Results from similar investigations for five years on the soil at Lundgard are shown in Figures 3 and 4 for NO_3-N and NH_4-N respectivley. The content of NO_3-N and NH_4-N did not vary much from the first to the fifth year, although NO_3-N content tended to be greater after application of slurry for three - five years than after one year. With 50 and 100 t ha^{-1} of slurry given annually for 5 years there was a tendency for

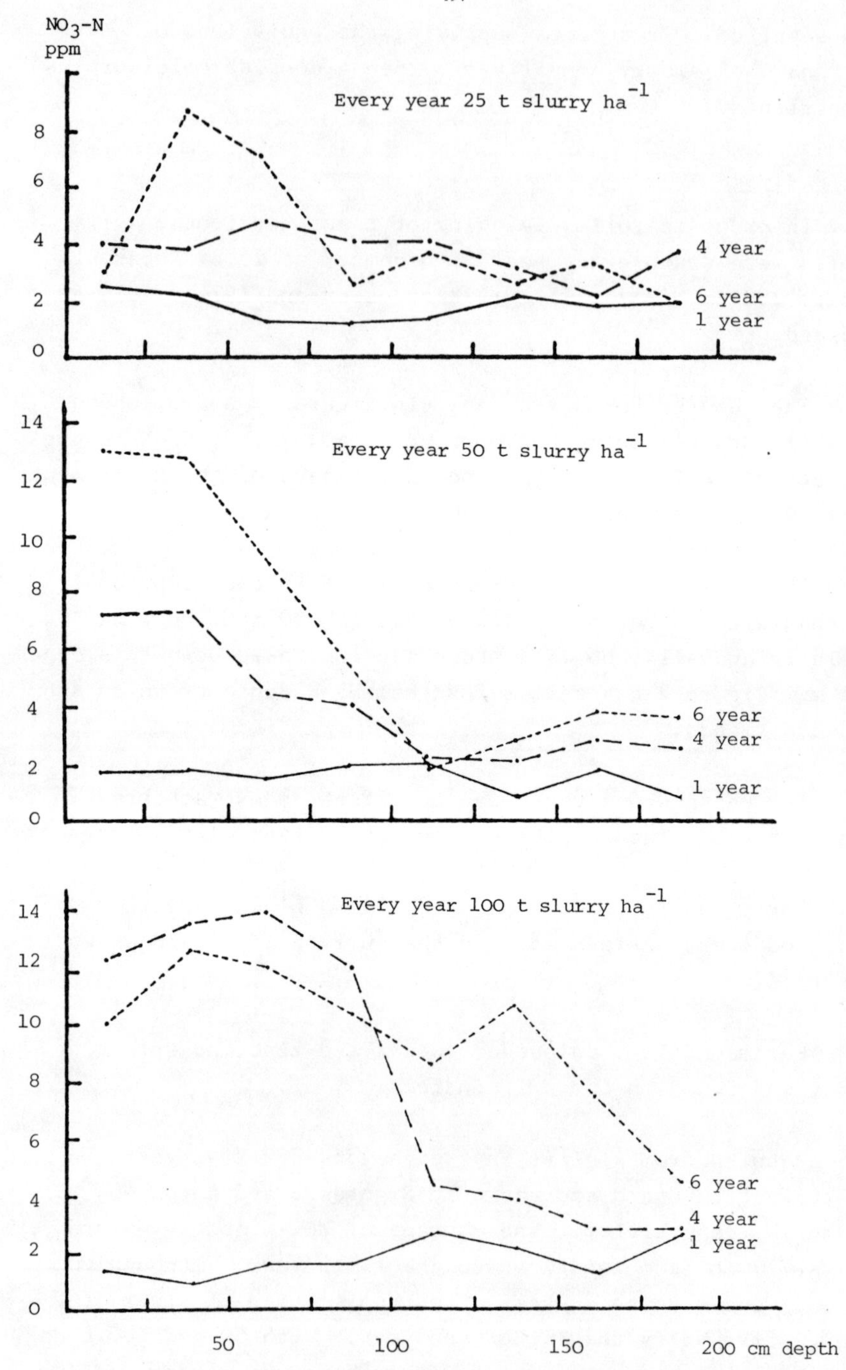

Fig. 1. Content of NO_3-N at various depths in loam soil, Askov 1973-78.

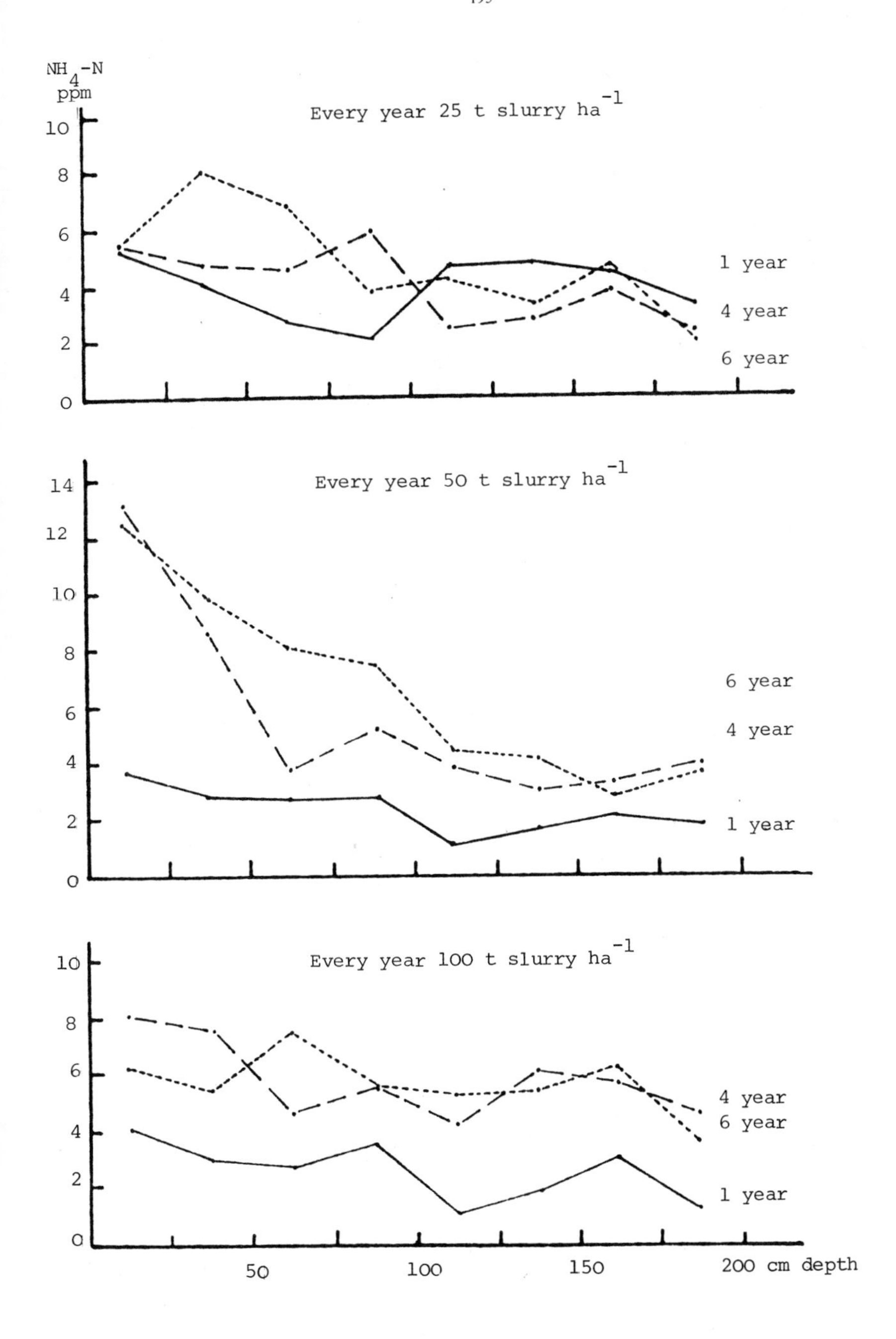

Fig. 2. Content of NH_4-N at various depths in loam soil, Askov 1973-78.

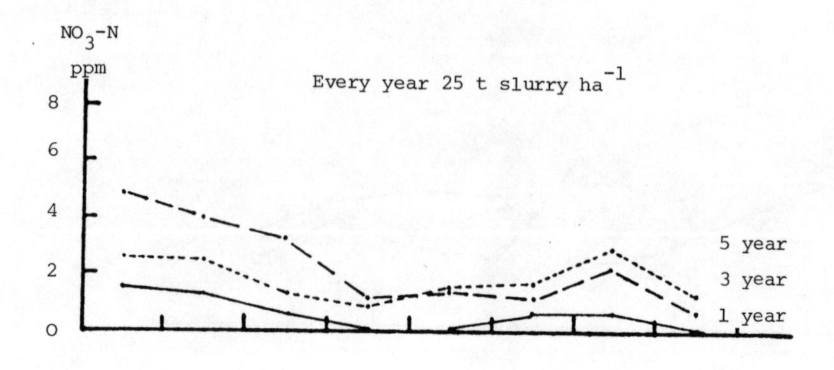

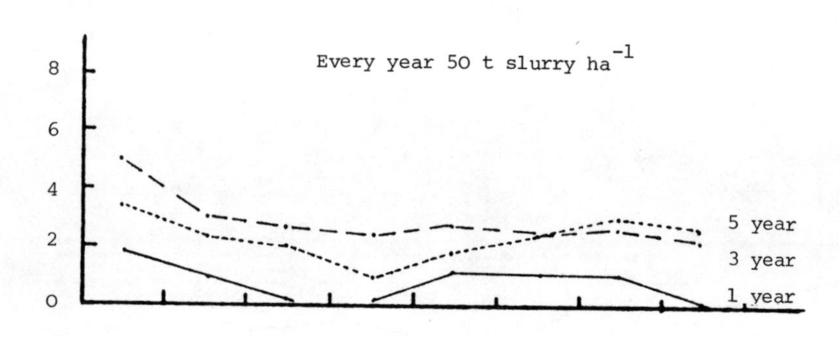

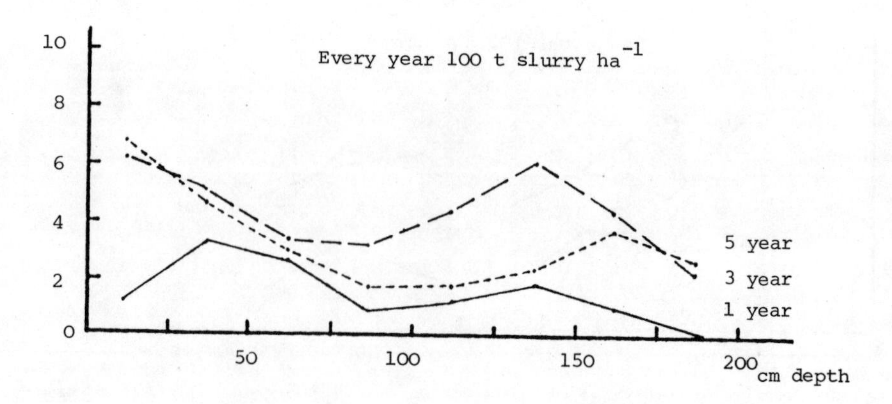

Fig. 3. Content of NO_3-N at various depths in sandy soil - Lundgård 1974-78.

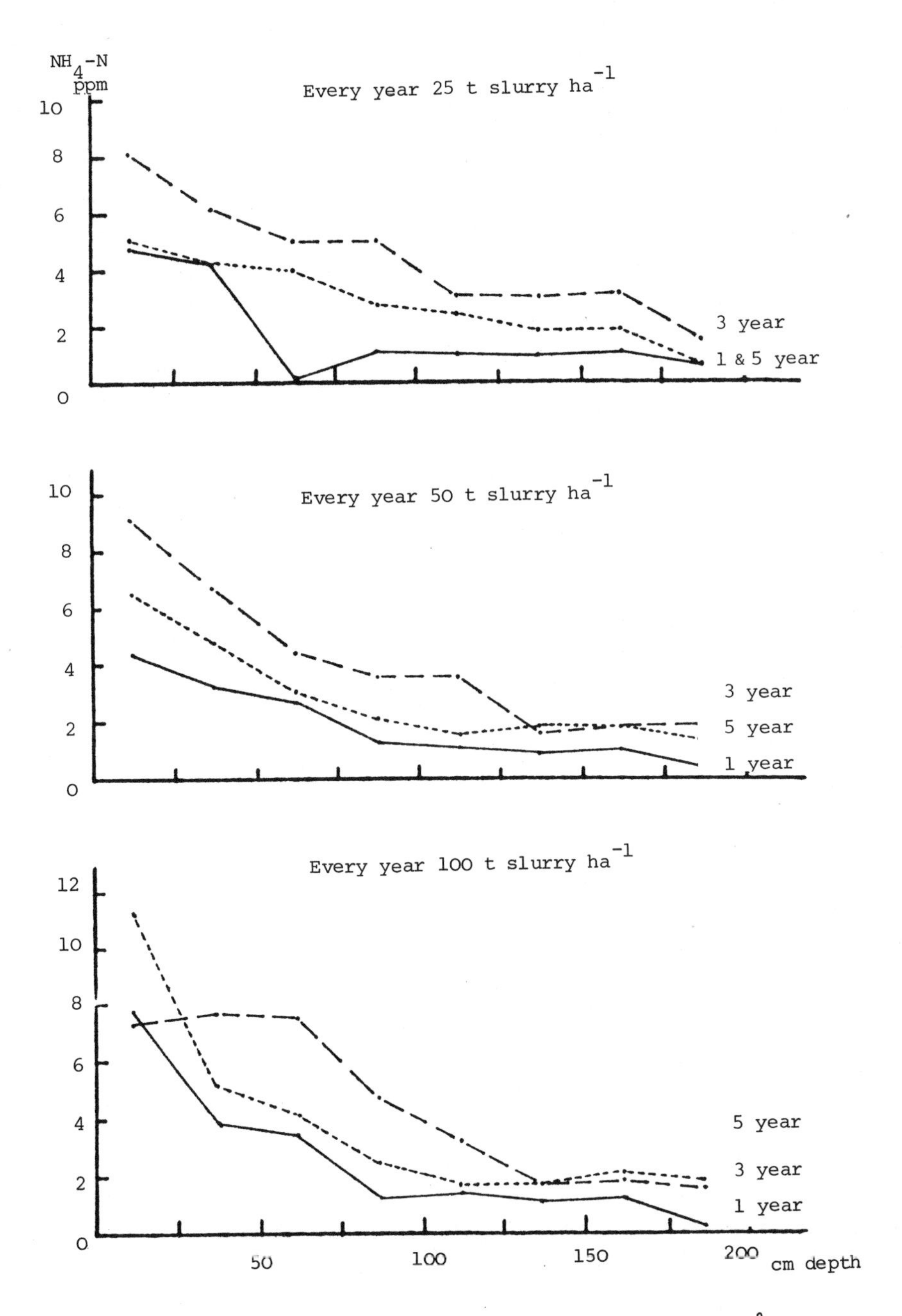

Fig. 4. Content of NH_4-N at various depths in sandy soil, Lundgård 1974-78

extractable-P to increase in the lower soil layers, and for extractable-K to increase in the top 50 cm of soil.

EXPERIMENTS WITH FARMYARD MANURE (FYM)

Analyses of FYM and amounts supplied

The FYM used was analysed each year and Table 8 gives the average composition for the years 1973 - 1977. Table 9a shows the amounts of nutrients applied for the four year period at Askov; and Table 9b, the amounts at Lundgård.

TABLE 8

COMPOSITION OF FYM, AVERAGE FOR YEARS 1973-77

	%
Dry matter	21.48
Sand	5.74
Total-N	0.55
NH_3-N	0.15
P	0.18
K	0.32
Na	0.05
Ca	0.25
Mg	0.09
	ppm
Cu	10
Mn	54
Zn	34

Crop yields

Barley: In the period of 1973 - 1978 barley was used as a test crop five times for both sites together. Table 10 shows the yields of grain.

TABLE 9

TOTAL AMOUNTS OF DIFFERENT NUTRIENTS, kg ha^{-1} APPLIED AS FERTILISER OR FYM

A) Askov for the four years 1973-76

		Total-N	NH_3-N	P	K
Fertiliser applied annually	1 NPK	240	-	190	343
	2 NPK	480	-	381	685
	4 NPK	960	-	761	1 370
FYM applied annually	25 t ha^{-1}	559	147	184	341
	50 t ha^{-1}	1 119	294	368	683
	100 t ha^{-1}	2 237	589	736	1 366
FYM applied every 2nd year	50 t ha^{-1}	614	183	209	390
	100 t ha^{-1}	1 227	366	418	781
	200 t ha^{-1}	2 454	731	836	1 562
FYM applied every 4th year	100 t ha^{-1}	604	159	272	446
	200 t ha^{-1}	1 208	318	544	892
	400 t ha^{-1}	2 416	636	1 088	1 784

B) Lundgård for the four years 1974-77

		Total-N	NH_3-N	P	K
Fertiliser applied annually	1 NPK	240	-	180	316
	2 NPK	480	-	360	631
	4 NPK	960	-	719	1 262
FYM applied annually	25 t ha^{-1}	552	164	180	316
	50 t ha^{-1}	1 105	328	360	631
	100 t ha^{-1}	2 209	656	719	1 262
FYM applied every 2nd year	50 t ha^{-1}	512	120	178	301
	100 t ha^{-1}	1 024	240	355	601
	200 t ha^{-1}	2 048	480	710	1 202
FYM applied every 4th year	100 t ha^{-1}	575	151	200	314
	200 t ha^{-1}	1 150	302	400	628
	400 t ha^{-1}	2 300	604	800	1 256

Results in Table 10 show that the best yields of grain were obtained when 100 t ha^{-1} of FYM was applied. Considering the utilisation of plant nutrients and their potential leaching from the soil, 50 t ha^{-1} of FYM is suggested for barley.

Beet: Table 11 gives the yields for the four test crops of beet. On average of the four experiments the best yield was obtained with 400 t ha^{-1} of FYM, although the last 200 t ha^{-1} only increased yield by 0.37 t of DM. Taking into account yield, quality and the environment, 75 - 100 t ha^{-1} of FYM for beet appears to be adequate.

Italian ryegrass and maize: Table 12 gives the results from one experiment with ryegrass at Askov and one experiment with maize at Lundgård.

For ryegrass, the FYM was given in three dressings, half just before the winter ploughing in December and one-quarter each after the first and second cuts. Although yields of dry matter were increased with up to 200 t ha^{-1} the maximum amount of FYM should not exceed 75 - 100 t ha^{-1}, in order to minimise the risk to the environment.

For maize, no generalisation could be made from a single experiment but other experience suggested that 100 t ha^{-1} of FYM is the largest amount that should be applied if the nutrients are to be reasonably used.

Average yields for the rotation

In the previous section the effect of FYM on single crops was discussed. However, the effect of single treatments on the average of a four year rotation has to be considered. Table 13 gives the effects of fertiliser and FYM given every year, alternate years and every four years, on yields of single crops and averages for the rotation at Askov. P and K have been supplied as equal amounts in fertiliser and FYM, whether

TABLE 10

BARLEY YIELDS OF GRAIN FOR THE YEARS 1973-78, t ha^{-1}

	Askov 3 crops	Lundgård 2 crops	Average 5 crops
Fertiliser			
1 NPK	3.97	2.04	3.29
2 NPK	4.52	2.67	3.78
4 NPK	4.04	2.91	3.59
FYM			
25 t ha^{-1}	3.87	1.96	3.11
50 t ha^{-1}	4.70	2.71	3.90
100 t ha^{-1}	4.79	3.23	4.17

TABLE 11

BEET YIELDS OF DRY MATTER, FOR THE YEARS 1973-78, t ha^{-1}

	Askov, 2 crops		Lundgård, 2 crops		Average of 4 crops	
	root	tops	root	tops	root	tops
Fertiliser						
1 NPK	8.39	3.21	8.21	2.59	8.30	2.90
2 NPK	10.04	3.92	9.11	2.91	9.58	3.42
4 NPK	11.00	4.79	9.29	3.93	10.15	4.36
FYM						
25 t ha^{-1}	6.06	2.81	5.89	1.54	5.98	2.18
50 t ha^{-1}	7.45	3.24	7.97	2.33	7.71	2.79
100 t ha^{-1}	8.25	3.76	10.57	3.53	9.41	3.65
200 t ha^{-1}	9.43	3.87	12.52	4.52	10.98	4.20
400 t ha^{-1}	10.06	4.28	12.63	5.15	11.35	4.72

TABLE 12

YIELDS OF DRY MATTER, t ha^{-1}, OF ITALIAN RYEGRASS AT ASKOV 1975; AND MAIZE AT LUNDGÅRD, 1976.

		Ryegrass	Maize
Fertiliser	1 NPK	4.45	6.20
	2 NPK	6.28	5.89
	4 NPK	7.65	5.78
FYM	25 t ha^{-1}	3.72	5.35
	50 t ha^{-1}	4.59	6.30
	100 t ha^{-1}	6.79	6.63
	200 t ha^{-1}	8.49	6.08

TABLE 13

AVERAGE YIELDS OF GRAIN OR DRY MATTER, t ha^{-1} AND cu ha^{-1} (1 cu = 1 crop unit = 1 t barley)

	Beet		Barley	Italian ryegrass	Barley	Average cu ha^{-1}
	root	tops				
Fertiliser, annually						
1 NPK	6.19	2.55	4.44	4.45	3.44	5.16
2 NPK	7.69	3.23	4.95	6.28	3.96	6.42
4 NPK	9.12	4.06	6.28	7.65	2.65	7.29
FYM, annually						
25 t ha^{-1}	4.36	2.09	4.25	3.72	3.51	4.49
50 t ha^{-1}	5.14	2.36	5.17	4.79	3.77	5.29
100 t ha^{-1}	6.71	2.94	5.84	7.04	3.73	6.50
FYM, alternate years						
50 t ha^{-1}	5.26	2.56	2.68	4.38	32.2	4.35
100 t ha^{-1}	6.29	2.86	2.97	6.53	3.91	5.43
200 t ha^{-1}	6.77	2.96	3.95	8.49	3.57	6.20
FYM, every 4th year						
100 t ha^{-1}	6.21	2.82	3.15	3.09	2.17	4.20
200 t ha^{-1}	6.83	2.89	4.00	3.52	2.63	4.82
400 t ha^{-1}	7.67	3.27	5.14	4.79	3.58	6.00

more N was given in FYM than as fertiliser. Table 14 gives the corresponding results for Lundgård, except that maize was grown instead of ryegrass. Results are also shown as averages for the rotation expressed as crop units (cu), where 1 cu = 1 tonne barley.

TABLE 14

AVERAGE YIELDS OF GRAIN OR DRY MATTER, t ha^{-1} AND cu ha^{-1} AT LUNDGÅRD FOR THE YEARS 1974-77 (1 cu = 1 crop unit = 1 t barley)

	Beet		Barley	Maize	Barley	Average cu ha^{-1}
	root	tops				
Fertiliser, annually						
1 NPK	9.78	3.05	1.57	6.20	2.51	5.33
2 NPK	9.64	3.18	1.54	5.89	3.79	5.62
4 NPK	8.59	4.32	1.48	5.78	4.33	5.87
FYM, annually						
25 t ha^{-1}	7.54	1.77	1.76	5.35	2.16	4.32
50 t ha^{-1}	10.46	2.74	1.95	5.88	3.46	5.76
100 t ha^{-1}	12.88	4.07	1.93	7.33	4.52	7.27
FYM, alternate years						
50 t ha^{-1}	10.35	2.67	1.28	6.72	0.91	4.93
100 t ha^{-1}	12.19	3.79	1.57	5.93	1.16	5.57
200 t ha^{-1}	12.94	4.90	1.78	6.08	1.73	6.23
FYM, every 4th year						
100 t ha^{-1}	11.80	3.99	1.53	5.02	0.82	5.24
200 t ha^{-1}	13.46	4.91	1.96	5.84	1.10	6.18
400 t ha^{-1}	13.37	5.39	2.24	7.33	1.48	67.6

At Askov (sandy-loam) best yields were obtained with fertiliser but at Lundgård (sandy-soil) FYM was better than fertiliser except for crops given 25 t ha^{-1} of FYM.

Yields of crops with 25 t ha^{-1} of FYM annually, 50 t ha^{-1} of FYM every second year and 100 t ha^{-1} of FYM every fourth year, can be compared over four years because all received 100 t ha^{-1} of FYM. Results in Table 13 show yields of 4.49, 4.35 and 4.20 cu ha^{-1} respectively. Moderate manuring every year was better than a heavier dressing at less frequent intervals. The same trend is seen for the rest of the treatments.

Table 14 gives similar results for Lundgård.

As for slurry, FYM increased yields more on the light sandy soil at Lundgård than on the loamy soil at Askov, except when 25 t ha^{-1} was applied annually.

Residual effects of FYM

The residual effects of FYM were calculated and showed that even after the largest amounts of FYM the residual effects in the first and second years were smaller than the increases from the smallest amount of fertiliser and in the third year practically no effect was measured.

Crop analyses

Every crop was analysed for its content of different nutrients. Composition followed the same trends as crop receiving slurry.

Removal of nutrients in crops

As for slurry, in general, more of all nutrients had been supplied in FYM than was removed in the crops, except that more N and K were removed with the least amounts of fertiliser and FYM than was applied. For crops receiving the equivalent of 100 t ha^{-1} $year^{-1}$ a considerable surplus of nutrients was applied.

Soil analyses

In order to follow leaching of plant nutrients, soil samples were taken every year in December at 25 cm intervals down to 200 cm, from 1973 - 1978 at Askov and 1974 - 1978 at Lundgård.

The results for N are shown in Figures 5 and 6, for Askov, where lines are drawn showing the content of NO_3-N and NH_4-N in the soil after the first, fourth and sixth years of the experiment with treatments given FYM every year. After six years with 50 and 100 t ha^{-1} of FYM given every year, both NO_3-N and NH_4-N contents were increased down to a depth of 1 m.

Figures 7 and 8 show results from similar investigations for the five years 1974 - 1978 in the soil at Lundgård. Both the contents of NO_3-N and NH_4-N increased a little from the first to the fifth year.

For phosphorus, there was no effect on extractable-P even from the large surplus of organic-P given with FYM, at either Askov or Lundgård.

For potassium, the largest dressing of FYM increased the available-K in the top 50 cm of soil at both Askov and Lundgård.

None of the other elements measured were affected.

EFFECTS ON PORE SIZE AND CONTENT IN THE SOIL

At Askov, in co-operation with the experimental station at Højer, some soil physical measurements were made on samples from five of the plots receiving FYM and one plot with fertiliser. Soil samples were taken on 12th October 1976. Table 15 shows the total pore space and the distribution between sizes.

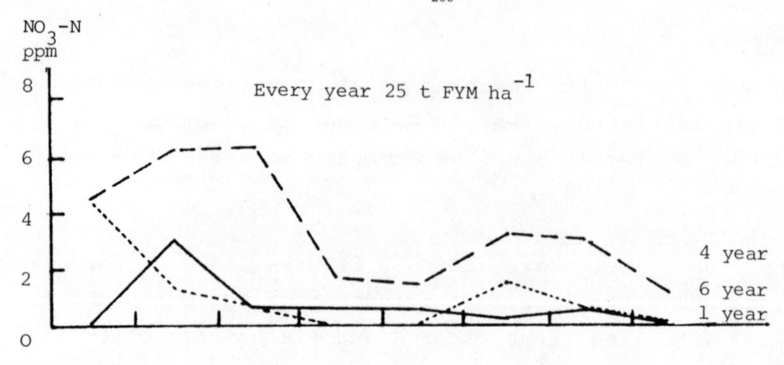

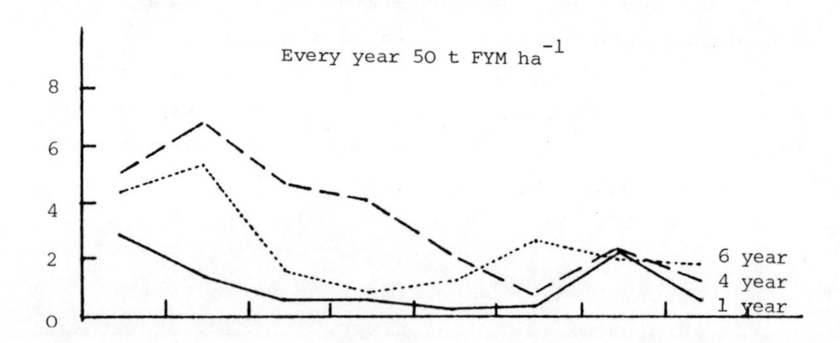

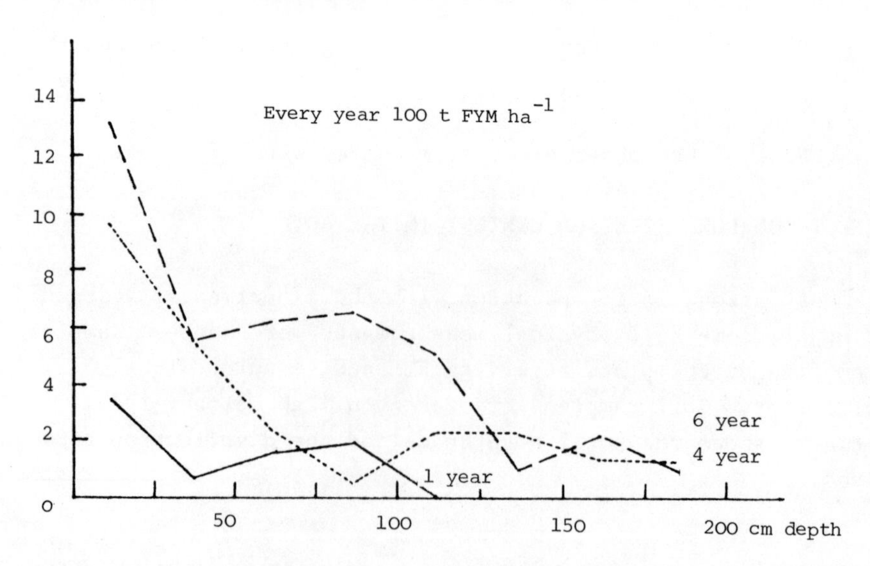

Fig. 5. Content of NO_3-N at various depths in loam soil - Askov 1973-78

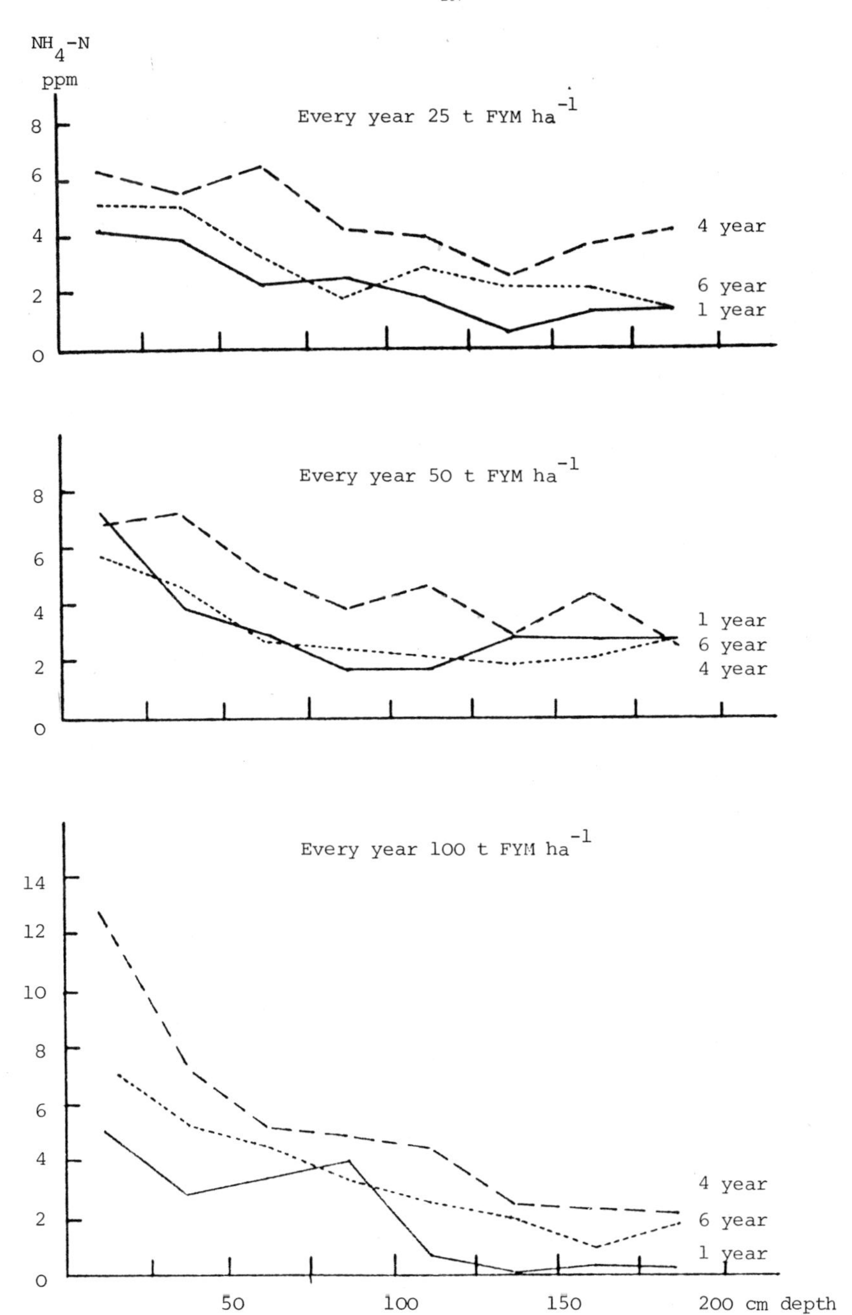

Fig. 6. Content of NH_4-N at various depths in loam soil - Askov 1973-78

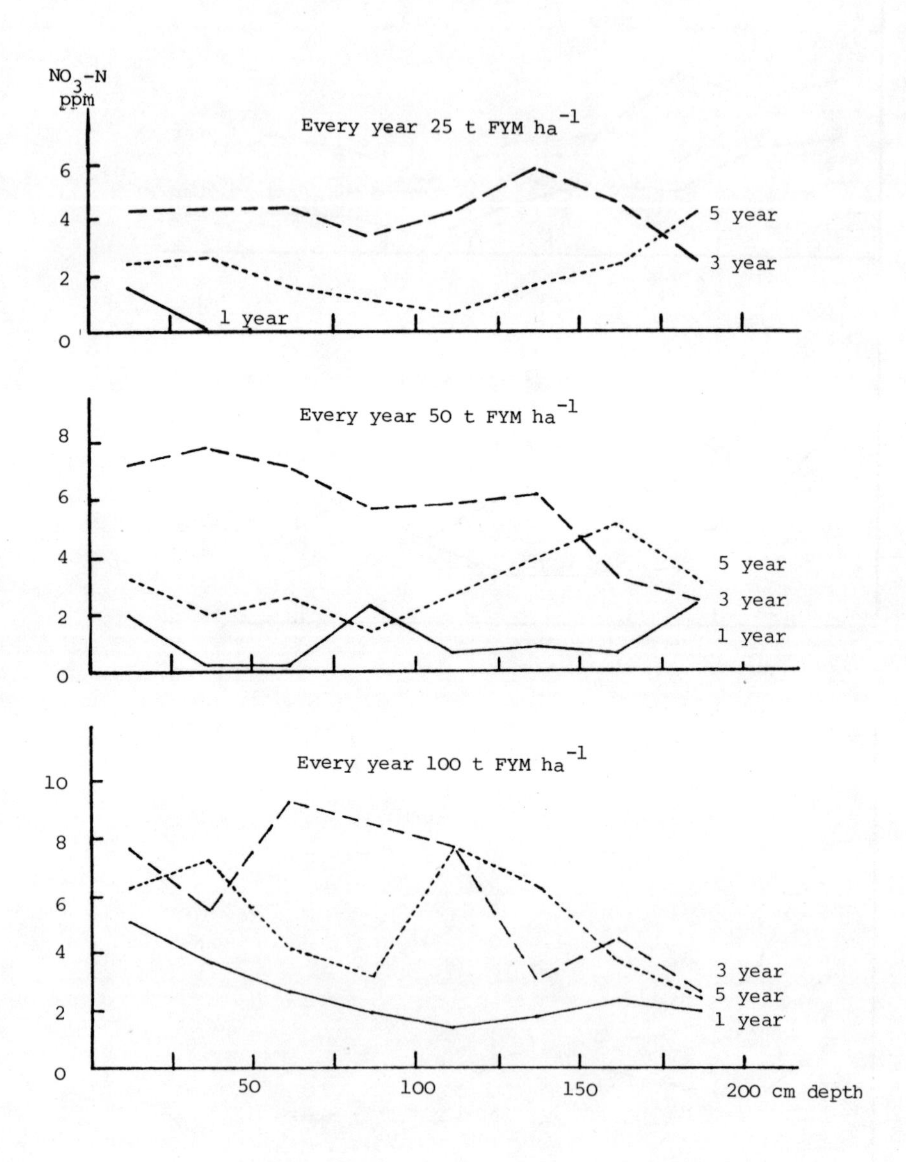

Fig. 7. Content of NO_3-N at various depths in sandy soil - Lundgård 1974-78.

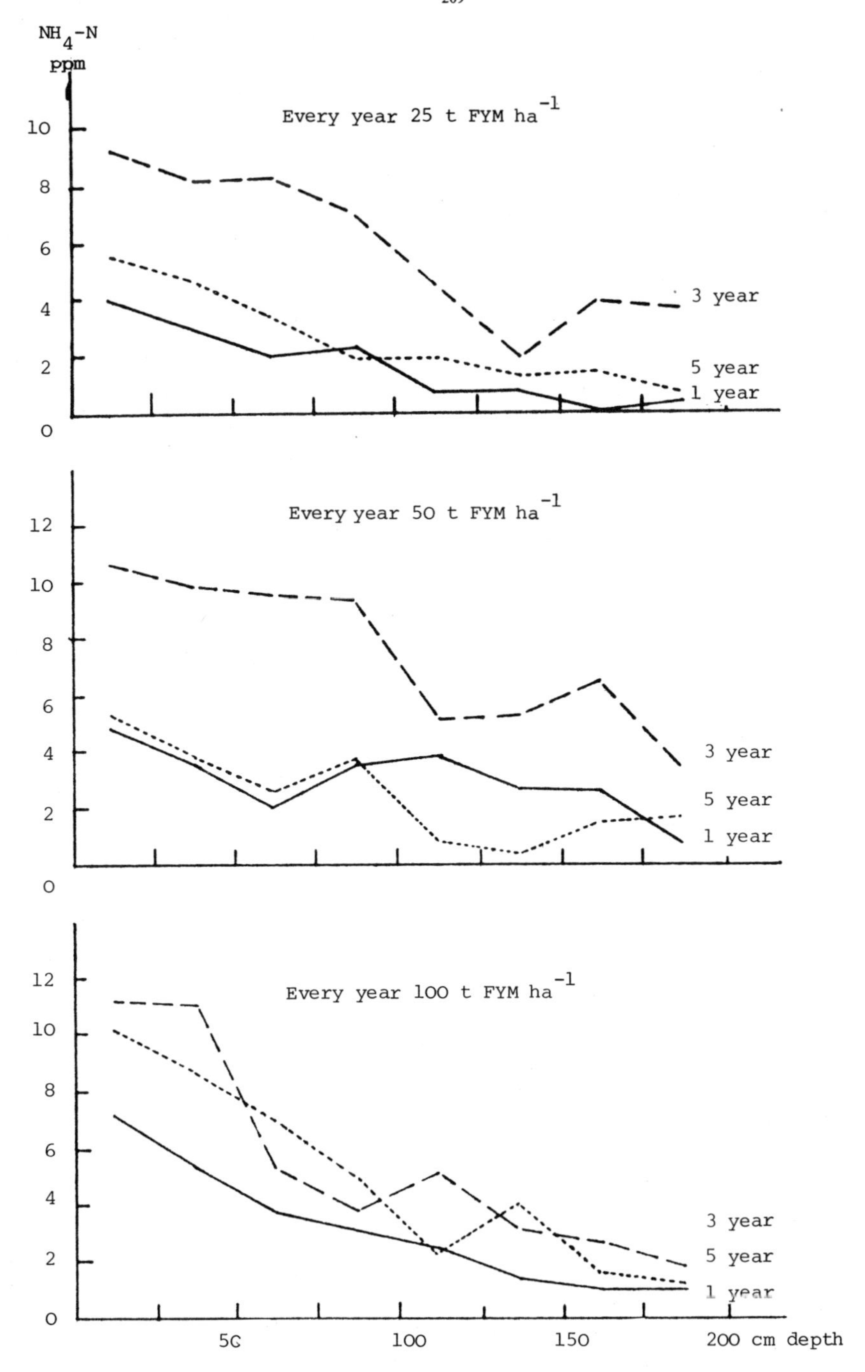

Fig. 8. Content of NH_4-N at various depths in sandy soil - Lundgård 1974-78

The soil had the least content of pores when fertiliser was given and the most after 100 t ha^{-1} of FYM annually. There was a slight tendency for the percentage of large and small pores to increase with FYM applied annually.

INVESTIGATIONS ON POPULATIONS OF EARTHWORMS

In co-operation with the zoological institute of the Royal Veterinary and Agricultural University in Copenhagen, earthworms were collected from soils treated with fertiliser, slurry and FYM and the samples obtained were analysed for size and age of the worms together with the species composition of the population. Earthworms were sampled in three periods - October 1976, April 1977 and October 1977.

The number of worms was nearly the same from the slurry and FYM plots but fewer were found in the plots with fertiliser. The number of worms increased with increasing amounts of organic manures except when 400 t ha^{-1} were applied in the first year.

The weight of worms varied less than their numbers and irregularly, and was similar after fertiliser and on average of the rates of slurry applied. Application of heavy regular dressings of slurry tended to increase the weight of worms. After weighing and counting, the worms were separated into five species - *A. longa, A. caliginosa, A. rosea, A. chlorotica* and *L. terrestris*.

A. caliginosa dominated populations from the slurry plots. This species was also the largest single group in soil from the FYM plots but here *A. longa* and *L.terrestris* are more numerous than the same species in the slurry plots.

COMPARISON BETWEEN SLURRY AND FARMYARD MANURE (FYM)

In most comparisons, crops with slurry have yielded better than those with FYM. Slurry contains more ammonium-N than FYM does, per unit weight, and since ammonium-N is most important for the N effect in the first year after spreading organic manure, this difference in yields is to be expected when slurry and FYM are compared on a weight basis.

Figure 9 illustrates this by showing the relation between yields after applications of slurry and of FYM on the basis of the amounts of NH_4-N supplied by the two types of manure.

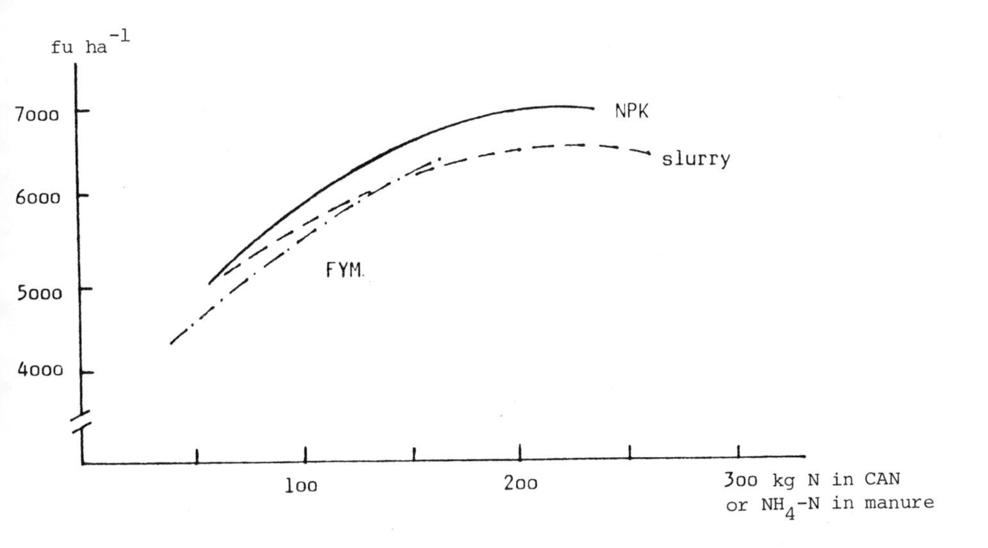

Fig. 9. Effect of FYM and slurry compared on NH_4-N basis.

When compared in this way, slurry and FYM are seen to be equally efficient.

LYSIMETER EXPERIMENTS

Following the same design as the field experiments, lysimeter experiments were carried out at Askov using light sandy-soil from Lundgård and loam soil from Rønhave experimental station. Slurry was used on both soil types, FYM on the loam soil only. Table 16 shows the texture of the soils used together with the pH.

TABLE 15

SOIL PORE SIZE DISTRIBUTION IN ASKOV LOAM

	Total pore size %	Pore size, mm		
		> 3.0 %	3.0 - 0.2 %	< 0.2 %
Fertiliser, annually				
1 NPK	43.2	12.8	20.4	10.0
FYM, annually				
25 t ha^{-1}	45.8	17.0	18.6	10.2
50 t ha^{-1}	45.4	14.7	20.2	10.5
100 t ha^{-1}	48.0	17.4	19.6	11.0
FYM, alternate years				
200 t ha^{-1}	46.3	15.5	19.2	11.6
FYM, every 4th year				
400 t ha^{-1}	44.4	14.4	19.0	11.0

TABLE 16

SOILS USED FOR THE LYSIMETER EXPERIMENT

	Analyses of texture				
	clay	silt	fine sand	coarse sand	pH
Rønhave					
0 - 25 cm	13	18	46	23	7.4
25 - 40 cm	17	16	48	19	7.2
40 - 100 cm	21	19	33	21	7.1
Lundgård					
0 - 25 cm	7	3	23	67	6.0
25 - 40 cm	6	3	29	62	6.4
40 - 100 cm	6	3	23	68	6.0

ANALYSES OF SLURRY AND FYM

The animal manures used were analysed every year and the average composition is given in Tables 17 and 18 as averages for the four year period.

Table 19 gives the total amount of nutrients supplied in four years as fertiliser, slurry and FYM. The small variations were due to variations in the composition of manures from year to year.

YIELDS

Table 20 gives the average yields of single crops for the years 1974 - 1977. The results obtained do not differ appreciably from those from the field experiments except that there is a higher level of production from the lysimeters.

NUTRIENT UPTAKE

Crops and leachate were analysed for their contents of different nutrients, and the removal of nutrients in crops and water was calculated. Where less N and P were given some N and P was taken from the soil. Spreading of larger amounts of slurry and FYM resulted in a temporary accumulation of N and P in the soil. Normally more K, Na and Ca were removed than were supplied.

ANALYSES OF LEACHATES

Leachates were analysed for NO_3-N, P, K, Na, Ca, Mg, Cl and SO_4-S. Increasing manuring resulted in increasing concentration of nitrate, calcium, magnesium and after fertiliser, sulphate. The loss of nitrate by leaching was increased with increasing amounts of manure applied and varied for both soil types from 3 to 10 g NO_3-N m^{-2}. The largest loss was from an annual application of 10 kg m^{-2} of manure.

TABLE 17

COMPOSITION OF SLURRY ON AVERAGE OF THE YEARS 1974-77

	%
Total-N	0.49
NH_3-N	0.30
P	0.08
K	0.38
Na	0.06
Ca	0.11
Mg	0.06
	ppm
Cu	3
Mn	29
Zn	14

TABLE 18

COMPOSITION OF FYM ON AVERAGE OF THE YEARS 1974-77

	%
Total-N	0.58
NH_3-N	0.20
P	0.17
K	0.33
Na	0.05
Ca	0.31
Mg	0.10
	ppm
Cu	15
Mn	63
Zn	59

TABLE 19

TOTAL AMOUNTS OF DIFFERENT NUTRIENTS, g m^{-2} APPLIED AS FERTILISER, SLURRY OR FYM IN LYSIMETER EXPERIMENTS AT ASKOV FOR THE FOUR YEARS 1974-77

	Total-N	NH_3-N	P	K
Fertiliser, annually				
1 NPK	24	-	10	38
2 NPK	48	-	19	76
4 NPK	96	-	38	150
Slurry, annually				
2.5 kg m^{-2}	49	29	9	38
5 kg m^{-2}	98	58	19	75
10 kg m^{-2}	196	116	38	150
Slurry, alternate years				
5 kg m^{-2}	49	30	8	39
10 kg m^{-2}	98	60	16	78
20 kg m^{-2}	196	120	32	156
Slurry, every 4th year				
10 kg m^{-2}	48	29	9	38
20 kg m^{-2}	96	58	18	76
40 kg m^{-2}	191	116	36	152
Fertiliser, annually				
1 NPK	24	-	17	31
2 NPK	48	-	33	62
4 NPK	96	-	67	123
FYM, annually				
2.5 kg m^{-2}	53	20	17	33
5 kg m^{-2}	116	39	34	56
10 kg m^{-2}	233	78	68	133
FYM, alternate years				
5 kg m^{-2}	58	15	16	28
10 kg m^{-2}	105	30	32	56
20 kg m^{-2}	210	59	65	112
FYM, every 4th year				
10 kg m^{-2}	53	15	16	29
20 kg m^{-2}	105	29	31	58
40 kg m^{-2}	210	58	62	114

TABLE 20

YIELDS OF CROPS FROM THE LYSIMETER EXPERIMENTS AT ASKOV FOR THE YEARS 1974-77 g m^{-2}

	Fertiliser			Organic manure kg m^{-2}								
	Annually			Annually			Alternate			Every 4th year		
	1	2	4	2.5	5	10	5	10	20	10	20	40
Sandy soil							Slurry					
Beet, root	911	1 488	1 647	1 156	1 602	2 335	1 624	1 980	2 535	1 887	2 614	2 733
Beet, top	332	470	689	291	415	459	565	655	646	943	981	1 481
Barley, grain	380	556	586	363	484	611	251	259	328	279	312	450
Italian ryegrass	344	557	782	352	498	733	525	658	916	220	247	306
Barley, grain	368	523	751	371	503	539	266	322	381	231	210	274
Loam soil							Slurry					
Beet, root	1 702	1 729	2 081	1 643	1 943	2 091	1 757	2 325	2 705	2 181	2 389	2 733
Beet, top	614	731	1 022	667	573	861	669	915	1 256	918	1 300	1 743
Barley, grain	650	808	882	693	748	816	534	546	561	540	569	683
Italian ryegrass	572	629	986	510	730	922	674	892	1 167	367	340	347
Barley, grain	454	636	880	494	672	928	384	358	495	340	352	366
Loam soil							FYM					
Beet, root	2 076	1 852	2 228	2 008	1 854	1 943	2 020	1 939	2 428	1 957	2 167	2 733
Beet, top	684	892	1 068	642	798	766	703	733	995	853	990	1 348
Barley, grain	721	807	932	627	724	836	550	598	667	595	648	741
Italian ryegrass	609	785	1 060	484	568	820	535	737	1 072	419	411	468
Barley, grain	494	662	860	519	582	765	390	467	596	324	397	433

CONCLUSIONS

Field experiments

1) Application of slurry gave greater yields of crops than did the same weights of FYM. However, when compared on the basis of their contents of ammonium-N, slurry and FYM were equally effective for increasing yields.

2) Annual application of slurry or FYM had, in most cases, given better yields than corresponding weights of the manure applied as a single dressing or two dressings during the four year period, when average yields over this period are compared.

3) In all cases more N and P was supplied by manure than was removed in the crops. With a few exceptions the same was true for K also.

4) The movement of some plant nutrients was limited to the upper 200 cm of the soil. No downwards movement was found for the elements, Mg, Na and Cl.

For nitrogen, some increase was found from the first to the sixth year especially in the upper 100 cm, when 100 t ha^{-1} manure was applied annually.

For phosphate and potassium, a slight increase in the upper 50 cm was noted.

5) If animal manure is given as the only fertiliser to the same field for many years there may be some difficulties in balancing the supply of the three major plant nutrients. For many crops supplementary nitrogen fertilising is necessary.

Lysimeter experiments

Nitrate nitrogen lost annually by leaching was 5 - 10 g m^{-2} in loam soil and 5 - 7 g m^{-2} in sandy soil. Greatest losses were from 10 kg m^{-2} of slurry given annually for four years.

DISCUSSION

H. Tunney *(Ireland)*

You showed phosphorus concentrations at different depths in the profile, to a depth of 2 m, with three rates of slurry 25 t $year^{-1}$, 50 t $year^{-1}$ and 100 t $year^{-1}$. There seems to be very little difference in the values or in the shape of the curves between the three rates. So that 100 t containing four times as much P as 25 t gives similar values for soil-P. Is the soil test not a good indicator, or what is the explanation?

A. Dam Kofoed *(Denmark)*

Part of the phosphorus goes into the soil, certainly not all that is applied from organic manure. The comparison was a valid one, and the lack of difference was surprising and is unexplained.

J.R. O'Callaghan *(UK)*

I would like to ask what do you do about your N, P and K balance in the long term?

A. Dam Kofoed

For phosphorus we are not worried at present about putting on more than is removed, because although the phosphorus index has been raised over the last twenty years, a higher value is desirable. For potassium we do not wish to increase supplies too much, so we look at phosphate and potash and balance the nutrient requirements with fertiliser for nitrogen.

H. Vetter *(West Germany)*

We must now finish our discussion, thank you Dr. Dam Kofoed.

INFLUENCE OF DIFFERENT SLURRY DRESSINGS ON THE YIELD AND QUALITY OF PLANTS, AND THE NUTRIENT CONTENTS OF THE SHALLOW GROUNDWATER AND OF THE SOIL

H. Vetter and G. Steffens
Landwirtschaftskammer Weser-Ems, Mars la Tour Strasse 4,
2900 Oldenburg, West Germany.

Since 1974 our Institute has been intensively engaged in the problems of slurry application on arable land. Two series of field trials, in co-operation with farmers in the vicinity of Oldenburg, deal with the following problems:

1. Influence of different quantities of slurry on plant growth and nutrient contents of the shallow groundwater and of the soil.
2. Influence of different dates of application of slurry on plant growth and nutrient contents of the shallow groundwater and of the soil.

All the field trials were with pig slurry on arable land. The following report is about the most important results of both series of field trials.

1. INFLUENCE OF DIFFERENT QUANTITIES OF SLURRY ON THE YIELD AND QUALITY OF PLANTS AND THE NUTRIENT CONTENTS OF THE SHALLOW GROUND WATER AND OF THE SOIL

1.1 Experimental layout

In this first series of field trials, running since September 1974, 0, 30, 60 and 90 m^3 ha^{-1} of pig slurry have been spread in autumn every year; Table 1 shows the treatments and the nutrient amounts. In case the nitrogen requirement of the plants could not be covered by the slurry dressings, all the treatments have been subdivided and fertilised with 0, 15 and 30 kg ha^{-1} of mineral nitrogen at the beginning of the growing season.

TABLE 1

TREATMENTS AND AVERAGE NUTRIENT AMOUNTS SPREAD WITH THE SLURRY EVERY YEAR

Treatment	kg N ha^{-1}	kg P_2O_5 ha^{-1}	kg K_2O ha^{-1}
Without slurry	0	0	0
30 m^3 pig slurry ha^{-1}	180	150	100
60 m^3 pig slurry ha^{-1}	360	300	200
90 m^3 pig slurry ha^{-1}	540	450	300

Table 2 shows the soil characteristics of the different fields. Two trials were on loamy sand, two on sand; the trials on loamy sand and sand were established on fields, which had been fertilised with high amounts of slurry in previous years, the two other trials were established on fields which had received lower amounts of slurry in the previous years. In Table 2 the differences between high and low slurried fields can be characterised in the first instance by the P_2O_5-contents in the soil depths of 0 - 30 cm and 30 - 60 cm. The P_2O_5 content of the high slurried loamy sand was 86 mg P_2O_5 100 g^{-1} soil (DL) in the top soil, whilst of the sandy soil it was 40 mg P_2O_5 100 g^{-1} soil.

All the field trials have a boulder clay layer at 2 - 3 m depth, on which the ground water lies in the autumn and winter periods. After spreading the slurry, we established small wells in each plot by drain filter tubes, 2 - 3 m deep, out of which a sample of water was taken.

1.2 Results

1.2.1 Yields

Table 3 shows the yields of cereals as an average of the 4 year field trials. In all cases the yields increase with increasing amounts of slurry. Nevertheless in all the field trials the application of 90 m^3 pig slurry ha^{-1} spread in autumn, was not enough to provide the plants with nitrogen in

TABLE 2

SOIL CHARACTERISTICS OF THE FIELD TRIALS WITH INCREASING QUANTITIES OF SLURRY

Trial No.		% < µ	pH KCl	mg 100 g^{-1} P_2O_5	mg 100 g^{-1} K_2O	mg 100 g^{-1} Mg	ppm Cu	% C	% N	C/N
1. lS, nutrient	topsoil	7.7	5.9	86	39	5.8	5.3	1.36	0.11	12
level high	subsoil	16.1	4.4	14	23	5.7	1.7	0.45	0.04	11
2. lS, nutrient	topsoil	7.2	5.1	32	21	5.6	14.0	1.62	0.11	15
level low	subsoil	11.9	4.0	6	15	3.8	3.2	0.61	0.04	15
3. hS, nutrient	topsoil	6.4	5.1	40	14	7.0	9.3	2.94	0.14	21
level high	subsoil	3.1	4.4	20	5	2.8	1.4	1.72	0.08	22
4. hS, nutrient	topsoil	5.3	4.6	24	11	2.8	2.3	2.17	0.10	22
level low	subsoil	4.9	4.4	7	6	0.7	0.5	0.90	0.04	23

TABLE 3

YIELD OF CEREALS, AVERAGE OF 4 YEARS

	Trial 1 lS	Trial 2 lS	Trial 3 hS	Trial 4 hS
Without slurry, without min. N	26.4	32.4	37.1	22.9
" " , + 15 kg " N	26.8	36.8	43.4	25.6
" " , + 30 kg " N	30.6	40.5	47.5	30.6
30 m^3 slurry, without min. N	33.6	34.0	40.4	27.2
" " , + 15 kg " N	35.7	38.2	44.9	29.9
" " , + 30 kg " N	37.8	40.8	49.3	32.7
60 m^3 slurry, without min. N	38.0	36.0	45.7	28.9
" " , + 15 kg " N	39.6	40.8	47.5	34.5
" " , + 30 kg " N	42.0	43.7	50.2	37.7
90 m^3 slurry, without min. N	40.1	37.5	48.2	31.1
" " , + 15 kg " N	41.2	42.4	49.4	36.1
" " , + 30 kg " N	43.5	45.0	50.2	38.7

TABLE 4

INFLUENCE OF INCREASING AMOUNTS OF SLURRY ON THE NITRATE CONTENT OF CATCHCROPS (% NO_3 IN DRY MATTER), DETERMINED IN NOVEMBER 1976

Trial	Catch crop	Slurry application	Treatments 1 0	2 30	3 60	4 90	Relative treatment 4 : 1
			m^3 slurry ha^{-1}				
1. lS	Bird rape	16.8.1976	0.11	0.59	1.96	2.62	24 : 1
2. lS	Bird rape	11.9.1975	0.19	0.31	0.33	0.52	2.7 : 1
4. hS	Turnip like rape	6.10.1975	2.10	3.71	3.29	5.07	2.4 : 1

an optimal way. Additional mineral N-fertilisations of 15 and 30 kg N ha^{-1} at the beginning of the growing period increased the yields on all the slurry treatments. This means that the effectiveness of the slurry nitrogen was very low when the slurry was spread in autumn, and the effectiveness of the slurry nitrogen decreased as the precipitation rates increased in autumn and winter.

1.2.2 Plant quality

It is well known, that high N-dressings can diminish the quality of plants. In 1976, 3 of the field trials were sown with catch crops. In trial No. 1 the slurry was spread directly before sowing, in trial No. 2 and trial No. 3 the spreading of the slurry took place one year earlier. Table 4 shows the nitrate contents of the plants in relation to the amount and the application time of the slurry.

In all trials the nitrate contents are clearly influenced by the fertilisation, with the greatest influence in trial No. 1, where the slurry was spread directly before sowing. The nitrate content of the plants from the plots with 90 m^3 pig slurry ha^{-1} was 24 times higher than from the plots without slurry and reached a content of 2.26 percent which is too high if the acceptable limit of the nitrate content in the dry matter of the plants is 0.5 percent, an amount exceeded by the dressing of 30 m^3 pig slurry ha^{-1}.

In trials No. 2 and 3 the nitrate contents were 2.7 and 2.4 times higher than in the plots without slurry but the dressing of 90 m^3 pig slurry ha^{-1} was applied one year earlier. The after-effect was due to the weather conditions in 1976, when a relatively warm and moderately wet autumn favoured the mineralisation of the soil-N during the growth of the catch crops.

1.2.3 Nutrient contents of the shallow ground water

The risks of downward movement of nutrients into deeper soil layers and into the groundwater are very high when the slurry is spread in autumn, especially for nutrients, which have low binding affinities with the soil- and humus-particles, e.g. N, Na, Cl.

Figure 1 shows the average nitrogen contents (NO_3-N + NH_4-N) of the shallow groundwater of the 4 field trials. The following points should be noted.

- There was a considerable increase in the nitrogen contents of the shallow groundwater as the amounts of slurry increased.
- Nitrogen contents of water were higher on the sandy soils than on the loamy soils.

Table 5 shows the average K, Na, Cl, P and C-contents as well as the conductivity rates in the shallow groundwater of the different field trials. Just as the N-contents increased so also the Na- and Cl-concentrations and the conductivity rates in the shallow groundwater increased with increasing slurry amounts in all the field trials. There was a considerable increase of the K-contents only on the sandy soils. The P- and C-concentrations in the groundwater were not influenced by the different slurry dressings.

1.2.4 Nutrient contents in the soil

At the end of February 1979 soil samples were taken from all the fields, and the light soluble nitrogen content (N_{min}) determined; Table 6 shows the nitrate-N-contents at depths of 0 - 30, 30 - 60, 60 - 90 and 0 - 90 cm. In all the field trials only very little was found at depths of 0 - 30 and 30 - 60 cm (10 - 20 kg N ha^{-1} in the treatments with 90 m^3 pig slurry ha^{-1}). The nitrate-N contents in the soil are well related to the yields and groundwater measurements. The small increases in yields and the nitrogen losses to the shallow groundwater can be explained very well in this way.

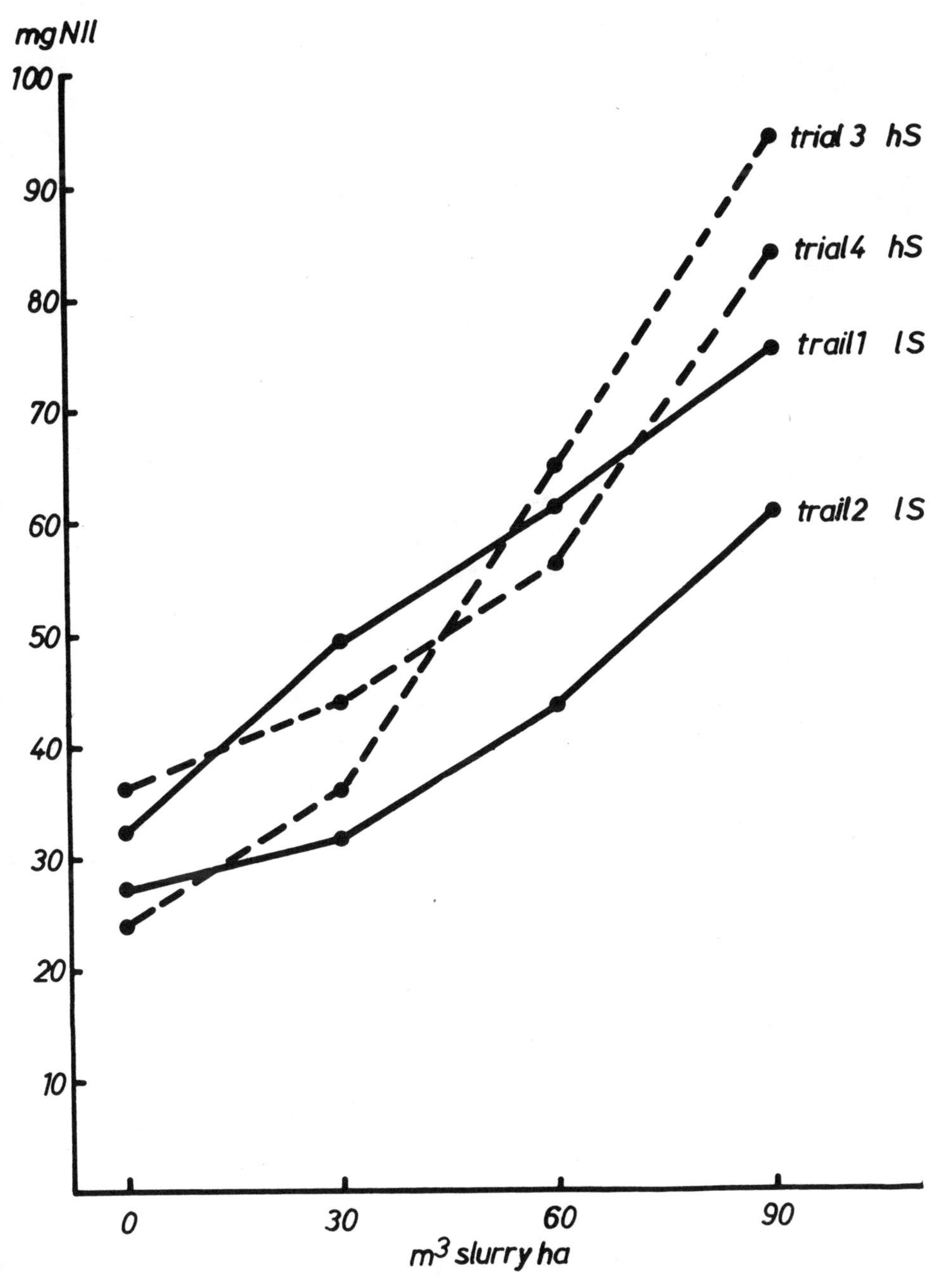

Fig. 1. Average N-contents in the shallow ground water of sandy and loamy soil after spreading increasing amounts of slurry in autumn.

TABLE 5

AVERAGE K, Na, Cl, P, C CONTENTS (mg l^{-1}) AND CONDUCTIVITY RATES (mg KCl l^{-1}) IN THE SHALLOW GROUND WATER (AVERAGE OF 3 YEARS)

	K	Na	Cl	P	C	Conductivity rates
Trial 1, lS						
without slurry	2.0	23.5	27.5	0.13	18.9	267
30 m^3 "	7.3	26.8	36.6	0.12	19.1	305
60 m^3 "	5.9	28.7	46.5	0.16	18.2	364
90 m^3 "	10.8	33.7	49.1	0.06	17.9	422
Trial 2, lS						
without slurry	6.1	23.9	28.1	0.10	16.0	217
30 m^3 "	6.3	24.5	28.5	0.13	15.0	224
60 m^3 "	5.9	28.7	36.6	0.08	16.3	273
90 m^3 "	6.9	33.0	42.7	0.08	14.3	331
Trial 3, hS						
without slurry	22.3	15.8	19.1	0.11	26.5	206
30 m^3 "	29.0	22.2	27.8	0.20	20.9	262
60 m^3 "	37.2	28.1	51.0	0.18	21.5	382
90 m^3 "	44.8	34.4	61.0	0.09	22.0	543
Trial 4, hS						
without slurry	49.7	15.1	17.6	0.06	18.7	208
30 m^3 "	52.7	16.4	28.5	0.07	17.7	275
60 m^3 "	62.6	19.1	30.6	0.09	20.2	320
90 m^3 "	80.2	23.6	45.3	0.09	20.8	418

Table 7 shows the changes in the P_2O_5, K_2O, Cu and Zn contents of the soil after spreading 90 m^3 pig slurry ha^{-1} four years earlier. In all the field trials there was a considerable increase of the P_2O_5, K_2O, Cu and Zn contents of the soil. Although the fields have not been ploughed deeper than 30 cm,

TABLE 6

NITRATE-N CONCENTRATIONS (kg ha^{-1}) IN THE SOIL AT THE BEGINNING OF THE GROWING SEASON (END OF FEBRUARY 1978)

	0-30 cm	30-60 cm	60-90 cm	0-90 cm
Trial 1. lS				
without slurry	21.3	11.3	19.5	52.1
30 m^3 "	26.8	11.3	25.8	63.9
60 m^3 "	22.2	22.4	43.1	87.7
90 m^3 "	18.8	22.2	53.5	94.5
Trial 2. lS				
without slurry	11.7	7.8	14.3	33.8
30 m^3 "	13.2	17.5	33.1	63.8
60 m^3 "	13.0	21.9	44.8	79.7
90 m^3 "	13.8	18.5	72.4	104.7
Trial 3, hS				
without slurry	9.9	12.6	10.1	32.6
30 m^3 "	11.9	12.6	8.6	33.1
60 m^3 "	19.2	24.1	13.6	56.9
90 m^3 "	19.9	25.9	16.6	62.4
Trial 4, hS				
without slurry	10.7	6.8	3.7	21.2
30 m^3 "	11.7	7.6	6.1	25.4
60 m^3 "	12.6	8.7	8.7	30.0
90 m^3 "	31.3	19.5	17.8	68.6

the P_2O_5, K_2O and Cu-contents of the soil increased more in the 30 - 60 cm zone than at the depth of 0 - 30 cm. An enrichment of the Zn-contents of the soil took place mainly in the depth of 0 - 30 cm. There was little increase in the P_2O_5-contents of the soil at depths of 60 - 90 cm.

TABLE 7

THE CHANGE OF THE NUTRIENT CONTENTS IN THE SOIL
A: CONTENTS AT THE BEGINNING OF THE FIELD TRIALS
B: CONTENTS AFTER 4 YEARS' APPLICATION OF 90 m^3 PIG SLURRY ha^{-1}

	mg P_2O_5 100 g^{-1} DL		mg K_2O 100 g^{-1} DL		ppm Cu		ppm Zn	
	A	B	A	B	A	B	A	B
Trial 1, lS								
0-30 cm	86	84	(39)	30	5.3	6.1	13.5	22.0
30-60 cm	14	26	23	26	1.7	3.1	5.1	8.0
60-90 cm	3	4	17	16	0.9	0.8	2.4	1.2
Trial 2, lS								
0-30 cm	32	35	21	33	14.0	15.5	7.2	13.5
30-60 cm	6	9	15	20	3.2	7.3	3.7	5.7
60-90 cm	2	3	14	13	1.5	1.4	1.4	1.7
Trial 3, hS								
0-30 cm	40	54	14	15	9.3	10.5	-	-
30-60 cm	20	24	5	8	1.4	4.1	-	-
60-90 cm	3	4	4	5	0.3	0.2	-	-
Trial 4, hS								
0-30 cm	24	31	11	12	2.3	5.3	14.0	17.0
30-60 cm	7	18	6	8	0.5	2.0	6.2	6.5
60-90 cm	2	3	5	5	0	0.1	1.5	1.5
The change in average of all field trials								
0-30 cm	+ 5.5		(+ 1)		+ 1.6		+ 5.9	
30-60 cm	+ 7.5		+ 3		+ 2.4		+ 1.7	
60-90 cm	+ 1.0		± 0		- 0.1		- 0.3	

2. INFLUENCE OF DIFFERENT DATES OF APPLICATION OF SLURRY ON THE YIELD OF THE PLANTS AND ON THE NUTRIENT CONTENTS OF THE SHALLOW GROUNDWATER AND THE SOIL

2.1 Experimental layout

The series of field trials with different application dates ran from August 1976. In August, October, December and February/March 30 m^3 of pig slurry ha^{-1} were spread on the fields every year. In addition, 15 m^3 pig slurry ha^{-1} was applied to an untreated plot in February/March. Table 8 shows the different treatments and the average nutrient amounts applied.

TABLE 8

TREATMENTS AND AVERAGE NUTRIENT AMOUNTS IN PIG SLURRY IN THE FIELD TRIALS AT DIFFERENT APPLICATION DATES

		N kg ha^{-1}	P_2O_5 kg ha^{-1}	K_2O kg ha^{-1}
Without slurry		-	-	-
30 m^3 slurry ha^{-1}	August	210	140	120
30 m^3 "	October	210	140	120
30 m^3 "	December	210	140	120
30 m^3 "	Febr./March	210	140	120
15 m^3 "	Febr./March	105	70	60

Just as in the field trials with different amounts of slurry the plots were subdivided and fertilised with 0, 20 and 40 kg mineral-N ha^{-1} at the beginning of the growing period. Two of the field trials were on humus sand, two on clayey silt. Table 9 shows the soil characteristics. In these field trials there was also a boulder-clay layer at a depth of 2 - 3 m. Small wells for the collection of the shallow groundwater were made.

TABLE 9

SOIL CHARACTERISTICS OF THE FIELD TRIALS

	Soil % <2 μ	pH $CaCl_2$	P_2O_5	K_2O	Mg	Cu ppm	C %	N %	C/N
			mg 100 g^{-1}						
Trial 1, hS									
topsoil	4.0	4.9	30	8	4.2	3.9	2.77	0.17	16
subsoil	3.0	4.6	12	5	2.5	0.7	1.62	0.09	19
Trial 2, hS									
topsoil	4.2	4.9	28	11	4.9	5.2	2.05	0.15	14
subsoil	6.4	5.0	10	7	2.9	1.0	1.55	0.09	17
Trial 3, tU									
topsoil	16.0	5.7	31	34	8.3	8.3	1.45	0.12	12
subsoil	15.3	5.7	12	23	7.0	2.5	0.69	0.07	10
Trial 4, tU									
topsoil	15.5	5.0	31	25	5.0	4.2	1.54	0.16	10
subsoil	16.9	5.0	14	29	4.3	1.9	1.08	0.10	11

2.2 Results

2.2.1 Yields

Figure 2 shows the yields, as an average for 2 years, of cereals on the humus sands, and Figure 3 shows those on clayey silts. The following points should be noted:

- On both soil types the effectiveness of the N in the slurry was very low, when the slurry was spread in August or October; it was better by spreading in December and best by spreading the slurry in February/ March.

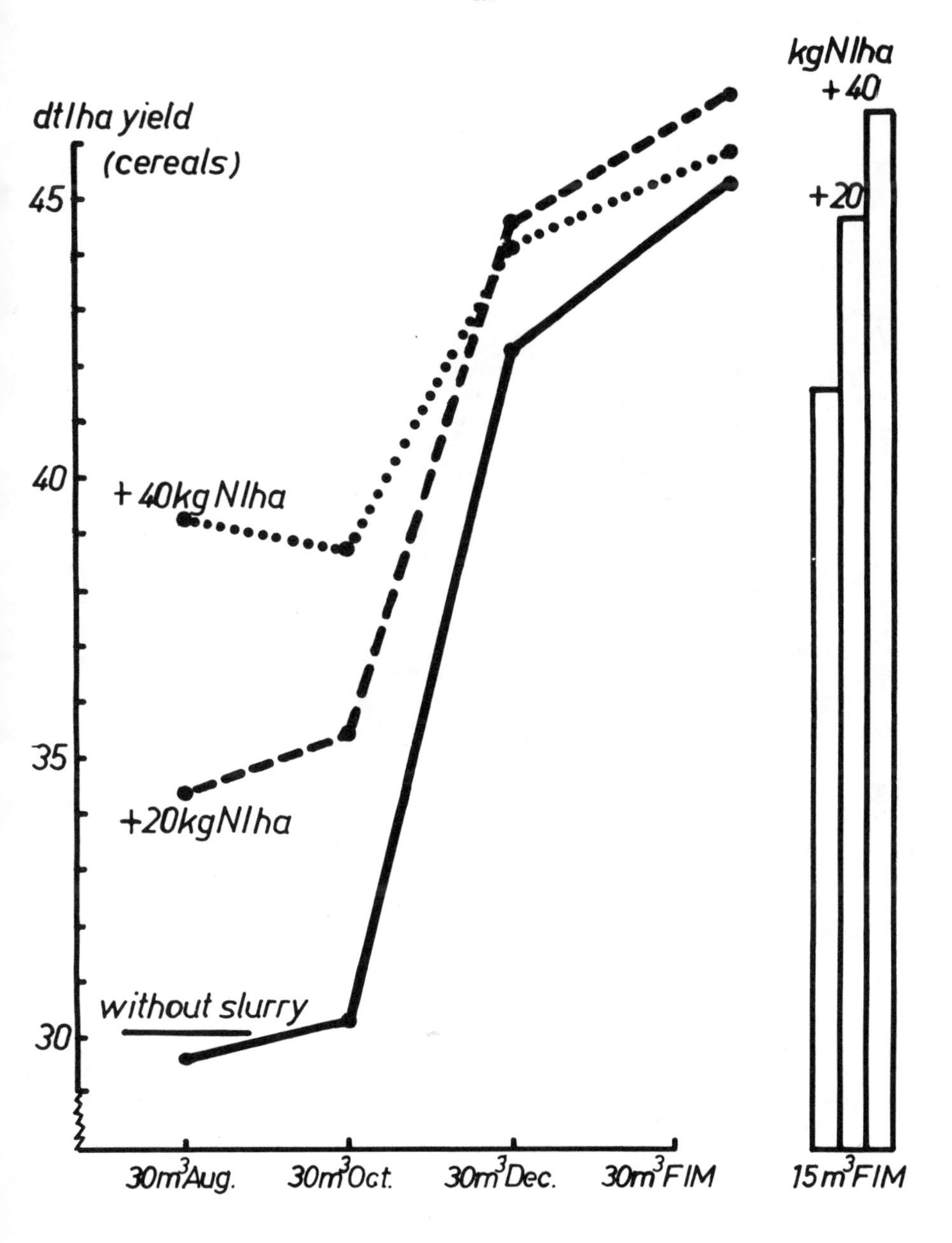

Fig. 2. Average yields on humus sand (1977, 1978), spreading the slurry (m^3 ha^{-1}) at different dates.

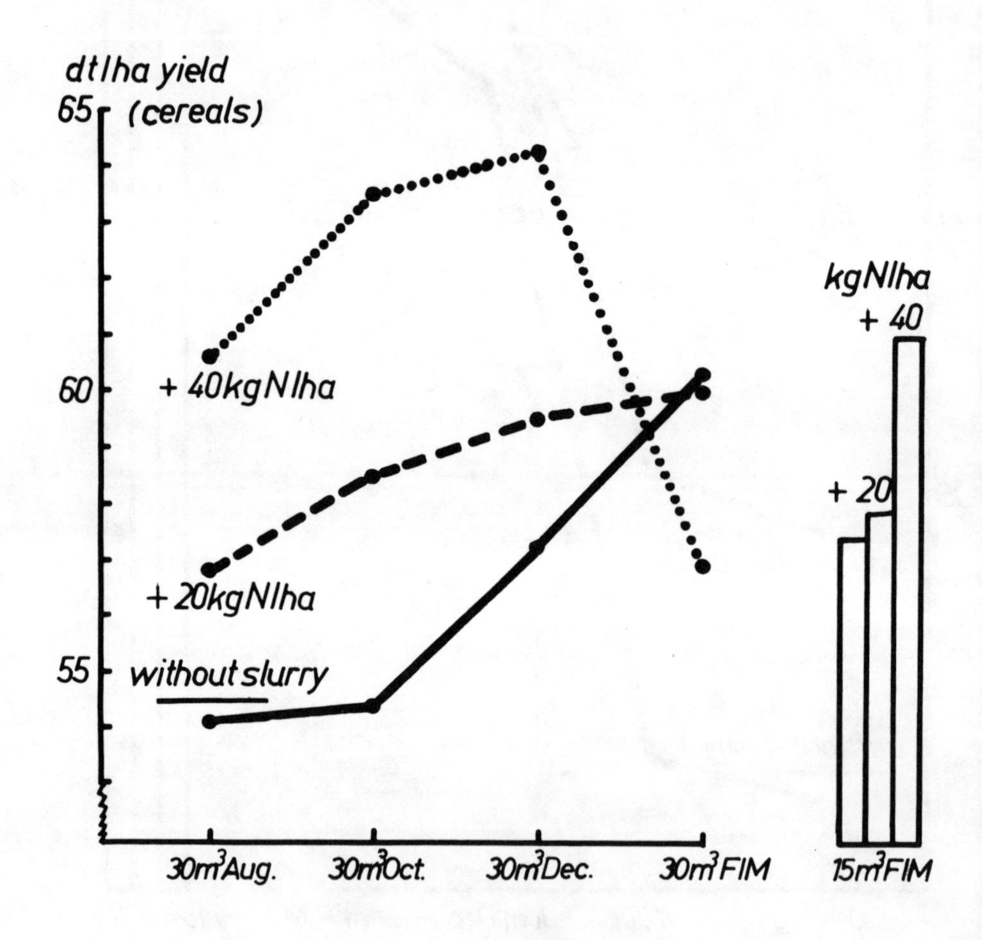

Fig. 3. Average yields on clayey silt (1977, 1978), spreading the slurry (m^3 ha^{-1}) at different dates.

- The yield increases were much higher on the sandy soils than on the clayey soils, due probably to the higher fertility of the clayey silts.
- On the sandy soils the highest yields were obtained by spreading 30 m^3 of pig slurry ha^{-1} in February/March with an additional mineral N-fertilisation of 20 kg ha^{-1}; 40 kg N ha^{-1} resulted in a yield decrease, caused by over-fertilisation.
- On the clayey soils the optimal N fertilisation was exceeded by spreading 30 m^3 of pig slurry ha^{-1} in February/March. Additional mineral N-applications of 20 and 40 kg ha^{-1} decreased the yields by over-fertilisation.
- On the sandy soils it was satisfactory to apply only 15 m^3 of pig slurry ha^{-1} in spring with little additional mineral N-fertiliser.

2.2.2 Nutrient contents in the shallow groundwater

Figure 4 shows the N-contents of the shallow groundwater. The N-contents are the sum of NO_3-N + NH_4-N; the NH_4-contents were always very low (< 5 mg NH_4-N l^{-1}).

In the field trials on clayey silt the N-contents of the shallow groundwater were above the average of other groundwater measurements. As the rate of the N-leaching is lower on heavier soils than on sandy soils, the higher N-contents of the shallow groundwater in the field trials on clayey silt may be due to the higher slurry dressings in the years before starting the field trials. After spreading the slurry in August, October and December the N-concentrations in the shallow groundwater were much higher than after spreading the slurry in February/March.

Table 10 shows the average K, Na, Cl and P-contents as well as the conductivity rates of the shallow groundwater as an average of the two years. The differences between the treatments are relatively small.

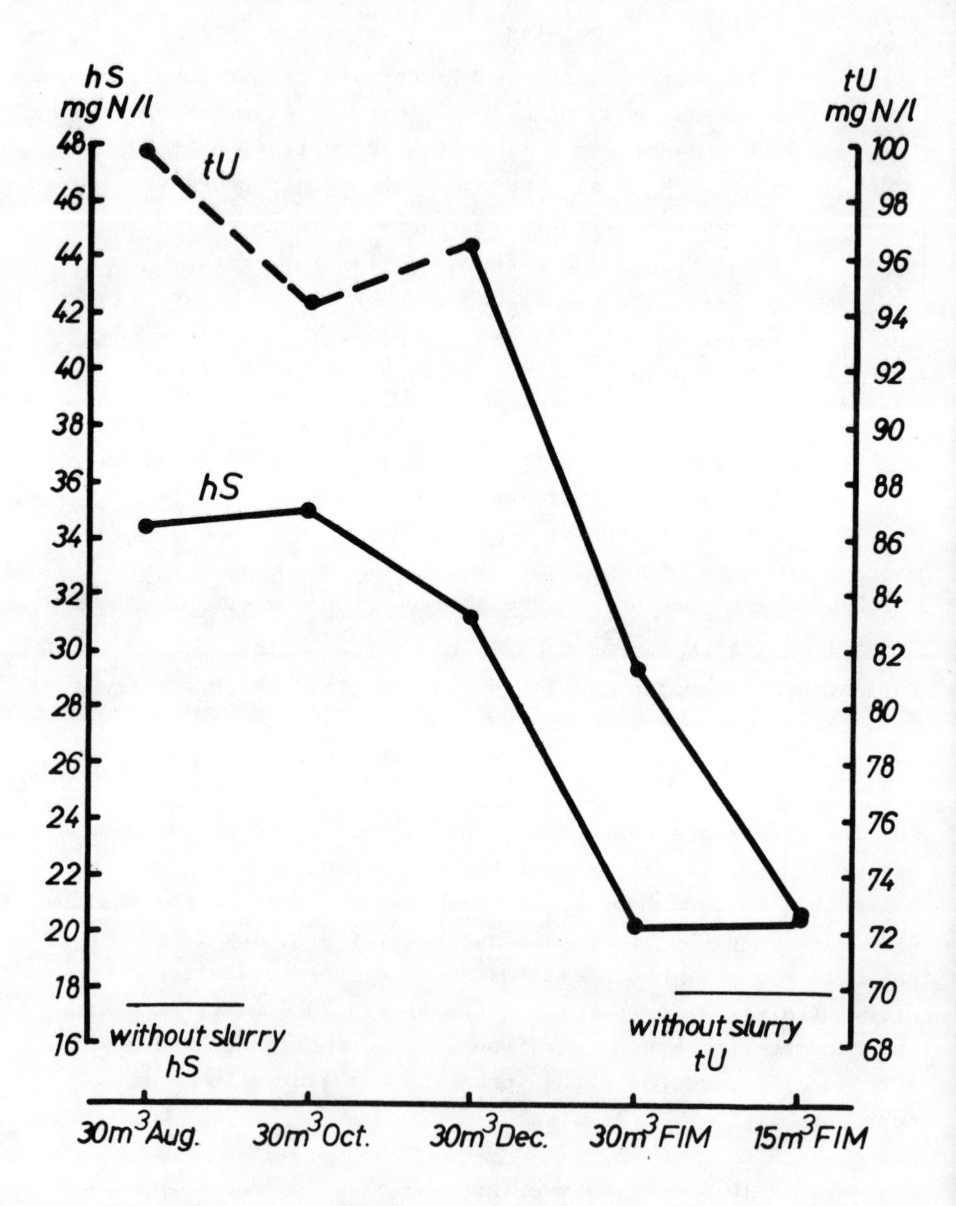

Fig. 4. Average N-contents (NO_3-N + NH_4-N) in the shallow ground water, spreading the slurry at different dates.

TABLE 10

AVERAGE K, Na, Cl, P-CONTENTS (mg l^{-1}) AND CONDUCTIVITY RATES (mg KCl l^{-1}) IN THE SHALLOW GROUND WATER (AVERAGE OF 2 YEARS)

	K	Na	Cl	P	Conductivity rates
Trial 1 and 2, hS					
without slurry	16.3	16.2	63.1	0.06	481
30 m^3 slurry August	33.8	23.7	62.7	0.06	546
30 m^3 " October	20.5	25.3	69.7	0.07	607
30 m^3 " December	22.4	24.9	70.4	0.08	559
30 m^3 " Febr./March	12.4	24.2	61.3	0.08	506
15 m^3 " Febr./March	14.4	21.1	60.1	0.06	465
Trial 3 and 4, tU					
without slurry	7.8	41.6	65.9	0.06	741
30 m^3 slurry August	6.3	42.1	84.7	0.08	810
30 m^3 " October	11.1	42.4	71.0	0.12	804
30 m^3 " December	4.0	48.5	86.3	0.11	874
30 m^3 " Febr./March	6.1	45.8	83.4	0.10	825
15 m^3 " Febr./March	3.9	44.6	73.4	0.09	767

2.2.3 Nutrient contents in the soil

As in the first series of field trials, soil samples were taken to determine the effects of different application times on the nitrate-N contents of the soil. Table 11 shows the nitrate-N contents, in kg ha^{-1} and the following points should be noted:

- In the soil profile, especially in the 60 - 90 cm zone, the nitrate-N contents of the clayey silts were much higher than the N-contents of the sandy soils.
- Trial No. 2 (hS) was cropped with rape as a catch crop in autumn 1977. The lower N-losses of catch crops becomes evident by comparing the two field trials on sandy soils. In trial No. 2 the nitrate-N contents in the soil were higher than in trial No. 1.

TABLE 11

NITRATE-N-CONTENTS (kg ha^{-1}) IN THE SOIL, END OF FEBRUARY 78, BEFORE SLURRYING THE PLOTS IN FEBRUARY/MARCH

	Depths			
	0-30 cm	30-60 cm	60-90 cm	0-90 cm
Trial 1, hS				
without slurry	8.6	4.3	6.2	19.1
30 m^3 slurry, August	6.9	5.0	4.0	15.9
30 m^3 " , October	6.7	3.7	4.6	15.0
30 m^3 " , December	48.0	16.5	6.3	70.8
30 m^3 " , Febr./March	10.1	5.2	3.6	18.9
15 m^3 " , Febr./March	6.2	3.8	3.5	13.5
Trial 2, hS				
without slurry	17.3	7.1	5.5	29.9
30 m^3 slurry, August	33.6	13.3	5.1	52.0
30 m^3 " , October	27.8	23.5	22.3	73.6
30 m^3 " , December	111.9	45.6	7.9	165.4
30 m^3 " , Febr./March	22.7	20.8	5.8	49.3
15 m^3 " , Febr./March	23.2	9.8	7.0	40.0
Trial 3, tU				
without slurry	14.3	27.8	47.3	89.4
Aug., 2nd year, without slurry	7.9	35.3	58.7	101.9
30 m^3 slurry, October	5.5	35.1	81.5	122.1
30 m^3 " , December	26.1	80.8	92.0	198.9
30 m^3 " , Febr./March	5.3	37.4	90.8	133.5
15 m^3 " , Febr./March	11.7	28.6	66.6	106.9
Trial 4, tU				
without slurry	29.7	25.6	22.9	78.2
30 m^3 slurry, August	22.2	25.0	34.5	81.7
30 m^3 " , October	45.4	56.8	40.9	143.1
30 m^3 " , December	139.9	41.3	30.5	211.7
30 m^3 " , Febr./March	33.2	28.3	30.3	91.8
15 m^3 " , Febr./March	29.8	23.7	18.3	71.8

- In all the field trials the highest nitrate-N contents were associated with spreading the slurry in December.
- Spreading the slurry in August or October gave nitrate-N contents in the soil almost the same as in the plots without slurry.
- The low nitrate-N contents in the plots spread in February/March can be explained by the fact that soil sampling took place before spreading the slurry.

3. SUMMARY AND CONCLUSIONS

The most important results can be summarised as follows:

- The effectiveness of N in slurry applied in autumn is different on different soils, but on the whole very low for plant growth in spring. On the heavier soils (1S, first series of field trials) 540 kg slurry-N ha^{-1} were not enough to provide the plants with nitrogen in an optimal way.

  The effectiveness is much improved by spreading the slurry in spring. Table 12 summarises all the results in mineral N-fertiliser-equivalents. Whereas the effectiveness of the slurry nitrogen, depending on weather conditions and soil type, is 5 - 50 percent after spreading the slurry in autumn in comparison to mineral N-fertilisation in spring, the effectiveness of the slurry nitrogen is 70 - 100 percent after spreading the slurry in spring, depending on the dry matter content of the slurry and the weather conditions.

  Spreading the slurry in autumn directly to catch crops, reduces the N losses into the deeper soil layers and improves the N-effectiveness for the following crop.

- High amounts of slurry, spread in autumn to catch crops, can produce high nitrate contents in the plants.

TABLE 12

THE EFFECTIVENESS OF NITROGEN ON DIFFERENT SOILS BY SPREADING THE SLURRY AT DIFFERENT APPLICATION DATES
100 KG SLURRY-N = x KG MINERAL FERTILISER-N

Application date	S	lS	lS-L very fertile
August/September	5 - 10	20 - 30	40 - 50
October/November	15 - 25	30 - 40	50 - 60
December/January	50 - 60	50 - 65	60 - 70
February/March	70 - 80	70 - 80	70 - 90
Beginning of the growing season	70 - 100	70 - 100	70 - 100

- Intensive slurry dressings in autumn cause nitrogen enrichments, on the sandy soils also K-enrichments, of the shallow groundwater, which can reach the surface waters through drain pipes. The N- and K-leachings increase with increasing slurry amounts. The N-leaching is higher on sandy soils than on loamy soils and higher on nutrient-rich soils than on nutrient-poor soils. The N-leaching is very low when the slurry is spread in spring.

- While none of the field trials gave higher P-concentrations in the shallow groundwater, a 4-year dressing with 90 m^3 pig slurry ha^{-1} increased the P-contents in the soil by considerable amounts. On average, the P-enrichment in the soil depth at 0 - 30 cm was 5.5 at 30 - 60 cm 7.5 and at 60 - 90 cm was 1 mg 100 g^{-1} soil (DL). A considerable downward movement in the soil could also be found in fundamental researches of our Institute, from soils which had been dressed with high amounts of slurry many years ago (Table 13). Here P-enrichments in the soil of 57 mg could be determined at a depth of 0 - 30 cm; 17 mg at 30 - 60 cm; and of 3 mg 100 g^{-1} soil, at 60 - 90 cm. When the calculated time for P-enrichments in the soil or P-downward movements into deeper soil layers and into the shallow groundwater is considered not over 5 or 10 years, but over 50 or 100 years, it is likely that still higher P-levels will be found in the shallow groundwater, if high slurry dressings have been put on over many years. Fertilising recommendations with mineral P-fertilisers are 1.5 times the amount of the plant withdrawal; which means about 100 kg P_2O_5 ha^{-1} and it is not obvious why much higher amounts should be applied with organic manures.

TABLE 13

THE P_2O_5-INCREASE IN THE SOIL BY AN APPLICATION OF ABOUT 2 200 m^3 slurry ha^{-1} IN 22 YEARS (22 FIELDS WITH SLURRY, 12 FIELDS WITHOUT SLURRY)

	mg P_2O_5 100 g^{-1} soil (DL)		
Soil depth	without slurry	with slurry	increase
0 - 30 cm	17	74	57
30 - 60 cm	5	22	17
60 - 90 cm	2	5	3

- Besides the considerable increase in the P-content of the soil there was also a considerable increase of the Cu and Zn-contents in the upper soil layers after 4-years application of 90 m^3 pig slurry ha^{-1}.

In the short term, nitrogen is the factor which limits the quantity of a slurry application. When the autumn dressings were higher than 30 m^3 pig slurry ha^{-1} (180 - 200 kg N ha^{-1}), without catch cropping, considerable N-enrichments of the shallow groundwater were found. On the other hand with catch cropping, the nitrate contents in the plants were too high. The spreading of 30 m^3 of pig slurry ha^{-1} in spring was enough to meet the nutrient requirements of the plants.

From the long term point of view, a 4 years field trial is too short for:

- Quantifying the disadvantages of high slurry applications for plant growth and water eutrophication.
- Specifying the slurry quantities which can be spread without harm to soil fertility and to the quality of the water.

Only long term field trials are suitable for examining these problems. Nevertheless the results indicate that phosphate can be leached into the shallow groundwater, when large amounts of slurry have been applied to fields over many years, and that the Cu- and Zn-contents of the soil can reach amounts which are toxic for plants.

To provide the plants with nitrogen, phosphate and copper in an optimal way the spreading of 30 m^3 of pig slurry ha^{-1} (180 - 200 kg N ha^{-1}, 130 - 150 kg P_2O_5 ha^{-1}, 1.2 - 1.5 kg Cu ha^{-1}) will be enough, when the slurry is applied in spring. If these quantities are exceeded P, Cu- and Zn-enrichment in the soil will occur. If animal manures are applied at the correct times, and in utilisable amounts the needs of both plant growth and environmental protection will be served.

SPREADING OF PIG AND CATTLE SLURRIES ON ARABLE LAND: LYSIMETER AND FIELD EXPERIMENTS

P. Spallacci and V. Boschi
Istituto Sperimentale Agronomico, Sezione di Modena,
Modena, Italy.

PART A LYSIMETER EXPERIMENTS: Influence of pig slurry application rates on yield and quality of crops, on soil properties and nutrient balance in relation to soil type.

A.1 OBJECTIVE

This study aims to determine the rates of slurry application that, according to the type of soil, would enable good yields of crops without causing either pollution of the drainage waters or dangerous variations of the physical and chemical properties of the soil. In addition, it will allow the balance of nutrient and polluting elements to be drawn up and the capacity of soils as a purification system to be evaluated.

A.2 MATERIALS AND METHODS

The experiment was carried out in lysimeters (size 1 x 1 x 1 m) containing 4 types of soil: sand, sandy loam, sandy clay and clay.

The trials were started in spring 1976, with applications of 4 increasing rates of pig slurry, in factorial combination with 4 types of soil. In spring 1977 a comparison with a normal rate of fertiliser was added (20 treatment combinations replicated three times, i.e. a total of 60 lysimeters).

The following crops were grown: forage sorghum in 1976, followed by Italian ryegrass in 1976 - 77, utilising the residual fertility of the previous dressings; grain maize in 1977 and soft wheat in 1977 - 78.

TABLE 1

LYSIMETER EXPERIMENTS, MODENA 1976 - 78. ANALYSIS OF PIG SLURRY

		1976 (a)		1977 (b)		1977 - 78 (b)	
		In fresh matter	In DM	In fresh matter	In DM	In fresh matter	In DM
pH		7.3	-	6.9	-	6.9	-
Dry matter	%	1.921	100.0	3.004	100.0	4.039	100.0
COD	mg kg^{-1}	-	-	27200	-	39100	-
BOD_5	mg kg^{-1}	6770	-	4390	-	2940	-
Total N	%	0.275	14.31	0.273	9.09	0.312	7.72
Organic N	%	0.109	5.67	0.104	3.46	0.132	3.27
NH_3-N	%	0.166	8.64	0.169	5.63	0.180	4.46
NO_3-N	%	0	0	0	0	0	0
Ash	%	0.685	35.66	0.898	29.89	0.897	22.21
P	%	0.079	4.11	0.088	2.93	0.100	2.48
K	%	0.238	12.39	0.289	9.62	0.200	4.95
Na	%	0.091	4.74	0.095	3.16	0.060	1.49
Ca	%	0.053	2.76	0.073	2.43	0.105	2.60
Mg	%	0.030	1.56	0.044	1.46	0.053	1.31
SO_4	%	0.039	2.03	0.041	1.36	0.037	0.92
Cl	%	0.140	7.29	0.147	4.89	0.093	2.30
Cu	ppm	6.41	333.85	11.51	383.27	14.92	369.48
Zn	ppm	8.88	462.26	11.86	394.81	16.18	400.68

a) Mean of 4 dressings; b) Mean of 3 dressings (analysis of 2 samples per dressing).

TABLE 2

LYSIMETER EXPERIMENTS, MODENA 1976-78. SUMMARY OF NUTRIENT AMOUNTS, kg ha^{-1}, APPLIED TO DIFFERENT CROPS WITH VARIOUS RATES OF PIG SLURRY

	1st period (+)		
	1976 (forage sorghum) (++)		
	rate 1 160 m^3ha^{-1}	rate 2 320 m^3ha^{-1}	rate 3 480 m^3ha^{-1}
Dry matter	3074	6148	9222
Organic matter	1978	3956	5934
Total N	440	880	1320
Organic N	174	348	522
NH_3-N	266	532	798
Ash	1096	2192	3288
P	126	252	378
K	381	762	1143
Na	146	292	438
Ca	85	170	255
Mg	48	96	144
SO_4-S	21	42	63
Cl	224	448	672
Cu	1.02	2.04	3.06
Zn	1.42	2.84	4.26
	2nd period (+)		
	1977 (grain maize)		
	rate 1 100 m^3ha^{-1}	rate 2 200 m^3ha^{-1}	rate 3 300 m^3ha^{-1}
Dry matter	3033	6066	9099
Organic matter	2107	4214	6321
Total N	280	560	840
Organic N	100	200	300
NH_3-N	180	360	540
Ash	926	1852	2778
P	89	178	267
K	311	622	933
Na	96	192	288
Ca	75	150	225
Mg	45	90	135
SO_4-S	14	28	42
Cl	149	298	447
Cu	1.31	2.62	3.93
Zn	1.37	2.74	4.11

(+) Periods fixed for nutrient balances: the former from April 1976 to March 1977, the latter from April 1977 to June 1978.

(++) Slurry applied to sorghum followed by Italian ryegrass utilising the residual effect.

TABLE 2 (Continued)

	2nd period (+)		
	1978 (wheat)		
	rate 1 70 m^3ha^{-1}	rate 2 140 m^3ha^{-1}	rate 3 210 m^3ha^{-1}
Dry matter	2827	5654	8481
Organic matter	2199	4398	6597
Total N	218	436	654
Organic N	92	184	276
NH_3-N	126	252	378
Ash	628	1256	1884
P	70	140	210
K	140	280	420
Na	42	84	126
Ca	72	144	216
Mg	37	74	111
SO_4-S	26	52	78
Cl	65	130	195
Cu	1.04	2.08	3.12
Zn	1.13	2.26	3.39

(+) Periods fixed for nutrient balances: the former from April 1976 to March 1977, the latter from April 1977 to June 1978.

The slurry dressings were applied in divided amounts during the growing period: May, June, July and August for sorghum; May, June and July for maize; October, February and April for wheat.

The chemical composition of slurry is shown in Table 1.

The amounts of nutrients applied in different rates of slurry are shown in Table 2.

The leachates were collected and analysed once or twice monthly, according to season and amount of rainfall.

The nutrient balances were made between amounts applied in slurry (or in fertiliser) and amounts recovered in crops and leachates, for periods covering a complete cycle of cropping, manuring and leaching: from April 1976 to March 1977 and from April 1977 to June 1978.

The possible accumulation of nutrients in soils will be included in the balance only after a period longer than one or two years.

Soil samples were taken before the start of the trials (April 1976), after the first period of cropping (April 1977) and again after the second one (June 1978).

A.3 RESULTS

A.3.1. Crop yields

The yields of crops grown during the three-year experiments are summarised in Figure 1.

a) Sandy loam and sandy clay soils produced the highest yields of forage sorghum.

The smallest amounts of slurry increased the yield on all soils; the highest rates increased it only on clay soil.

b) Like sorghum, Italian ryegrass produced the highest yields on sandy loam and sandy clay soils. The residual effects of slurry increased the yield linearly up to maximum rate.

c) Sandy loam and sandy clay soils produced the highest grain yields of maize.

The first slurry rate gave a very high increase of yield on all soils; the second one increased the yield on sand and clay soils and the third one only on sand soil. Both grains ear^{-1} and 1000-grain weight components of yield were increased.

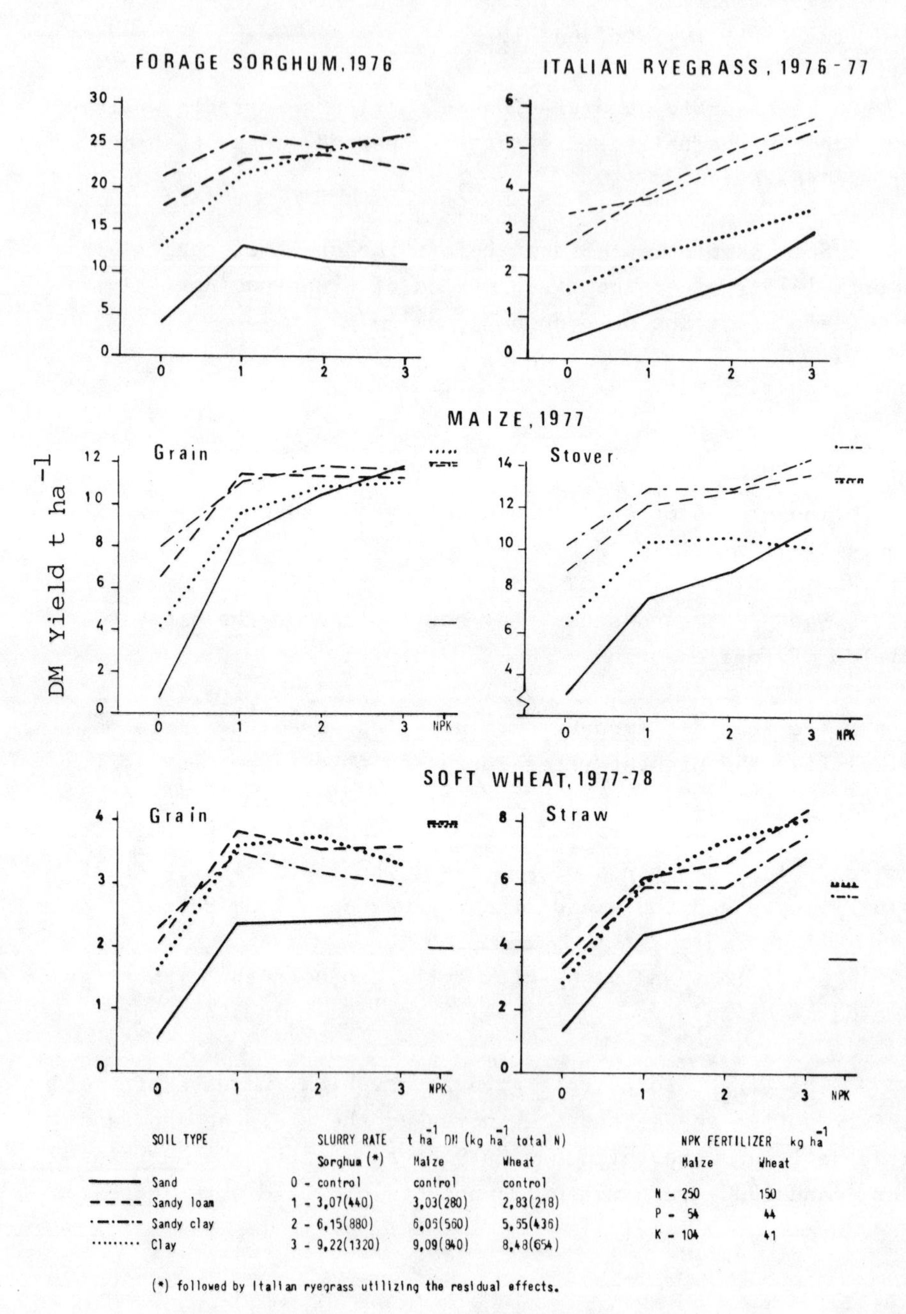

Fig. 1. Lysimeter Experiments, Modena 1976 - 78. Influence of soil type and slurry rate on yields, t ha^{-1}, of the crops.

Greater differences in stover yields were obtained than in grain yields. The slurry applications, up to the highest rates, gave considerable increases on all soils except clay.

d) The yields of wheat grain were very similar for the different soils, except for the sandy one, which gave a lower yield.

The lowest slurry rate increased the yield on all soils. The highest rates were of no use or even harmful; only clay soil can tolerate the second rate. The yield was increased only by grains ear^{-1}; the 1000-grain weight decreased for applications of slurry greater than the lowest rates.

The straw yield increased progressively up to the highest slurry rate.

A.3.2. Nutrients in the crops

The nutrient contents influenced by slurry application are shown in Figures 2 and 3.

a) The soil type showed significant effects on nitrate, calcium, magnesium and ash contents of forage sorghum. The slurry rates increased Kjeldahl and nitrate N, potassium and magnesium: but the highest rate only affected nitrate. It is remarkable that nitrates reached dangerous levels with the second application rate, mainly at first cut and on light soils.

b) The soil type showed some differences on all nutrients of Italian ryegrass except for nitrate content which changed very much.

Applying slurry increased mainly Kjeldahl and nitrate N contents. The nitrate increased on sandy loam and sandy clay more than on the other soils. Dangerous levels of nitrate were reached only at first cut in the autumn.

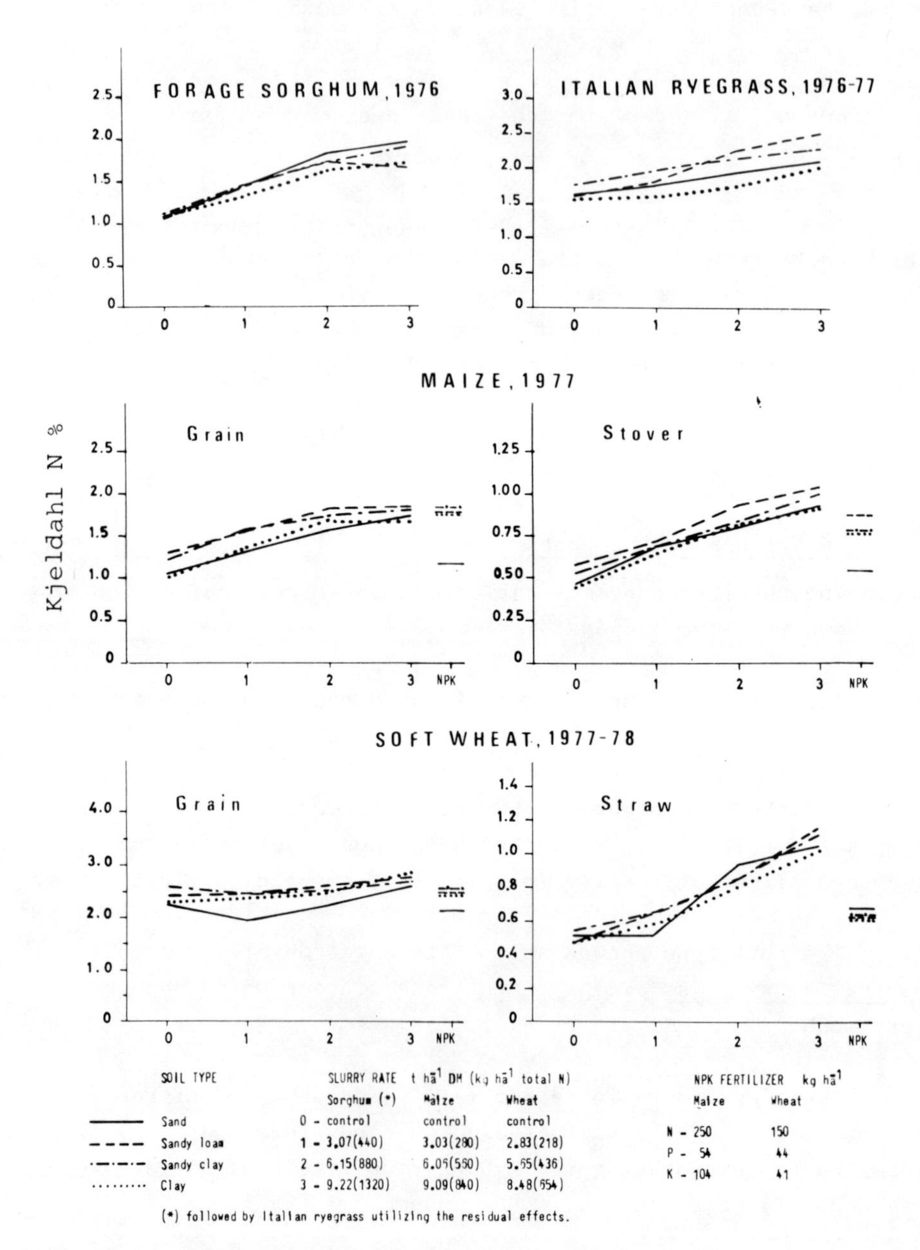

Fig. 2. Lysimeter experiments, Modena 1976 - 78. Influence of soil type and slurry rate on Kjeldahl N contents, % in DM, in the crops.

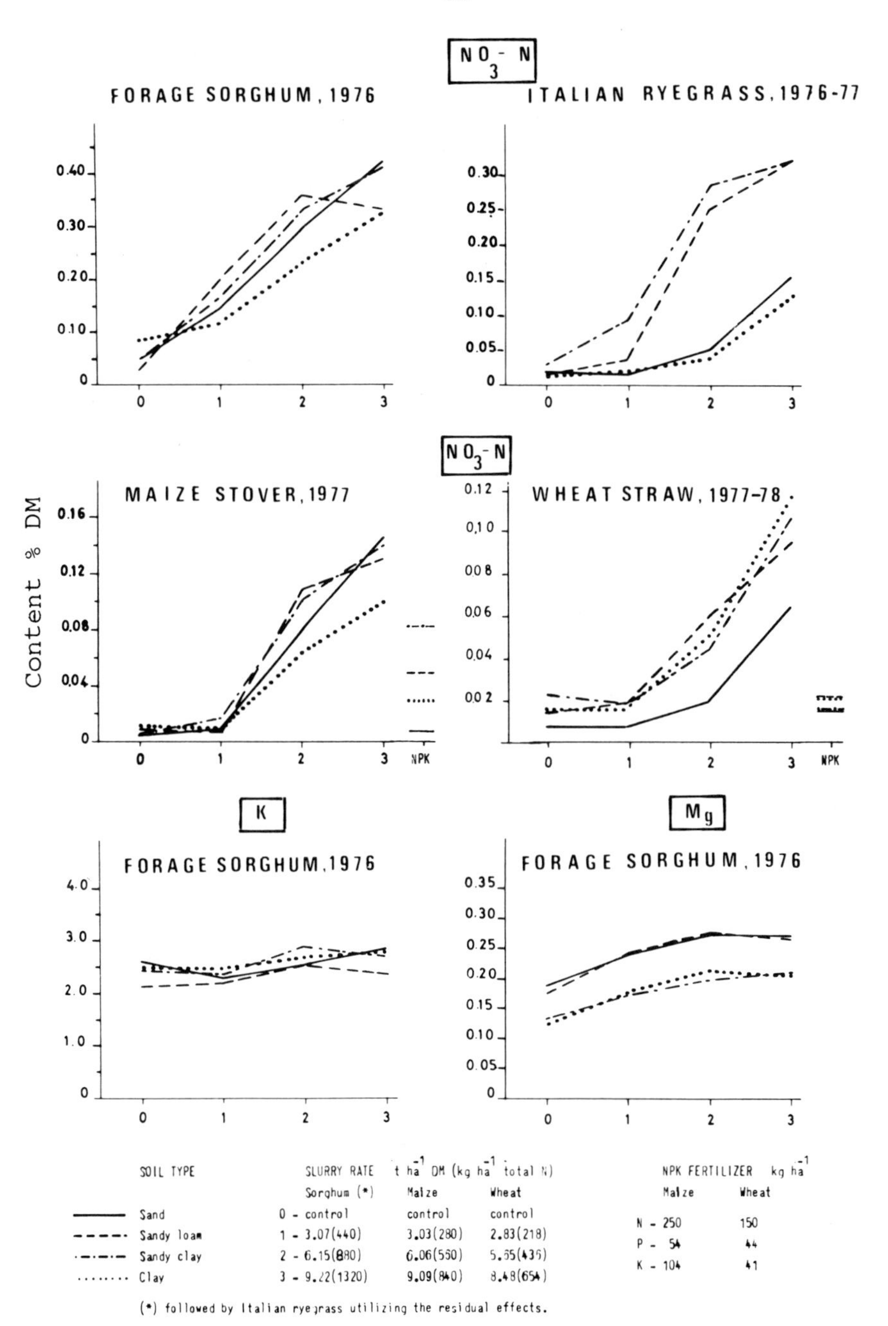

Fig. 3. Lysimeter experiments, Modena 1976 - 78. Influence of soil type and slurry rate on some nutrient contents, % in DM, in the crops.

c) The N content of maize grain differed with the soil type, but mainly due to the slurry rates: nitrogen increased up to the second rate, where it levelled off. All the other determinations (ash, P, K, Na, Ca, Mg) did not change.

The soil type influenced all nutrient contents in the stover, except for calcium. The slurry rates increased considerably Kjeldahl and nitrate N and K contents, but they decreased phosphorus; the other elements did not show signficant variations.

d) The N content of wheat grain differed with the soil type and with the rate of slurry application. Among the other nutrients, the slurry rates increased K and Ca contents slightly while Mg showed a little decrease.

The soil type influenced all nutrient contents in the straw, but the change of Kjeldahl N was very small. The slurry rates gave high increase of Kjeldahl and nitrate N; also P, K, and Ca contents increased, while Mg decreased slightly.

A.3.3. Chemical composition of leachates

The examination of analytical data is in progress, for the whole of the experimental period. The following determinations were made: pH, suspended matter, filtrable residue at $105^{o}C$ and $550^{o}C$, conductivity at $20^{o}C$, DO, BOD_5, COD, organic N, NH_3, NO_2, NO_3, total and organic P, orthophosphate, K, Na, Ca, Mg, HCO_3, SO_4, Cl.

A.3.4. Nutrient losses in leachate waters

The amounts of nutrient losses in leachate waters are shown in Figure 4 (April 1976 - March 77) and in Figure 5 (April 1977 - June 78).

The effects of the soil type and of the slurry application rates were quite similar in the two periods under consideration. However, more nutrients were lost in the second period

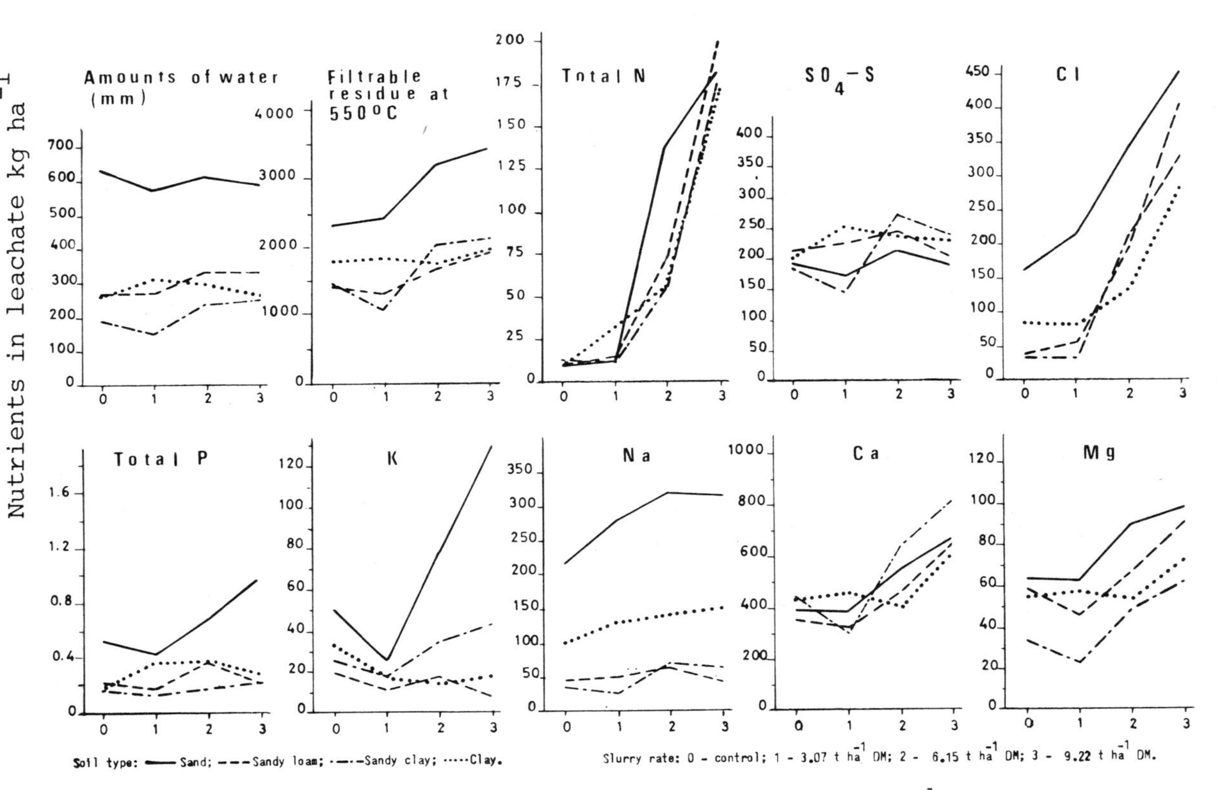

Fig. 4. Lysimeter experiments, Modena 1976 - 77. Losses of nutrients, kg ha^{-1}, in leachate waters as influenced by soil type and slurry rate (April 1976 - March 77).

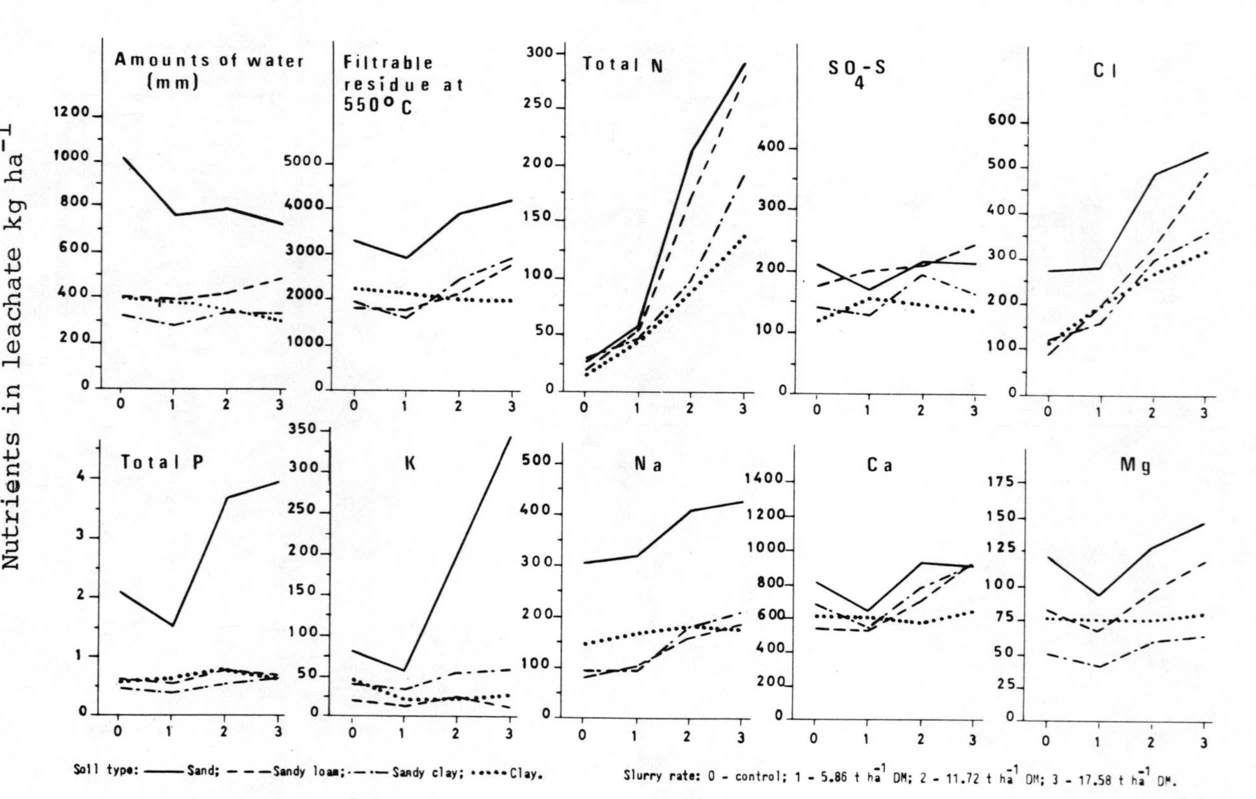

Fig. 5. Lysimeter experiments, Modena 1977 - 78. Losses of nutrients, kg ha^{-1}, in leachate water as influenced by soil type and slurry rate (April 1977 - June 78).

than the first one, due to its greater length, the larger amounts of nutrients applied and the residual effects of the preceding period.

The nutrient losses were higher on the light soils, especially those treated with high slurry applications. The clay soil showed a considerable resistance in modifying the amounts of losses; however this resistance was less for nitrogen and chloride.

The leached N was almost entirely nitrate (90 - 98%); increases in the other forms of nitrogen were obtained only at the end of winter or in spring and with the highest slurry applications.

The phosphorus and potassium losses were very low, with little dependence on the slurry application rates except on the sandy soil.

As a consequence the lowest rate of application of slurry was tolerated by all the soils, the second rate only by sandy clay and clay soils, while the third rate was always too high.

A.3.5. Nutrient balance

The nutrient balances, expressed as differences between the amounts in the slurry dressings and those removed by crops and leaching, are shown in Tables 3 and 4.

Most N accumulated in sandy soil because of low crop uptakes and in spite of the high losses by leaching. The lowest slurry application was enough to obtain equilibrium on sandy loam and sandy clay soils and it did not give a dangerous surplus even on the other soils. The other rates of application should be considered every time, in regard both to the accumulation on the different types of soil and to the contents of leachate waters.

TABLE 3

LYSIMETER EXPERIMENTS, MODENA 1976 - 78. INFLUENCE OF SOIL TYPE AND SLURRY RATE ON BALANCES OF N, P AND K (AMOUNTS APPLIED - AMOUNTS REMOVED BY CROPS AND BY LEACHING)

Soil type	Slurry rate or fertiliser	Nitrogen kg ha^{-1} N		Phosphorus kg ha^{-1} P		Potassium kg ha^{-1} K	
		A	B	A	B	A	B
Sand (S)	0	-59.7	-68.7	-10.2	-17.2	-170.3	-135.8
	1	+196.5	+198.5	+96.0	+103.7	+16.7	+188.8
	2	+466.5	+429.4	+221.7	+249.0	+346.0	+413.3
	3	+798.8	+726.1	+340.0	+391.0	+594.4	+580.7
	Mean	+350.5	+321.3	+161.9	+181.6	+196.7	+261.7
Sandy loam (Sl)	0	-245.4	-216.1	-45.5	-60.0	-479.0	-248.7
	1	-28.3	+34.2	+66.7	+73.1	-270.5	+98.8
	2	+183.3	+321.9	+185.4	+233.7	-37.6	+445.0
	3	+516.8	+626.5	+312.8	+385.0	+399.7	+883.8
	Mean	+106.6	+191.6	+129.8	+158.0	-96.9	+294.7
Sandy clay (Sc)	0	-321.8	-263.9	-49.6	-63.1	-663.7	-339.4
	1	-78.2	+54.3	+62.8	+80.6	-373.7	+0.8
	2	+211.7	+430.2	+187.5	+243.7	-146.4	+371.0
	3	+389.6	+738.6	+301.7	+388.8	+195.2	+774.2
	Mean	+50.3	+239.8	+125.6	+162.5	-247.1	+201.6
Clay (C)	0	-181.3	-138.9	-32.4	-37.8	-398.0	-241.5
	1	+55.5	+130.3	+78.5	+87.2	-234.6	+78.2
	2	+315.1	+463.5	+193.5	+241.6	+3.2	+485.3
	3	+542.3	+867.7	+314.4	+398.4	+283.2	+860.6
	Mean	+182.9	+330.7	+138.5	+172.4	-86.5	+295.6

TABLE 3 (Continued)

Soil type	Slurry rate or fertiliser	Nitrogen kg ha^{-1} N		Phosphorus kg ha^{-1} P		Potassium kg ha^{-1} K	
		A	B	A	B	A	B
S	NPK fertiliser	-	+8.9	-	+67.0	-	-12.3
Sl		-	-174.0	-	+23.6	-	-156.9
Sc		-	-174.7	-	+19.2	-	-322.0
C		-	-108.5	-	+16.0	-	-393.3
				Amount of nutrient applied			
	0	0	0	0	0	0	0
	1	440	498	126	159	381	451
	2	880	996	252	318	762	902
	3	1320	1494	378	477	1143	1353
	NPK	-	400	-	98	-	145

A) 1st period: from April 1976 to March 1977 (sorghum and Italian ryegrass)
B) 2nd period: from April 1977 to June 1978 (maize and wheat)

TABLE 4

LYSIMETER EXPERIMENTS, MODENA 1976-78. INFLUENCE OF SOIL TYPE AND SLURRY RATE ON BALANCES OF Na, Ca, Mg AND MINERAL MATTER (AMOUNTS APPLIED - AMOUNTS REMOVED BY CROPS AND BY LEACHING).

Soil type	Slurry rate or fertiliser	Sodium kg ha^{-1} Na		Calcium kg ha^{-1} Ca		Magnesium kg ha^{-1} Mg		Mineral matter kg ha^{-1} ash	
		A	B	A	B	A	B	A	B
Sand (S)	0	-217.9	-307.7	-414.8	-829.3	-71.8	-133.4	-2763	-3602
	1	-138.4	-184.1	-358.1	-547.9	-49.1	-48.1	-2541	-2215
	2	-32.1	-134.4	-440.1	-697.0	-28.5	-10.4	-2191	-1866
	3	+116.6	-15.0	-479.4	-549.8	+8.5	+46.3	-1426	-1139
	Mean	-67.9	-160.3	-423.1	-656.0	-35.2	-36.4	-2230	-2206
Sandy loam (Sl)	0	-50.4	-87.2	-435.6	-596.3	-91.7	-124.3	-3357	-3235
	1	+85.2	+24.8	-357.6	-471.3	-60.2	-43.4	-2630	-2015
	2	+213.0	+111.3	-446.0	-498.5	-45.3	+13.3	-2360	-1015
	3	+374.6	+221.1	-526.6	-568.3	-16.5	+68.3	-1361	-123
	Mean	+155.6	+67.5	-441.5	-533.6	-53.4	-21.5	-2427	-1597
Sandy clay (Sc)	0	-40.7	-100.0	-556.6	-780.1	-66.6	-85.8	-3919	-3452
	1	+110.9	+38.0	-354.9	-482.4	-25.4	-5.2	-3111	-1954
	2	+210.6	+93.7	-611.4	-576.4	-8.3	+60.7	-3038	-1336
	3	+363.3	+199.3	-716.9	-581.6	+17.3	+127.9	-2404	-476
	Mean	+161.0	+57.7	-559.9	-605.1	-20.7	+24.4	-3118	-1805
Clay (C)	0	-103.3	-151.5	-498.5	-657.6	-72.8	-97.1	-3298	-3211
	1	+12.4	-33.1	-469.8	-538.3	-50.7	-34.1	-3128	-2384
	2	+146.5	+89.2	-351.0	-375.7	-14.9	+42.5	-2431	-815
	3	+280.3	+233.4	-469.0	-287.1	+12.5	+121.6	-1871	+612
	Mean	+84.0	+34.5	-447.1	-464.7	-31.5	+8.2	-2682	-1450

TABLE 4 (Continued)

Soil type	Slurry rate or fertiliser	Sodium $kg\ ha^{-1}$ Na		Calcium $kg\ ha^{-1}$ Ca		Magnesium $kg\ ha^{-1}$ Mg		Mineral matter $kg\ ha^{-1}$ ash	
		A	B	A	B	A	B	A	B
		Amount of nutrient applied							
	O	O	O	O	O	O	O	O	O
	1	146	138	85	147	48	82	1096	1554
	2	292	276	170	294	96	164	2192	3108
	3	438	414	255	441	144	246	3288	4662

A) 1st period: from April 1976 to March 1977 (sorghum and Italian ryegrass)
B) 2nd period: from April 1977 to June 1978 (maize and wheat)

The P gains increased progressively, according to the slurry application rates. The differences due to the soil type were small.

The K balance in the first period was positive only with the maximum rate of application on all types of soil, except for the sandy one; in the second period there was a gain with the second slurry application rate.

The Na balance was positive with the second and third slurry application rates on all types of soils, except the sandy one.

The Ca and Mg balances resulted nearly always in loss, with some exceptions for Mg, mainly in the second period of balance.

The mineral matter (ash) balances resulted in more loss in the first period than in the second one.

Figure 6 shows the N amounts taken up by crops by comparison with the loss by leaching.

Crop uptakes increased linearly while the losses were very low, also with the lowest rate of slurry application; with highest application rates the losses increased rapidly. From these different shapes it was possible to pick out some points, where the highest difference between the uptake curve and the leaching one shows an 'optimum utilisation' of nitrogen applied in slurry (i.e. a high yield without pollution).

The optimum points corresponding to different N amounts for the various soils are: 350 for sandy loam; 470 for sandy clay; 570 for clay (kg ha^{-1} of total N applied during each period). On sandy soil, where this calculation was not possible, the observation of the N balance suggests that the optimum point can be about 200.

As each period includes two crops, the amounts shown should be shared proportionally to the N needs of the single crop.

The data shown in Figure 6, compared also with the nitrate contents in forages and in leachates, will give useful indications about the highest rates of nitrogen from slurry that can be applied on different soils.

A.3.6. Chemical and physical properties of soils

The data for soils taken at the start of the trials are shown in Tables 5 and 6 and by comparison with the data shown in Table 7, after the first year of treatments, show the following effects of slurry rates:

a) lowering of pH in sandy soil;

b) no increase in total N and organic matter with the lowest rate and some increases with the higher rates of application, though changing in relation to the type of soil; C/N ratio tends to increase in sandy soil and to decrease in the others;

c) no significant increases in total and extractable P with the lowest rate and progressive increase with the other rates;

d) exchangeable and soluble K variations similar to those for phosphorus; no significant variations in exchangeable and soluble Na.

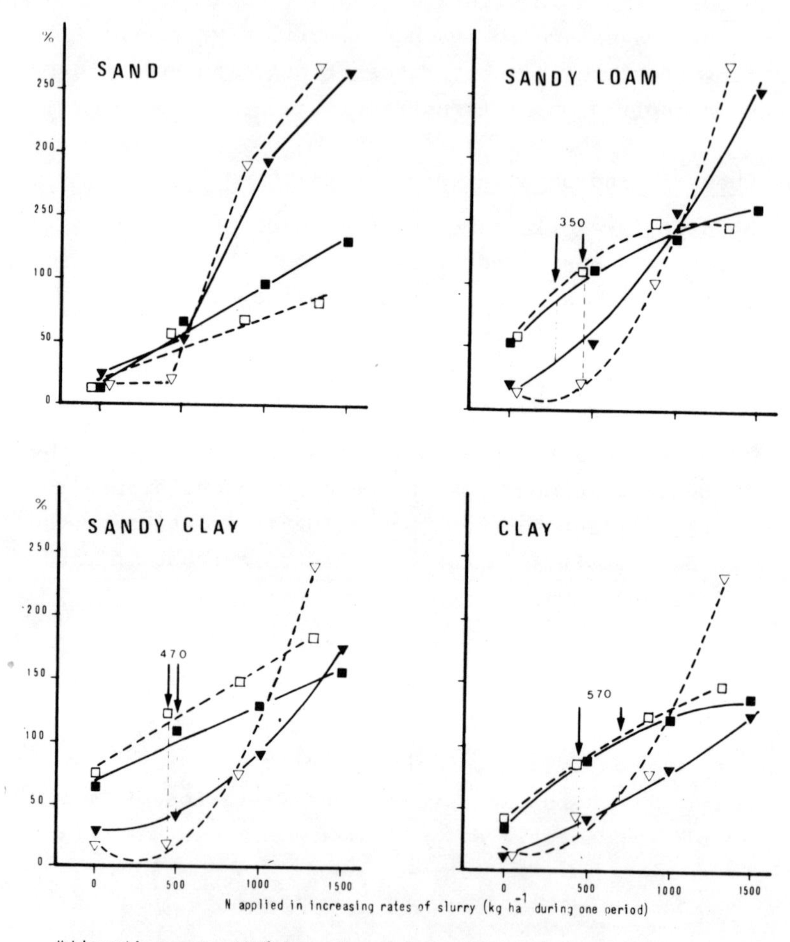

Uptakes and losses expressed in % of respective general means of each of the following periods:

	Average amounts of total N kg ha^{-1}			
	Supplied	Taken up	Leached	Balance
1) from April 1976 to March 77	660	414.6	72.8	+ 172.6
2) from April 1977 to June 78	747	365.2	110.9	+ 270.9

The arrows indicate the points with the maximum differences between uptakes and losses; the number is the average of the N rates corresponding to the arrows.-□-1st and-■-2nd period uptakes;--▽-1st and-▼-2nd period losses.

Fig. 6. Lysimeter experiments, Modena 1976 - 78. Relationship between nitrogen applied in slurry and nitrogen removed by crops and by leachate waters.

TABLE 5

LYSIMETER EXPERIMENTS, MODENA 1976. CHEMICAL ANALYSIS OF SOILS BEFORE THE START OF THE TRIALS ON SLURRY DRESSING*

Soil type	Slurry rate	pH in H_2O	pH in 1N KCl	Conductivity at 20°C (1:2.5 ratio) mmho cm^{-1}	Mineral matter %	Total $CaCO_3$ %	Active $CaCO_3$ %	Kjeldahl N ‰	Organic matter %	C/N ratio
Sand	0	7.9	7.6	0.084	98.7	5.6	0.4	0.13	0.06	2.9
	1	8.1	7.9	0.080	99.6	6.0	0.3	0.08	0.05	3.5
	2	8.2	7.9	0.082	99.2	4.9	0.3	0.09	0.05	3.5
	3	7.8	7.6	0.073	99.1	5.6	0.3	0.11	0.05	2.6
	Mean	8.0	7.8	0.080	99.2	5.5	0.3	0.10	0.05	3.1
Sandy loam	0	7.4	6.6	0.162	94.2	0.9	0.7	1.26	1.65	7.6
	1	7.0	6.1	0.129	94.4	0.3	0.6	1.30	1.58	7.0
	2	7.1	6.2	0.172	94.8	0.5	0.6	1.27	1.66	7.6
	3	7.2	6.3	0.133	95.2	0.5	0.6	1.18	1.52	7.5
	Mean	7.2	6.3	0.149	94.7	0.6	0.6	1.25	1.60	7.4
Sandy clay	0	7.9	7.3	0.170	94.9	14.0	3.1	1.18	1.27	6.3
	1	7.9	7.3	0.180	94.4	13.6	3.0	1.17	1.26	6.3
	2	8.0	7.5	0.175	94.3	14.7	3.2	0.98	1.10	6.5
	3	7.9	7.3	0.175	94.4	15.3	3.4	1.08	1.19	6.4
	Mean	7.9	7.3	0.175	94.5	14.4	3.2	1.11	1.21	6.4
Clay	0	8.1	7.3	0.199	91.0	14.3	9.9	1.41	1.66	6.8
	1	8.0	7.4	0.207	90.6	14.8	10.3	1.49	1.71	6.6
	2	8.1	7.4	0.196	91.2	14.6	9.9	1.41	1.62	6.7
	3	8.0	7.3	0.210	90.6	14.5	9.8	1.46	1.74	6.9
	Mean	8.0	7.4	0.203	90.9	14.6	10.0	1.44	1.68	6.8

* Samples taken in April 1976, at 0 - 25 cm depth. Data expressed on air dry basis and averaged from three replications.

TABLE 5 (Continued)

Soil type	Slurry rate	Total P	Extractable P (0.5M $NaHCO_3$)	Exchangeable K (1N NH_4OAc)	Soluble K (1:2.5 ratio)	Exchangeable Na (1N NH_4OAc)	Soluble Na (1:2.5 ratio)	CEC
		‰ P_2O_5	ppm P_2O_5	ppm K_2O	ppm K_2O	ppm Na_2O	ppm Na_2O	meq $100g^{-1}$
Sand	0	0.60	12	7	14	0	7	5.3
	1	0.68	12	13	9	0	8	5.2
	2	0.60	8	5	6	1	5	5.6
	3	0.55	11	11	8	5	4	6.4
	Mean	0.61	11	9	9	2	6	5.6
Sandy loam	0	1.26	41	141	20	18	17	20.2
	1	1.26	29	111	8	21	19	20.6
	2	1.34	41	136	11	17	20	20.0
	3	1.16	31	119	10	23	16	19.8
	Mean	1.26	36	127	12	20	18	20.1
Sandy clay	0	1.13	14	169	28	0	19	19.7
	1	1.17	11	175	30	0	21	20.2
	2	1.18	9	184	33	1	22	20.4
	3	1.24	12	178	28	0	22	20.4
	Mean	1.18	12	177	30	0	21	20.2
Clay	0	1.39	13	359	16	35	29	25.2
	1	1.33	21	377	20	43	32	26.2
	2	1.29	11	299	14	21	29	25.2
	3	1.34	18	364	19	35	29	26.0
	Mean	1.34	16	350	17	34	30	25.7

TABLE 6

LYSIMETER EXPERIMENTS, MODENA 1976. PHYSICAL ANALYSIS OF SOILS BEFORE THE START OF THE TRIALS ON SLURRY DRESSING*

Soil type	Slurry rate	Sand (2-0.02 mm) %	Silt (0.02-0.002 mm) %	Clay (<0.002 mm) %	Bulk density g cm^{-3}
Sand	0	95.7	0.6	3.7	-
	1	98.6	0.5	0.9	-
	2	98.8	0.9	0.3	-
	3	97.2	2.7	0.1	-
	Mean	97.6	1.2	1.2	1.45
Sandy loam	0	57.3	31.7	11.0	-
	1	59.7	28.6	11.7	-
	2	54.0	34.4	11.6	-
	3	55.2	35.3	9.5	-
	Mean	56.5	32.5	10.9	1.32
Sandy clay	0	63.7	21.0	15.3	-
	1	61.0	22.0	17.0	-
	2	66.4	19.2	14.4	-
	3	63.9	19.7	16.4	-
	Mean	63.7	20.5	15.8	1.26
Clay	0	11.3	45.7	43.0	-
	1	11.8	48.9	39.3	-
	2	10.6	49.3	40.1	-
	3	9.8	46.3	43.9	-
	Mean	10.9	47.5	41.6	1.25

* Samples taken in April 1976, at 0 - 25 cm depth. Data expressed on oven-dry basis and averaged from three replications.

TABLE 6 (Continued)

Soil type	Slurry rate	Particle density	Total porosity	Water infiltration rate (in field)	Water retentivity (Richards)		Moisture of air-dry soil
					pF 2.54 **	pF 4.19	
		g cm^{-3}	%	cm h^{-1}	%	%	%
Sand	0	-	-	-	2.3	1.5	0.15
	1	-	-	-	2.5	1.5	0.17
	2	-	-	-	2.1	1.3	0.14
	3	-	-	-	2.5	1.5	0.17
	Mean	2.58	43.8	201.0	2.3	1.4	0.16
Sandy loam	0	-	-	-	21.4	12.3	1.72
	1	-	-	-	21.3	11.9	1.69
	2	-	-	-	22.5	12.2	1.68
	3	-	-	-	22.1	12.3	1.76
	Mean	2.50	47.2	20.7	21.8	12.2	1.71
Sandy clay	0	-	-	-	21.3	12.7	1.61
	1	-	-	-	22.2	13.8	1.80
	2	-	-	-	21.4	13.3	1.62
	3	-	-	-	21.3	12.7	1.68
	Mean	2.48	49.2	16.1	21.5	13.1	1.68
Clay	0	-	-	-	34.0	23.1	3.45
	1	-	-	-	34.9	22.5	3.30
	2	-	-	-	33.4	21.7	3.28
	3	-	-	-	34.2	20.9	3.37
	Mean	2.62	52.4	19.0	34.1	22.1	3.35

** pF = 0.01 for the sand.

TABLE 7

LYSIMETER EXPERIMENTS, MODENA 1977. CHEMICAL ANALYSIS OF SOILS AFTER THE FIRST PERIOD OF CROPPING AND MANURING*

Soil type	Slurry rate	pH in H_2O	pH in 1N KCl	Conductivity at 20°C (1:2.5 ratio) mmho cm^{-1}	Total $CaCO_3$ %	Active $CaCO_3$ %	Kjeldahl N ‰	Organic matter %	C/N ratio
Sand	0	7.8	7.4	0.072	5.2	0.6	0.13	0.09	4.0
	1	7.8	7.4	0.067	6.5	0.4	0.20	0.11	3.1
	2	7.5	7.2	0.073	5.3	0.5	0.28	0.22	4.6
	3	7.4	7.2	0.091	5.4	0.5	0.27	0.23	5.0
	Mean	7.6	7.3	0.076	5.6	0.5	0.22	0.16	4.2
Sandy loam	0	7.5	6.4	0.157	1.2	1.1	1.29	1.58	7.1
	1	7.1	6.0	0.142	1.0	1.0	1.32	1.51	6.7
	2	7.2	6.2	0.170	1.1	1.1	1.31	1.69	7.5
	3	6.9	6.0	0.149	1.1	1.0	1.34	1.61	7.0
	Mean	7.2	6.2	0.155	1.1	1.1	1.32	1.60	7.1
Sandy clay	0	7.7	7.2	0.180	13.5	2.9	1.26	1.35	6.2
	1	7.8	7.2	0.191	13.3	2.8	1.30	1.45	6.4
	2	8.0	7.4	0.197	13.6	3.0	1.32	1.38	6.1
	3	8.0	7.3	0.202	13.9	3.1	1.40	1.46	6.1
	Mean	7.9	7.3	0.193	13.6	2.9	1.32	1.41	6.2
Clay	0	8.1	7.3	0.217	13.9	10.4	1.48	1.53	6.0
	1	8.0	7.3	0.205	13.9	10.4	1.60	1.83	6.7
	2	8.1	7.3	0.216	13.8	10.4	1.56	1.73	6.4
	3	8.0	7.2	0.231	14.2	10.4	1.70	1.83	6.2
	Mean	8.1	7.3	0.217	14.0	10.4	1.59	1.73	6.3

* Samples taken in April 1977, at 0 - 25 cm depth. Data expressed on air-dry basis and averaged from three replications.

TABLE 7 (Continued)

Soil type	Slurry rate	Total P	Extractable P (0.5M $NaHCO_3$)	Exchangeable K (1N NH_4OAc)	Soluble K (1:2.5 ratio)	Exchangeable Na (1N NH_4OAc)	Soluble Na (1:2.5 ratio)	CEC
		‰ P_2O_5	ppm P_2O_5	ppm K_2O	ppm K_2O	ppm Na_2O	ppm Na_2O	meq $100g^{-1}$
Sand	0	0.50	6	12	6	3	7	5.4
	1	0.61	39	28	7	1	6	5.4
	2	0.81	107	33	8	0	7	6.4
	3	0.86	124	33	10	1	9	6.7
	Mean	0.70	69	27	8	1	7	6.0
Sandy loam	0	1.26	33	101	12	1	63	19.9
	1	1.27	33	89	9	29	75	20.4
	2	1.39	75	187	17	18	60	19.6
	3	1.64	138	348	31	15	62	19.8
	Mean	1.39	70	181	17	16	65	19.9
Sandy clay	0	1.10	13	172	16	0	56	20.4
	1	1.22	42	196	23	0	74	20.7
	2	1.36	43	198	27	0	66	20.9
	3	1.79	119	337	43	0	61	21.2
	Mean	1.37	54	226	27	0	64	20.8
Clay	0	1.40	13	347	21	34	32	25.8
	1	1.53	62	443	22	40	36	26.0
	2	1.76	61	433	24	39	39	25.6
	3	2.00	88	602	46	38	42	25.7
	Mean	1.67	56	456	28	38	37	25.8

PART B FIELD EXPERIMENTS: Effects of applying pig slurry on maize, sorghum and wheat; and of cattle slurry on silage maize in relation to soil type.

B.1 OBJECTIVE

Field trials, of which the most significant results are reported, were carried out to ascertain:

1) the effects of pig slurry, with rates applied for several years, on yields and qualities of silage maize and forage sorghum;
2) yields and qualities of soft and durum wheats given increasing applications of pig slurry;
3) the influence of different times, methods and rates of pig slurry dressing on the infection by *Ustilago maydis;*
4) the effects of increasing rates of cattle slurry in different soil types on the yield of silage maize.

B.2 MATERIALS AND METHODS

The trials were carried out during several years with pig slurry on clay soil only, and with cattle slurry on three types of soil: sandy loam, sandy clay and clay.

Table 1 (part A) gives the composition of pig slurry, Table 8 the composition of cattle slurry, Table 5 (part A) the analysis of soils. Further data on the employed methods are illustrated in Figures 7 to 10.

B.3 RESULTS

B.3.1. Pig slurry

Figure 7 gives the effects of increasing rates of pig slurry and extra NPK application on DM yields and crude protein

TABLE 8

CHEMICAL COMPOSITION OF CATTLE SLURRY APPLIED TO EXPERIMENTAL FIELDS, 1978

	Dressing 1 2/6	Dressing 2 8/7	Dressing 3 28/7
pH	7.49	7.73	7.70
Dry matter (+)	5.462	4.181	5.226
Organic matter	4.328	3.085	4.007
BOD_5	4128	2096	1672
Total N	0.301	0.299	0.327
Organic N	0.140	0.145	0.150
NH_3-N	0.161	0.154	0.177
Ash	1.134	1.096	1.219
Total P	0.051	0.044	0.058
Soluble P	0.0022	0.0010	0.0016
Total K	0.296	0.256	0.305
Soluble K	0.134	0.155	0.161
Ca	0.113	0.095	0.109
Mg	0.041	0.045	0.047
Chloride	0.100	0.108	0.101
Na	0.065	0.070	0.065
CO_3	0.042	0.037	0.040
SO_4	0.052	0.046	0.054
Cu	6.22	7.80	4.40
Zn	14.80	11.28	10.60

(+) % on w/w; BOD mg kg^{-1}; Cu and Zn ppm.

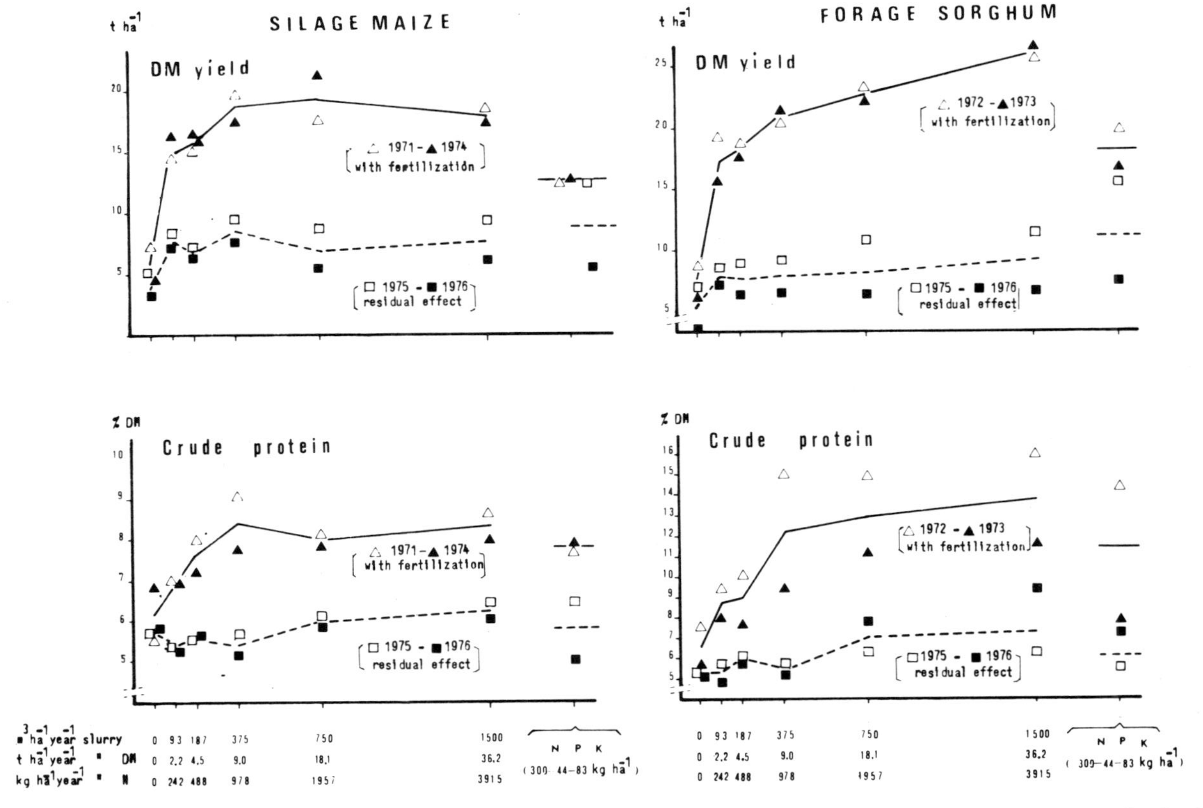

Fig. 7. Pig slurry dressings, up to very high rates, and extra N P K fertilisation, repeated successively for 5 years on the same plots: effects on yields and crude protein contents of silage maize and forage sorghum and influence of residual effect during the following 2-year period without fertilisation, 1970 - 76.

contents of silage maize and forage sorghum. The dressings were applied on the same plots during 5 successive years (1970 - 74).

We have also reported the yields obtained without slurry application during the following 2-year period (1975 - 76), to emphasise the residual effect of 5-year manuring.

a) DM yield of silage maize increased significantly up to application rates of 9 to 18 t ha^{-1} slurry DM, while DM yield of forage sorghum increased even with maximum rates;

b) crude protein contents of forages increased significantly, more in sorghum than in maize;

c) the extra NPK fertilisation was less effective, mainly in maize;

d) in the 5-year period no cumulative effect was found harmful to the crops, even with maximum rates;

e) in the following 2-year period without manuring, the residual effect produced an average increase of 3.7 t ha^{-1} DM yield of maize and 2.8 of sorghum, in comparison with unmanured control plots. This increase was more marked on plots supplied with NPK.

The analyses of 5-year-treated soils showed significant increases in organic matter and nutrient contents (total and available P, exchangeable K).

Nutrients from faeces and urine, mainly nitrogen, were removed during the year, while the least decomposable materials (feeds, hair) have been accumulating in the soil.

The effects of increasing quantities of pig slurry on yields and qualities of soft wheat and durum wheat are illustrated in Figure 8.

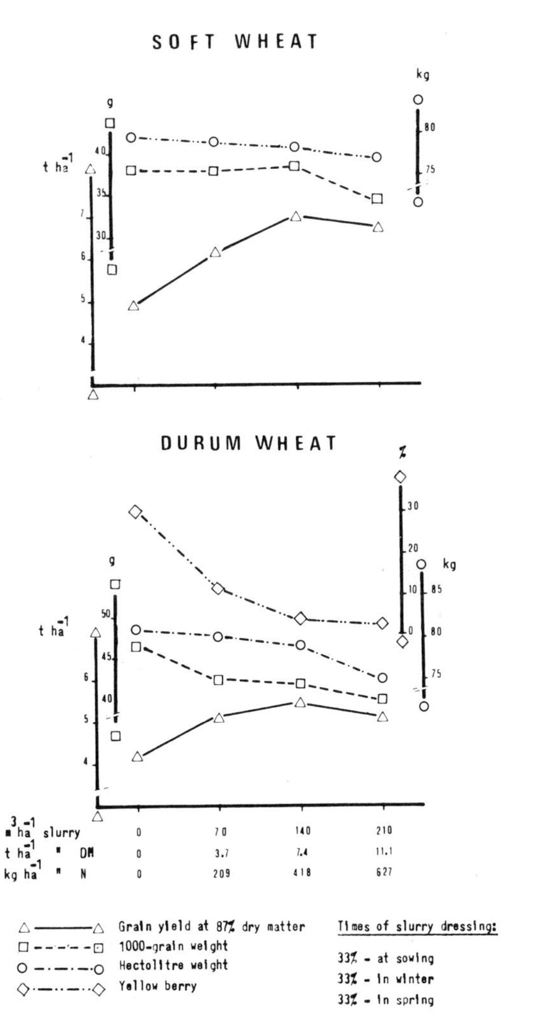

Fig. 8. Effects of increasing rates of pig slurry on yields and qualities of winter wheats (soft and durum), 1977 - 78.

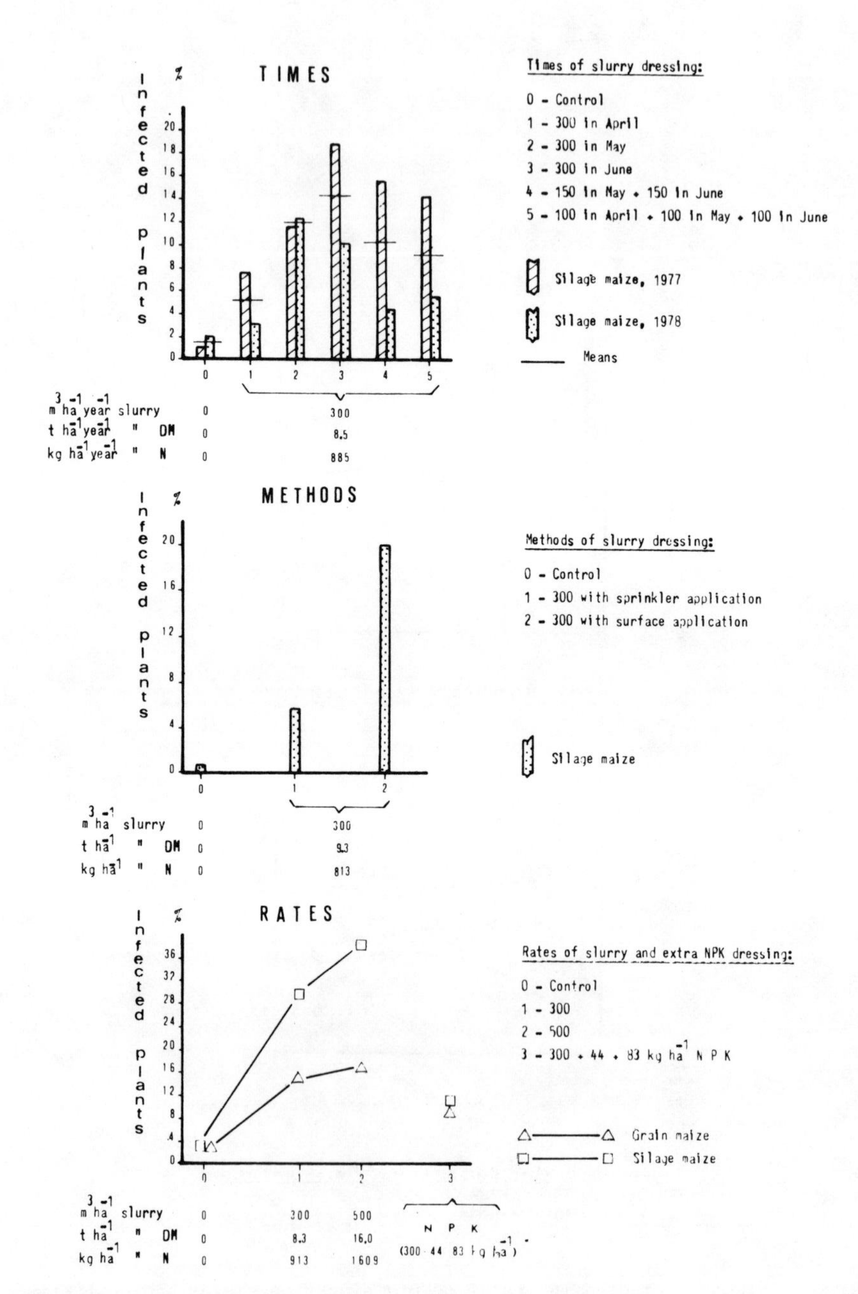

Fig. 9. Effects of different times, methods and rates of pig slurry dressings on the infection by *Ustilago maydis* in silage maize and grain maize. 1977 - 78.

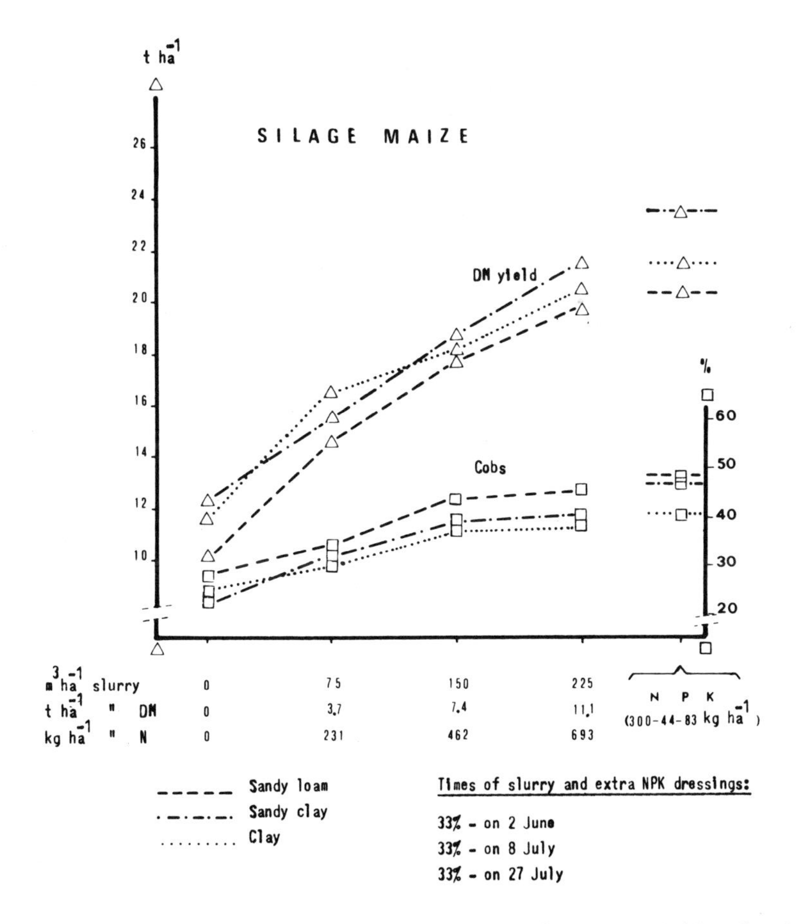

Fig. 10. Effects of increasing rates of cattle slurry on yield and cob percentage of silage maize, 1978.

a) Yields of wheats increased with slurry, up to the second application rate of 7.4 t ha^{-1} slurry DM, and decreased with higher rates;

b) plant density was unaffected by spring dressing, even when applied on young plants;

c) 1000-grain weight and hectolitre weight decreased over the second application rate, in both wheats;

d) yellow berry decreased with increasing rates, up to the second application rate.

Figure 9 shows the effect of various times, methods and rates of pig slurry application on the infection by *Ustilago maydis*;

a) the smut infection was light when the total amount of slurry was applied at sowing, higher when applied in May and the highest in June;

b) surface application produced infection three times higher than sprinkler;

c) the infection was more marked with increasing rates of application, more in silage maize than in grain maize.

B.3.2. Cattle slurry

Figure 10 shows DM yield of silage maize grown on different types of soil (sandy loam, sandy clay, clay) given increasing applications of cattle slurry.

a) on all soils DM yields increased significantly with increasing applications, up to a maximum of 11.1 t ha^{-1} slurry DM, and reached the highest values on plots given NPK fertiliser;

b) the percentage of cobs also increased markedly with increasing rates, mainly with 3.7 - 7.4 t ha^{-1} slurry DM, and was higher with NPK fertiliser.

DISCUSSION

A. Maton *(Belgium)*

What is the reason for the low BOD and COD values shown in Table 4?

P. Spallacci *(Italy)*

The values are low because of the dilution of the slurry giving DM content from 1.9 to 4 percent. The slurry was dilute because the pigs were fed on whey from a cheese factory and by washing water.

H. Vetter *(West Germany)*

Thank you Dr. Spallacci for your report.

SUMMARY OF SESSION 1

H. Vetter

In the EEC countries, animal husbandry is becoming concentrated in fewer and bigger units, which means that there are now regions with much larger amounts of animal manures than formerly.

Against this background, three questions have been raised which should be answered by the EEC research programme.

A. How much manure can be applied without harm to the vegetation?

B. How much manure can be applied without risk to the quality of water?

C. How much manure can be applied economically?

The answers to these questions lead to conclusions and recommendations.

Although our present state of knowledge does not allow these questions to be fully answered, I will summarise the results obtained and try to give answers to the following aspects:

1. Duration of effects with different rates of application
2. Soil conditions
3. Climatic conditions
4. Application times of the manure
5. Origin and composition of the manure
6. Types of crop.

A. HOW MUCH MANURE CAN BE APPLIED WITHOUT HARM TO THE VEGETATION?

1. Duration of effects with different rates of application

We have to distinguish between long and shorter term fertilising effects. With increasing amounts of manure applied over periods of 10 to 20 years, the nitrogen content of the manure will limit the dressings which can be used without harm to the plants. However, the application of large quantities of manure over many decades will be limited by the content of other nutrients such as phosphate, copper and calcium, or the organic matter.

Too much nitrogen leads to the following difficulties: with cereals, lodging and/or decreasing yield; with sugar beet, decreasing sugar content in the roots; with pototoes, decreasing starch content in the tubers; with forage plants, harmful nitrate content in the plants (Dam Kofoed; Vetter and Steffens); with grassland, scorching of the grass (Smilde); with forests, too rapid decomposition of the humus and litter (Smilde). In the long term, other nutrients can also have a negative influence on the growth of plants and plant quality, and their significance becomes evident when we look at the nutrient levels which are applied to the soil with specific quantities of nitrogen. Assumming that 240 kg N ha^{-1} are applied as cow slurry, pig slurry, poultry slurry and poultry manure, Table 1 gives the amounts of other nutrients that are applied.

Comparing these quantities with the nutrient uptake by plants in a standard rotation of crops, the amounts of P applied with pig and poultry slurry are 3.5 times more than the uptake; and the amounts of Cu, 16 times more. The nutrient content of slurries and the relation between amounts applied and plant uptake will vary somewhat, but in general, with pig and poultry manure, too much phosphate will be applied. Poultry manure also supplies excess Ca; pig manure too much Cu; and cow slurry too much K. The enrichment of the soil with P and Cu after only four years' application of 90 m^3 pig slurry ha^{-1} shows that with continued application, the P and Cu contents of the soil can reach critical levels. Furthermore, high organic loadings can produce unfavourable physical top soil conditions because the soil can become too porous.

TABLE 1

THE AMOUNT OF OTHER NUTRIENTS SUPPLIED BY SLURRY SUPPLYING 240 kg N ha^{-1} FOR THE REGION WESER-EMS AND THE AMOUNTS REMOVED IN CROPS

	N	P_2O_5	K_2O	CaO	MgO	Cu	t manure or slurry	DM %
	kg ha^{-1}							
Cow slurry	240	130	340	157	36	0.29	57	10
Pig slurry	240	214	131	160	29	1.1	37	10
Poultry slurry	240	225	115	372	21	0.21	25	10
Poultry manure	240	240	159	320	41	0.25	9	50
Removal on average of a crop rotation kg ha^{-1}	200	70	220	70	30	70 g		
Ratio applied: removed		(2-3.5)				(3-16)		
Results of Maton, Belgium								
Pig slurry	240	190	-	-	-	1.6	36	10

2. Soil

On heavier soils, more nutrients can be applied than on sandy soils.

3. Climatic conditions (no comments)

4. Time of application

The amounts of slurry applied can be increased with a longer time between the application of the slurry and the beginning of crop growth.

5. Origin and composition of the manure

The composition of slurries from different animals is also influenced by different feeding regimes. In Belgium and the Netherlands obviously the pig feed contains more Cu, than in Germany (Maton). With continued application, excessive amounts of both slurry and farmyard manure can be harmful to plants (Dam Kofoed).

6. Types of crop

The optimal amounts of slurry increase from least for cereals to most for grassland with root crops, potatoes to maize, being intermediate.

B. HOW MUCH MANURE CAN BE APPLIED WITHOUT RISK TO SURFACE AND GROUNDWATER?

1. Duration of effects with different rates of application

For the year of application nitrogen limits the amount of slurry that can be applied without risk to water quality. With continued application, nitrogen leaching into deeper soil layers increases with the number of dressings, as also does the amount of phosphate leached (Dam Kofoed; Spallacci; Vetter and Steffens). Particularly if large slurry dressings are used for two or three decades.

2. Soil

Nitrogen concentrations in the ground waters of heavy soils are similar, or sometimes a little higher, than in sandy soils, but the amount of nitrogen leached from heavier soils is generally less than from sandy soils, because there is less drainage water. More phosphate moves downwards into deeper soil layers on sandy soils than on heavier soils (Vetter and Steffens; Spallacci).

3. Climatic conditions

In wet regions, more N is leached than in drier regions with low precipitation rates; dry summers with high evapotranspiration rates lead to more leaching in autumn and winter than after wetter summers.

4. Time of application

The N content of the ground water is increased less when slurry is applied nearer to the start of crop growth (Vetter and Steffens). When the same amount of nitrogen is applied, more N is leached from organic manure than from fertiliser. With continued applications similar amounts of N are leached from farmyard manure and from slurry (Dam Kofoed). More N is leached from organic manures than from fertilisers, because mineralisation of organic-N cannot be calculated, and because the time of application is not always related to the time of nutrient demands by the plants.

5. Origin of the manure (no comments)

6. Types of crop

The risk of loss of nutrients by leaching is lessened when growing crops are present. For example, on grassland, where the crop is present all the year, less nutrients are lost by leaching than from arable land. The different authors obtained similar results, and over a prolonged period leaching from farmyard manure would be expected to be the same as from slurry. A very important point to note is that the order of magnitude is almost the same in all the experiments.

C. HOW MUCH MANURE CAN BE APPLIED ECONOMICALLY?

1. Duration of effects with different rates of application

Applying large amounts of organic manures causes the amount of nutrients used from the slurry to decrease from year to year, because of the residual effects of the earlier slurry dressings.

2. Soil

The nutrients in the slurry are utilised better on heavier soils than on sandy soils for the first few years.

3. Climatic conditions (no comments)

4. Time of application

The amounts of slurry required for crop production are smaller, the nearer the application time of the slurry to the start of crop growth.

5. Origin and composition of the manure

For the first few years, more N can be applied as farmyard manure than as slurry, although the amounts of available-N as ammonium are the same for farmyard manure and slurry (Dam Kofoed). With application for several decades, the K_2O content of both cattle slurry and farmyard manure can limit the optimal rates. The same is true for P and Cu for pig slurry, and for P and Ca for poultry manure and slurry.

6. Types of crop

Least N is required for cereals, increases for potatoes and sugar beet, and most is need for grassland and maize.

D. CONCLUSIONS

1. Water quality is not adversely affected when slurry is spread at optimal rates for crops at the best times.
2. In all the countries of the EEC the total nutrient requirements of crops are greater than amounts produced as animal manures. In West Germany two and one-half times more nutrients in the form of animal manures could be applied for crop production.
3. The amounts of organic manure giving optimal crop yields are also economically viable. The value of the nutrients in the animal manures is sufficient to make storage

profitable, allowing slurry to be spread at the best time. Also transport is justified over some distance from regions with overproduction to regions requiring nutrients.

4. The most important reasons why slurry is not spread at optimal application times are:

   a) insufficient storage capacity

   b) the lack of slurry tankers able to distribute slurry on to the land evenly and exactly

   c) too much damage to the soil by the big wheels of heavy slurry tankers

   d) odour emission during slurry spreading.

The most important objectives for further scientific research are:

1. The determination of the limits for long term applications of slurry, which requires the continuation of the existing field and lysimeter trials.

2. To examine the influence of different times of slurry application on the composition of plants and on the leaching of nutrients to the ground water or the drainage water.

3. The development of slurry tankers which allow even distribution of slurry to land, at exact rates and with less less damage by the wheels of the tankers.

4. The determination of the fertilising value of slurry treated either aerobically or anaerobically.

RECOMMENDATIONS FOR ADMINISTRATIVE DECISIONS

The more intensive concentration of animal holdings in particular regions gives less incentive to transport slurry to regions requiring nutrients, because of transport costs which have to be set against the value of the slurry as fertiliser.

At a time when there is serious concern about the lack of raw materials and shortage of phosphate, the waste of nutrient phosphate, as happens in many regions with intensive husbandry is not responsible behaviour. Similarly, nitrogen should also be used to best advantage.

In other economic areas the trend is to re-cycle waste and the same must apply to slurry by finding ways for optimal use while preventing environmental problems.

GENERAL DISCUSSION

There was general support for the conclusions formulated by the Chairman for the attention of the Commission. The following specific points were raised in discussion.

A. Dam Kofoed *(Denmark)*

The three questions should be integrated and damage to the soil possibly added as a fourth one, in order to provide some specific recommendations as guidelines for farmers.

J.K.R. Gasser *(UK)*

This last point is relevant to an earlier comment when the evidence presented suggested that the time which is most convenient for the farmer to apply slurry does not allow maximum protection of the environment, nor does it allow the best use of the nutrients. There are two opposing requirements which have to be reconciled and this is a topic which should receive much attention in the future.

A. Aumaitre *(France)*

Attention should be drawn to the effects of aeration on the N content of slurry, which causes the loss of half of the nitrogen content of the manure and adversely affects the balance of nutrients for crop growth. The same is true for copper. The transport of pig slurry was unlikely to be economic over distances greater than 20 km.

H. Vetter *(West Germany)*

Now I must help Dr. Aumaitre; that is only true for phosphate.

SESSION II

THE USE OF SLURRY FOR GRASSLAND AND FORAGE CROPS

Chairman: H. Tunney

COMPARISON OF THE LEACHING PATTERNS OF NUTRIENT ELEMENTS FROM MINERAL FERTILISERS AND LIQUID MANURE*

F. Van de Maele and A. Cottenie
Laboratorium voor Analytische en Agrochemie,
Faculteit van de Landbouwwetenschappen,
Rijksuniversiteit Gent, Gent, Belgium.

1. INTRODUCTION

In Western Europe, a number of soils have received very heavy fertiliser dressings during recent years, especially as a consequence of the intensification of livestock raising, often practised in circumstances of land shortage (Van de Maele, 1975). This results in a general overloading of NPK, causing nutrient element accumulation in the upper layers or losses in surface or ground water.

Percolation experiments were carried out with undisturbed soil cores in order to compare the incorporation and leaching patterns of nutrients after applications of liquid manure and mineral fertilisers.

2. MATERIAL AND METHODS

Plastic percolation tubes of 60 cm height and 19.5 cm diameter were tapped into the soil to a depth of 50 cm. After an equilibration period of 4 weeks, the tubes were dug out and transported to the laboratory.

Chemical soil characteristics are given per 10 cm layer in Table 1. Beneath the plough layer (30 - 50 cm) an accumulation zone of NO_3-N was noticed and the salt content in this zone was also higher.

*Research subsidised by I.W.O.N.L. (Instituut voor Aanmoediging van het Wetenschappelijk Onderzoek in Nijverheid en Landbouw) Brussels.

The three following treatments were applied (4 replications):

1. Control
2. 100 tons of pig slurry ha^{-1}, containing 1096 kg total nitrogen were mixed with the 10 cm upper layer.
3. Mineral fertilisation with the same K, P and Mg amounts: 114 kg P_2O_5, 435 kg K_2O and 316 kg MgO ha^{-1} were mixed with the upper 10 cm layer.

TABLE 1

CHEMICAL CHARACTERISTICS OF THE DIFFERENT SOIL LAYERS (EXTRACTION WITH 0.5 n Am-Ac + 0.02 M EDTA (pH 4.65)

Depth, cm	0 - 10	10 - 20	20 - 30	30 - 40	40 - 50
pH H_2O	6.32	5.90	5.20	4.85	4.90
pH KCl	5.02	4.42	3.62	3.45	3.62
conductivity in mmho/cm	0.70	0.78	0.78	0.72	1.35
Na mg/100 g	1.6	1.8	1.8	1.4	1.6
K "	15.6	20.7	14.8	10.9	8.6
Ca "	96.0	75.0	18.6	12.8	11.4
Mg "	6.1	6.1	3.2	2.2	2.2
P_2O_5 mg/100g	44.2	40.0	22.5	19.3	10.5
NO_3-N "	0.78	0.78	0.89	1.05	2.40
Cl "	1.45	1.45	2.61	1.45	1.45
CEC meq/100g	5.00	5.00	4.00	3.50	3.00
% C	1.01	0.89	0.55	0.52	0.46

Furthermore, a nitrogen dressing of 200 kg N ha^{-1} (as NH_4NO_3) was given to each tube receiving mineral fertilisers. During 28 days, the soils were leached daily with a total quantity of water corresponding to about 400 mm rainfall, i.e. twice a day, 200 ml of distilled water was given to each tube. Before each addition of water, the quantity percolated as a result of the preceding addition was collected and measured in order to prepare a mixed sample.

3. RESULTS AND DISCUSSION

3.1. The percolates

The changes in some chemical characteristics of the percolates as a function of the volume of percolate is given in Figure 1. These curves show a leaching pattern of several nutrient elements depending on their origin.

The following general tendency was noticed: the highest concentrations of nutrient elements, except P_2O_5, were found in the first percolates while an equilibrium concentration was reached after percolation of 6 l water.

Taking into account that the same quantities of P, K and Mg were added to the tubes as mineral and organic fertilisation, the following phenomena were observed.

The pH of the percolates from the soil cores treated with mineral fertilisers decreased rapidly but became equal to the value of the control before the end of the experiment. The pH minimum corresponded with a maximum of electrical conductivity (soluble salt content). Earlier investigations (Cottenie and Van de Maele, 1976) showed that the electrical conductivity of the saturation extract of the soil is a practical method for diagnosing the general effect of excessive use of livestock effluents and for indicating whether more detailed analysis is required.

K, Ca and Mg followed the same leaching pattern after mineral fertilisation: their concentrations in the percolates were highest between 2 and 4 l of percolate. After fertilisation with liquid manure their concentrations in the percolation solutions showed little variation. This means that the leaching of nutrients originating from mineral fertilisers was more pronounced than from liquid manure.

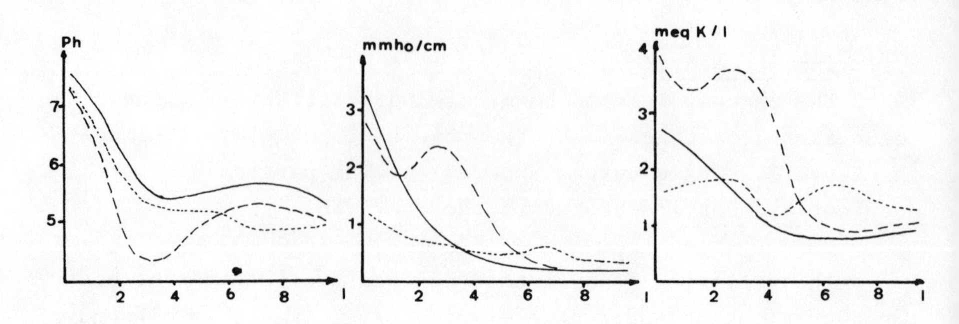

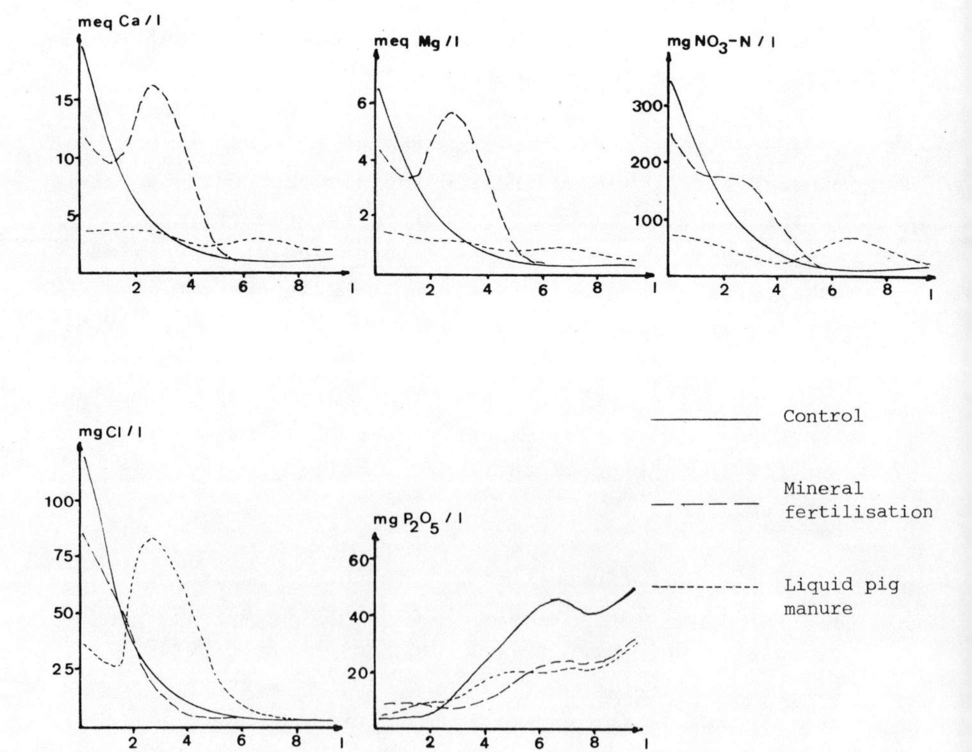

Fig. 1. Changes in some chemical characteristics of the percolates as a function of the volume leached.

In spite of the fact that the quantity of nitrogen added as liquid manure was more than the 200 kg N in mineral form (treatment No.3), much more nitrate percolated from the mineral treated tubes during the test. The organic manure nitrogen caused only a slight nitrate increase in the corresponding percolates during the second part of the test. The cumulative percolation curves show that more nitrate leaching can be expected in the latter case (Figure 2).

Analysis of the saturation extract of the 40 - 50 cm soil layers after the percolation experiment confirmed the analytical results of the last percolates (Figure 3).

Much chloride was leached after treatment with organic manure (Figure 1).

3.2. The percolated soil

After percolation, each 10 cm layer of soil was analysed in order to determine the quantity of available nutrients (0.5 N NH_4Ac + 0.02 N EDTA, pH 4.65 extraction). Knowing the original and residual quantities of each available nutrient, the amount added before percolation, and the amount leached during the test, a balance-sheet was drawn up for each treatment (see Table 2 and Figure 4).

In each diagram the first block represents the originally present and the added amount of the element under consideration. The second block shows the retained and leached quantities. Comparison of both indicates whether retention or release has taken place during the experiment.

In all cases K was retained, as well as Mg and Cl after organic manuring. From Table 2 one can conclude as follows:

Potassium: The leaching of this element increased slightly after treatment with liquid manure, while addition of the same quantity of K as a mineral fertiliser almost doubled

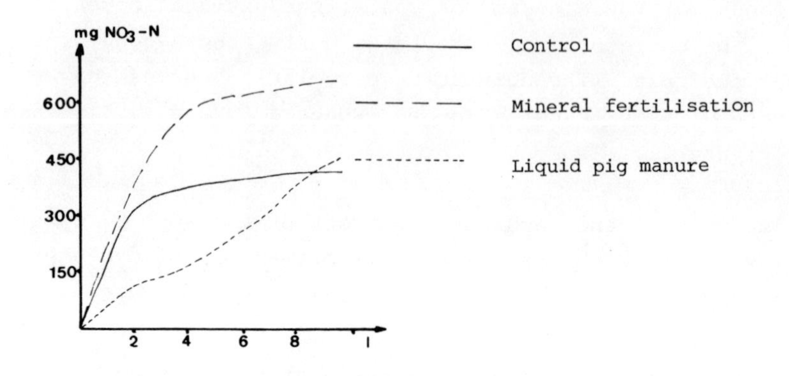

Fig. 2. Cumulative curve of the total quantity of NO_3-N leached in function of the 3 treatments.

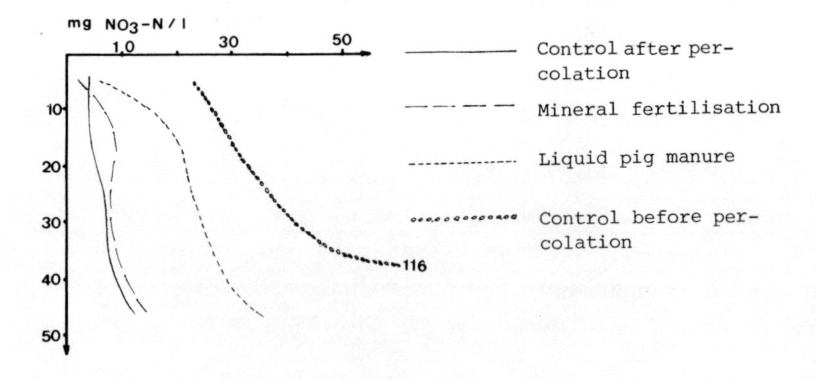

Fig. 3. NO_3-N content of the saturation extracts of the soil profile in function of depth and treatment.

TABLE 2

BALANCE-SHEET OF THE PROFILE

Treatment		a	b	c	d	e
K	B	3170.5	0	2680.1	426.3	+ 64.1
	F	3170.5	509.4	2767.6	760	+ 152.3
	F-B			87.5	333.7	
	M	3170.5	509.4	2802.7	545.5	+ 331.7
	M-B			122.6	119.2	
Mg	B	882.5	0	884.4	101.0	- 102.6
	F	882.5	264	1029.6	235.7	- 118.6
	F-B			145.2	134.7	
	M	882.5	268	958.8	103.5	+ 88.2
	M-B			74.4	2.5	
Ca	B	9399	0	11370.2	555.6	- 2526.6
	F	9399	44.9	11603.2	966.6	- 3125.7
	F-B			233.-	411	
	M	9399	908.0	11424.-	445	- 1562
	M-B			53.8	- 110.6	
P_2O_5	B	6117.4	0	6854.4	272.-	- 1009
	F	6117.4	158	7835.5	162.-	- 1722.1
	F-B			981.1	- 110.-	
	M	6117.4	161	8167.-	162.7	- 2051.3
	M-B			1312.6	- 109.3	
Cl	B	376.8	0	283.6	136.1	- 42.9
	F	376.8	0	320.3	121.7	- 65.2
	F-B			36.7	- 14.4	
	M	376.8	220.6	306.9	251.9	+ 38.
	M-B			23.3	115.8	
NO_3 (N)	B	264.3	0	62.3	415.2	- 213.2
	F	264.3	186.7	71.2	642.4	- 262.6
	F-B			8.9	227.2	
	M	264.3	15.0	164.9	416.8	- 302.4
	M-B			102.6	1.6	

B : control (blanc)
F : mineral fertilisation
F-B : treatment F minus control (blanc)
M : liquid pig manure
M-B : treatment M minus control (blanc)

a : mg available per tube before percolation
b : mg added per tube
c : mg available per tube after percolation
d : mg percolated per tube
e : retention (+) or release (-) : $\{(a + b) - (c + d)\}$

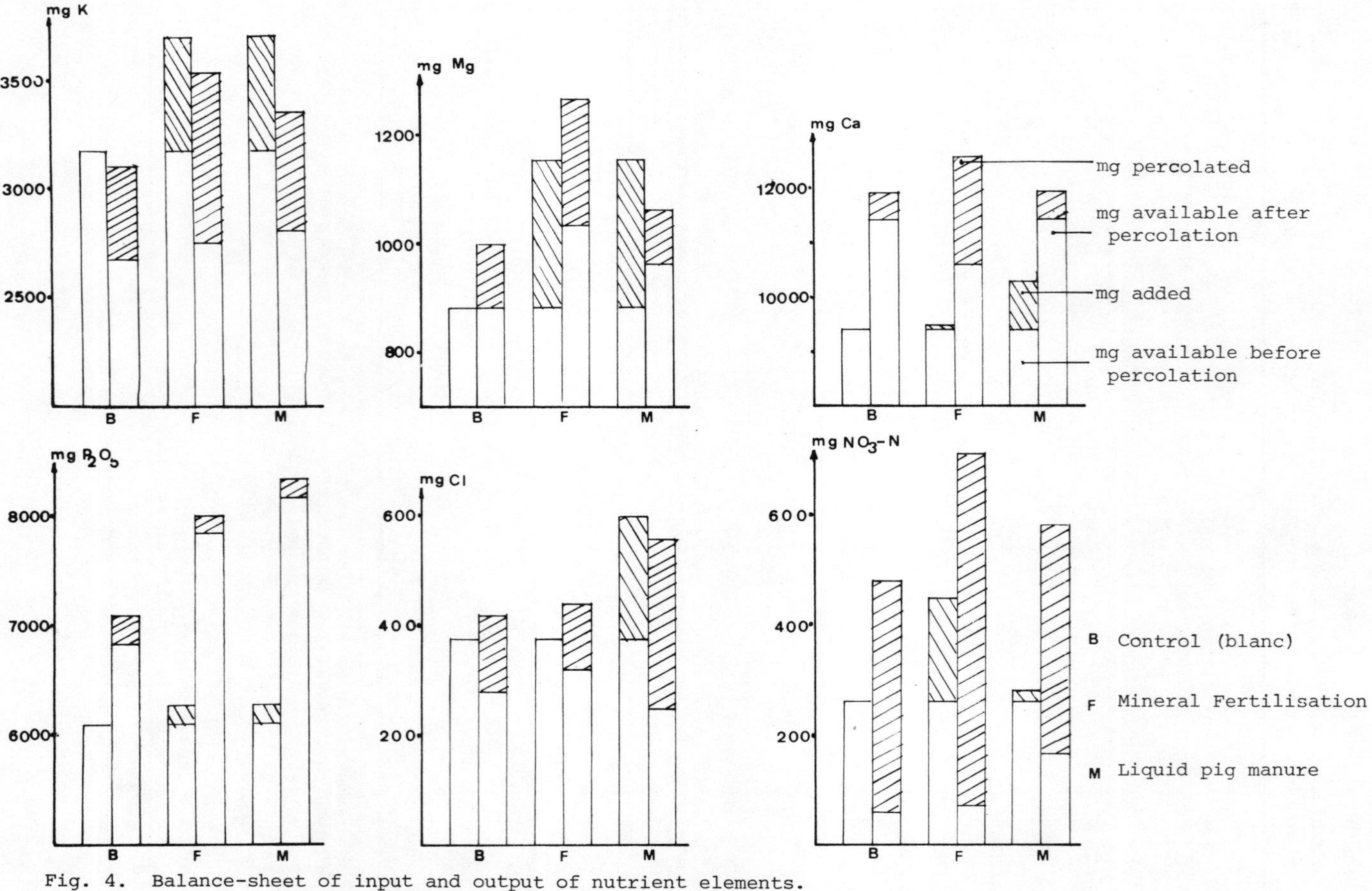

Fig. 4. Balance-sheet of input and output of nutrient elements.

the quantity leached in comparison with the control. After the three treatments, no clear difference in available K in the soil was seen.

The retention efficiency (%) was calculated as =(quantity retained/quantity added) x 100. After correction for the control 34% retention was found in the soil with mineral fertilisation and 76% in the soil with organic manure.

Magnesium: The retention efficiency of Mg reached values of 49 and 99%. This means that all Mg added in the organic form was retained in the upper 50 cm of the soil.

Calcium: Ca added in liquid manure was much higher than after mineral treatment, but the quantity leached was only half of that after mineral fertilisation.

Phosphorus: No difference was noticed for the amounts of P_2O_5 leached after the two treatments. The control released more P_2O_5 than the other tubes.

Chlorides: The remaining chloride contents of the soils after percolation were almost equal in all treatments, but the leached quantities after organic fertilisation were higher so that soil water enrichment with chloride can be expected.

Nitrates: After percolation, the nitrate content in the soil treated with liquid manure was twice as high as in the control. However, the percolated quantities in both treatments were comparable. This results from the favourable conditions in which the test took place is also related to the soil used for the experiment (last fertilisation: 650 kg NH_4NO_3, 26% and 40 tonnes of liquid pig manure ha^{-1}).

For a total N input of 1546 mg $tube^{-1}$ after organic manuring and of 282 mg after mineral fertilisation, about 80 percent of the latter leached out, while no increased nitrate percolation took place after organic manuring in comparison to the control.

4. SUMMARY AND CONCLUSION

By means of percolation tests with undisturbed soil cores treated with liquid pig manure or with an equivalent amount of mineral fertiliser, the leaching patterns of nutrient elements from mineral and organic origin were compared.

Chemical analysis of the percolating liquid showed that K, Ca, Mg and NO_3-N leached faster and to a greater extent after mineral fertilisation in comparison with organic manuring.

A balance-sheet of nutrients input and output indicated that retention of K also took place in all treatments and this was highest after organic manuring, while Ca, Mg and NO_3-N were not retained.

The available quantities of nitrates and to a certain extent also phosphates, were higher after percolation when liquid pig manure was applied than after mineral fertilisation.

Finally, chloride enrichment of the percolates after treatment with organic manure was rather important.

REFERENCES

Cottenie, A. and Van de Maele, F., 1976. Soil, water, plant relationship as influenced by intensive use of effluents from livestock. EEC Seminar on the Utilisation of Manure for Landspreading, Modena. Ed. J.H. Voorburg, EUR 5672e, p 225-246.

Van de Maele, F. 1975. Fumures pratiqués dans les exploitations d'élevage intensif. Revue de l'Agriculture No.6, novembre-décembre, p 1487-1494.

DISCUSSION

J.H. Voorburg *(The Netherlands)*

You were leaching for 28 days with 400 mm of simulated rain, that is almost 15 mm daily which is far more than normal. Were your samples aerobic or anaerobic with the possibility of denitrification; and was ammonium leached?

F. Van der Maele *(Belgium)*

Little ammonium was leached. The tubes received 200 ml of water twice a day, in the morning and the evening. In the early stage the water percolated slowly and by the following percolation almost all the water that was given before had leached. Some anaerobic conditions may have been created by this, with the possibility of denitrification because the room temperature was 25°C.

H. Vetter *(West Germany)*

Your slurry contained ten times as much nitrogen as phosphate. Was this pig slurry or a mixture with cow slurry, or was the slurry taken from the top of the tank without mixing?

F. Van der Maele

The slurry applied was pure pig slurry which was mixed before taking the sample from the pit. For sampling we use a small beaker with a heavy base which sinks to the bottom allowing a mixed sample to be obtained.

R. de Borger *(Belgium)*

The results from research depend very much on the prevailing conditions. We analysed the water from lysimeters during the season 1977 - 1978 and we had the opposite results. The leaching of calcium, magnesium and chloride was increased by applying slurry instead of mineral fertilisers. The lysimeters were uncropped and fertiliser potassium was applied as the chloride.

P. Sequi *(Italy)*

You compared the addition, to the soil, of solid fertilisers with liquid manure, equivalent to 100 t ha^{-1}, which corresponds to about 100 mm of rain. Do you think that, at the beginning of your incubation time, this could cause percolation? Do you think it is comparable with the other results?

F. Van der Maele

First, 100 t ha^{-1} of pig slurry is a commonly used dressing. Second, the liquid manure was put on the surface and then it was mixed with the top 10 cm of soil so you have absorption of the water in it. There was no percolation to the next layer of soil.

Th. M. Lexmond *(The Netherlands)*

Would you comment on the very high concentrations of phosphate in your leachates. Also leachates from the control treatment contained large concentrations of phosphate, could the large amounts of demineralised water have dispersed the organic matter?

F. Van der Maele

We had previously analysed soils coming from bio-- industrial farms and the saturation extract of these soils - the soil-water contained much phosphate. This soil is being used, and has been for a long time, as an acceptor of liquid manure. Therefore I think it originates from the background. The question of dispersion of soil organic matter was not studied.

EFFLUENTS FROM INTENSIVE LIVESTOCK UNITS
FERTILISER EQUIVALENTS OF CATTLE SLURRY FOR GRASS AND FORAGE MAIZE

B.F. Pain and Lesley T. Sanders
National Institute for Research in Dairying,
Shinfield, Reading, RG2 9AT, UK.

EXPERIMENTAL

The experiment had plots without treatment and with two rates of cow slurry and ammonium nitrate with and without PK fertiliser applied to grass and forage maize plots in a factorial arrangement. The total amounts of slurry and fertiliser were applied to grass in three parts during each season and the grass was harvested in late May, July and September. The treatments were applied to the soil for maize shortly before sowing in late April - May and the crop was harvested for silage in October.

The fate of applied nitrogen was investigated by measuring:

(i) the amount removed in the harvested crops,

(ii) loss of N from freshly applied slurry by ammonia volatilisation, using boric acid to absorb free ammonia,

(iii) nitrate concentrations in the soil profile by extracting NO_3-N in 10 or 15 cm subsections to 60 cm depth followed by analysis,

(iv) evolution of nitrous oxide as an indicator of denitrification using gas chromatography to analyse the gases. The soil atmosphere under grass and maize was sampled by installing diffusion equilibrium reservoirs at different depths. In addition, the flux of nitrous oxide through the soil surface was measured by trapping gases evolved from small grass plots.

Slurry and fertiliser treatments were applied either for one year only or for two consecutive years. The 'residual' effects of the previous year's treatments were measured when slurry and fertiliser were not applied.

The experiments commenced in 1976 but grass failed because of prolonged drought, necessitating the establishment of a second series of grass and maize plots in 1977 on an adjacent field; hence data for maize only is available for 1976.

Other experiments were carried out on grassland, using the liquid fraction produced by mechanically separating cow slurry together with liquid formulation N fertiliser.

The soil in the field experiments was a surface-water gley with a clay-loam texture. Soil to a depth of 15 cm at the beginning of the experiments in 1977 had pH 6.7, 0.21% total N, 3.9% organic matter and 68, 235, mg ℓ^{-1} of P and K respectively. Rainfall was 560 mm in 1976, 659 mm in 1977 and 606 mm in 1978.

EFFECTS ON CROP YIELDS

Grass

In 1977, total DM yields were increased by ammonium nitrate at each rate of slurry applied, but by slurry only when no N fertiliser was applied (Table 1 a). Re-applying the treatments in 1978 gave similar results (Table 1 b) although slurry also increased yields slightly when ammonium nitrate was applied at 150 kg N ha^{-1}. Averaged over the season, slurry was 25 to 30 per cent as efficient as ammonium nitrate on an equal N basis for increasing grass yields in each year, but relative efficiencies differed at each cut. For example, in 1978 slurry applied in March was 50 percent as efficient as N fertiliser in increasing yields at the first harvest in late May but only 15 to 20 percent when re-applied in June or July for later cuts. In 1978, Table 1 c shows the residual effects of both

TABLE 1

YIELDS OF DRY MATTER OF GRASS (kg ha^{-1}) IN 1977 AND 1978

a) 1977

Rate of slurry (kgN ha^{-1})	Rate of inorganic N (kgN ha^{-1}) 0	150	300	Linear effect of N. (kg^{-1} ha^{-1}) (S.E. ± 1.71)
	Yield (kg ha^{-1})			
0	11827	15259	16415	15.3***
210	13038	15815	16603	11.9***
420	13236	15328	15865	8.6***
Linear effect of slurry (kg^{-1} N ha^{-1}) (S.E. ± 1.22)	3.4**	0.2	-1.3	

b) 1978 Slurry and fertiliser applied in 1977 and 1978

Rate of slurry (kgN ha^{-1})	Rate of inorganic N (kgN ha^{-1}) 0	150	300	Linear effect of N (kg^{-1} ha^{-1}) (S.E. ± 1.60)
	Yield (kg ha^{-1})			
0	3270	8855	10336	23.6***
174	5098	9967	10302	17.3***
348	6994	9946	10311	11.1***
Linear effect of slurry (kg^{-1} N ha^{-1}) (S.E. ± 1.38)	10.7***	3.1*	-0.1	

c) 1978 Slurry and fertiliser applied in 1977 only

Rate of slurry (kgN ha^{-1})	Rate of inorganic N (kgN ha^{-1}) 0	150	300	Linear effect of N (kg^{-1} ha^{-1}) (S.E. ± 1.18)
	Yield (kg ha^{-1})			
0	2938	4236	5144	7.4***
210	3784	4776	5306	5.1***
420	4355	5277	6157	6.0***
Linear effect of slurry (kg^{-1} N ha^{-1}) (S.E. ± 0.84)	3.4***	2.5**	2.4**	

Note i) Levels of significance: *, $P < 0.05$; **, $P < 0.01$: ***, $P < 0.001$

ii) Standard errors have 43 d.f.

the slurry and ammonium nitrate applied in 1977, which were greater for fertiliser than for slurry and occurred at the first and second cuts.

PK fertiliser did not increase yields of dry matter, either with or without slurry application.

Forage maize

Treatments did not affect maize yields during the drought conditions of 1976, but high rates of slurry and fertiliser decreased plant density. The results for maize in 1977 and 1978 were similar to those obtained for grass. Ammonium nitrate increased yields of dry matter at each rate of slurry application but slurry only in the absence of N fertiliser (Tables 2 a and 2 b). The residual effects were significant for ammonium nitrate only (Table 2 c). In both years, the best yields were obtained with combinations of slurry and fertiliser. The relationship between DM yield of ears and treatment were similar to that for whole crop yield. Slurry and fertiliser did not influence DM or ears as a proportion of the whole crop. PK fertiliser did not affect yield.

EFFECTS ON CROP COMPOSITION

Values in Table 3 shows that slurry and fertiliser treatments affected the nutrient composition of grass more than that of maize. Ammonium nitrate increased the N, P, K, Mg and Ca content of grass more than slurry, although grass generally contained most percentage of nutrients when high rates of slurry and N fertiliser were applied. In contrast to grass, increasing the amounts of slurry or ammonium nitrate for maize tended to decrease the P percent and K percent in the whole crop; the crop had the largest percentages with no treatment.

TABLE 2

YIELDS OF DRY MATTER OF FORAGE MAIZE (kg ha^{-1}) IN 1977 AND 1978

a) 1977

Rate of slurry (kgN ha^{-1})	Rate of inorganic N (kgN ha^{-1}) 0	60	120	Linear effect of N (kg^{-1} ha^{-1}) (S.E. ± 6.39)
	Yield (kg ha^{-1})			
0	6289	8526	9157	23.9***
138	8135	8244	10070	16.1*
276	7700	9055	9414	14.3*
Linear effect of slurry (kg^{-1} N ha^{-1}) (S.E. ± 2.78)	5.1	1.9	0.9	

b) 1978 Slurry and fertiliser applied in 1977 and 1978

Rate of slurry (kgN ha^{-1})	Rate of inorganic N (kgN ha^{-1}) 0	60	120	Linear effect of N (kg^{-1} ha^{-1}) (S.E. ± 7.42
	Yield (kg ha^{-1})			
0	6161	11266	13024	57.2***
87	8435	10038	15070	55.3***
174	8836	11887	13500	38.9***
Linear effect of slurry (kg^{-1} N ha^{-1}) (S.E. ± 5.12)	15.4**	3.6	2.7	

c) 1978 Slurry and fertiliser applied in 1977 only

Rate of slurry (kgN ha^{-1})	Rate of inorganic N (kgN ha^{-1}) 0	60	120	Linear effect of N (kg^{-1} ha^{-1}) (S.E. ± 6.03)
	Yield (kg ha^{-1})			
0	5432	5762	7074	13.7*
138	5700	5966	8119	20.2**
276	6293	6814	7027	6.1
Linear effect of slurry (kg^{-1} N ha^{-1}) (S.E. ± 2.62)	3.1	3.8	-0.2	

Note i) Levels of significance: *, $P < 0.05$; **, $P < 0.01$; ***, $P < 0.001$

ii) Standarderrors have 43 d.f.

TABLE 3

PERCENTAGE NUTRIENTS IN GRASS AND FORAGE MAIZE IN 1978

% in DM	Grass					Maize				
		Rate of slurry (m^3 ha^{-1})[a]		Rate of Inorganic N (kgN ha^{-1})			Rate of slurry (m^3 ha^{-1})[b]		Rate of Inorganic N (kgN ha^{-1})	
	Nil	50	100	150	300	Nil	30	60	60	120
N	1.11	1.31	1.51	1.70	2.10	1.12	1.10	1.26	1.14	1.22
P	0.26	0.30	0.32	0.33	0.35	0.27	0.23	0.25	0.22	0.21
K	2.53	2.84	3.07	3.15	3.32	1.33	1.27	1.25	1.26	1.28
Mg	0.12	0.14	0.14	0.15	0.18	1.44	1.31	1.29	1.32	1.13
Ca	0.44	0.51	0.51	0.53	0.59	0.25	0.22	0.26	0.24	0.24

Note a 10 m^3 contained 35, 7 and 25 kg of N, P and K respectively.

b 10 m^3 contained 33, 6 and 17 kg of N, P and K respectively.

APPARENT RECOVERY OF APPLIED NUTRIENTS IN THE HARVESTED CROPS

In 1976 the apparent recovery of applied N in the maize crop was very low (less than 10 percent) but in the second series of experiments in 1977 averaged 51 percent for ammonium nitrate N and 17 percent for slurry N. The corresponding figures for grass were 75 and 16 percent respectively. The apparent recovery of slurry P and K averaged 20 and 44 percent respectively for grass and 21 and 10 percent for maize. Analyses of the 1978 crops showed that the pattern of results was similar to the previous year.

AMMONIA VOLATILISATION

Most ammonia was volatilised from the surface during the first 3 days following slurry application and losses were negligible after 10 days. Maximum losses (up to 5 kgN ha^{-1}) were after spreading slurry on grass during the summer. In most instances, less than one percent of slurry-N was lost in this way.

NITRATE CONCENTRATIONS IN THE SOIL PROFILE

Figure 1 illustrates how the concentrations of nitrates in the soil profile varied greatly from year to year as well as with treatment. The poor response of the 1976 maize crop to applied N was due to high soil nitrate concentrations. The amounts of nitrate leached from maize plots during the winter of 1976 - 1977 ranged between 60 kgN ha^{-1} (no treatment) up to about 304 kgN ha^{-1} from soil with ammonium nitrate supplying 120 kgN ha^{-1}+ slurry N supplying 192 kgN ha^{-1}. Concentrations were generally much smaller between 1977 and 1978 in the second series of experiments. As in 1976, the highest values were recorded where large amounts of both slurry and ammonium nitrate were applied together. In spite of the low apparent recovery of slurry N in the crops, more N appeared to be lost by leaching as nitrate from ammonium nitrate than from slurry. Soil nitrate

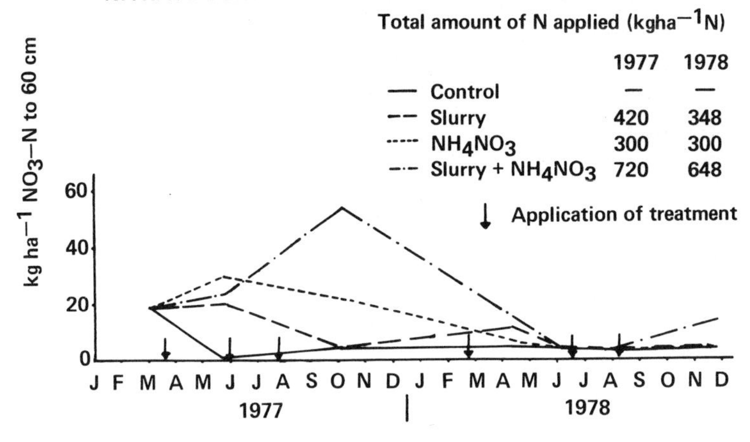

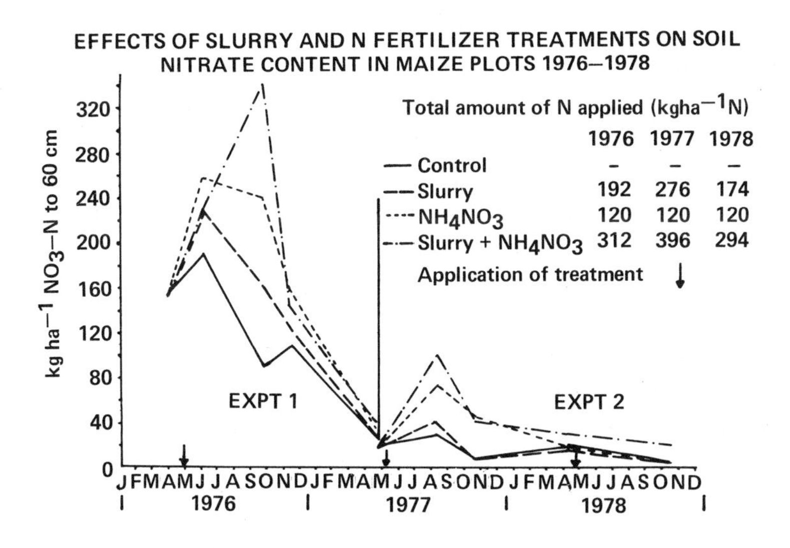

Fig. 1.

Effects of slurry and N fertiliser treatments on soil nitrate content in grass and maize plots 1976 - 1978

concentrations under grass tended to be smaller than those under maize. The very low values for all treatments under both crops in November 1978 suggested that the potential for winter leaching was minimal.

DENITRIFICATION

Rates of nitrous oxide production under grass and maize

Nitrous oxide was evolved sporadically throughout the year and similar amounts were produced with slurry as with ammonium nitrate under the conditions of the field experiments. Nitrous oxide was produced even at low soil temperatures and low soil nitrate concentrations but production was enhanced by wet soils. The largest concentrations were generally detected where the largest amounts of N were applied as slurry with N fertiliser.

Nitrous oxide fluxes from small grass plots

The technique used in the above experiments allowed rates of nitrous oxide production to be compared but did not enable the losses from the soil to be quantified. Other experiments were designed to measure the flux of nitrous oxide through the soil surface by trapping gases evolved from small plots of grass and Figure 2 shows the fluxes from ammonium nitrate and slurry applied in August. Little was generally lost but more from ammonium nitrate than from slurry. Furthermore, nitrous oxide production commenced more rapidly after applying N fertiliser than after slurry. Maximum fluxes were recorded in mid-summer when it was estimated that 2.8 percent of the N applied as ammonium nitrate was lost as nitrous oxide and 0.44 percent of slurry N over a period of 10 days. These values were equivalent to 2 and 0.5 kgN ha^{-1} respectively.

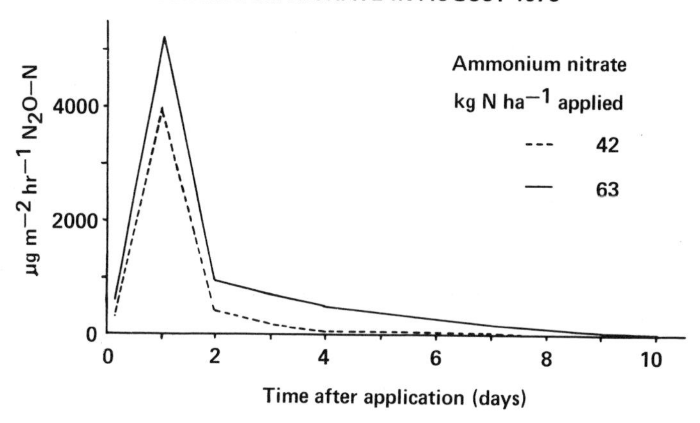

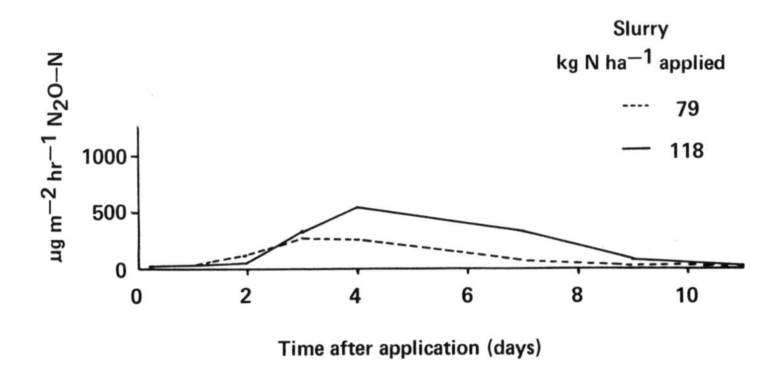

Fig. 2

Loss of N as N_2O from grass plots treated with either ammonium nitrate or cow slurry in August 1978.

FERTILISER EQUIVALENT OF MECHANICALLY SEPARATED SLURRY

The liquid fraction produced by mechanically separating cow slurry was applied to grassland at 3 rates together with 3 rates of liquid formulation N fertiliser in a factorial arrangement. Calculated volumes of N fertiliser were added to the separated slurry before application to the land.

Averaged over the season, separated slurry was up to 45 percent as efficient as fertiliser N in increasing yields of grass when applied with or without additional N. For example, applying a mixture of N fertiliser equivalent to 50 kgN ha^{-1} and separated slurry at 40 000 l ha^{-1} containing 100 kg total-N, gave yields similar to those obtained with 100 kg ha^{-1} N fertiliser.

Applying separated slurry with added N fertiliser to areas of grass of 0.2 ha with a 10 m wide boom mounted on the tanker gave similar increases in yields to those obtained from small plots.

CONCLUSIONS

Crop yields clearly responded little to cow slurry from 1976 - 1978 and only a small proportion of slurry N was utilised by the crops. In spite of the poor apparent recovery of slurry N in grass and maize during the first season of application, the 'residual' effects of slurry N in the second year were smaller than those obtained from ammonium nitrate. Furthermore, there was no evidence that losses of N from the slurry applications by nitrate leaching or as nitrous oxide were greater than those from ammonium nitrate. Large amounts of slurry N were unaccounted for, especially where the treatments were applied for two consecutive years. Possibly, a large proportion of the slurry N is immobilised in the soil, and will be released to form nitrate under suitable conditions. For example, large amounts of nitrates were leached from maize plots during a wet autumn following an exceptionally hot, dry summer in 1976.

Separated liquid slurry was a more effective fertiliser for grassland than unseparated material. Preliminary results suggest that it may be feasible to produce a more balanced fertiliser material by fortifying the separated slurry with artificial fertiliser before application to the land.

DISCUSSION

H. Vetter *(West Germany)*

Why are yields increased more with separated slurry?

B.F. Pain *(UK)*

We can only speculate at the moment. There is some evidence to suggest that a greater proportion of the nitrogen in separated slurry is in a plant available form because, during the separation process, water soluble compounds are squeezed out. Also separated liquid does not smother or shade or cause mechanical damage to the sward.

M. Sherwood *(Ireland)*

With no added slurry and no fertiliser in 1977, about 12.0 t ha^{-1} of dry matter were produced. In 1978 it was only about 3.0 t ha^{-1}. Had the site been fallowed for a year?

B.F. Pain

No. We started on a fairly fertile site, and by the second year it was obviously mineralising less soil-N. It was the same plot, and with no fertiliser for two years, the fertility decreased. Secondly, the sward deteriorated over that period as well. It did not contain clover.

J. Brogan *(Ireland)*

Would you comment a little more on the applicator for the separated liquid. It seemed to be a low-cost machine. Could you apply more quickly than with the untreated slurry? Would there be an economic advantage in the rate of spreading?

B.F. Pain

Yes, I think there was an economic advantage. The first aim in producing that machine was to provide a tool to enable separated liquid to be applied to large scale experiments. At the same time it demonstrated how easily, accurately and evenly, separated liquid could be applied to grassland, compared with unseparated slurry which would obviously block such a machine. Using separated slurry offers a potential for reducing odour because the boom can be mounted very close to the ground and decrease spray and wind-drift during application compared with conventional slurry spreaders.

EFFECTS OF LANDSPREAD ANIMAL MANURES ON THE FAUNA OF GRASSLAND

J.P. Curry, D.C.F. Cotton, T. Bolger and V. O'Brien*
Department of Agricultural Biology,
University College, Dublin, Ireland

INTRODUCTION

The principal objective of the study was to assess the influence of animal manures applied as slurry on the invertebrate fauna of grassland. The main field plot investigations were carried out at sites maintained by the Agricultural Institute at Johnstown Castle and Kilmore, Co. Wexford, and at Grange, Co. Meath. The effects of gross soil contamination by pig slurry on earthworms and arthropods were studied at Celbridge, Co. Kildare. The arthropod fauna involved in the decomposition of cattle slurry on the soil surface was studied at Grange.

1. EFFECTS OF SLURRY ON POPULATIONS OF SOIL FAUNA

Earthworms (Lumbricidae)

Considerable attention was given to earthworms because they are important in the recycling of dead organic matter and the maintenance of soil fertility in grassland. The main study (Cotton and Curry, 1979a) was carried out at Johnstown Castle where cattle and pig slurry were being compared with inorganic fertiliser in a large plot (0.4ha) silage production experiment. The slurry was applied at 80 - 100 t ha^{-1} yr^{-1}. On average, a moderate increase in both numbers (25 - 31 percent) and biomass (23 - 38 percent) was recorded in slurry treated plots but the results of statistical analysis of the data summarised in Table 1 reveal considerable inter-block and inter-sampling date variability in regard to treatment effects. Community structure was largely unaffected by slurry apart from an increase in the proportion of *Satchellius* (= *Dendrobaena*) *mammalis* (Sav.) Two species recorded from this site, *Allolobophora*

*Paper presented by D.B.R. Poole

TABLE 1

SUMMARY OF THE EFFECTS OF ANIMAL MANURES ON NUMBERS AND BIOMASS OF EARTHWORMS IN GRASSLAND AT JOHNSTOWN CASTLE

		*Log_{10} numbers			*biomass		
Block I	Nov. 76	C	F	P	F	C	P
	Feb. 77	F	C	P	F	P	C
	April 77	P	F	C	F	P	C
	Oct. 77	F	P	C	P	F	C
Block II	Feb. 77	C	F	P	C	F	P
	April 77	F	C	P	F	C	P
	Oct. 77	P	F	C	F	C	P
Block III	April 77	F	C	P	F	C	P
	Oct. 77	P	C	F	F	P	C

Treatments are ranked in ascending order from left to right and those not underscored by the same line differ at the 95% level of significance.

F = Inorganic fertiliser C = Cattle slurry P = Pig slurry

* Overall average numbers were F = 310, C = 388 and P = 407 worms m^{-2}. Corresponding values for biomass were 129, 159 and 178g m^{-2}.

TABLE 2

MEAN EARTHWORM NUMBERS AND BIOMASS RECOVERED FROM PLOTS TREATED THREE TIMES ANNUALLY WITH PIG SLURRY AT VARIOUS SITES

	Rates of slurry application (m^3 ha^{-1} yr^{-1})			
	0	69	138	345
	Sandy loam site at Johnstown Castle			
Numbers m^{-2}	516	475	390	300
Biomass (g m^{-2})	196	224	225	208
	Sandy site at Kilmore Quay			
Numbers m^{-2}	39	62	47	56
Biomass (g m^{-2})	32	37	26	56
	Silty site at Kilmore Quay			
Numbers m^{-2}	51	73	59	23
Biomass (g m^{-2})	15	26	20	12

limicola (Michaelsen) and *A. tuberculata* (Eisen) were first records from Ireland (Cotton, 1978).

The effects of increasing rates of pig slurry application on earthworms were studied on grassland sites on sandy loam, silty loam and sandy soils at Johnstown Castle and Kilmore (Cotton and Curry, 1979b). The results are summarised in Table 2. Site variation and low population densities obscured effects for the sandy soil but at the other two sites the highest level of application studied, 345 t ha^{-1} yr^{-1} of slurry significantly reduced earthworm population density. Immature worms in general and adults of *S. mammalis* and *Eiseniella tetraedra* (Sav.)were the groups affected.

Effects of gross pollution arising from the dumping of large quantities of pig slurry in a quarry were studied at the University farm, Celbridge, Co. Kildare (Curry and Cotton, 1979). In wet weather slurry frequently overflowed from the quarry, contaminating a strip approximately 10 m wide and 100 m downhill of the quarry. There were very few earthworms in the area adjacent to the quarry in April, 12 months after the last major spill (Figure 1). The populations had substantially recovered by November, but a high proportion of worms adjacent to the quarry were surface dwelling, pigmented forms. The typical grassland non-pigmented species (*Allolobophora* and *Aporrectodea* spp.) remained scarce. Soil copper levels up to 100 ppm were recorded at the quarry edge but these declined rapidly with increasing distance from the quarry. A corresponding gradient in copper levels present in earthworm bodies was noted although levels in worm bodies tended to be lower in November than in the previous April.

Soil Arthropods

The effects of various rates of application of cattle slurry on soil arthropods were studied at Grange (Bolger and Curry, 1979). The treatments were:

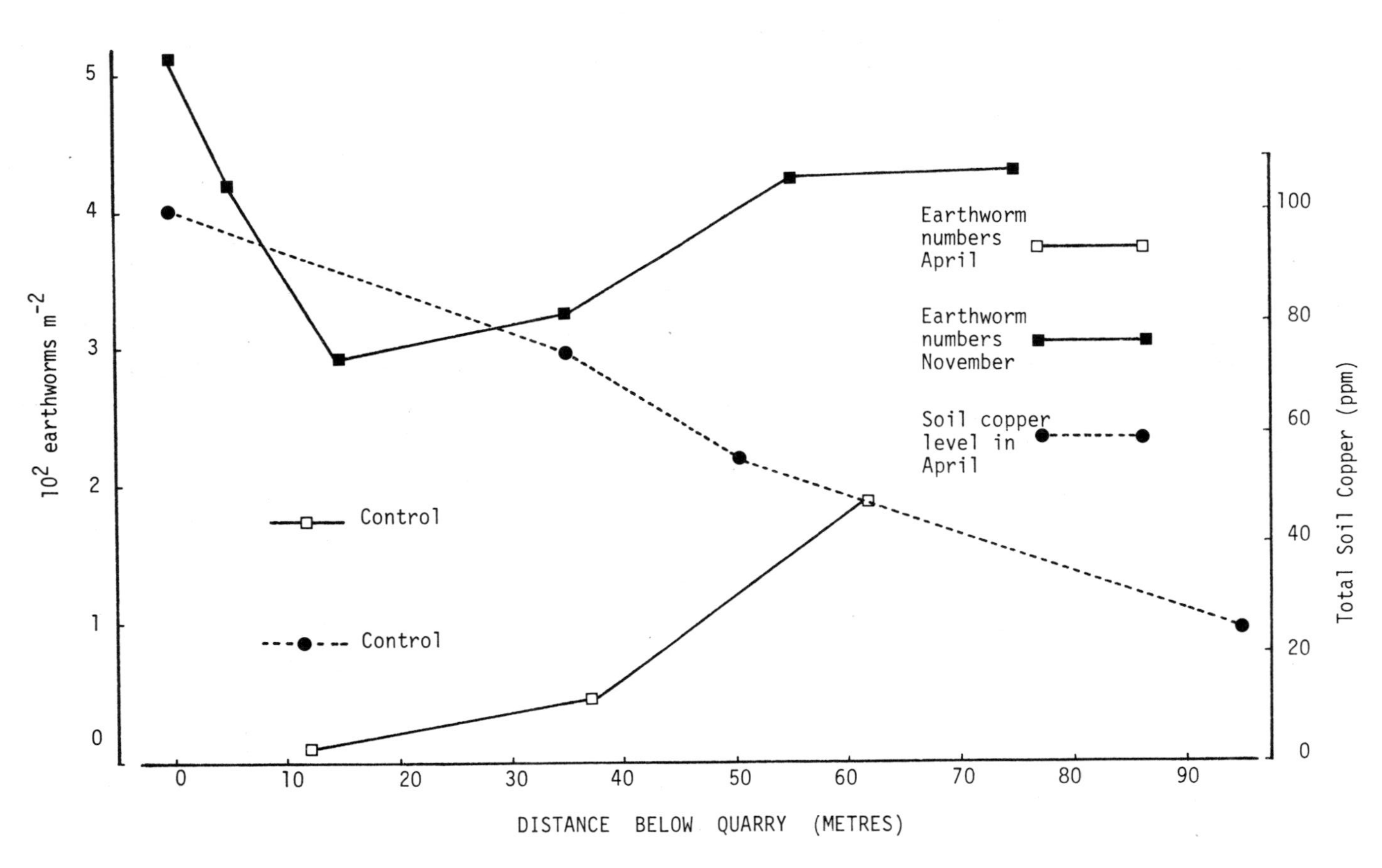

Fig. 1. Effects of gross pig slurry contamination at Celbridge, Co. Kildare.

(1) untreated control,

(2) a single application of 55 t ha^{-1} of cattle slurry,

(3) six applications of 110 t ha^{-1}within a period of 21 months,

(4) two applications each of 550 t ha^{-1} separated by an interval of 14 months.

The highest rate of application significantly depressed populations of Acari and Collembola (Table 3). Substantial recolonisation by haemiedaphic species occurred within 9 months but numbers of euedaphic species were still low.

TABLE 3

EFFECTS OF THE CATTLE SLURRY TREATMENTS ON ACARINE AND COLLEMBOLAN POPULATIONS AT GRANGE, CO. MEATH.

Treatment	Numbers m^{-2} with standard errors	
	Collembola	Acari
Untreated control	28 313 ± 7 453	21 600 ± 10 914
55 t ha^{-1}	46 600 ± 26 665	17 181 ± 6 875
6 applications at 110 t ha^{-1}	45 894 ± 10 919	10 713 ± 3 535
2 applications at 550 t ha^{-1}	25 006 ± 10 782	7 844 ± 3 649

Numbers of both groups were significantly depressed by 2 applications at 550 t ha^{-1} ($P < 0.01$) and remained so for at least 9 months following application of the treatment.

The major effects of cattle and pig slurry at 80 - 100 t ha^{-1} yr^{-1} in silage production experiments at Johnstown Castle are indicated in Table 4. Acarine and collembolan numbers were depressed by the slurry treatments, particularly by pig slurry, at various times during the study. The most seriously affected species were the collembolans *Friesea mirabilis* (Tulbg.) and *Isotoma viridis* (Bourl.) and the predatory mite *Rhodacarellus silesiacus* (Willm.).

TABLE 4

MEAN NUMBERS OF ACARI AND COLLEMBOLA PER SAMPLE WITH STANDARD ERRORS IN SOIL AT JOHNSTOWN CASTE

	No slurry	Pig slurry	Cattle slurry
		Acari	
Oct. '76	37.1 ± 5.8	17.1 ± 3.7	33.7 ± 6.5
Feb. '77	46.5 ± 16.9	12.3 ± 4.1	29.6 ± 6.2
Apr. '77	47.1 ± 9.8	25.9 ± 3.9	35.9 ± 4.5
Jul. '77	42.7 ± 6.6	69.6 ± 15.3	72.3 ± 16.8
Nov. '77	49.7 ± 13.2	31.6 ± 4.0	42.6 ± 5.2
		Collembola	
Oct. '76	52.0 ± 7.6	17.3 ± 4.5	22.3 ± 4.9
Feb. '77	79.2 ± 25.6	22.6 ± 9.1	36.5 ± 10.1
Apr. '77	77.6 ± 20.2	31.8 ± 7.4	47.4 ± 8.8
Jul. '77	99.2 ± 23.8	73.7 ± 27.0	72.7 ± 15.2
Nov. '77	91.3 ± 16.1	123.5 ± 26.3	94.6 ± 13.7

The effects of increasing amounts of pig slurry on soil arthropod populations were studied at two sites, a sandy loam at Johnstown Castle and a sandy site at Kilmore. Population levels at Johnstown Castle were low and no significant effects of slurry treatments on Acari or Collembola were detected. Acari and Collembola were considerably more abundant at Kilmore (Table 5). Significant effects of treatments on population density were noted on the first sampling date (Jan. 1978) but not subsequently. On that occasion numbers of Collembola were significantly depressed by the heaviest rate of slurry application (345 t ha^{-1} yr^{-1}) while numbers of Acari were significantly higher in plots which had received the lighest slurry application (69 t ha^{-1} yr^{-1}). Changes in the relative abundance values of many of the dominant species were recorded in both sites. These effects appeared to be related to the phenological state of the various species populations at the time of slurry application, to the rates of slurry application, and to the intervals between applications and sampling.

TABLE 5

MEAN NUMBERS OF ACARI AND COLLEMBOLA m^{-2} WITH STANDARD ERRORS IN SANDY GRASSLAND SOIL AT KILMORE

	Acari	Collembola
Jan. '78	83 495 ± 15 668	57 005 ± 11 918
May '78	144 485 ± 22 778	63 125 ± 10 168
Sept.'78	90 550 ± 13 699	22 150 ± 2 486

Invertebrate pests

Slug (Mollusca) and leatherjacket (Tipulidae) populations in the plots which had received 80 - 100 t ha^{-1} yr^{-1} pig or cattle slurry at Johnstown Castle were sampled in April 1977. Populations of leatherjackets in all plots were low and no significant effects of manurial treatments were observed. The distribution of slugs in the site appeared to be affected mainly by factors such as the proximity of hedgerows and local drainage characteristics. No relationships between populations and slurry treatments were established.

Populations of soil nematodes were sampled in February 1977. They were largely unaffected by slurry, apart from a small increase in numbers of the plant parasitic *Paratylenchus* spp.

2. EFFECTS OF SLURRY ON ARTHROPOD POPULATIONS ON THE VEGETATION

Populations of microarthropods on the vegetation at Johnstown Castle were sampled by Tullgren funnel extraction and macroarthropods by vacuum net suction. Microarthropod populations were quite variable in the site as a whole

(Table 6). Numbers of the acarid mite *Tyrophagus longior* (Gervais) were increased by pig slurry application but tarsonemids were reduced.

TABLE 6

MEAN NUMBERS OF ACARI PER PLOT WITH STANDARD ERRORS EXTRACTED BY TULLGREN FUNNEL FROM SMALL VEGETATION SAMPLES AT JOHNSTOWN CASTLE

	No slurry	Pig slurry	Cattle slurry
Oct. '76	140 ± 35	101 ± 22	111 ± 30
Feb. '77	44 ± 8	43 ± 11	76 ± 25
Apr. '77	6 ± 3	4 ± 1	7 ± 2
May '77	14 ± 3	13 ± 3	9 ± 2
June '77	10 ± 3	21 ± 9	14 ± 5
Aug. '77	127 ± 40	107 ± 17	95 ± 23

Suction sampling indicated that mean collembolan numbers were drastically reduced by pig and cattle slurry applications (Table 7). Numbers of Acari and Insecta were not affected by slurry in any consistent way. The detritivorous collembolan species *Sminthurinus aureus* (Lub.) and Isotomidae spp. were drastically reduced whereas the phytophagous *Sminthurinus viridis* (L.) was less affected. Laboratory and field studies are in progress to determine which components of slurry are responsible for this toxicity.

TABLE 7

MEAN NUMBERS OF COLLEMBOLA m^{-2} WITH STANDARD ERRORS CAUGHT BY SUCTION SAMPLING AT JOHNSTOWN CASTLE

	No slurry	Pig slurry	Cattle slurry
Apr. '77	658 ± 208	175 ± 54	207 ± 43
May '77	1 120 ± 302	371 ± 83	490 ± 73
June '77	1 600 ± 251	355 ± 87	511 ± 86
Aug. '77	987 ± 412	650 ± 239	618 ± 242

3. THE FAUNA COLONISING CATTLE SLURRY

The arthropod fauna colonising cattle slurry on the soil surface at Grange was studied in some detail (Curry, 1978). Arthropod numbers ranged from 20 to 40 g^{-1} with Collembola, Acari, dipterous larvae and Coleoptera being abundant (Table 8). A total of 144 species were recorded including 71 Acari and 25 Collembola. The community was remarkably stable and there was little evidence of arthropod succession related to the age of the slurry. This stability was attributed to resource renewal through periodic addition of slurry.

TABLE 8

MEAN NUMBERS OF ARTHROPODS EXTRACTED FROM CATTLE SLURRY APPLIED (a) IN SINGLE ANNUAL APPLICATIONS OF 550 t ha^{-1} AND (b) PERIODICALLY AT THE RATE OF 110 t ha^{-1}.

	Numbers 10 g^{-1} dry cattle slurry with standard errors	
	(a)	(b)
Collembola	197.1 ± 46.5	217.9 ± 76.8
Acari	62.6 ± 11.9	95.2 ± 27.9
Diptera	42.8 ± 17.0	14.1 ± 3.7
Coleoptera	2.3 ± 1.2	5.4 ± 4.6
Araneae	+	+
Thysanoptera	+	
Hymenoptera	+	
Hemiptera	+	+
Psocoptera	+	+
Pauropoda	+	+
Neuroptera	+	
Total Arthropoda	305.1	331.1

Populations of enchytraeid and nematode worms were sampled twice. Nematodes were abundant, particularly in slurry applied periodically at the rate of 110 t ha^{-1}. Enchytraeids were only present in October 1973, and only in the 110 t ha^{-1} treatment (Table 9).

TABLE 9

MEAN NUMBERS OF ENCHYTRAEID WORMS AND NEMATODES EXTRACTED FROM CATTLE SLURRY, (a) AND (b) AS IN TABLE 8.

	Numbers 10 g^{-1} dry cattle slurry with standard errors	
	Enchytraeids	Nematodes
Oct. '73 (a)	0	32 750 ± 2 100
(b)	84 ± 48	43 720 ± 3 700
June '73 (a)	0	15 830 ± 4 850
(b)	0	24 200 ± 2 100

CONCLUSIONS

The main conclusion from these studies is that earthworms are only adversely affected by very high levels of slurry which might arise from the dumping of large quantities of unwanted slurry. Such adverse effects are transitory and populations appear to recover within 12 - 15 months. High levels of soil copper from pig slurry may create a hazard under conditions of continous dumping over a long period of time. Much of this copper is apparently bound in organic-copper chelates and further work is required to determine how much is actually absorbed by earthworms. Moderate applications result in enhanced earthworm populations in soils where worms are food-limited.

It is difficult to assess the implications of the effects noted on soil arthropods for soil fertility and plant growth. Earthworms were by far the dominant element in the decomposer fauna at Grange and Johnstown Castle and in these situations it is likely that soil arthropods play only a minor role in organic matter transformation. Disruption of arthropod communities resulting from slurry application was relatively minor, except following the very heavy applications of cattle slurry (550 t ha^{-1} yr^{-1}) at Grange. Arthropods were relatively more important in decomposition and mineralisation processes in the sandy soil at Kilmore because there were few earthworms. However, it is doubtful whether the relatively minor changes induced by slurry in arthropod community structure would have any seriously detrimental effects on decomposition processes. Decaying animal manure provides a rich substrate for an abundant and diverse invertebrate fauna. This may have implications for the amelioration of sites where the decomposer fauna is impoverished and also for systems of crop production where, because of reduced plant diversity, animal diversity is low. Such situations are characterised by high instability and pest outbreaks and might benefit from slurry application.

REFERENCES

Bolger, T. and Curry, J.P., 1979. Effects of cattle slurry on soil arthropods in grassland. Pedobiologia. In press.

Cotton, D.C.F., 1978. A revision of the Irish earthworms (Oligochaeta: Lumbricidae) with the addition of two species. Ir. Nat. J. 19, 257-260.

Cotton, D.C.F. and Curry, J.P., 1979a. The effects of cattle and pig slurry fertilisers on earthworm (Oligochaeta, Lumbricidae) in grassland managed for silage production. Pedobiologia 19 in press.

Cotton, D.C.F. and Curry, J.P., 1979b. The response of earthworm populations (Oligochaeta, Lumbricidae) to high applications of pig slurry. Pedobiologia. In press.

Curry, J.P., 1978. The arthropod fauna associated with cattle manure applied as slurry to grassland. Proc. R. Ir. Acad. 79, 15-27.

Curry, J.P. and Cotton, D.C.F., 1979. Effects of heavy pig slurry contamination on earthworms in grassland. In: Proceedings of the VII ISSS Soil Zoology Colloquium, Syracuse, N.Y., July 1979.

DISCUSSION

G. Steffens *(West Germany)*

You worked on grassland, but have you also experience of arable land? Many farmers have asked us if the earthworm population is decreased because after spreading slurry many earthworms were on top of the soil.

D.B.R. Poole *(Ireland)*

I am not aware of any work that was done by this team on arable land.

H. Tunney *(Ireland)*

I know that they are planning some work on this - but mostly with sugar beet. I can add an additional comment though, in grassland we also noted earthworms on the surface after application. Some of the work done by Dr. Curry, indicates that you tend to kill a very small percentage of earthworms; for lack of oxygen perhaps, or maybe high concentrations of ammonia. However, the longer term effect is to increase the earthworm numbers because the amount of available food is increased. Therefore, earthworm numbers do not decrease with average rates of application. If you put on very large amounts like 1 000 m^3 ha^{-1} you may have problems.

EFFECTS OF CATTLE SLURRY, PIG SLURRY AND FERTILISER ON YIELD AND QUALITY OF GRASS SILAGE

H. Tunney, S. Molloy and F. Codd
Agricultural Institute,
Johnstown Castle, Wexford, Ireland.

1. INTRODUCTION

In the nine EEC countries most of the farm land is devoted to grass production. The greatest proportion of animal manures is produced by dairy and beef animals that are fed primarily on grass. This situation is perhaps most noticeable in Ireland where over 90 percent of farmland is devoted to grass and where silage or hay account for most of the winter feed. Intensive dairy and beef farms often have no tillage on the farm and therefore the manure must be spread on grassland.

The recommendation is to spread manure where possible, on the area conserved for hay or silage. This means that the plant nutrients in the winter feed are recycled in the slurry to produce the winter feed for the next winter. In addition spreading slurry on the conserved area minimises the risk of spread of disease.

The continuing world energy shortage emphasises the importance of efficient utilisation of plant nutrients in manures. Chemical fertiliser prices will increase with increasing oil prices. At present fertiliser prices, manures in the EEC countries are probably worth more than £1 000 million sterling per annum. On a wider scale, it has been estimated that the annual world manure production from farm animals contains in the region of 54, 12 and 52 million tonnes of N, P and K respectively (Tunney, 1979). This compares with the 1976 estimated world chemical fertiliser production of 44, 11 and 20 million tonnes of N, P and K respectively (FAO, 1977). It is clear that the efficient utilisation of manures is important for world food production.

In this paper the results of two separate but related experiments on the use of slurry on grass are summarised for the three years 1976, 1977 and 1978. In the first (Experiment A) fertiliser, cattle slurry and pig slurry were compared for grass silage production on a total area of about 12 ha. The grass from the three treatments was ensiled in three separate pits for both Cut 1 and Cut 2 and fed to three groups of beef animals over the winter to study effects on animal performance. In the second (Experiment B) incremental rates of slurry were applied in factorial combination with incremental rates of fertiliser nitrogen on a total area of about 1 ha. The primary aim was to study the efficiency of slurry nitrogen relative to fertiliser nitrogen for grass silage production.

The experiments were carried out at Johnstown Castle Research Centre in the south east corner of the country. This area has a maritime climate with mild damp winters and cool summers. The annual rainfall is 800 mm and precipitation deficiencies are less than 50 mm. The field plots were on moderately well drained loam (24% clay) over fine loamy subsoil, derived from shale and schist till.

2. EXPERIMENT A - EFFECTS OF SLURRY ON GRASS PRODUCTION AND ANIMAL PERFORMANCE

2.1 Materials and methods

This experiment consisted of 30 randomised field plots of 0.4 ha each. There were three treatments namely cattle slurry, pig slurry, and chemical fertiliser, with 10 plots per treatment. The sward was predominantly ryegrass with some white clover. There were two silage cuts in each of the three years and the treatments were applied about 7 weeks before cutting. Fertiliser plots received 100 kg N ha^{-1} for Cut 1 and 60 kg N ha^{-1} for Cut 2. The slurry plots received no fertiliser nitrogen. Details of the cattle and pig slurry treatments for the three years are summarised in Table 1.

TABLE 1

DETAILS OF QUANTITIES OF SLURRY APPLIED AND NUTRIENTS PRESENT FOR 1976, 1977, AND 1978

		Slurry		kg ha^{-1} applied in slurry			
		t ha^{-1}	% DM	N	P	K	Mg
				1976			
Cattle	Cut 1	44	7	115	23	136	17
	Cut 2	40	11	178	32	211	21
Pig	Cut 1	44	7	186	64	78	35
	Cut 2	40	5	136	39	68	23
				1977			
Cattle	Cut 1	55	8	128	17	232	15
	Cut 2	44	8	149	18	162	22
Pig	Cut 1	55	8	220	92	50	46
	Cut 2	44	3	120	21	55	15
				1978			
Cattle	Cut 1	44	9	166	24	196	18
	Cut 2	44	9	154	22	176	18
Pig	Cut 1	44	3	132	29	55	16
	Cut 2	44	5	151	52	57	26

There was a basal dressing of 500 kg ha^{-1} of a fertiliser containing 7% phosphorus and 30% potassium annually on the fertiliser plots. Cattle slurry plots received supplemental fertiliser P at 16 kg ha^{-1} in 1976 and pig slurry was supplemented with fertiliser K at 56 kg ha^{-1} in 1976 and 112 kg ha^{-1} in 1977.

Silage Cut 1 was taken at the end of May and Cut 2 at the end of July. Grass yield from all plots was measured and sampled for analysis at each harvest.

The grass from the three treatments was ensiled in three pits for both Cut 1 and Cut 2. These silages were fed to three groups of beef animals over the winter feeding period for each of the three years. There were 20 animals for each treatment (total = 60). Cut 2 was fed first between November and January, and Cut 1 was fed between January and April to the same group of animals without re-randomisation. The mean 24 hour fasted weight at the start of the experiment was 370, 445 and 459 kg per animal for 1976, 1977 and 1978 respectively. Animal weights were recorded at the start of the experiment and again after feeding the first and second cut silage. The daily liveweight gain was then calculated. Silage intake was measured by weighing in the silage over a one week period and weighing out the residue at the end of the week. Silage samples were analysed to determine silage quality and preservation. The animals were fed silage ad lib plus 2 kg meals head^{-1} day^{-1} in a slatted floor house. The slurry from this house was used in the field plot experiment.

2.2 Results and discussion

The cattle slurry was usually visible on the grass surface for 2 weeks or longer in dry weather after spreading. Fragments of cattle slurry were often visible on the grass at time of cutting, especially at the second cut. Pig slurry could not be seen on the grass at time of cutting. Visual observations indicated that grass growth on cattle slurry plots was less than on the other two treatments. This visual difference tended to decrease as cutting date approached. In general, grass always appeared healthy and the crop looked good at time of cutting. There was more clover in the cattle slurry plots especially at the second cut.

2.2.1. Effects on grass yields

The grass yields for the three treatments are summarised in Table 2.

TABLE 2

GRASS YIELDS (t DM ha^{-1}) AT EACH CUT 1976 - 1978 (EACH VALUE IS THE MEAN OF 10 REPLICATES)

	Cattle slurry	Pig slurry	Fertiliser	Standard error
		1976		
Cut 1	5.19	5.55	6.29	0.18
Cut 2	2.47	3.29	4.05	0.17
Total	7.66	8.84	10.34	
		1977		
Cut 1	4.04	4.77	4.79	0.18
Cut 2	1.45	2.47	3.33	0.12
Total	5.49	7.24	8.12	
		1978		
Cut 1	3.68	4.31	4.31	0.07
Cut 2	2.42	3.00	3.72	0.12
Total	6.10	7.31	8.05	

TABLE 3

THE MEAN COMPOSITION OF THE GRASS AT HARVEST 1976 - 1978

		% in plant dry matter			
	% DM	N	P	K	Mg
		Cut 1			
Cattle slurry	19.8	2.15	.35	3.17	.21
Pig slurry	18.7	2.33	.42	3.03	.24
Fertiliser	17.6	2.50	.40	3.00	.23
		Cut 2			
Cattle slurry	23.9	1.97	.37	2.76	.23
Pig slurry	23.0	2.02	.39	2.58	.25
Fertiliser	20.5	2.31	.36	2.03	.27

The mean total yield for the two cuts over the three years for cattle slurry, pig slurry and fertiliser was 6.42, 7.80 and 8.83 t DM ha^{-1} $year^{-1}$. The results in Table 2 show that on average the fertiliser treatment gave over 2 t DM ha^{-1} more than the cattle slurry treatment, and about 1 tonne more than the pig slurry treatment.

2.2.2. Effects on grass composition

The average composition of the grass is shown in Table 3.

The results of plant analyses showed that, over the 3 years, grass from cattle slurry plots had a statistically significantly higher dry matter and potassium content than that from the fertiliser plots. Grass from pig slurry plots had intermediate levels. The nitrogen content of the grass from the cattle slurry plots was generally significantly lower than the fertiliser plots.

Cattle slurry plots usually had lower magnesium and pig slurry plots had higher phosphorus levels but these differences were not statistically signficant.

The main conclusions to be drawn from these results is that cattle slurry gave lower yields and higher percentage dry matter and potassium than fertiliser, while pig slurry gave intermediate results.

2.2.3. Effects on soil analyses

Soil analyses showed that cattle slurry plots had the highest potassium levels and pig slurry plots had the highest phosphorus levels. These results reflect the levels of nutrients applied in the slurry as shown in Table 1. In addition the results indicate a low efficiency for cattle slurry nitrogen relative to pig slurry and fertiliser nitrogen.

2.2.4. Effects on animal liveweight gain

The results of animal liveweight gain on the three treatments are summarised in Table 4.

TABLE 4

LIVEWEIGHT GAIN (kg animal^{-1} day^{-1}), FOR ANIMALS FED EACH OF THE THREE SILAGES

	Cut 1 (75 days)	Cut 1+2 (135 days)
1976		
Cattle slurry	.67	.74
Pig slurry	.74	.76
Fertiliser	.70	.77
Standard error	.045	.035
% Coefficient of variation	28.3	28.1
1977		
Cattle slurry	.85	.37
Pig slurry	.92	.49
Fertiliser	.99	.62
Standard error	.032	.030
% Coefficient of variation	15.4	26.8
1978		
Cattle slurry	0.86	.69
Pig slurry	.90	.76
Fertiliser	.93	.78
Standard error	0.053	.037
% Coefficient of variation	43.5	22.1

In all cases, for Cut 1, Cut 2 and Cut 1+2, the silage produced from plots treated with cattle slurry gave lower live-weight gains than that from the pig slurry or fertiliser treatments. However, only in one case, Cut 1 in 1977, was the difference statistically significant.

2.2.5. Effects on silage composition

The silage analyses were similar for 1976 and 1977, however in 1978 the dry matter and other parameters tended to be higher as a result of wilting. No additive was used on the silage. The average silage results for Cut 1 and Cut 2 1976 and 1977 are summarised in Table 5.

TABLE 5

AVERAGE SILAGE ANALYSES FOR CUT 1 AND CUT 2 IN 1976 AND 1977

	pH	% dry matter	% crude protein	% dry matter digestibility
			Cut 1	
Cattle slurry	4.1	21.0	12.8	70.5
Pig slurry	4.1	19.5	13.1	70.1
Fertiliser	4.1	19.8	13.4	69.4
			Cut 2	
Cattle slurry	4.2	20.9	10.9	60.7
Pig slurry	4.2	24.5	10.8	61.9
Fertiliser	4.1	22.1	12.8	61.2

The results of silage analyses in Table 5 do not explain the difference in animal performance shown in Table 4. The pH of the silage indicates no difference in preservation between the treatments. Silage intake was in the region of 40 kg animal^{-1} day^{-1} or 9 kg silage dry matter. The intake studies did not show a consistent difference between treatments. However, the approach of feeding in silage for a week, to 20 animals, and then weighing out uneaten material may not be sensitive enough to detect real differences. During the intake studies it was clear that a proportion of the silage was pulled into the pens and some of this fell down between the slats. This silage would be recorded as silage intake. It is suggested that more sensitive intake studies are required, perhaps with fewer animals in special pens, to establish conclusively if the cattle slurry treatment reduces silage intake.

3. EXPERIMENT B - EFFECT OF SLURRY AND FERTILISER NITROGEN ON GRASS PRODUCTION

3.1 Materials and methods

This was a factorial experiment with incremental rates of slurry and fertiliser nitogen. In 1976 and 1977 the cattle slurry rates were 0, 28, 42 and 84 t ha^{-1}. In addition there was one pig slurry treatment at the medium rate of 42 t ha^{-1}. The fertiliser nitrogen treatments were 0, 30, 60 and 90 kg N ha^{-1}.

In 1978 the slurry and nitrogen rates were higher. The slurry rates were 0, 40, 80, 120 t ha^{-1}. The fertiliser nitrogen levels were 0, 40, 80 and 120 kg N ha^{-1}.

There were, therefore, five slurry treatments in factorial combination with four fertiliser nitrogen treatments giving a total of 20 treatments. For each treatment there was a washed and unwashed treatment. The washing treatment consisted of washing the slurry from the grass surface with 120 t water ha^{-1} shortly after application. This gave a total of 40 plots, and there were six replications giving a grand total of 240 plots. Each plot was 8 m x 3 m.

All plots received a basal dressing of P and K at the start of the experiment to ensure high fertility. The sward was an old pasture with a mixture of grass species. The sward was sprayed to eliminate clover in order to avoid the effects of biological nitrogen fixation. Fertiliser and slurry were applied once each year in April and first harvest (Cut 1) was taken in early June. A second harvest was taken in September to study the residual effects of the treatments on grass growth. Yields were recorded at each harvest and representative samples were collected for analyses.

3.2 Results and discussion

Observations at the time of spreading showed that cattle slurry, particularly the higher rates, coated the grass for about two weeks after spreading. New grass growth then pushed through the slurry on the surface. The washing treatment was effective in removing slurry from the grass surface. The effect of pig slurry was much less noticeable on the grass surface and was usually not evident a week after spreading. Rainfall shortly after spreading tended to wash the cattle slurry off the grass. This was most noticeable in the spring of 1977 when rain fell the day after spreading. The relationships between the grass yield and fertiliser nitrogen applied for the different slurry treatments for Cut 1, 1977, are summarised in Figure 1. The corresponding results for Cut 1, 1978, are summarised in Figure 2. These results are for the unwashed treatments only, the effects of washing are discussed later.

It is clear from Figures 1 and 2 that there was a significant positive yield response to slurry. In both years, and also in 1976, response to pig slurry was superior to cattle slurry. A striking feature of the results is the small yield difference between the three rates of cattle slurry. This suggests that the low rate of cattle slurry was almost as effective as the high rate. The nitrogen efficiency of cattle slurry at the low rate was much better than at the high rate. The nitrogen efficiency of the low rate of cattle slurry was in the region of 37% in 1977 and 23% in 1978. The low efficiency of nitrogen in cattle slurry is probably due to loss of nitrogen by ammonia volatilisation. The effect of cattle slurry on the grass surface must also have contributed to the low yield response.

In 1978 there was a higher maximum yield from pig slurry than from fertiliser nitrogen even at 120 kg N ha^{-1}. This suggests that some factor, other than nitrogen, supplied by the pig slurry may have limited yield. Alternatively the form of nitrogen in the pig slurry may have been more effective than the fertiliser nitrogen (Calcium ammonium nitrate with 26% N).

In 1976 many of the treatments with fertiliser nitrogen alone gave better yields than with slurry plus the same amount of fertiliser nitrogen. So that with the exception of controls

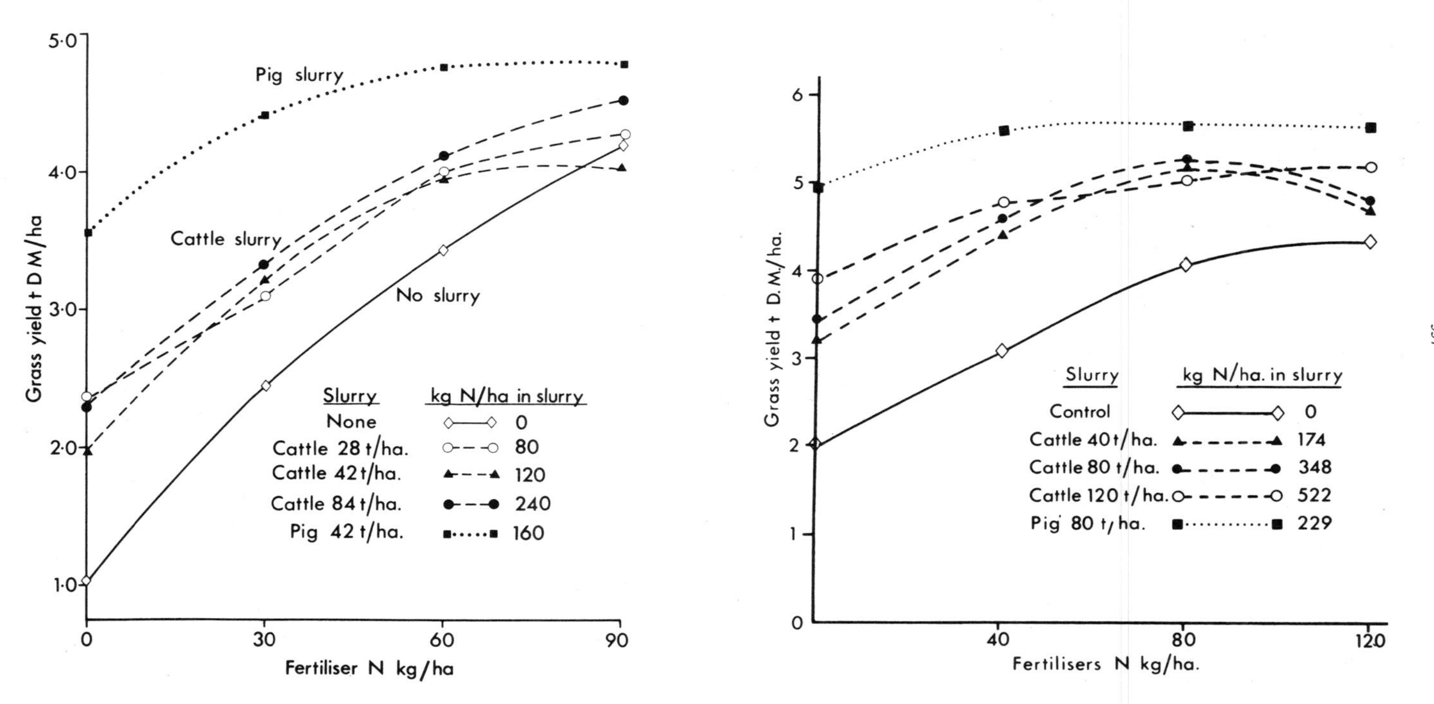

Fig. 1. Experiment B grass yield response to slurry and fertiliser nitrogen, Cut 1, 1977 (LSD = 0.7)

Fig. 2. Experiment B grass yield response to slurry and fertiliser nitrogen, Cut 1, 1978 (LSD = 0.6)

where no fertiliser nitrogen was applied, slurry tended to depress grass yields for Cut 1 in 1976. However on the washed treatments there was a yield response to slurry, relative to fertiliser alone.

The mean yields over the three years for Cut 1 are summarised in Table 6.

TABLE 6

MEAN GRASS YIELD (t DM ha^{-1}) FOR CUT 1 OVER THE THREE YEARS 1976 - 1978

Fertiliser nitrogen	None	Low	Medium	High
		Not washed		
No slurry	2.08	3.50	4.32	4.77
Cattle slurry - low	2.89	3.98	4.73	4.78
Cattle slurry - medium	2.76	4.27	4.78	4.79
Cattle slurry - high	3.27	4.08	4.73	4.76
Pig slurry - medium	4.25	4.94	5.36	5.27
		Washed		
No slurry	1.94	3.64	4.05	5.10
Cattle slurry - low	3.55	4.58	5.04	5.16
Cattle slurry - medium	3.21	4.64	5.05	4.88
Cattle slurry - high	4.12	5.18	5.33	5.38
Pig slurry - medium	4.74	5.41	5.52	5.74
Standard error of means = 0.23				

The results in Table 6 show that overall there was a yield response to washing on all slurry treatments. However, this difference was not statistically significant in all cases.

Other results showing yield on washed and unwashed treatments are summarised in Table 7.

TABLE 7

SOME EFFECTS OF WASHING ON GRASS YIELD (t DM ha^{-1}) 1976 - 1978

	1976	1977	1978
		Overall total yield (Cut 1+2)	
Not washed	5.56	3.99	6.43
Washed	6.31	4.16	6.64
Standard error	.144	.071	.062
		Overall yield - Cut 1	
Not washed	4.65	3.50	4.48
Washed	5.46	3.65	4.74
Standard error	.141	.073	
		Cattle slurry high - no nitrogen	
Not washed	3.58	2.35	3.90
Washed	4.61	2.75	4.75
Standard error	.33	.23	.20
		Cattle slurry high - high nitrogen	
Not washed	4.64	4.27	5.12
Washed	6.17	4.37	5.54
Standard error	.33	.23	.20

The Cut 2 yields were low by comparison with Cut 1. There was, however, generally a yield response due to the residual effects of slurry treatments. The mean yield for Cut 2 was 0.88, 0.50 and 1.93 t DM ha^{-1} in 1976, 1977 and 1978 respectively. There was a tendency for residual effects of slurry to be highest on the high fertiliser nitrogen treatment. There was no residual effect with fertiliser nitrogen only. Residual effects of slurry accounted for up to 50% of Cut 2 yield.

Slurry and nitrogen treatments had significant effects on the composition of the grass. These effects were noticeable in Cut 2. As expected cattle slurry increased the potassium and nitrogen content of grass relative to the control. The nitrogen treatments of course significantly increased the nitrogen content of the grass. One interesting effect of slurry was that it significantly reduced the manganese content of the grass. The manganese content of slurry treated plots was usually in the region of 50% of the level in the non slurry treated plots. The non slurry plots had grass with manganese levels of about 200 micrograms per gram whereas the slurry plots had in the region of 100. This reduction in manganese occurred at Cut 1 and Cut 2 in the three years. The fertiliser nitrogen treatments also reduced the manganese levels in grass but to a lesser extent than the slurry treatments.

It can be concluded from this work that pig slurry can give yields as good or better than fertiliser nitrogen, but a supplementation of about 30 kg fertiliser N ha^{-1} may be necessary to ensure maximum yield. Cattle slurry has a very low nitrogen efficiency on grassland and should be supplemented with 60 to 90 kg fertiliser N ha^{-1} to ensure maximum yield. The main benefit from cattle slurry is the phosphorus, potassium and other nutrients present.

4. CONCLUSIONS

The results of the two experiments discussed in this paper indicate that slurry can be used for grass silage production but there are, however, associated problems.

The main advantage of the slurry is that it can supply, by recycling, the P and K requirements of the crop. It can also supply some nitrogen and other plant nutrients.

Cattle slurry at rates as low as 44 t ha^{-1} applied 7 weeks before cutting silage can decrease animal performance. In general, pig slurry appears to be more suitable for use on grassland. In recent years there has been increased emphasis on early and frequent (6 week intervals) silage cutting to ensure high digestibility. This approach makes it even more difficult to use cattle slurry effectively on silage land. Where land is not grazed between November and the first silage cut in May, it is not advisable to spread cattle slurry in March or April as the grass growth is generally too advanced.

It appears that autumn application of cattle slurry after second or third cut silage may have some advantages for grassland. However, it also presents problems as stored slurry tends to form a crust and dry out during summer so it is more difficult to agitate and spread. If spreading is postponed to October a sudden weather change may make the land too wet and tanks cannot be emptied without serious damage to the land. The risks of water pollution are also increased. It is proposed to continue to study the effects of cattle slurry on silage quality, but with adequate supplementation with fertiliser nitrogen to ensure maximum yield. This may help to reduce the adverse effects of slurry indicated in the present study.

It is important to spread slurry on bare ground shortly after grazing or silage cutting to ensure minimum grass contamination. When supplementing cattle slurry with nitrogen it is advisable to apply the nitrogen before slurry is spread. In dry weather fertiliser granules can be held by the slurry and prevented from entering the soil. There is an absence of information on the supplementation of slurry with urea, but it is possible that slurry may contribute to the volatilisation of ammonia from urea.

This work does not solve the special problems of using slurry on grassland. It does, however, help to quantify the effects. It also indicates how slurry can be integrated into management of a grassland farm with maximum benefits and minimum risks. There is clearly a need for more research on the use of cattle slurry on grassland.

5. ACKNOWLEDGMENTS

The authors wish to thank staff of the Analytical Laboratories at Johnstown Castle Research Centre for soil and plant analyses and at Dunsinea Research Centre for silage analyses. Thanks are also due to Mr. M. Clancy for manure analyses and to Mr. M. O'Keefe for statistical analyses of results.

The practical help received from many of the staff of Johnstown Castle Research Centre with this work and the valuable suggestions and discussions are gratefully acknowledged.

REFERENCES

FAO, 1977. Annual Fertiliser Review 1976. FAO Rome.

Tunney, 1977. Agricultural Wastes as fertiliser. Chapter 1 in 'Foodstuffs and Fertilisers from Agricultural Wastes'. Ed. B.M. Bewick, Van Mostrand Reinhold, N.Y. (In Press).

THE USE OF ANIMAL MANURES ON PASTURE FOR GRAZING

D.P. Collins*
The Agricultural Institute,
Grange, Co. Meath, Ireland.

INTRODUCTION

Animal manures have traditionally been a source of plant nutrients in farming, especially for cereal or root crops where their organic matter content is also of benefit to soil structure and water holding capacity. The advent of cheap inorganic fertilisers, which can be easily and accurately applied, corrected much nutrient deficiency in crops and indirectly devalued organic manures. However, more intensive farming, initially with poultry and later with pigs and cattle, produced environmental and mechanical difficulties in the disposal of animal manures.

When animal manures became a nuisance to farmers there was a renewal of interest in methods of disposal and, in turn, in their fertiliser value. Application to arable land, which is immediately ploughed, reduces or removes odours, prevents evaporation losses of nitrogen and may reduce the spread of pathogens. The optimum application rate is related to the nitrogen (N) level because excess N causes lodging in cereals and reduces the sugar content in beet. Many farms have either insufficient arable land or none at all, for disposal of animal manures. Application to grassland is the alternative, especially on areas cut for silage or hay. The limiting factor with large applications to pasture is high potassium (K) because it can be a predisposing factor to grass tetany (Walshe and Conway, 1974). Ultimately it is necessary to dispose of some animal manures on pasture which may subsequently be grazed by cattle.

The present experiment was designed to study the behaviour of cattle grazing on pasture dressed with slurry and to detect any disadvantage of different rates of application.

* Paper presented by H. Tunney.

EXPERIMENTAL

Treatments

Details are given in Table 1 of the different types of animal manures used, the rates of application, and the number of applications per year.

TABLE 1

ANIMAL MANURE TREATMENTS AND RATES APPLIED (t ha^{-1})

Treatment	Type of Manure	t ha^{-1}	No. Applications
A (68 kg N ha^{-1})	Control Fertiliser Nitrogen	-	1
B	Cattle manure	32.5	1
C	Cattle manure	65.0	1
D	Pig manure	65.0	1
E	Cattle manure	97.5	1
F	Cattle manure	97.5	3 (32.5 t)
G	Cattle manure	97.5	2 (65.0 t : 32.5 t)
H	Cattle manure	97.5	3 (32.5 t)
I	Poultry manure	32.5	1
J (204 kg N ha^{-1})	Fertiliser Nitrogen	-	4

Each treatment was replicated four times on a Perennial ryegrass *(Lolium perenne)* White clover *(Trifolium repens)* sward growing on a moderately well drained Brown Earth soil. The total area of each treatment (0.4 ha) was sufficient to provide grazing for at least three animals at a stocking rate of 7.4 animals ha^{-1}. Single applications of manures were applied in early 1976 and in the third week of March in 1977 and 1978. In the case of Treatment G the second application was after a silage cut in early June and the three applications, each of 32.5 t ha^{-1} (Spring, early June and Autumn) were on continuous grazing in Treatment F and grazing plus two silage cuts in Treatment H.

Animal manures

Cattle manure was collected from finishing cattle which were fed grass silage plus barley and were housed in a slatted floor building. The June and Autumn applied cattle manure was diluted with wash water. Pig manure came from fattening animals and was stored in overground tanks. The poultry manure came from laying hens in cages. It was stored in an overground tank for a few days and diluted with water to facilitate spreading. All the manures were well agitated before spreading with a vacuum tanker.

Animals

Twenty seven Friesian steers 15 - 18 months old, were used to graze the treatments. The mean liveweight at the start of the grazing season was 321.5 kg ± 12.0 in 1976, 271.3 kg ± 4.9 in 1977 and 298.3 kg ± 4.7 in 1978. They were divided into three groups of nine animals according to liveweight. One animal from each group was allocated randomly to each treatment. The animals were moved from each block of plots to the next at the same time. A stocking rate of 7.4 animals ha^{-1} was used and the grazing cycle was 24 days.

MEASUREMENTS

Herbage production

Prior to grazing each treatment and again when stock were removed, two 4.5 m^2 areas were cut from the swards with a powered rotary mower at a height of 1 - 2 cm above soil level. In 1977 and 1978 the animals were confined for one day to one-sixth of each treatment replication to measure intake. This procedure helped to make the determinations more accurate. Herbage from the sub plots was weighed and two 100 g samples used for dry matter determinations. Pre-grazing samples were used to estimate the herbage presented to the animals while the post grazing samples allowed determinations of herbage intake and rejection. Digestibility analysis of the herbage as presented was undertaken to indicate the quality of the sward.

Animal behaviour

Vibracorders were used to monitor the grazing behaviour of the animals on the different treatments. In this technique the grazing time is automatically recorded through a pendulum and stylus on to an eight day recording chart. Work elsewhere (Castle et al., 1975; Werk et al., 1974) has shown that vibracorders are simple, reliable and accurate instruments for measuring grazing behaviour and can save many tedious hours of day and night observations. The times grazing started and ended, plus the times spent grazing, were recorded.

Animal liveweight gain

At the end of each grazing cycle the cattle were weighed to assess performance and daily liveweight gain.

RESULTS

In 1976 dry warm weather followed the application of the manure causing the cattle manure, especially the 97.5 t ha^{-1} application, to dry on the soil surface and retard grass growth. A similar effect occurred again in 1977 when dry cold weather followed the manure applications. However, in 1978, 24 mm of rainfall occurred during the week the manures were spread and this, plus 40 mm precipitation in April, helped to wash the manures quickly into the base of the sward. There was, therefore, no smothering of the grass swards as in the previous years.

Chemical analysis of the manures

Details of the chemical analysis of the manures are given in Table 2.

There was definite variation in the cattle manures applied in the spring of 1976 and 1977. This was because some of the manure came from animals fed meals plus silage while some was from animals fed only silage. There was also some dilution by water from drinking bowls. The cattle manure applied in

TABLE 2

CHEMICAL COMPOSITION OF MANURE SAMPLES

Year	Type of Manure		Analysis			
			% DM	% N	% P	%K
1976	Cattle:	Spring	10.5 ± 0.5	0.50 ± 0.02	0.08 ± 0.04	0.53 ± 0.09
		Spring	7.85 ± 1.22	0.36 ± 0.08	0.07 ± 0.02	0.31 ± 0.04
		Summer	5.9 ± 1.4	0.27 ± 0.09	0.06 ± 0.01	0.06 ± 0.03
	Pig		8.8 ± 6.7	0.32 ± 0.13	0.16 ± 0.14	0.10 ± 0.03
	Poultry		9.8 ± 1.9	0.53 ± 0.24	0.23 ± 0.09	0.26 ± 0.10
1977	Cattle:	Spring	12.1 ± 0.58	0.54 ± 0.03	0.10 ± 0.01	0.59 ± 0.22
		Spring	9.2 ± 0.56	0.49 ± 0.04	0.10 ± 0.22	0.36 ± 0.05
		Summer	5.7 ± 0.80	0.26 ± 0.06	0.06 ± 0.03	0.21 ± 0.04
	Pig		13.3 ± 3.7	0.53 ± 0.08	0.34 ± 0.14	0.28 ± 0.20
	Poultry		12.6 ± 4.7	0.98 ± 0.44	0.36 ± 0.13	0.34 ± 0.24
1978	Cattle:	Spring	9.0 ± 1.7	0.39 ± 0.04	0.08 ± 0.01	0.48 ± 0.12
		Summer	5.8 ± 1.3	0.25 ± 0.08	0.04 ± 0.02	0.29 ± 0.04
	Pig		11.0 ± 1.0	0.47 ± 0.07	0.24 ± 0.02	0.25 ± 0.07
	Poultry		13.6 ± 1.6	0.96 ± 0.11	0.27 ± 0.02	0.73 ± 0.20

TABLE 3

ESTIMATED AMOUNTS OF NITROGEN APPLIED IN MANURES

	Manure t ha^{-1}		N kg ha^{-1}			Mean
			1976	1977	1978	
A	Control	0	68	68	68	68
B	Cattle	32.5	140	149	124	138
C	Cattle	65.0	280	336	228	281
D	Pig	65.0	208	345	380	311
E	Cattle	97.5	419	504	380	434
F	Cattle	97.5	316	319	286	318
G	Cattle	97.5	368	384	309	354
I	Poultry	32.5	172	319	312	268
J	Fertiliser N	0	204	204	204	204

the summer of all three years was diluted with wash water and consequently had a lower nutrient level than spring applied manures. The pig and poultry manures used in 1976 were generally of lower nutrient content than in 1977 and 1978. This was due to more water, from rain and washing, gaining access to the pig manure and greater dilution with water of the poultry manure for spreading purposes.

The amount of nitrogen (N) applied in the manures is of greatest interest in regard to grass production and animal performance. Estimates of the amounts applied ha^{-1} $year^{-1}$ are given in Table 3. There was variation in the quantity of N applied, depending on the type, quantity and time of manure application. There was also variation from year to year, especially for pig and poultry manures in 1976 compared with 1977 and 1978. Very high rates of N ha^{-1} were applied in the pig and poultry manures in 1977 and 1978 and at the high rate of application of cattle manure in all three years.

Herbage production

Grazing of the swards was delayed until the first week of May each year, i.e. about 5 weeks after spreading the manures. Some estimates of herbage dry matter production for the three years are given in Table 4. Treatment H is discarded because the grass was cut for silage. Drought in the summer of 1976 reduced herbage production much below normal and curtailed the autumn grazing season. Cold weather in April and May plus a mid season drought retarded growth in 1977, while 1978 was a very good grass growing season. The cattle manure treatments produced either a small increase in herbage yields or reduced it slightly as with Treatment G in 1977 and Treatment B in 1978. Pig manure and more particularly poultry manure significantly increased herbage yields in all three years. This suggests that the N content of poultry manure was more effective than that of cattle or pig manure.

TABLE 4

THE EFFECT OF MANURE APPLICATIONS ON HERBAGE DRY MATTER PRODUCTION (100 kg ha^{-1})

Year	Treatments									SE
	A	B	C	D	E	F	G	I	J	
1976	42.1	47.0	47.5	50.2	48.2	40.1	46.1	57.0	54.4	1.78
1977	60.5	62.4	61.1	68.3	65.9	64.1	54.5	79.8	70.7	2.46
1978	62.4	54.3	80.9	82.4	79.7	70.2	60.5	72.2	85.9	9.6

TABLE 5

INFLUENCE OF MANURE APPLICATIONS ON HERBAGE INTAKE (kg DM 100 kg^{-1} liveweight)

Manure	t ha^{-1}	1976		1977		1978	
		May 7 July 20	July 20 Oct. 12	May 3 July 25	July 25 Oct. 19	May 2 July 26	July 26 Oct. 9
Control	0	1.41	0.67	1.82	0.82	1.60	0.97
Cattle	32.5	1.17	0.66	1.77	0.84	1.56	0.66
Cattle	65.0	1.39	0.33	1.70	1.04	1.94	1.26
Pig	65.0	1.33	0.60	1.70	1.07	2.30	1.11
Cattle	97.5	1.12	0.58	1.57	1.12	2.07	1.06
Cattle	97.5	1.03	0.73	1.72	1.07	1.80	0.95
Cattle	97.5	0.93	0.81	1.58	0.74	1.65	0.72
Poultry	32.5	1.18	0.95	1.66	1.41	1.97	0.71
Fert. N	0	1.22	0.95	2.03	0.88	2.17	1.25
SE		0.35	0.13	0.10	0.14	0.18	0.16

TABLE 6

INFLUENCE OF CATTLE MANURE ON GRAZING BEHAVIOUR (MEAN OF FIRST GRAZING 1976, 1977 AND 1978)

	Start	Finish	Morning	Afternoon	Daily total
Control	05.48 hrs	21.33 hrs	145 mns	327 mns	472 mns
Cattle manure 95.7 t ha^{-1}	05.10 hrs	22.05 hrs	196 mns	370 mns	566 mns

Herbage intake

The 1976 applications of animal manures had the biggest influence on herbage intake. In the first grazing period of that year there was a reduction in intake which was very evident with the high rate of cattle manure application i.e. 97.5 t ha^{-1}. Poultry manure also reduced the intake in 1976. Drying of the manures on the surface of the herbage following their application was responsible for the reduced intakes. A similar but smaller effect occurred at the first grazing in 1977, while in 1978 the treatments did not affect intake. Some details of mean intakes are given for the first and second halves of the grazing seasons in Table 5.

In the second half of all three years there was a marked reduction of herbage intake. This was due to drought restricting grass growth in 1976 and 1977 and overstocking in 1978. Whilst the differences were small between treatments there was generally a lower intake on cattle manure treatments relative to the other treatments in the May to July period. In the July to October period intake was usually highest on the poultry manure and high N treatment.

Digestibility analysis of the herbage presented to the grazing animals in 1976 showed a steady decline from a mean DMD percent in May of 71.3 to 67.9 in June, 60.5 in July, 53.0 in August, 44.0 in late August - September and 35.3 in September - October. This showed the steady progressive effect of drought in that year. In 1977 herbage quality was also poor due to drought, varying from 58.8 percent DMD in May to 53.2 in June, 55.0 in July, 61.2 in August, 66.3 in late August - September and 61.0 in late September - October. In 1978, however, herbage quality was high, usually in excess of 70 percent DMD, throughout the season.

Animal behaviour

Over the season the grazing behaviour of the animals was very much influenced by the time of sunrise and sunset. In May grazing started at 5.30 ± 15 minutes and finished at

21.50 ± 15 minutes but in late May and most of June grazing started 30 minutes earlier and finished 30 minutes later. The total time spent grazing of 540 ± 20 minutes varied very little in May to August period. In the long-day period of May and June the animals tended to rest more both in the morning and afternoon. From September onwards grazing started later at 6.30 ± 20 minutes approximately and ended at 20.30 ± 40 minutes, while in October grazing finished at 17.50 to 18.15. There were, therefore, fewer rest periods between the start and the finish of grazing each day.

Application of animal manures modified the grazing behaviour. This occurred particularly in the first and second grazing cycles of each season, where the highest rate of cattle manure (97.5 t ha^{-1}) was applied (Table 6).

Animals commenced grazing earlier and finished later on the plots treated with animal manure. This was also evident in 1978 when the manures were more thoroughly washed into the base of the swards than in the previous two seasons. Evidently the animals spent more time selecting herbage and grazing was more intermittent than on the control or nitrogen treated swards. There was little or no difference in the grazing behaviour of the animals after the second grazing cycle on most swards or after the third cycle on the 97.5 t ha^{-1} cattle manure treatment.

Grazing during the night was very rare except for the occasional 20 to 30 minutes which occurred on most treatments. In October 1977, there was a complete breakdown in the grazing behaviour when grass was scarce, and animals regularly grazed for 60 to 90 minutes most nights. In general, however, one third of all grazing time took place in the morning with the remaining two thirds taking place beween noon and nightfall. The most concentrated grazing occurred in the three to four hours before sunset.

Animal performance

Any detrimental effects on animal performance arising from the manurial treatments might be expected to occur in the first or second grazing cycles immediately following application. Manure rates as high as 65 t ha^{-1} did not reduce animal performance in the first grazing cycles of 1976 and 1977, but there was a slight reduction of 5 to 6 kg $animal^{-1}$ in 1978. The single high rate of 97.5 t ha^{-1} cattle manure reduced markedly animal performance at the first cycle in 1976, i.e. liveweight gain 14 kg $animal^{-1}$ in contrast to 36 kg $animal^{-1}$ on the control treatment. This reduction in performance was partly compensated for by better liveweight gains in the second and third grazing cycles. Details are given in Table 7 of the mean animal performance for the three years over five grazing cycles.

Reduction in animal performance at the first grazing using the single high rate of cattle manure was overcome somewhat by splitting the application. There was only a small benefit, however, from such a high rate of application. Over all of the grazing cycles the best performance usually occurred with the poultry manure treatment.

Details of the liveweight gains for the whole of the seasons in 1976, 1977 and 1978 and the mean annual production for the three years are given in Table 8.

Significant reductions in animal liveweight gains occurred in the first year, mainly on the cattle and pig manure treatments. In 1977 and 1978 these same treatments increased production. The mean for the three years showed that all the animal manures produced a positive effect with the exception of the 32.5 t ha^{-1} cattle manure. The pig manure and the single or double application of 97.5 t ha^{-1} cattle manure were significantly effective ($P < 0.05$) while the poultry manure and the 204 kg N ha^{-1} increased animal gains very significantly ($P < 0.01$). The same trends were also evident in the liveweight production ha^{-1}.

TABLE 7

INFLUENCE OF ANIMAL MANURES ON LIVEWEIGHT GAIN (kg) DURING 24 DAY GRAZING CYCLES (MEAN OF 3 YEARS)

Treatment	Rate of manure application ($t\ ha^{-1}$)	Grazing cycle (24 days)				
		1	2	3	4	5
A. Control	0	39.7	32.3	16.3	14.3	3.0
B. Cattle	32.5	36.3	24.3	15.7	22.0	-4.0
C. Cattle	65.0	39.0	31.7	15.3	19.0	1.3
D. Pig	65.0	39.0	30.3	17.3	22.3	3.7
E. Cattle	97.5	31.0	30.0	23.0	19.0	4.3
F. Cattle	97.5	36.6	35.7	15.3	19.3	-1.0
G. Cattle	97.5	34.0	27.3	21.3	20.7	11.0
I. Poultry	32.5	42.7	29.3	24.7	33.3	5.3
J. High N	0	33.7	31.7	21.7	21.7	8.7

TABLE 8

EFFECT OF MANURES ON ANIMAL LIVEWEIGHT GAIN AND PRODUCTION

Treatment	Rate of manure application ($t\ ha^{-1}$)	Liveweight gain ($kg\ animal^{-1}\ season^{-1}$)				Production 1976-1978 ($kg\ animal^{-1}\ ha^{-1}$)
		1976	1977	1978	1976-78	
Control	0	123.3	87.6	88.0	99.7	739
Cattle	32.5	112.0	96.3	77.0	95.1	705
Cattle	65.0	102.3	120.3	87.6	103.4	766
Pig	65.0	91.3	130.6	102.3	108.1	801
Cattle	97.5	115.3	112.6	93.3	107.1	794
Cattle	97.5	110.0	110.0	101.6	107.2	794
Cattle	97.5	100.3	122.0	80.6	101.0	748
Poultry	32.5	138.6	122.3	128.0	129.7	961
Fert. N	0	132.0	106.3	105.6	114.7	850
	SE	9.16	7.58	13.39	3.70	
	DF	16	16	16	48	

SOIL ANALYSIS

Phosphorus (P) - analysis

In 1976 all the animal manure treatments, exept 32.5 ha^{-1} cattle manure, increased the soil P level (0 - 7.5 cm depth). The pig and poultry manure raised the readings from 4 to 5 mg kg^{-1} to 8 mg kg^{-1} in the first year and by the end of 1978 the levels had risen to 13 - 14 mg kg^{-1}. The 65 t ha^{-1} cattle manure raised the P level to 6.5 mg kg^{-1} at the end of 1978 while the 97.5 t ha^{-1} raised it to 8 - 9 mg kg^{-1}. There was only a small increase of the soil P level with the 97.5 t ha^{-1} cattle when the sward was cut for silage.

Potassium (K) - analysis

Where cattle manure application only slightly increased soil P levels, it resulted in very high levels of K in the soil. The 32.5 t ha^{-1} cattle manure increased the K level from 57 mg kg^{-1} in 1976 to 154 mg kg^{-1} in late 1978. The 97.5 t ha^{-1} cattle manure doubled the soil K level in the first year and produced almost a five fold increase by the end of 1978 (254 mg kg^{-1}). Where silage was cut on the high cattle manure treatment there was a slightly lower soil K level (194 mg kg^{-1}). Somewhat smaller K levels of 118 mg kg^{-1} and 84 mg kg^{-1} were recorded with the pig and poultry manures at the end of the third year.

Soil pH

All the animal manures increased the soil pH readings. The cattle manures raised the pH from 6.2 to 6.4 - 6.6, while pig manure gave mean readings of pH 6.6 and poultry manure of pH 6.8.

Botanical composition

There was no deleterious effect of the animal manures on the sward composition. The single high dressing of cattle manure (97.5 t ha^{-1}) increased the *L. perenne* content from

58 percent to 69 percent while reducing slightly the content of *Poa trivialis* and *Agrostis spp*. It did, however, reduce the sward density from 993 plants m^{-2} in the control to 844 plants m^{-2}. Continuous grazing in association with the manures helped to maintain the *L. perenne* content. When swards were cut for silage and then grazed or vice versa, it was reduced to 30 and 43 percent respectively. Both the pig and poultry manures increased the *P. trivialis* contents marginally while reducing the *L. perenne* 5 to 7 percent.

DISCUSSION

Herbage production generally showed a positive relationship to the rate of cattle and pig manure applied. In the first grazing period each year, and sometimes the second, the higher rates of manure (65 and 97.5 t ha^{-1}) produced some negative effects which were offset by positive effects in later grazings. Poultry manure produced the highest herbage yields in two out of the three seasons, thus reflecting its high content of readily available N. An effect of both the poultry manure and the high rate of cattle manure was evident in the autumn grazing period.

Correlation of herbage yields with herbage intakes was made by comparing pre and post grazing samples each time the treatments were grazed. In 1976 and 1977 there was some reduction of intake in the May to July period on animal manure treatments but in 1978 there was an increase in intake. The dry weather pattern of 1976 and 1977 was mainly responsible for this result. In the first grazing of 1976, the single high dressing of cattle manure reduced the grass yield. This caused reduced intake which in turn was reflected in reduced animal performance.

Approximately five weeks elapsed between spreading the manures and the commencement of grazing. With the exception of the 97.5 t ha^{-1} cattle manure in 1976 and 1977, this period was adequate to allow the rain to wash the sward and for the grass to grow through the manure. If grazing occurred earlier it is likely that the effects recorded on the high cattle manure treatment would occur also on the other treatments. In most situations, delay in the start of grazing is a major loss because it shortens the grazing season. This is one disadvantage of spreading manures on pasture.

Rejection of herbage contaminated by dung droppings is a well known feature of pastures and is influenced by the stocking rate (Collins, 1967). This type of rejection arises because animals have a choice between clean and fouled grass. However, when spread on pasture, there is little opportunity for selection and animal grazing behaviour is modified. The degree of change in behaviour is related to the rate of manure applied. With low rates of 32.5 t ha^{-1} grazing is somewhat more intermittent because more time is spent selecting grass. At the higher rates of manure application, especially with 97.5 t ha^{-1}, animals spent more time selecting their feed. This can extend the grazing time by as much as 20 percent. Such an effect is usually most marked in the first grazing cycle and unimportant after the second cycle. This modification in grazing is less evident when rain follows manure application.

Short term reductions in grass growth and modifications in grazing behaviour which reduced herbage intake had only a transitory influence on animal performance.

The manures, except the poultry manure, reduced animal liveweight gain in 1976, but increased it in 1977 and 1978. The mean results for the three seasons showed a positive effect of the manures and in particular the poultry manure ($P < 0.01$). Some of the poor animal performance in both 1976 and 1977 arose because of drought which accentuated the bad effects of the manures on sward regrowth. Reduced animal performance in a grazing cycle, which is compensated for by increased performance in later grazing periods, is acceptable with beef cattle, but not with dairy cows. If milk production is reduced, it is difficult to increase again to the former level. Also, dairy cows have less time for grazing than beef cattle because a proportion of the day is spent in milking. Therefore, the extra grazing time required to compensate for the time lost in selecting feed may not be available. For these reasons it is advisable to spread animal manure at either low rates or on pasture grazed with beef cattle.

The effectiveness of the N supplied in the manures was low, except for poultry manure. In the experiment the fertiliser N produced a mean response of 0.82 kg liveweight gain kg^{-1} N applied. Based on this result the effectiveness of the N in the different manures was as follows: (a) Cattle - mean 26 percent; (b) Pig - 38 percent; (c) Poultry - 100 percent. When residual effects of cattle manure are determined a substantial increase is expected in the effectiveness of the N supplied at the higher rates of application.

The conclusions from the findings are that pasture is an outlet for the disposal of animal manures, but there are restrictions. When moderate rates (32.5 t ha^{-1}) of cattle manure are applied no difficulties arise and animal performances comparable to that of swards treated with fertiliser at 68 kg N ha^{-1} can be expected. Problems of animal grazing behaviour and performance are only encountered at high (65 t ha^{-1}) or excessively high (97.5 t ha^{-1}) rates of application which should not be used in practical farming. Besides the immediate effect of animal manures on herbage utilisation, the more long term effect of K build-up in the soil must be considered. Even the low rate of cattle manure (32.5 t ha^{-1}) can substantially increase the soil K level within three years. Hence, if swards are continuously grazed cattle manure should be applied at lower annual rates.

ACKNOWLEDGMENT

The author wishes to thank Mr. P. Collins for technical assistance in undertaking the experiment, Miss B. Wheeler for chemical analysis of manures, and Mr. J. Dwyer for digestibility analysis.

REFERENCES

Castle, M.E., MacDaid, E. and Watson, J.N., 1975. The automatic recording of the grazing behaviour of dairy cows. J. Br. Grassl. Soc., Vol. 30, 161-163.

Collins, D.P., 1968. Dung distribution by cattle and sheep. Research Report, Soils Div., An Foras Taluntais, Dublin 1967, pp. 96 and 90.

Walshe, M.J. and Conway, A., 1960. Hypomagnesaemia in ruminants. Proc. Eight Int. Grassl. Congr., Reading, 5B/2, 548-553.

Werk, O., Finger, H. and Ernst, P., 1974. Investigations in feed times of grazing cattle. 5th Gen. Meet. Eur. Grassld. Fed., Uppsala, Vaxtodling, Plant Husbandry (1974), No. 29, pp. 193-6.

DISCUSSION

J.K.R. Gasser *(UK)*

Could I offer a comment on the effectiveness of your poultry slurry on grass, which may be related to animal nutrition because poultry manure can be used directly as a feed supplement for ruminants. In applying poultry manure to the grass, is a direct effect on animal nutrition obtained from the poultry manure as well as an indirect effect through the grass, particularly as cattle can take up to 5 percent or 10 percent of ingested dry matter as soil and if the poultry manure slurry is on the soil they will obviously ingest it.

H. Tunney *(Ireland)*

I agree, that may be a part of the explanation.

J.R. O'Callaghan *(UK)*

On feeding the silage made from grass with slurry to the livestock, the performance really dropped catastrophically, so that there appears to be some effect on the quality of the silage which is adversely affecting the performance.

H. Tunney

The paper gives some silage quality results, which show that slurry did not adversely affect silage quality, neither the cattle slurry nor the pig slurry, relative to chemical fertiliser. The pH was the same. The dry matter digestibility was the same. The measurement of intake was not accurate enough to allow a definite decision to be made whether intake was lower with silage from slurry treated grass - it may be, but we must test that.

TIME AND RATE OF APPLICATION OF ANIMAL MANURES

P.V. Kiely*
The Agricultural Institute,
Johnstown Castle Research Centre, Wexford, Ireland.

INTRODUCTION

The intensification of pig and cattle produced in larger units and the continuing trend towards slatted floor housing have resulted in the production of large quantities of semi-liquid slurry, which has to be disposed of. Although spreading slurry on land is one of the cheapest methods of disposal, a major cost item in both pig and cattle housing is the provision of storage facilities for slurry, which should therefore be kept as small as is practicable. This would be helped by the ability to spread slurry throughout the year particularly during the winter period when most cattle slurry is produced.

Research on landspreading of animal slurry has been aimed principally at determining maximum amounts which can be applied without causing water pollution or damage to the sward (O'Callaghan et al., 1973; Cromack, 1974; Dodd et al., 1975). As adequate areas of grassland are available in Ireland for the disposal of slurry the objective is to make optimum use of it as a fertiliser source, because animal manures are a valuable source of nutrients for crop production (Tunney, 1975). The rapid and large increases in the costs of fuel and mineral fertilisers in recent years have aroused a new interest in the fertilising value of animal manures, particularly slurry. Studies on the availability of nutrients in slurry have shown that all of the K is available for plant growth in the year of application, because the K in animal manures is all water soluble. Most research indicates that the P in animal manures is as effective as that in mineral fertilisers (Kofoed, 1976; Kolenbrander and de la Lande Cremer, 1976; Vetter, 1973; May and Martin, 1966). However, some studies suggest that P

*Paper presented by J.C. Brogan

in manures is only half as effective as fertiliser P in the year of application (Anon, 1976). In the short term pig slurry was only one-tenth as effective as fertiliser P in increasing the P content of herbage (Adams, 1974) suggesting that the release of P from slurry is a gradual process.

Nitrogen in animal manures appears to vary very much in availability ranging from zero to almost complete uptake and to be affected by time of application, weather conditions, dry matter content of the slurry and whether it is incorporated into the soil or left on the surface. In England from 50 to 75 percent of the nitrogen in both pig and cattle slurry was available in the year of application (Anon., 1976), in Scotland and Northern Ireland, from 44 to 84 percent of the nitrogen in pig slurry was available (Castle and Drysdale, 1966); in West Norway, from 60 to 80 percent of the nitrogen in cattle slurry was available (Naess and Myhr, 1976); in Denmark, 40 percent of the nitrogen in animal manures was estimated to be available (Kofoed, 1976); in Germany, 50 percent of the nitrogen in cattle slurry was estimated to be effective (Tietjen, 1976); and in Canada, 50 percent of the nitrogen in cattle manure was available in the year of application (Baldwin, 1975). In each case, availability is compared with uptake from mineral fertilisers and is not based on the total nitrogen content of the slurry.

The effect of time of application on N availability has received little attention. In England,Cromack (1974); Anon. (1976) estimated that approximately 16 percent of the N in pig and cattle slurry was available from slurry applied in December, 25 percent from that in January and 50 percent from that in March.

This study aimed to determine the effects of time and rate of application of pig and cattle slurry on nutrient loss (especially nitrogen), yield of herbage, sward composition, residual value of slurry nitrogen and damage to the soil by machinery.

MATERIALS AND EXPERIMENTAL

Experiments were established in spring 1976 on grassland on both a free draining and and an impeded soil type. Two amounts of pig slurry and two amounts of cattle slurry were each applied at one of four times during the season, late November, late February, late March and early June. These dressings were compared with three amounts of N applied as fertiliser and with grass without any treatments.

The rates and times of application of fertiliser-N were:

N_0 = no nitrogen

	Late March	Early June	August	
N_1 =	33.3	25	16.7	)
N_2 =	66.7	50	33.3	) kg N ha^{-1}
N_3 =	100	75	50	)

The rates of slurry application were approximately 30 and 45 m^3 ha^{-1}. The pig slurry used was from a fattening house and the cattle slurry from a slatted floor house where mature animals were fed grass silage and some concentrates.

An experimental layout of 20 treatments in a randomised block with 5 replications was employed. The experimental area at each site was approximately 2.5 ha.

Grass was cut in late May/early June, early August and October. Yield of dry matter and N, P, K and Mg contents of the herbage were determined. Soil samples were taken at the end of each season to measure changes in soil properties caused by the different treatments. A botanical analysis of the plots was carried out at the end of the third season to determine botanical changes caused by the successive slurry and fertiliser nitrogen treatments.

YIELDS OF DRY MATTER

As results varied from year to year due to differences in weather conditions after slurry application and other factors the results are more easily presented on a yearly basis:

1976

The experiment was not begun until spring, so that the slurry applications which should have been applied in November and February were not applied until late March with the late March application. Consequently only slurry applied in late March and June could be compared with the fertiliser N treatments for the 1976 season. Responses to N and to slurry were affected by the relatively dry season particularly on the free draining soil type.

On the impeded soil late March/early April applications of nitrogen as pig slurry supplying 112 and 153 kg N ha^{-1} were both superior to 100 kg N ha^{-1} as ammonium nitrate at the first cut (Table 1) and there was a small residual effect from 153 kg N ha^{-1} as slurry at the second cut in August. There was no response to similar levels of N applied as cattle slurry. A comparable application of nitrogen as pig slurry in June after the first harvest increased yields of grass by an amount equivalent to 12 to 30 percent of its N content while cattle slurry applied in June depressed yield, presumably by coating the sward.

On the free draining soil approximately 10 percent of the nitrogen applied as pig slurry in early April was available at the first cut whereas cattle slurry did not significantly increase yields of dry matter (Table 2). Pig slurry applied in June did not increase yields and cattle slurry applied in June depressed them.

TABLE 1

MEAN YIELDS OF DRY MATTER OBTAINED ON THE IMPEDED SOIL IN 1976

Treatment	Time of application	Amount of N applied kg N ha^{-1}	Yield of dry matter		
			Cut 1 May/June t ha^{-1}	Cut 2 August t ha^{-1}	Cut 3 October t ha^{-1}
SLURRY					
Cattle 1	Late March/ early April	104	6.11	2.25	0.77
Cattle 2		151	6.11	2.42	0.76
Pig 1		112	8.06	2.68	0.76
Pig 2		153	7.78	3.31	0.75
Cattle 1	June	103	6.20	2.26	0.75
Cattle 2	"	149	6.79	2.25	0.89
Pig 1	"	111	6.01	2.99	0.82
Pig 2	"	152	5.99	3.75	0.84
FERTILISER					
N_0	See text	0	6.71	2.71	0.86
N_1	"	75	6.42	3.25	0.86
N_2	"	150	7.48	3.56	1.23
N_3	"	225	7.47	4.41	1.71
Least significant difference ($P = 0.05$)			0.59	0.47	0.14

The application of pig slurry to the soil with impeded drainage was followed by rain whereas the other applications were followed by a dry period which might explain some of the differences in response obtained. Cattle slurry depressed Mg content of herbage on both soil types presumably because of its high K content. Potassium contents of soils were consistently larger in plots which received cattle slurry and pig slurry increased the content of soil P.

TABLE 2

MEAN YIELDS OF DRY MATTER OBTAINED ON THE FREE DRAINING SOIL TYPE IN 1976

Treatment	Time of application	Amount of N applied kg N ha^{-1}	Yield of dry matter Cut 1 May/June t ha^{-1}	Cut 2 August t ha^{-1}	Cut 3 October t ha^{-1}
SLURRY					
Cattle 1	Early April	106	3.13	1.47	0.39
Cattle 2		153	3.59	1.48	0.37
Pig 1	"	113	3.68	1.50	0.41
Pig 2	"	162	3.80	1.45	0.42
Cattle 1	June	104	3.30	1.27	0.42
Cattle 2	"	150	3.24	1.23	0.45
Pig 1	"	96	3.19	1.48	0.49
Pig 2	"	138	3.20	1.63	0.45
FERTILISERS					
N_0	See text	0	3.26	1.68	0.38
N_1	"	75	4.05	2.82	0.57
N_2	"	150	5.44	3.59	0.72
N_3	"	225	5.37	3.89	0.80
Least significant difference (P = 0.05)			0.52	0.34	0.17

The most notable result obtained in 1976 was that cattle slurry applied in late spring under dry weather conditions was of no value as a source of N for grass. Also cattle slurry applied in June in dry weather conditions after a first silage cut might depress grass growth presumably by coating the sward. Pig slurry applied in late spring under wet weather conditions was an excellent source of N and produced yields better than those from a similar amount of fertiliser N.

1977

On the impeded soil, slurry N applied in late November, late February and late March applications significantly increased yields of grass at the first cuts (Table 3). Cattle slurry increased yields most when applied in late November;

TABLE 3

MEAN YIELDS OF DRY MATTER OBTAINED ON THE IMPEDED SOIL IN 1977

Treatment	Time of application	Amount of N applied kg N ha^{-1}	Yield of dry matter Cut 1 May/June t ha^{-1}	Cut 2 August t ha^{-1}	Cut 3 October t ha^{-1}
SLURRY					
Cattle 1	November	98	7.66	1.57	1.23
Cattle 2	"	142	7.48	1.83	1.36
Pig 1	"	95	7.84	1.57	0.97
Pig 2	"	138	8.73	1.58	0.99
Cattle 1	Late February	178	7.81	1.62	1.26
Cattle 2		261	6.98	1.40	1.05
Pig 1	"	198	7.53	1.48	1.02
Pig 2	"	290	8.51	1.48	0.81
Cattle 1	Late March	131	6.88	1.69	0.97
Cattle 2	"	191	6.93	1.39	1.20
Pig 1	"	121	8.48	1.56	1.05
Pig 2	"	177	8.24	1.43	0.68
Cattle 1	June	114	6.41	1.35	1.09
Cattle 2	"	164	6.27	1.58	1.23
Pig 1	"	130	6.38	2.76	1.28
Pig 2	"	178	6.59	3.31	1.20
FERTILISER					
N_0	See text	0	5.76	1.34	1.18
N_1	"	75	6.92	1.80	1.08
N_2	"	150	7.93	2.79	1.42
N_3	"	225	8.78	3.39	1.93
Least Significant difference (P = 0.05)			0.69	0.31	0.16

up to 56 percent of N in cattle slurry was utilised. The smaller amount applied was better than the larger suggesting that application rates of more than 150 kg N ha^{-1} as cattle slurry were undesirable. Nitrogen in pig slurry was always more available than that in cattle slurry and yields of grass obtained with pig slurry were comparable with those obtained with 100 kg ha^{-1} as fertiliser N. As much as 70 percent of N from pig slurry was utilised. The late November and spring applications of pig slurry differed little in their effects on yields of grass (Table 3). Cattle slurry applied in June under wet soil and weather conditions increased yields little while pig slurry applied in June increased by an amount suggesting that about 40 percent of its N content was available.

On the free draining soil type largest yields and best utilisation of slurry N was obtained from slurry applied in late February when more than 70 percent of the N in pig slurry and 35 percent of that in cattle slurry were utilised (Table 4). The late November and late March applications gave similar increases when approximately 40 percent of the N in pig slurry and 20 percent of that in cattle slurry were utilised. Forty percent of the N content of pig slurry and 10 percent of that in cattle slurry applied in June under wet soil and weather conditions was utilised. The yields of grass obtained with pig slurry as a source of N in spring were greater than those from fertiliser N.

The 1977 results indicated that:

(a) pig and cattle slurry were good sources of N when applied in late winter or spring. Pig slurry was also a moderately good source of N when applied under wet soil and weather conditions in summer.

(b) the optimum time to apply cattle slurry for maximum utilisation of its N content was late winter on soils with impeded drainage and early spring on free draining soils.

TABLE 4

MEAN YIELDS OF DRY MATTER OBTAINED ON THE FREE DRAINING SOIL TYPE IN 1977

Treatment	Time of application	Amount of N applied kg N ha^{-1}	Yield of dry matter Cut 1 May/June t ha^{-1}	Cut 2 August t ha^{-1}	Cut 3 October t ha^{-1}
SLURRY					
Cattle 1	November	100	3.13	1.74	0.46
Cattle 2	"	146	3.85	1.97	0.46
Pig 1	"	140	5.08	2.26	0.52
Pig 2	"	203	5.73	2.17	0.59
Cattle 1	Late February	108	3.99	1.81	0.42
Cattle 2		158	5.20	2.04	0.44
Pig 1	"	142	6.00	2.00	0.49
Pig 2	"	207	6.54	2.04	0.59
Cattle 1	Late March	84	3.58	1.91	0.42
Cattle 2	"	123	4.03	1.88	0.42
Pig 1	"	150	5.19	2.08	0.47
Pig 2	"	218	5.15	2.33	0.47
Cattle 1	June	135	2.60	2.37	0.43
Cattle 2	"	195	3.05	2.33	0.51
Pig 1	"	150	2.68	3.76	0.91
Pig 2	"	216	2.92	4.54	1.06
FERTILISERS					
N_o	See text	0	2.66	2.08	0.43
N_1	"	75	4.39	2.47	0.66
N_2	"	150	5.51	3.68	0.80
N_3	"	225	5.69	4.73	1.16
Least significant difference (P = 0.05)			0.32	0.32	0.08

<u>1978</u>

On the impeded soil, swards responded little to applied N and to slurry partly because of the changes in botanical composition resulting from the different treatments. The optimum time to apply both pig and cattle slurry was late February (Table 5). Pig slurry applied in late spring was better than that applied in late November whereas for cattle slurry, late November application was better. Early spring application of pig slurry increased yields of herbage more than 100 kg N ha^{-1} as fertiliser N; 75 percent of N in pig slurry was utilised. Cattle slurry applied in June depressed yield of the second cut whereas almost 60 percent of the N in pig slurry applied in June was utilised.

The optimum time to apply pig slurry on the free draining soil type for a first cut for silage in late May was late February when 60 percent of N in pig slurry was utilised (Table 6). Approximately 40 percent of N in pig slurry was utilised from that applied in late November and 25 percent from late March application. Cattle slurry increased yields little whenever it was applied - only 15 to 30 percent of N in cattle slurry was utilised at the first silage cut. Cattle slurry applied in June did not increase yields, whereas for pig slurry 35 percent of the N applied in June was utilised at the second cut.

BOTANICAL COMPOSITION

A botanical analysis at the end of the third season indicated marked changes in botanical composition resulting from the different treatments. Clover content of the sward decreased with increasing amounts of N applied (Tables 7 and 8) on both soil types. Swards receiving cattle slurry had significantly more clover than those with pig slurry reflecting the greater availability of N in pig slurry. On the impeded soil, Cocksfoot (*Dactylis glomerata*) was the dominant grass species on plots with the N_2 and N_3 fertiliser treatments and on plots

TABLE 5

MEAN YIELDS OF DRY MATTER OBTAINED ON THE IMPEDED SOIL TYPE IN 1978

Treatment	Time of application	Amount of N applied kg N ha^{-1}	Yield of dry matter Cut 1 May/June t ha^{-1}	Cut 2 August t ha^{-1}	Cut 3 October t ha^{-1}
SLURRY					
Cattle 1	Late November	87	6.57	1.54	1.75
Cattle 2		129	6.81	1.24	1.53
Pig 1	"	115	6.53	1.03	1.50
Pig 2	"	184	7.49	1.08	1.53
Cattle 1	Late February	117	6.76	1.15	1.65
Cattle 2		164	6.90	1.08	1.43
Pig 1	"	135	7.02	1.12	1.38
Pig 2	"	197	8.05	1.02	1.11
Cattle 1	Late March	118	6.14	1.28	1.32
Cattle 2		172	6.27	1.21	1.61
Pig 1	"	106	7.06	1.28	1.20
Pig 2	"	154	7.47	1.10	1.30
Cattle 1	June	88	6.21	0.93	1.54
Cattle 2	"	129	6.23	1.03	1.73
Pig 1	"	87	5.77	1.63	1.36
Pig 2	"	149	6.18	1.88	1.36
FERTILISERS					
N_0	See text	0	6.35	1.06	1.29
N_1	"	75	6.86	1.17	1.30
N_2	"	150	6.97	1.37	1.25
N_3	"	225	7.22	1.99	1.74
Least significant difference (P= 0.05)			0.67	0.32	0.25

TABLE 6

MEAN YIELDS OF DRY MATTER OBTAINED ON THE FREE DRAINING SOIL TYPE IN 1978

Treatment	Time of application	Amount of N applied kg N ha^{-1}	Yield of dry matter Cut 1 May/June t ha^{-1}	Cut 2 August t ha^{-1}	Cut 3 October t ha^{-1}
SLURRY					
Cattle 1	November	75	3.36	1.14	1.11
Cattle 2	"	120	3.19	0.82	0.92
Pig 1	"	99	4.20	0.74	0.63
Pig 2	"	144	3.87	0.63	0.36
Cattle 1	Late February	131	2.80	0.68	0.89
Cattle 2		170	3.48	0.92	1.07
Pig 1	"	128	4.49	0.77	0.44
Pig 2	"	186	5.10	0.85	0.46
Cattle 1	Late March	138	3.46	1.12	1.07
Cattle 2		200	3.37	1.11	0.99
Pig 1	"	151	3.45	0.90	0.52
Pig 2	"	219	3.91	0.97	0.47
Cattle 1	June	88	2.91	0.78	1.12
Cattle 2	"	129	3.32	0.68	1.20
Pig 1	"	132	3.04	2.05	1.33
Pig 2	"	192	3.36	2.42	1.75
FERTILISERS					
N_0	See text	0	2.83	0.87	0.90
N_1	"	75	3.56	1.37	0.81
N_2	"	150	4.18	2.01	1.63
N_3	"	225	5.20	2.64	2.37
Least significant difference (P = 0.05)			0.40	0.31	0.29

TABLE 7

BOTANICAL COMPOSITION OF THE SWARD WITH THE DIFFERENT TREATMENTS AT THE END OF THE THIRD SEASON ON THE IMPEDED SOIL TYPE

Treatment	Time of application	Component of sward		
		Grass	Clover % of sward	Weeds
SLURRY				
Cattle 1	Late November	88.0	20.0	0
Cattle 2		77.0	23.0	0
Pig 1	"	82.0	17.6	0.4
Pig 2	"	80.0	19.5	0.5
Cattle 1	Late February	82.5	17.5	0
Cattle 2		72.0	27.0	1.0
Pig 1	"	79.0	20.5	0.5
Pig 2	"	84.0	16.0	0
Cattle 1	Late March	76.0	24.0	0
Cattle 2		70.6	27.8	1.6
Pig 1	"	78.5	21.0	0.5
Pig 2	"	86.5	13.0	0.5
Cattle 1	June	76.0	22.0	2.0
Cattle 2	"	70.0	30.0	0
Pig 1	"	85.0	15.0	0
Pig 2	"	85.0	14.0	1
FERTILISERS				
N_0	See text	75.0	25.0	0
N_1	"	87.5	12.4	0.1
N_2	"	91.5	7.9	0.6
N_3	"	98.3	0.3	1.4

TABLE 8

BOTANICAL COMPOSITION OF THE SWARD WITH THE DIFFERENT TREATMENTS AT THE END OF THE THIRD SEASON IN THE FREE DRAINING SOIL TYPE

Treatment	Time of application	Component of sward		
		Grass	Clover % of sward	Weeds
SLURRY				
Cattle 1	Late November	56	40	4
Cattle 2		48	47	5
Pig 1	"	73	21	6
Pig 2	"	85	4	11
Cattle 1	Late February	45	51	4
Cattle 2		64	33	3
Pig 1	"	81	9	10
Pig 2	"	92	4	4
Cattle 1	Late March	73	21	6
Cattle 2		67	28	5
Pig 1	"	63	31	6
Pig 2	"	88	7	5
Cattle 1	June	70	24	6
Cattle 2	"	66	28	6
Pig 1	"	83	15	2
Pig 2	"	93	4	3
FERTILISERS				
N_o	See text	64	31	5
N_1	"	90	5	5
N_2	"	97	1	2
N_3	"	100	0	0

with the larger amounts of pig slurry, while ryegrass was the dominant species on the plots with cattle slurry and smallest amounts of fertiliser N.

SUMMARY OF RESULTS

Pig slurry was a good source of N for grass particularly when applied in early spring. Up to 75 percent of N in pig slurry was utilised when less than 150 kg N ha^{-1} was applied at the first silage cut. Grass with pig slurry commonly out-yielded that receiving 100 kg N ha^{-1} as fertiliser. Pig slurry applied in late spring or summer in wet weather conditions was also a good source of N but pig slurry applied from April to June in dry weather was of little value as a source of N.

Nitrogen in cattle slurry was much less available than that in pig slurry and generally less than 30 percent of N in cattle slurry was utilised. The optimum time to apply cattle slurry for grass cut for silage in late May or early June was in early spring on free draining soils and late winter or early spring on soils with impeded drainage. Cattle slurry applied in June did not increase yields of grass and when applied in dry weather conditions in June it depressed yields.

Grass with the larger amounts of slurry applied, corresponding to approximately 150 - 200 kg N ha^{-1}, generally produced more dry matter than grass with the smaller amount except for cattle slurry applied in June, when the smaller amount increased yields more. More of the N applied in slurry was utilised with the smaller than with the larger amount applied. Slurry applied for the first cut of grass seldom increased yields at the second cut.

The botanical composition of the plots differed markedly at the end of the third year. Clover content decreased with increasing amounts of fertiliser N and swards with cattle slurry had more clover than those with pig slurry reflecting the greater availability of N in pig slurry. On the impeded soil,

Cocksfoot (*Dactylis glomerata*) was the dominant grass species on plots with the N_2 and N_3 treatments and on those with the larger amount of pig slurry; while ryegrass was the dominant grass on plots with least fertiliser-N and with cattle slurry.

Soil tests at the end of each season indicated that it contained more K when plots had received cattle slurry and more P when treated with pig slurry.

REFERENCES

Adams, S.N., 1973. The response of pasture in Northern Ireland to N, P and K fertilisers and to animal slurries. II Effects of mineral composition. J. agric. Sci., Camb. 81, 419-428.

Anon., 1976. In: 'Studies on Farm Livestock Wastes'. Agricultural Research Council, London.

Baldwin, C., 1975. Manure as a fertiliser. Manure Management Seminar, Ontario, Ministry of Agric. and Food, Canada.

Castle, M.E. and Drysdale, A.D., 1966. Liquid manure as a grassland fertiliser v the response to mixtures of liquid manure (urine) and dung. J. agric. Sci. Camb. 67, 397-404.

Chambley, C., 1975. Profitable utilisation of livestock manures. Leaflet 171, Ministry of Agriculture, Fisheries and Food, UK.

Cromack, H.T., 1974. Landspreading of slurry. Basic Asset, Winter 1973-74.

Dodd, V.A., Mulqueen, J. and Grubb, L., 1975. The management of animal manures in the catchment area of Lough Sheelin. Report No. 2, An Foras Taluntais, Dublin.

Kofoed, A. Dam., 1975. Farmyard manure and crop production in Denmark. Proc. EEC Seminar, 'Utilisation of Manure by Landspreading', Modena, Italy, EUR 5672e.

Kolenbrander, G.L. and de la Lande Cremer, L.C.N., 1967. Stalmest en Gier. H. Veenham & Zonen N.V. Wageningen.

May, D.M. and Martin, W.E., 19766. Manures are a good source of phosphorus. Calif. Agric., 20 (7), 11-12.

Naess, O. and Myhr, K., 1975. Slurry manuring for lea in West Norway. Report No. 20, State Agric. Exp. Station, Fureneset.

O'Callaghan, J.R., Dodd, V.A. and Pollock, K.A., 1973. The long term management of animal manures. J. Agric. Engng. Res., 18, 1-12.

Tietjen, C., 1976. The yield efficient nitrogen portion in treated and untreated manure. Proc. EEC Seminar, 'Utilisation of manure by Landspreading, Modena, Italy, EUR 5672e.

Tunney, H., 1975. Fertiliser value of livestock wastes. Proc. 3rd Intern. Symposium on Livestock Wastes, Amer. Soc. Agric. Engineers.

Vetter, H., 1973. 'Mist und Gulle'. DLG Verlag, Frankfurt.

DISCUSSION

J.H. Voorburg *(The Netherlands)*

Should more attention be paid to the amount of nitrogen in mineral form - that is as ammonia - and the amount of nitrogen in organic form, which can explain most of the differences in the results reported by several speakers.

J. Brogan *(Ireland)*

The mineral nitrogen was approximately 55 percent of the total-N in both the cattle and the pig slurry in most years, so the amount of mineral-N applied would have been approximately 60 - 70 kg N ha^{-1}. However, recoveries were calculated on total-N and not on mineral-N. Therefore in some years when recoveries were up to 70 - 75 percent, and in one year 80 percent, it suggests that some of the organic-N becomes available and is being recovered when it is applied at the correct time and in favourable circumstances.

K.W. Smilde *(The Netherlands)*

The speaker said that the rain improved the N use. Could Dr. Tunney tell us what was the effect of washing the grass with water after the application of slurry?

H. Tunney *(Ireland)*

Washing improved growth in two years out of three. In the third year, when rain fell within one day of application, the response was smaller, and it does not appear to be an effect of the water applied.

J. Brogan

Our results agree with Dr. Tunney's report earlier. In a number of years the pig slurry was superior, because although the nitrogen response curve was reaching a plateau, yields from grass with pig slurry were larger, suggesting that it had an extra beneficial effect in addition to its nitrogen content.

THE EFFECTS OF LANDSPREADING OF ANIMAL MANURES ON WATER QUALITY

Marie T. Sherwood
The Agricultural Institute,
Johnstown Castle Research Centre, Wexford, Ireland.

INTRODUCTION

In Ireland, increasing numbers of both cattle and pigs are housed in units with slatted floors for ease of effluent collection and storage and the best method for disposal of such effluent is to spread it on the land. Unfortunately disposal on land can pose a hazard to both surface and underground water resources and each country must evolve its own recommendations for the management of animal manures depending on its soil types, topography, land use practice and climate, to protect its water resources from pollution.

Such recommendations can only be reliably made if sufficient information is available, and this project was undertaken to increase our knowledge of the effects of land spreading of animal manures on water quality.

EXPERIMENTAL

Small plots of grass (2 m x 30 m) on two different soil types - a free draining loam (Hoarstone) and an impermeable silty-clay soil with gleying (Castlebridge) - were equipped with gutters to collect surface runoff water. Ceramic probes were installed at depths of 15, 30, 60 and 100 cm to sample infiltrating soil water. The particle size distribution of the two soils is given in Table 1.

Pig slurry was applied at three different volumes (22.5, 45 and 90 m^3 ha^{-1} approximately) using a long extension pipe fitted to the outlet of a slurry tanker with a helical rotor. The applications were made to each plot three times each year. Constant volumes were applied so that the weight of nutrients varied with the concentration. The dates of treatment applications and the approximate nutrient applications in the 90 m^3 ha^{-1} pig slurry treatments are indicated in Table 2, which shows the weight of N applied varied from 330 to 736 kg N ha^{-1}. The medium and low rate treatments received approximately a half and a quarter of these nutrients respectively. Fertiliser supplying 120 kg N ha^{-1} was also applied, together with P & K and there were plots with no treatment.

TABLE 1

PARTICLE SIZE DISTRIBUTION (%)

	0 - 15 cm		45 - 60 cm	
	Castlebridge	Hoarstone	Castlebridge	Hoarstone
Coarse sand	15	18	4	11
Fine sand	18	32	2	28
Silt	41	32	42	37
Clay	26	18	52	24

Runoff water was sampled after every major rainfall and infiltrating soil water was sampled every two weeks. To facilitate more detailed studies of the transformations in nitrogen fractions, the largest volume of pig slurry was applied to extra grass and fallow plots (2 m x 15 m) at each site on the same dates, from August 1977 onwards. A fresh plot was used on each application date and soil samples at various depths were taken on days 0, 1, 2, 4, 7, 14, 21 (approximately) after application and ammonium and nitrate nitrogen were measured.

RESULTS

Volatilisation of Ammonia

Little nitrogen was recovered from slurry applied in 1976 indicating substantial loss of nitrogen. Soils analyses from

TABLE 2

NUTRIENTS APPLIED (KG HA^{-1})

Date of application	Slurry volume $90m^3 ha^{-1}$								Fertiliser		
	Hoarstone				Castlebridge				Hoarstone and Castlebridge		
	Total N	NH_4^+-N	P	K	Total N	NH_4^+-N	P	K	N	P	K
June 30, 1976	401	-	186	170	391	-	172	195	120	42	51
Dec. 3, 1976	558	263	206	233	547	227	189	210	120	42	51
Mar. 24, 1977	736	355	156	365	541	243	115	270	120	42	51
Aug. 18, 1977	346	287	40	313	404	304	43	235	120	42	51
Nov. 23, 1977	349	239	55	137	463	220	119	174	120	42	51
Mar. 28, 1978	330	183	81	123	380	283	101	133	120	42	51
June 14, 1978	503	339	102	227	520	344	110	225	120	42	51
Nov. 20, 1978	550	299	147	318	583	317	163	331	120	42	51
April 25, 1979	359	166	72	181	566	242	121	250	120	42	51

grass and fallow plots showed ammonium-N disappeared rapidly for 5 - 7 days following slurry application, with no corresponding rise in the NO_3^--N content (Figure 1). The loss was similar in both grass and fallow plots and was interpreted as mainly due to volatilisation of ammonia being comparable with the losses described by Lauer et al. (1976) in their field trials in the USA. Figure 2 shows that the percentage loss varied with season and more was generally lost from the wet Castlebridge soil than from the Hoarstone soil. The reason why more should have been lost in March than in August was not known but was not related to any single climatic parameter.

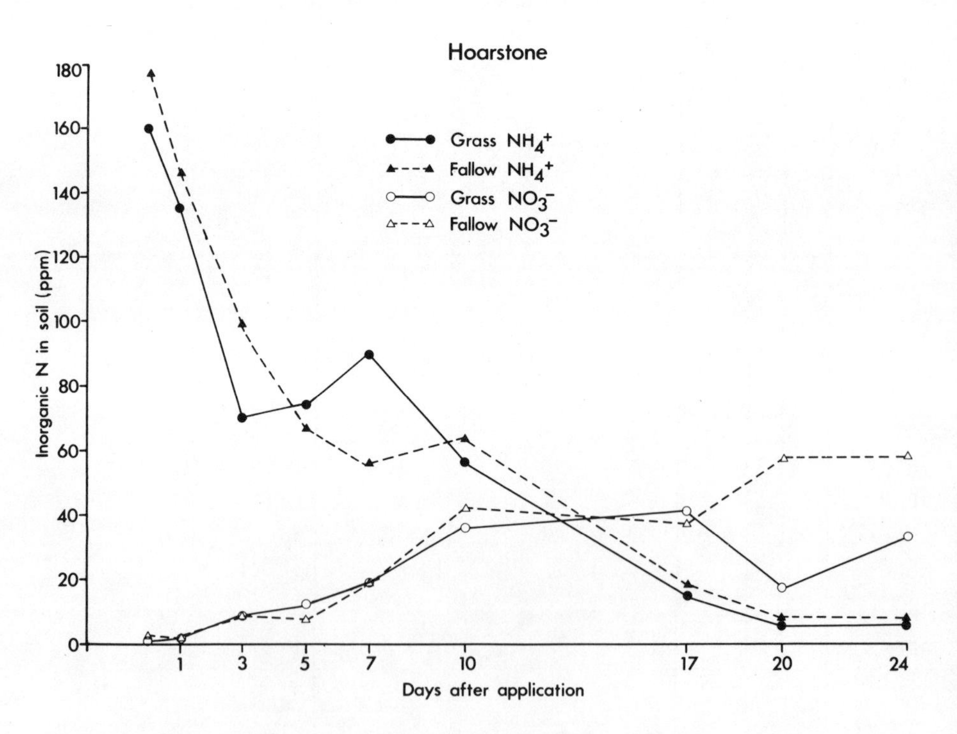

Fig. 1. Changes in ammonium and nitrate concentrations in soil following slurry applications to both grassland and fallow soil in August 1977.

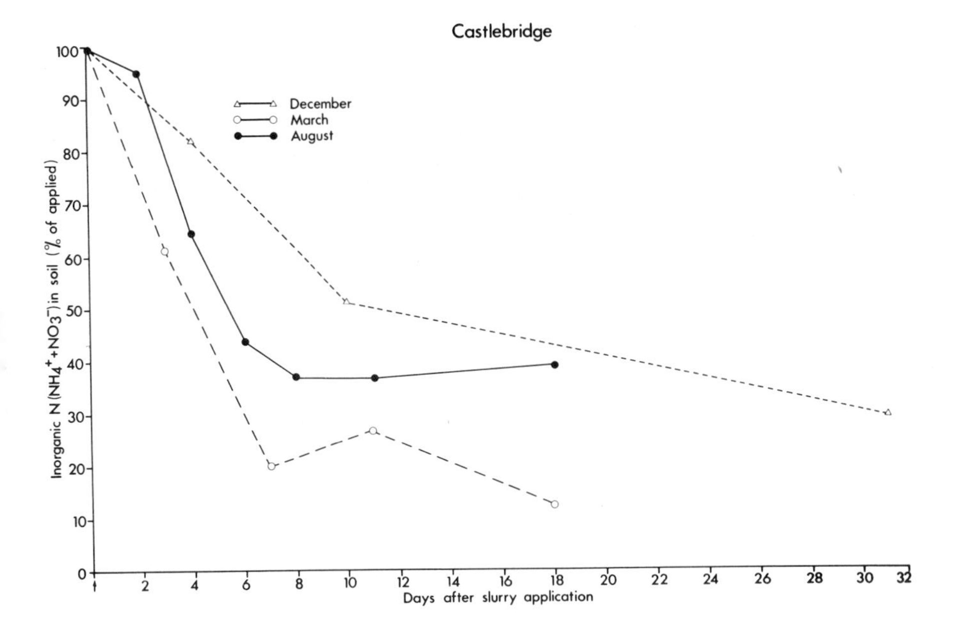

Fig. 2. Decline in inorganic N in a heavy soil following slurry application.

Runoff

The experiment suffered because volume of runoff was not uniform from plot to plot within a site and these differences were not related to treatments. Klausner et al. (1976) described a similar experience in their much more intensive study. Despite this disadvantage, the runoff experiment showed interesting trends which are described briefly. Nutrient concentrations in the runoff water from plots receiving $90m^3$ ha^{-1} pig slurry at Castlebridge for the three winters of the experiment are shown in Figure 3, which also shows the volumes of runoff water collected.

The concentrations of phosphate, ammonium-N and BOD were largest immediately after slurry application and all decreased steadily with time regardless of the volume of runoff. In the first winter, the concentrations of phosphate and BOD decreased more slowly than in other years. As a result, runoff water contained high phosphate and BOD on January 17 (i.e. six weeks after slurry application) and phosphate-P was still greater than 3 ppm on February 11th. The phosphate concentration at the Hoarstone site also decreased with time, except in January 1977 when snow fell on frozen ground and on melting increased the phosphate concentration.

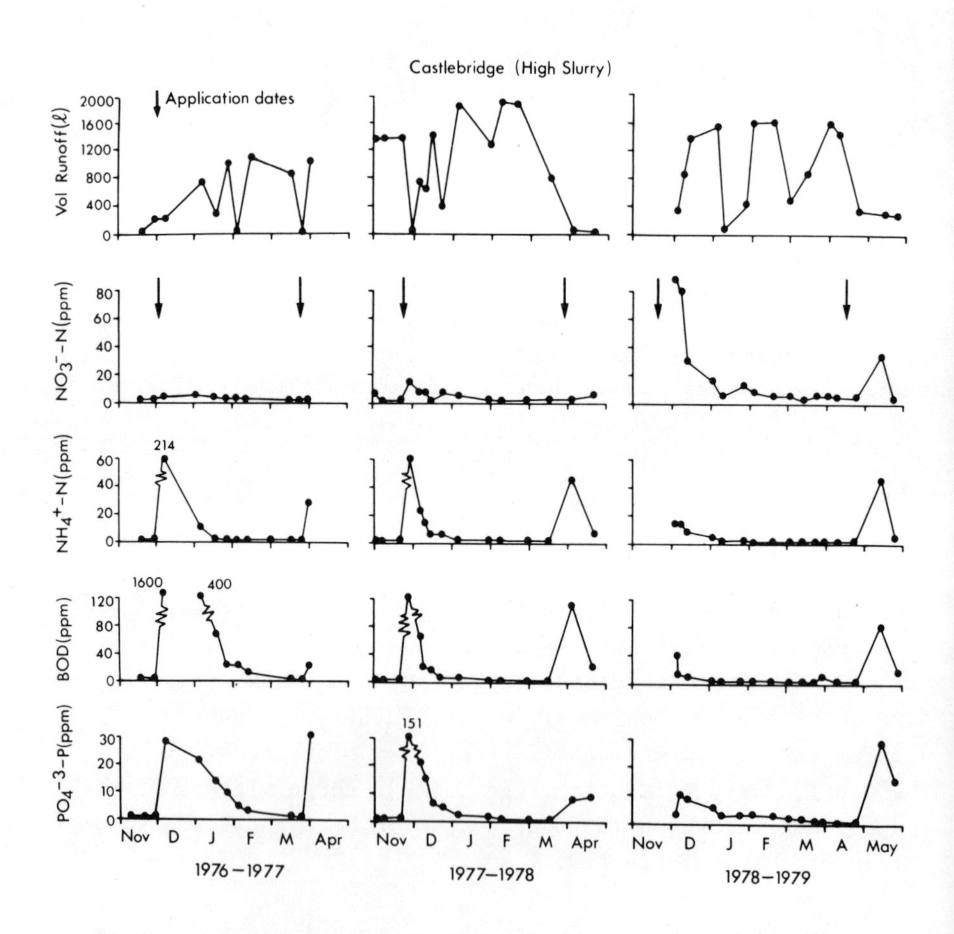

Fig.3. Volume of runoff water and nutrient concentrations in it from plots receiving $90m^3$ ha^{-1} pig slurry in Castlebridge.

Phosphate concentrations were comparatively low throughout the second winter, but small amounts of P were applied in the treatments (see Table 2). There was a substantial (greater than 25 mm) soil moisture deficit when the slurry was applied on 17 November, 1978. Nevertheless, runoff water contained 11 and 9 ppm of phosphate-P, which resulted from storms on December 7 and 12, which was 3 - 4 weeks after application and runoff water on February 19 had 2.5 ppm of phosphate-P.

Application during the growing season in March - April did not prevent runoff of phosphate when rain fell soon after application. In 1979, runoff contained 30 and 15 ppm of phosphate-P on May 14 and 25, which was 3 - 4 weeks after application, but it was an unusually wet and cold spring.

There was some evidence that except where rainfall immediately after slurry application resulted in direct runoff of diluted slurry (for example, in Castlebridge in November 1977), otherwise the concentration of phosphate-P in runoff water never exceeded 30 ppm. There is insufficient data to confirm if this was a coincidence or if 30 ppm is actually an upper limit of the solubility of some P fraction of the slurry which remains on the grass surface.

Runoff from the fertiliser treatment always contained less phosphate than that from the slurry treatments.

In summary, the results suggest that runoff water may contain phosphate for 6 - 8 weeks following slurry application and the hazard is not prevented either by spreading when there is an accumulated soil moisture deficit or during the growing season.

Infiltration of nitrate

The NO_3^--N concentrations in the soil water at 30 and 100 cm depths of land receiving 22.5 and 90 m^3 ha^{-1} pig slurry applications at both sites are shown in Figure 4. At the Hoarstone site nitrification proceeded normally in all three winters and the nitrate was subsequently leached. For the first two winters, however, very little nitrate was leached at Castlebridge. Soil analyses showed that the NH_4^+-N disappeared with time, so that nitrification appeared to take place in the surface soil where aeration was adequate but that the nitrate was denitrified as soon as it reached a waterlogged zone, agreeing with the findings of Guenzi et al. (1978), who showed that incubation with cattle manure resulted in almost simultaneous nitrification and denitrification.

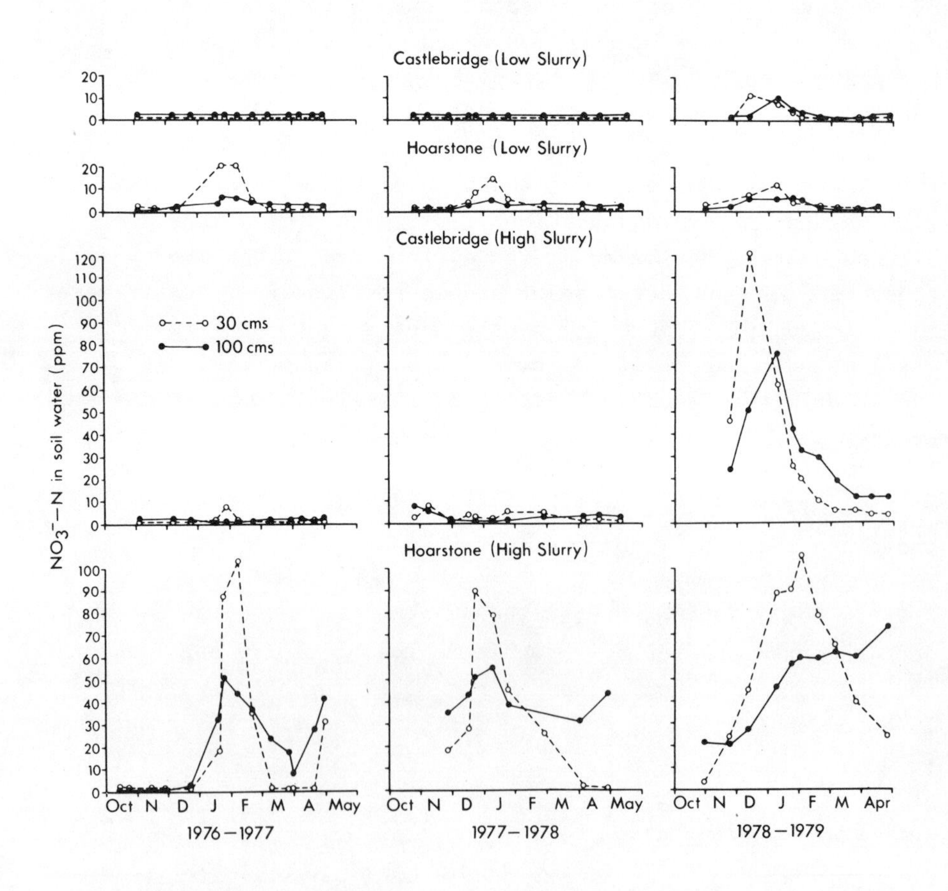

Fig. 4. NO_3^- concentrations of infiltrating soil water at 30 and 100 cm depths under plots receiving 22.5 and 90 m^3 ha^{-1} of pig slurry three times each year at Hoarstone and Castlebridge.

In the third winter there was a 25 mm soil moisture deficit when the slurry was applied. Considerable nitrification took place before the Castlebridge soil became waterlogged as was shown both by soil analyses and also by the high concentration of nitrate in the first runoff water (see Figure 3). As the soil rewet much of this nitrate was carried down with the water, resulting in large peaks in the concentration of NO_3-N in the infiltrating water at all depths. However, the nitrate peak at 100 cm depth was smaller than that at 30 cm depth and the data indicated that denitrification took place once the soil became waterlogged.

In the first winter, the fertiliser was applied as calcium ammonium nitrate. The nitrate leached at both sites, but at Castlebridge the concentration in the soil water decreased significantly with depth probably indicating denitrification even in the absence of slurry. In the second and third winters ammonium sulphate was applied and less nitrate was leached. Soil analyses indicated that the ammonium applied as fertiliser nitrified slower than the ammonium from slurry and this was also apparent in some incubation studies in the laboratory. The reasons for this are unexplained.

The suitability of pig slurry as a carbon source for denitrification was tested in incubation studies in the laboratory, where it was compared with glucose under waterlogged conditions at both 5° and $20^{o}C$. The results for the Castlebridge soil are shown in Figure 5, and although the slurry contained about ten times as much organic carbon as was present in the glucose, slurry is nevertheless a readily available source of carbon for denitrifying organisms, even at $5^{o}C$. In farming practice, these findings suggest that where slurry is used for its nitrogen content, but needs to be supplemented with fertiliser nitrogen as, for example, for first cut silage, the fertiliser should not contain nitrate, which might be liable to denitrification if the soil should become waterlogged.

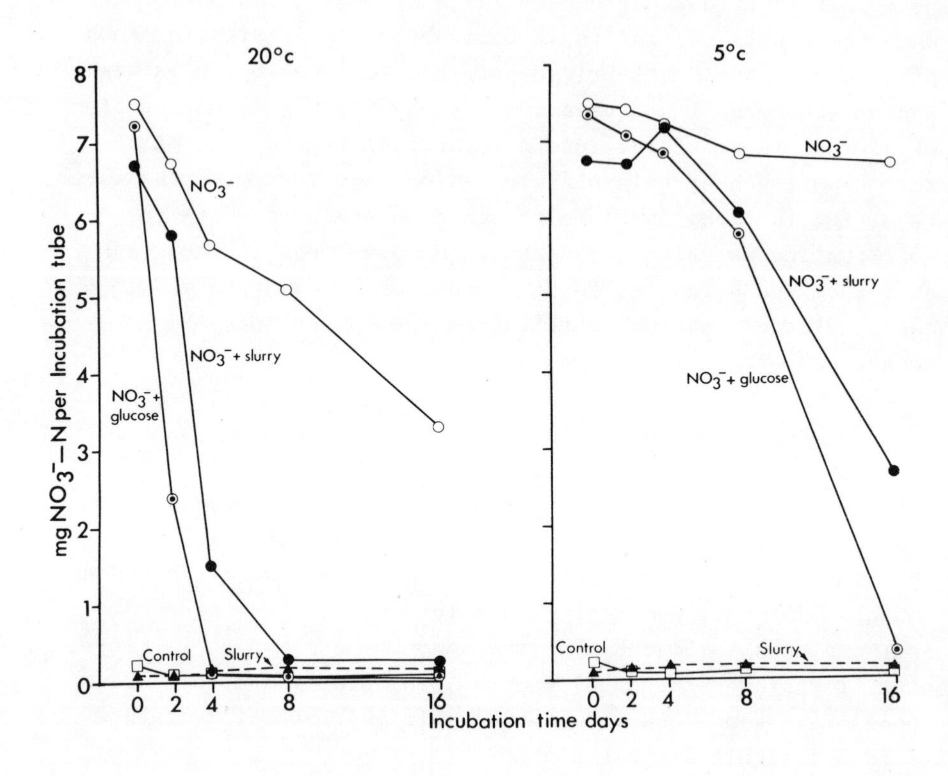

Fig. 5. Effect of slurry as a carbon source for denitrification at 5°C and 20°C in incubation studies using waterlogged Castlebridge soil.

Recoveries of nitrogen and phosphorus

Table 3 gives the percent nutrient recovered in the grass from treatment plots up to the end of 1978. Little nitrogen was recovered, due to losses by the volatilisation of ammonia, leaching and denitrification, of nitrate. Our results suggest that the smallest pig slurry treatment supplying three dressings of 22.5 m^3 ha^{-1} $year^{-1}$ might be the upper limit for disposal on grassland on a permeable soil if nitrate pollution of ground water is to be avoided. More nitrogen can be dissipated in a waterlogged soil, because of denitrification, although nitrous oxide will be formed.

TABLE 3

NUTRIENT RECOVERY IN GRASS 1976 - 1978 (%)*

Treatment	Castlebridge			Hoarstone		
	N	P	K	N	P	K
Slurry at 22.5 m^3 ha^{-1}	20	17	70	37	25	95
Slurry at 45 m^3 ha^{-1}	17	13	50	27	15	64
Slurry at 90 m^3 ha^{-1}	13	8	28	17	9	37
Fertiliser	43	19	104	48	18	116

*$\frac{\text{Nutrient recovered in treatment - nutrient recovered in control}}{\text{Total nutrient applied}}$ x 100

As we move into an era when efficient recycling of nutrients should be our interest, rather than disposal, future effort should probably be directed towards minimising the losses of nitrogen from animal manures.

The phosphorus distribution in the top 15 cm of soil after seven applications of slurry was examined and the results verified that the phosphorus was mainly retained in the top 2.5 cm, although a small amount had penetrated to the 10 - 15 cm zone.

REFERENCES

Guenzi, W.D., Beard, W.E., Watanabe, F.S., Olsen, S.R. and Porter, L.K., 1978. Nitrification and denitrification in cattle manure - amended soil. J. Envir. Qual. 7, 196-202.

Klausner, S.D., Zwerman, P.J. and Coote, D.R., 1976. In: Design Parameters for the Land Application of Dairy Manure. Publ. Environmental Research Laboratory, Athens, Georgia 30601.

Lauer, D.A., Bouldin, D.R. and Klausner, S.D., 1976. Ammonia volatilisation from dairy manure spread on the soil surface. J. Envir. Qual. 5, 134-141.

DISCUSSION

G. Steffens *(West Germany)*

Denitrification was found with slurry at $5^{o}C$. This denitrification is unusual because it is generally considered to take place at around $15^{o}C$. Have you any explanation for this?

M. Sherwood *(Ireland)*

We were surprised that it was so high because values in the literature suggest that at or below $10^{o}C$ it is not very important. In Ireland soil temperatures do not reach $10^{o}C$ until about May, and at that time the soils begin to dry out so that it seemed the denitrification would not be a problem. Since then I have seen other papers suggesting that denitrification can occur at a slower rate down to $0^{o}C$. This is important for giving advice to the farmer. In Ireland, there are many wet soils and the return of nitrogen from slurry, for grassland, is poor. Farmers are advised to spread slurry in the spring, and to supplement it with fertiliser-N. If they use calcium ammonium nitrate, with slurry, can you have denitrification of the fertiliser as well as losing some of the nitrogen from the slurry? It is a question which we have to clarify for ourselves.

B.F. Pain *(UK)*

In our field experiments, nitrous oxide was produced, again at quite low soil temperatures. However, soil moisture seemed to be more important than soil temperature. Wet soil conditions tended to enhance N_2O production.

H. Tunney

Would you comment on the amount of phosphate lost by runoff, which you suggested was about 8 kg P ha^{-1}, compared with losses reported in the literature of about 1 to 2 kg P ha^{-1}.

M. Sherwood

We are using small plots, 2 m wide and 30 m long, with only 1 m at the end of the plot left untreated. The water has only to traverse 1 m before it is collected. In a practical situation a farmer would obviously leave a much wider margin round the field, because he is using a tractor. Our conditions are conducive to maximum runoff because in farming practice this water sometimes enters drains, usually through a layer of soil which will filter much of the phosphorus out of it. In our experiments, the water ran over the surface, and was collected after only 1 m of clear grass. This over-emphasises the loss but it does show that with runoff of water, runoff of phosphorus can occur at any time. Season has little effect and moisture deficit makes some difference but not much.

SUMMARY OF SESSION 2

H. Tunney

The results of the three years work (1976 - 1978) on the utilisation of manures on grassland were presented at the meeting, which showed that a large proportion of the manures was still used on grassland in EEC countries.

There was reasonably good agreement between the results and the general conclusion can be made that slurry can be used effectively on grassland, but there are a number of problems and risks that need special attention. In the short term, slurry greatly in excess of crop needs can be applied. In the medium term, problems of water pollution will arise, and in the longer term, crop nutritional and animal health problems may develop with increased amounts of phosphorus and potassium.

Nutrients can be lost from grassland both by leaching and surface runoff. Nitrate losses by leaching are of primary importance, and results indicate that in general losses from grassland will be smaller than from arable soils. Where surface runoff occurs, the major losses will be as BOD, ammonia and phosphate.

One consistent result of major importance was the low efficiency of nitrogen in cattle slurry when applied to grassland. The reasons for this are not clear though there are several possibilities such as the volatilisation of ammonia, denitrification, slow release of organic nitrogen, and coating of slurry on the grass surface. The efficiency of nitrogen in cattle slurry and percentage recovery were often less than 20 percent and in a few cases cattle slurry actually depressed yields of grass.

Grass usually responded well to the nitrogen in pig and poultry slurry, with an efficiency of 50 percent or more being obtained. This difference in the efficiency of nitrogen in the slurry from ruminants and from monogastric animals should be studied and explained.

Clearly in the EEC large quantities of nitrogen in cattle manure are being lost each year. Perhaps this expensive loss could be reduced if we had a better understanding of what is happening.

Of course, all the manures are valuable sources of phosphorus and potassium for maintaining grassland fertility although large amounts of cattle slurry can give high K percentages in the herbage and pig and poultry slurry can increase the content of phosphorus in plants. Application of slurry in summer increased yields of grass least.

Slurry can increase some species of soil fauna and decrease others. But the results do not indicate a positive or a negative effect of these changes on grassland production or utilisation. It would be interesting to study the effects of slurry on the fauna of soil under arable crops. The short duration of the experiments (3 years) could not show what the long term effects of high nutrient levels in the soil might be. In the absence of information on the long term effects of large applications of slurry on the soil and the environment, the suggestion is made that no more than double the amount of nutrient required by the crop should be applied in one year. This is proposed as a short term guide line until further information becomes available. Present results indicate that large concentrations of phosphorus from manure in the soil give rise to serious problems when they occur in the catchment areas of inland lakes and other waters.

Pasture used for grazing can only accept a limited amount of slurry because of the high proportion of nutrients that are re-cycled. Therefore it is proposed that most of the manure on grassland will be re-cycled to grass cut for hay and silage. However, the results of the work reported here showed that although slurry can be used on pasture for grazing, poor animal performance was obtained with cattle slurry. Good animal performance was obtained from pasture when poultry and pig slurry were applied. Silage produced from grass treated with slurry was significantly less good when measured by animal performance than silage made from grass with fertiliser; though silage quality was not affected. In this work the slurry was applied in spring and summer less than two months before cutting the grass for silage. Applicaction of slurry in the autumn may avoid this problem.

The results indicate, particularly with cattle slurry, that caution is necessary, when spreading slurry on the grass surface to avoid damage of the sward and contamination of the grass.

GENERAL DISCUSSION

During the discussion of the Chairman's summary and conclusions much attention was focussed on various aspects of problems arising from the application of phosphorus in slurries. The following points were noted:

i) There is benefit to the farmer when phosphate accumulates in soils which are deficient in available-P.

ii) Excess phosphate, however, can lead to environmental hazards and problems of animal health.

iii) Pig slurries can contain up to 80 percent of their phosphate content in inorganic form and therefore resemble fertilisers. Cattle slurries contain a larger proportion of organic-P.

iv) Turnover of P in soils can be rapid and more information is required on the dynamics of this process.

There was some agreement that for grass cut for hay or silage, twice the crop requirement for nutrients could be applied as slurry, but less is required for grazed pastures.

More information is needed on the reasons for the apparent inefficiency of nitrogen in slurry, and the effects of long-continued applications on the nitrogen economy of the soil, availability to the plant and losses by leaching.

There is an urgent need to look at long term problems and for research workers to be able to predict effects of continued application of slurries, as well as interpret short term results and advise the farmers.

SESSION III

PROBLEMS OF COPPER IN SLURRY; MODELLING; TREATMENT OF SLURRY

Chairman: J.K.R. Gasser

REAL OR POTENTIAL RISK OF POLLUTION OF SOILS, CROPS, SURFACE AND GROUNDWATER DUE TO LANDSPREADING OF LIQUID MANURE

K. Meeus-Verdinne, G. Neirinckx, X. Monseur and R. de Borger
Institute of Chemical Research, Ministry of Agriculture,
Museumlaan 5, 1980 Tervuren, Belgium.

INTRODUCTION

The objective of this part of the research programme was the evaluation of the risks of pollution of soils, crops and water by mineral elements and by specific organic compounds brought into the environment by landspreading of liquid manure.

The first step in this work was the analysis of the manure and the determination of the elements and compounds that could cause some difficulties. It soon became clear that only copper and zinc were present in concentrations that could raise problems. Some attention was also paid to the elements arsenic, selenium and mercury. Only the last occurred sometimes in higher concentrations, due to the use of disinfectants in the piggeries. The average data for the copper and zinc content of liquid manure from pigs, cows and poultry are given in Table 1. The standard deviation of the mean values indicates that the composition of the manures varies widely. This phenomenon has to be ascribed to the changing composition of the feed given to the animals and also to the use of varying quantities of water in cleaning the stables. Thus the study of the influence of landspreading of manure on soil, crops and water was limited to that of copper and zinc.

Concerning the specific organic compounds it is impossible to perform a complete analysis of the complex and heterogeneous mixture formed by the manure. Neither is it feasible to look for such compounds in the soil or in the crops. The only work that could be done was the analysis of ground and surface water that had been polluted by manure.

TABLE 1

AVERAGE COMPOSITION OF LIQUID MANURE

	PIGS					
	n	$\bar{x}$	s	ν %	min.	max.
DM	132	7.1	4.2	60	1.1	17.6
pH	96	7.1	0.4	5.9	6.4	8.1
Hg (ppb/DM)	99	100*	147.7	147	9	138
Cu (ppm/DM)	132	574	159	28	48	996
Zn (ppm/DM)	132	919	315	34	180	1686
As (ppm/DM)	9	2.3	1.6	71	0.7	5
Se (ppm/DM)	17	0.8	0.3	33.3	0.4	1.3
Cd (ppm/DM)	10	0.6	0.2	30	0.3	0.8
	CATTLE					
	n	$\bar{x}$	s	ν %	min.	max.
DM	6	4.9	1.5	30	2.8	7.1
pH	6	7.1	0.3	4.5	6.5	7.4
Hg (ppb/DM)	6	113	23	20	96	156
Cu (ppm/DM)	6	57	8	14	48	71
Zn (ppm/DM)	6	580	159	28	430	840
As (ppm/DM)	5**	1.5	0.9	58	0.7	2.9
Se (ppm/DM)	5**	0.1	0.1	110	0	0.2
Cd (ppm/DM)	-	-	-	-	-	-
	POULTRY					
	n	$\bar{x}$	s	ν %	min.	max.
DM	17	20.3	7.6	37	6.2	28.4
pH	4	7	0.1	0.9	6.9	7.1
Hg/ppb/DM)	4	60.5	19.5	32	36	81
Cu (ppm/DM)	17	59	28	47	32	120
Zn (ppm/DM)	17	495	230	46	276	940
As (ppm/DM)	3	<0.5-1	-	-	-	-
Se (ppm/DM)	3	0.5	-	-	-	-
Cd (ppm/DM)	-	-	-	-	-	-

* 15 figures varying between 150 and 700 ppb are not included in this average. They were due to the use of disinfectants in the stables.

** Figure for calf manure; in cow manure the figures oscillate between 1.2 and 1.7 ppm for As and around 0.2 ppm for Se.

TABLE 2

INFILTRATION OF LARGE QUANTITIES OF PIG SLURRY (H_2O = 92.5%, Cu = 390 ppm/105^oC, Zn = 612 ppm/105^oC) SAMPLING OCCURRED 10 DAYS AFTER LANDSPREADING

Quantity of slurry applied l ha^{-1}	Cu added* ppm	Cu found in the soil ppm/105°C				Zn added* ppm	Zn found in the soil ppm/105°C			
		0-15	15-30	30-45	45-60		0-15	15-30	30-45	45-60
Control	0	12.6	11.4	10.5	10.5	0	60.2	51.6	41.6	44
192 500	3.1	16	12	11.4	10.8	4.9	87.2	58.4	55.6	46.4
278 250	4.5	30	13	10.5	10.5	7.1	91.2	59.4	49.6	47.6
428 250	7	14.2	13.4	13.2	12.6	11	60.8	55.8	48	48
534 700	8.7	22.4	14.2	10.8	9.2	13.7	71.6	53	46	38

* If the slurry is located in the upper 15 cm of the soil.

SOIL

On the basis of the average manure composition one can calculate the amount of copper and zinc that will be brought into the upper layer of the soil by a normal application. For example, on spreading 100 000 l ha^{-1} of pig manure the copper and zinc content of the upper 30 cm of the soil will be increased by 1.5 and 2.5 ppm respectively. The heterogeneity of the soil and of the manure, and the rough technology of the landspreading mean that this increase can hardly be detected on analysing soil samples. Even the infiltration of very large quantities (see Table 2) of liquid manure gave, for the copper and zinc content of the soil, very divergent results.

In addition to the determination of the total copper and zinc content, some attention was paid to the changes in the copper and zinc, extractable by a mixture of NH_4AC - EDTA; an indication of assimilation by plants. The analysis of soil samples from different experimental fields indicates that the amount of assimilable copper and zinc increases with landspreading of manure. A typical example is given in Table 3. The increase is observable for several months after the landspreading and then disappears gradually. The heterogeneity of the soil and of the manure itself, means once more, that the results noticed are not always coherent. Anyhow the results obtained on analysing several series of soil samples indicate that landspreading of more than 100 000 l of liquid manure ha^{-1} gives a significant increase in the assimilable copper and zinc, varying from 35 percent to 80 percent.

Because of the poorly reproducible results obtained on the samples from the experimental fields, the change in the assimilable copper and zinc was studied in the laboratory by incubating soil samples with manure. Unfortunately here too the fluctuations were relatively large. From both experiments (field and laboratory) we can conclude that the increase in the amount of assimilable copper and zinc due to the landspreading of liquid manure is most strongly marked in light soil, for example the phenomenon is more marked in sandy-loam soil than in loam soil. The increase is directly related to the quantity of manure applied but its magnitude is also determined by such factors as the time of application and the weather conditions.

TABLE 3

Cu AND Zn CONTENT OF SOIL SAMPLES (0 - 30 cm UPPER LAYER)

Treatment	ppm Cu in air dry soil		ppm Zn in air dry soil	
	Total	Extraction EDTA pH 4.65	Total	Extraction EDTA pH 4.65
Control	5.5	1.9	58.8	9.2
60 000 l ha^{-1}	7.5	2.5	55.8	8.6
180 000 l ha^{-1}	6.5	4.2	59	12.4

The figures are means of 4 analyses

As a summary we may state that in the short term the land-spreading of liquid manure with the quantities used for normal fertilising hardly affects the copper and zinc status of the soil. The amount of assimilable copper and zinc is temporarily increased. Bearing in mind the heterogeneity of the soil and the manure, it is very difficult to predict the possible long term effects of normal manure applications. Three years is too short a period to gather enough data for making a good long term prediction.

WATER

The analysis of the water collected under lysimeters showed us, that compared with an equivalent amount of fertiliser, the application of pig slurry gives an increase in the amount of minerals collected in the drainage water. This was very clear in the case of nitrate, chloride, calcium and magnesium ions (Table 4). It was less obvious for sodium and copper ions. For zinc ions, it was not possible to draw clear conclusions.

TABLE 4

LYSIMETER EXPERIMENTS - EXPORTATION OF MINERALS BY PERCOLATION WATER. PERIOD 1977 - 1978

	NO_3^- kg ha^{-1}	Cl^- kg ha^{-1}	Na kg ha^{-1}	Ca kg ha^{-1}	Mg kg ha^{-1}	Cu g ha^{-1}	Zn g ha^{-1}	Amount of water collected litres/m^2
Under wheat:								
Landspreading Nov. 77								
- sandy loam soil								
control (min. fert.)	17.6	63.6	174	163	43.3	10.9	230	357
treated (90 000 l ha^{-1} pig slurry)	60.4	138	109	587	51.2	14.3	90	315
- loam soil								
control	27.7	-	160	-	-	10.7	108	256
treated (90 000 l ha^{-1} pig slurry)	137	247	96.7	354	46.2	9.1	113	212
Under beetroots:								
Landspreading Nov. 77								
- sandy loam soil								
control	69.2	105	248	514	43.1	17.9	240	395
treated (90 000 l ha^{-1} pig slurry)	151	339	338	862	80.2	21.2	481	396
- loam soil								
control	145	598	367	818	96.6	14.7	486	336
treated (90 000 l ha^{-1} pig slurry)	526	1117	472	1398	162	18.7	273	340

TABLE 4 (Continued)

	NO_3^- kg ha^{-1}	Cl^- kg ha^{-1}	Na kg ha^{-1}	Ca kg ha^{-1}	Mg kg ha^{-1}	Cu g ha^{-1}	Zn g ha^{-1}	Amount of water collected litres/m^2
Under winterbarley:								
Landspreading Sept. 77								
- sandy loam soil								
control	105	239	277	746	67.8	16.4	204	398
treated (90 000 l ha^{-1} pig slurry)	579	604	442	1389	124.6	21.7	212	376
- loam soil								
control	132	852	503	1130	142	16.7	213	370
treated (90 000 l ha^{-1} pig slurry)	576	987	557	1611	176	22.4	162	341

Slurry comp.: 88.15% water, 390 ppm Cu/105°C, 607 ppm Zn/105°C

The quantity collected is strongly influenced by the time of application and the crop. The later the spreading the less will be collected, at least during the first drainage season. The opposite phenomenon is noticed when there is no plant cover during the winter period (difference between growing winter wheat and beetroots). No difference in behaviour between the two soil types studied (sandy-loam and loam) was found.

As mentioned before, the checking of specific organic compounds originating from liquid manure can only be done in water. Using gas liquid chromatography we made 'fingerprints' of the volatile organic compounds present in well water, in surface water and in water from field drains. We found similar chromatograms on analysing drainage water coming from fields treated with manure, and on analysing water from wells polluted by animal wastes. They present very characteristic fingerprint patterns. Immediately after the landspreading of manure the typical chromatogram was obtained in the drainage water, it disappeared with abundant rainfall, and reappeared with a drought period six weeks later. With well water it was noticed that the amount of water pumped influenced the concentration of the typical components.

The fingerprints obtained on analysing surface water polluted by animal wastes were strikingly different from those obtained on groundwater, which could in some ways be compared with those of extremely diluted manure.

The conclusion of this work is that on using the gas-chromatography - fingerprint technique we are able to identify and to distinguish pollution by liquid manure of ground water and of surface water.

CROPS

Straw from wheat grown on a loam soil treated with 100 000 l pig manure hectare^{-1}, contained 30 to 50 percent more copper and zinc than the control. The same treatment on a sandy soil also resulted in an increase of 40 to 90 percent of the copper and zinc content of the grain, which was also the case with winter barley. No influence could be detected on several cuttings of ryegrass grown on loam soil. A positive effect was established on the peelings and shoots of potatoes and on the leaves of turnips grown on sandy-loam soil. Especially for the leaves of the turnips, we observed that the copper and zinc contents were increased as the amount of pig manure spread on the field was increased (see Table 5).

TABLE 5

TURNIPS: SANDY LOAM SOIL

	Cu (ppm/75°C)		Zn (ppm/75°C)	
	Leaves	Roots	Leaves	Roots
Control	5.31	4.65	52.9	32.1
+ 25 000 l ha^{-1} pig slurry	6.13	6.19	63.1	41.3
+ 50 000 l ha^{-1} pig slurry	7.29	4.79	65.1	45.1
+ 75 000 l ha^{-1} pig slurry	9.31	4.15	67.3	50.2

Slurry: 97.56% H_2O, 116 ppm Cu/105°C, 725 ppm Zn/105°C

However, the increased copper and zinc contents noted in the crops are far below toxic concentrations. There is no danger of hazardous amounts being found in the crops grown on land treated with pig manure.

BALANCE

With the help of lysimeter experiments we tried to evaluate the amount of copper and zinc removed in the crops and lost in the drainage water, for an application of 90 000 l pig manure $hectare^{-1}$. Notwithstanding the fact that an application of liquid manure increases the amount of assimilable copper and zinc, the quantities removed in the plants are very low. During a period of one year, the water and the crops together only remove from 2 to 10 percent of the applied copper and zinc. As the yields of crops in a lysimeter are markedly lower than those obtained under normal circumstances, we estimate that more than 90 percent of the copper and more than 80 percent of the zinc introduced by manure spreading, remain in the soil. In the light of these facts and considerations, accumulation could be a problem in the long term.

CONCLUSION

In the short term, we may conclude that the landspreading of liquid manure, in quantities used for normal fertilising, no matter its origin, has no appreciable influence on the total copper and zinc content of the soil.

Although the assimilable copper and zinc contents of the soil are increased temporarily by landspreading of slurry, the rise of the copper and zinc contents of the crops stays far below toxic values. The drainage and percolation water of treated fields contain only slightly increased amounts of these metals.

In the long term however, regularly repeated landspreading of relatively large amounts of liquid manure will give rise to accumulation problems. We conclude this from the fact that little copper and zinc is removed in the plants being only a few percent of the quantity introduced. The drainage and percolation water do not remove much of these metals.

In this connection it is very important to know the accumulation capacity of the soils, and to know what will be the reaction of the crops towards soils containing different amounts of accumulated copper and zinc salts. This problem could not be elucidated during the short period of the research programme and forms a topic for further research.

DISCUSSION

F.A.M. de Haan *(The Netherlands)*

You have shown us a number of gas chromatograms. Did you measure an increase in BOD or COD values in the groundwater, the surface water or in the drainage water, because it might be important to have some indication of the oxygen demand? We found an increase in the BOD and COD of groundwater when waste water from potato starch manufacture was applied.

R. de Borger *(Belgium)*

We did not measure either BOD or COD in the groundwater. We looked for the organic compounds which could come from the slurry - qualitatively, not quantitatively.

D. McGrath *(Ireland)*

If I understood correctly, Dr. de Borger said that when added to soil initially, the copper from pig slurry is largely extractable. Does less copper become extractable with time and were measurements made over a period of time?

R. de Borger

Yes, changes in extractable copper and zinc were followed for about one year. In the first three or four months there was an increase; afterwards it gradually diminished slightly and in one soil more quickly than in another, so that overall availability diminished with time.

D. McGrath

Thank you, that is the point I wanted to establish because it may have a bearing on the long term effects of copper in soil.

TOXICITY OF COPPER

Th. M. Lexmond and F.A.M. de Haan
Department of Soil Science and Plant Nutrition
Agricultural State University
Wageningen, The Netherlands.

1. INTRODUCTION

This paper describes the effects of copper concentrations in soil on plant growth.

2. MATERIALS AND METHODS

The growth of maize *(Zea mays* c.v. *Capella)* grown in pots was studied for different copper concentrations and varying soil conditions. The experimental conditions are summarised as follows:

1977 I

a sandy soil (0.85% organic-C) was used at 5 pH values - pH in $CaCl_2$ 4.30; 4.48; 4.69; 4.90 and 5.49; at each pH value four amounts of copper were added together with no added copper; each treatment was replicated three times.

1977 II

three sandy soils were used with organic-C as the main variable, with each soil at three pH values:

Organic-C %		pH in $CaCl_2$		
A	6.10	4.20;	4.60;	4.86
B	3.90	4.45;	4.68;	5.01
C	1.05	4.37;	4.83;	5.41

at each pH value four amounts of copper were added together with no added copper; each treatment was replicated three times.

1978

seven soils with widely varying properties were used:

	Organic-C %	pH in $CaCl_2$
sandy soil	1.05	4.26
sandy soil	6.10	4.07
sandy soil	2.30	4.76
sandy soil	3.40	4.92
loess soil	0.85	5.49
river basin clay soil	1.55	5.27
peat soil	8.20	4.80

four amounts of copper were added to each soil together with no added copper; for five soils the treatments were replicated three times, for the remaining two they were duplicated.

The copper was added as copper sulphate or copper nitrate. Pots were filled with treated or untreated soils and incubated for three weeks. Then maize seeds were planted and three plants per pot grown until early tasselling.

Plants were harvested, dried and weighed. Sub-samples were analysed for Cu, Zn, Mn, Fe, N, PO_4, K, Na, Ca, Mg, Cl, SO_4 and NO_3.

The soil was sampled, dried, and passed through a 2 mm sieve. Reversibly bound copper was measured by extraction with 0.43 M HNO_3. Copper extracted by strongly acidic cation exchange resin in the Ca-form from a suspension of soil in 0.01 M $CaCl_2$ was used as an estimate of copper intensity in the soil. pH was measured in the same suspension after settlement.

3. RESULTS AND DISCUSSION

3.1 Soil analysis

Results from a series of preliminary experiments showed that extracting soil with HNO_3 was not an appropriate method for assessing potential copper toxicity to plant growth, because HNO_3 extraction gives a measure of the amount of copper absorbed, whereas copper toxicity will depend much more on the activity of the copper.

Resin extractable copper (Cu_{res}) and copper extracted by HNO_3 were well related for all experiments by a Freundlich type equation. The effect of pH on this relationship could be expressed by including a pH-term in the equation.

The results for the different experiments were:

1977 I

$\log Cu_{res} = 2.085 + 1.694 \log Cu\text{-}HNO_3 - 0.805\ pH \quad R^2 = 0.947$

1977 II

$\log Cu_{res} = 0.266 + 1.676 \log Cu\text{-}HNO_3 - 0.725\ pH \quad R^2 = 0.986$
(soil A)

$\log Cu_{res} = 0.884 + 1.672 \log Cu\text{-}HNO_3 - 0.716\ pH \quad R^2 = 0.994$
(soil B)

$\log Cu_{res} = 1.616 + 1.800 \log Cu\text{-}HNO_3 - 0.743\ pH \quad R^2 = 0.981$
(soil C)

Assuming that organic matter in soil is the main adsorbent for copper and noting that the soils used in 1978 varied widely in their organic-C contents, the data for 1978 were not expressed as log $Cu\text{-}HNO_3$ but as log ($Cu\text{-}HNO_3$/% org-C). The following equation then described the relationship between Cu_{res} and ($Cu\text{-}HNO_3$/% org-C) for the seven soils used in the 1978 experiment, again including a pH-term:

1978

$\log Cu_{res} = 1.604 + 1.429 \log (Cu\text{-}HNO_3/\%\ org\text{-}C) - 0.632\ pH$
$R^2 = 0.859$

This equation is useful for estimating the amount of resin-extractable copper in soil from the amount of

HNO_3-extractable copper, the organic matter content and the pH. This is important because Cu_{res} was found to give a reliable measure of copper toxicity.

3.2 Plant response

The main effects of increased copper concentrations in soil on plant growth and composition were comparable for the three different experiments. They were:

- increased copper and manganese contents of plants
- decreased phosphate content
- decreased dry matter production
- increased organic-N, free NO_3^-N and potassium contents.

Injury to plants showed itself by poor development of the root system; this decreased the uptake capacity for phosphorus, resulting in phosphate deficiency symptoms in leaves and stems. The ratio P/Po was a useful index for copper toxicity where P is the phosphate uptake by plants with added copper and Po the phosphate uptake by plants without added copper.

Table 1 presents typical effects on yield and plant composition for copper added to a loess soil (0.85% org-C, pH 5.49, 1978) and a sandy soil (3.90% org-C, pH 4.68, 1977 II).

Figure 1 presents the values of P/Po as a function of log $Cu\text{-}HNO_3$ for all treatments of the three experiments. The wide scatter of points shows that values for HNO_3-extractable copper in soil provide insufficient information for the prediction of plant injury.

Figure 2 presents a plot of P/Po as a function of Cu_{res}. Although this leads to a considerable improvement compared to Figure 1, the values of Cu_{res} alone clearly cannot be used as an acceptable index for copper toxicity.

TABLE 1

EFFECTS OF COPPER ON YIELD AND CHEMICAL COMPOSITION OF MAIZE PLANTS

Soil	Cu added mg kg^{-1}	Yield of dm g per pot	Content in dm						Relative phosphate uptake
			mmol kg^{-1}				mg kg^{-1}		
			Org-N	NO_3-N	PO_4	K	Cu	Mn	
Loess	0	95.7	935	24	96	714	4.5	86	1
	40	93.3	941	19	98	727	8.7	125	0.99
	80	88.1	1010	20	84	726	11.9	161	0.81
	120	50.0	1605	247	89	968	23.7	254	0.48
	160	16.1	1846	560	61	1007	36.2	303	0.11
Sandy soil	0	71.3	1096	40	97	601	2.9	355	1
	100	68.1	1153	36	91	641	7.6	393	0.90
	200	59.5	1217	60	78	655	10.6	460	0.67
	300	47.1	1321	89	61	699	14.2	486	0.42
	400	26.9	1538	141	53	708	16.0	481	0.21

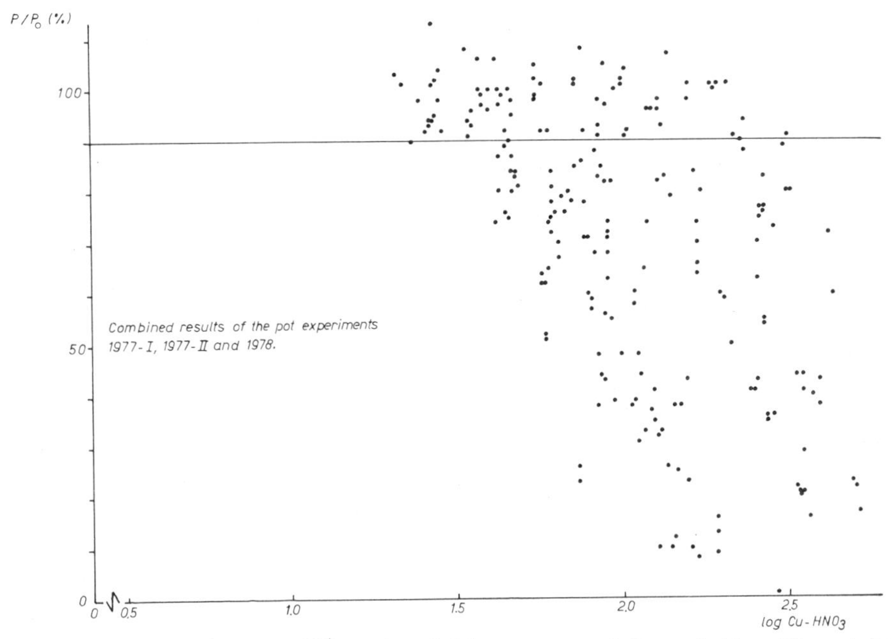

Fig. 1. Relation between HNO_3 extractable copper and the relative PO_4-uptake by maize.

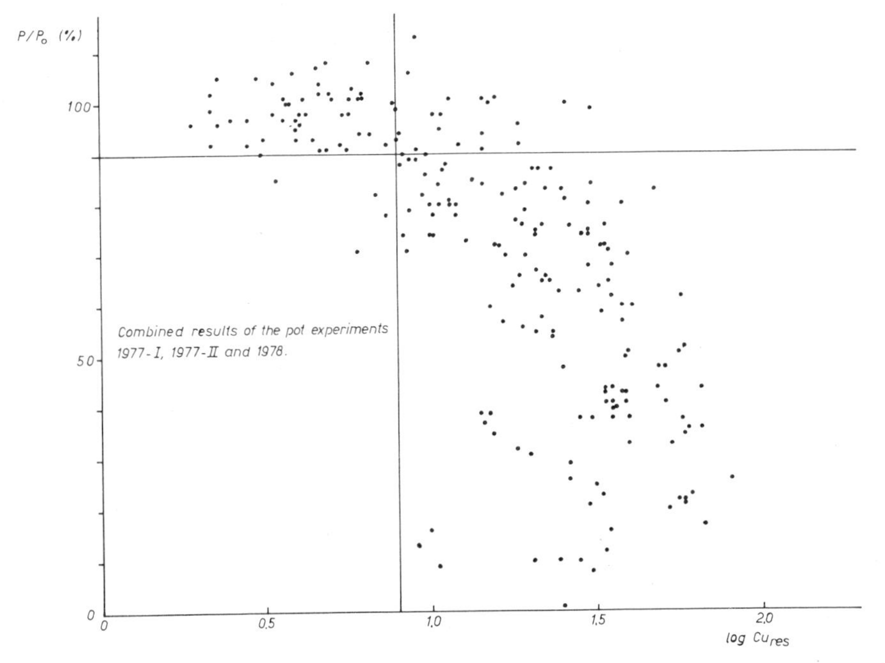

Fig. 2. Relation between resin extractable copper and the relative PO_4-uptake by maize.

pH was found to have a very significant effect on the relationship between P/Po and log Cu_{res} and, as Figure 3 shows, an almost complete separation between toxic and non-toxic treatments can be achieved when the copper status of the soil is expressed as log Cu_{res} + 0.33 pH. (The remaining variation is ascribed to the variation in phosphate supply by the different soils).

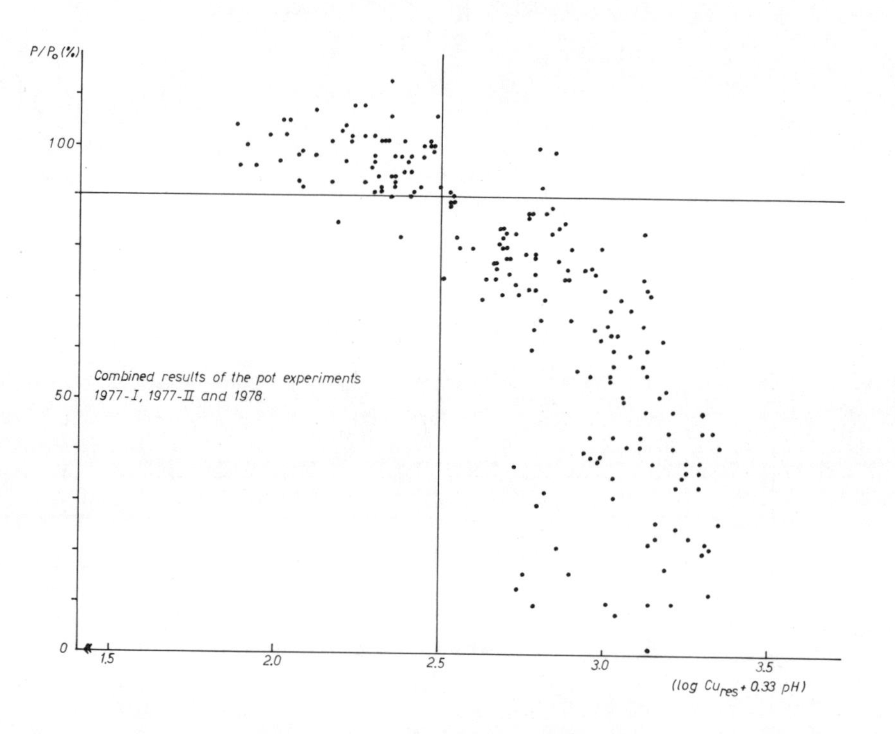

Fig. 3. Relation between resin extractable copper and pH and the relative pH and the relative PO_4-uptake by maize.

The effect of pH on the toxicity of copper to maize plants was studied in a separate water culture experiment by using nutrient solutions of different pH and copper concentrations. The increased toxicity of copper at increasing pH values, as found in the experiments with soil, was confirmed.

4. CONCLUSIONS

1) Using resin extractable copper as an intensity parameter and HNO_3 extractable copper as a quantity parameter, the quantity/intensity relationship for copper in acid soils was well described by a Freundlich equation.

2) The effect of pH on the quantity/intensity relationship in a particular soil could be described by including a pH-term in the Freundlich equation.

3) In the soils studied, organic matter was the main copper adsorbing component and a Freundlich equation was obtained which predicted reasonably well the resin extractable copper from Cu-HNO_3/% org-C (which served as a measure of the copper concentration in the organic matter) and pH.

4) The toxic effect of high copper concentrations increases with increasing pH. This has been shown in pot experiments using phosphate uptake as the measure of toxicity, and in water-culture using root development, also in the field using yield of dry matter.

5) The use of a toxicity index defined as $\log Cu_{res} + 0.33$ pH allowed toxic and non-toxic treatments to be distinguished in the pot experiments.

6) Comparing the magnitude of the pH effects on copper sorption and on the toxicity of copper ions in solution revealed that the former effect was greater, agreeing with the known beneficial effect of liming copper polluted soils.

7) In the pot experiments, where copper decreased plant uptake of phosphate, the magnitude of the effect could be predicted more accurately when available phosphate in the soil was included as a parameter. The presence of much available phosphate diminishes the toxic effect of copper.

5. RESEARCH NEEDS FOR PREDICTION OF LONG-TERM EFFECTS

The method described in the preceding sections provides a useful tool for predicting decreased productivity of soil as the result of increased copper activity. Further application of this approach in practice requires a means for the long-term prediction of copper contents of soil as a function of the copper application regime.

Transport and accumulation of copper in soil systems may be estimated by the use of simulation models, and one is now available in our laboratory for predicting the behaviour of phosphate in soil. This model should be explored for its value in modelling copper accumulation. This could eventually lead to the prediction of long-term effects of copper on soil productivity, and thus contribute to the development of measures required to maintain good soil conditions for crop production.

DISCUSSION

<u>J.K.R. Gasser</u> *(UK)*

How do the results from adding inorganic salts compare with the copper to be found in slurry, where it is in intimate contact with organic matter?

<u>Th. M. Lexmond</u> *(The Netherlands)*

That question cannot be answered yet. I think the first approach that should be made in order to answer it is to study the adsorption behaviour of copper in soils that have received slurry dressings for many years.

<u>L. Lecomte</u> *(Belgium)*

I think we have a good model for the long term application of copper in France where it has been used as a fungicide in vineyards for about 70 years. The amounts of copper which are distributed on each area of land are certainly much larger than the heaviest dressings of slurry would give.

<u>Th. M. Lexmond</u>

The largest dressings of slurry would introduce about 15 kg ha^{-1} of copper each year which is more than was introduced in the vineyards. So I think the magnitude of the accumulation problem is comparable in both situations.

<u>A. Maton</u> *(Belgium)*

Referring to Mr. Lecomte's comment, I would add that the same situation applies to hops, a crop grown in Bavaria, Belgium and the UK. After having used copper pesticides for many years, there does not seem to be any damage to the hop crop.

HAZARDS ARISING FROM APPLICATION TO GRASSLAND OF COPPER-RICH PIG FAECAL SLURRY

D. McGrath[1], D.B.R. Poole[2] and G.A. Fleming[1]
[1]The Agricultural Institute, Johnstown Castle, Wexford, Ireland.
[2]The Agricultural Institute, Dunsinea, Dublin, Ireland.

INTRODUCTION

In many countries copper supplementation of feed for fattening pigs gives slurry containing about 700 ppm ($\mu g\ g^{-1}$ DM) of copper. When such slurry is applied to grassland, the added copper is distributed between the herbage, the organic mat on the soil surface and the soil (Figure 1). Several features relating to the disposal of copper-rich pig slurry have previously been examined (Batey et al., 1972; Delgarno and Mills, 1975; Suttle and Price, 1976; Gracey et al., 1976) and serious, even fatal, (Feenstra and van Ulsen, 1973) consequences to animal health have previously been recorded.

The work reported studied (a) copper uptake by herbage grown in pots; (b) the influence of slurry on copper content of pasture; and (c) the effect of ingested slurry copper on copper status of animals. The ultimate objective was to define the rational use of slurry having regard to its copper component, so that its toxicity is minimised, while its value to plant and animal is maximised.

EXPERIMENTAL

Soils

Soils were selected to represent texturally different types ranging from a sand to a loam of high silt content (Table 1).

TABLE 1.

SOME PROPERTIES OF SOILS USED IN THE INVESTIGATION

Soil No.	pH	Organic carbon (%)	CEC (me $100g^{-1}$)	Extractable molybdenum (ppm)	Mechanical analysis		
					Sand (%)	Silt (%)	Clay (%)
1	6.9	1.62	12.3	0.32	89	7	4
1a	6.7	1.87	-	0.11	80	11	9
2	5.6	2.47	16.9	0.25	59	28	13
3	5.9	2.70	18.4	0.29	42	39	19
4	7.1	3.40	24.7	3.50	34	47	19

Field experiment

Slurry was spread on permanent pasture growing on Soils 1 - 4 using a vacuum tanker. Applications to supply 23, 46 and 115 m^3 ha^{-1} were made 3 times annually. Each treatment was replicated 5 times and there were plots without slurry. Herbage, predominantly perennial ryegrass, was generally sampled immediately before defoliating plots with a forage harvester and also at the end of each season.

Pot experiments

Soil from each soil type was used for a series of experiments. Except where stated otherwise, perennial ryegrass (*L. perenne*) S23, was used. Slurry was applied in three ways: (i) admixture of dried and ground slurry with soil, followed by sowing seed; (ii) slurry applied to herbage using a perforated container; (iii) slurry applied directly to soil surface.

Grazing experiments

Paddocks (0.54 ha) on Soils 1a (alternative to 1) and 3 were stocked with wether sheep for two grazing seasons, 1977 and 1978, while receiving pig slurry at the rate of 20 m^3 ha^{-1} three times annually. Herbage, faecal, and liver copper of selected animals were monitored. Liver copper was also measured at the end of both years following slaughter of all animals, and in June of the second year on animals removed from plots when stocking density needed to be reduced. In conjunction with the slurry experiment, soil ingestion patterns were measured for sheep at two stocking rates on both sites.

Feeding experiments

Three feeding trials were conducted with Scottish Blackface lambs; each treatment had 12 animals, penned in groups of 6. Initial liver copper was measured on samples following biopsy of all animals. Grass meal diets were enriched with sufficient copper compound to raise their copper contents from about 8 ppm to 23 ppm. Feeding was continued for a period of 8 - 11 weeks, after which all lambs were slaughtered and livers examined.

Analyses

Copper contents of plant, soil faeces and liver were estimated by atomic absorption. Soil content of faeces was calculated from their acid-insoluble residue content.

RESULTS

Figure 1 shows the different routes for copper from slurry to animal and is used in considering the results.

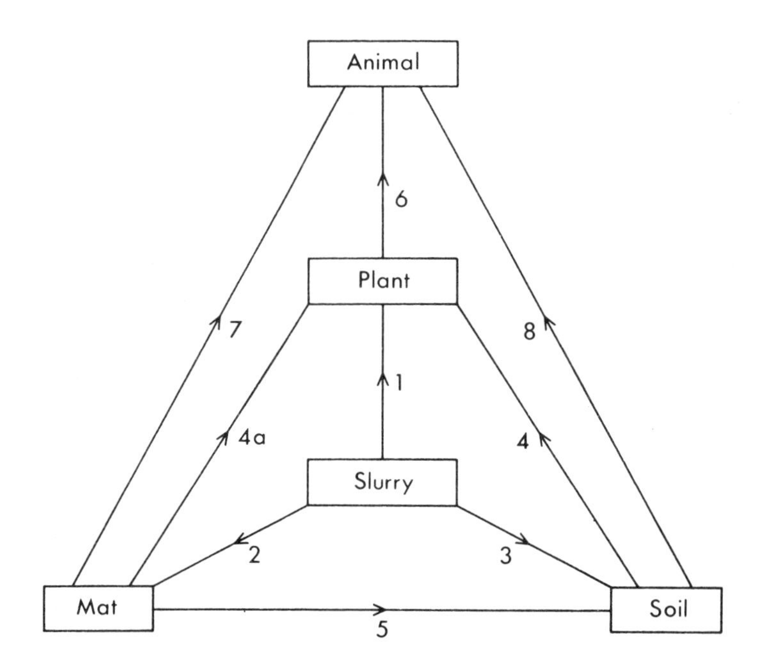

Fig. 1. Distribution of slurry-applied copper.

(1) Slurry-plant (plant contamination): Spreading slurry increased the copper content of herbage harvested for conservation. At the highest application rate this increase

varied from as little as 4 ppm to 100 ppm but was generally less than 15 ppm. Copper contents at harvesting were strongly influenced by rate of slurry application and sward height at time of spreading, smallest values being associated with application of low volumes to bare swards. Other factors identified as important, were the time interval between spreading and harvesting, the time interval between spreading and rainfall, and the intensity of the rainfall. There was also a plant species effect; in a pot experiment copper contents of finer grasses and clovers were increased by 10 - 25 ppm, while copper in coarser grasses including *Dactylis glomerata* and *Phleum pratense* was scarcely affected.

(2) Slurry-mat: Applications of slurry, especially when heavy and frequently repeated, caused the formation of a mat on the soil surface. Copper was lost from or diluted in mat material with time, because after 3 years, with all heavy application treatments, the average copper content of the mat was less than 399 ppm, corrected for ash to 10 percent, whereas the average copper content of the slurry applied had been 739 ppm. Mat formation would not be likely to occur to any extent under grazing conditions.

(3) Slurry-soil: After a cumulative slurry application of 1 000 m^3 ha^{-1} over three years, total copper in the 0 - 5 cm soil horizon (and including the mat) was increased by as much as 38 ppm. Increases of up to 6 ppm and 1.7 ppm were found for 5 - 10 cm and 10 - 15 cm horizons respectively. EDTA-extractable copper was increased similarly to total copper. After a cumulative application of 120 m^3 ha^{-1} of slurry over two seasons on land grazed by sheep, extractable copper was also increased in addition to other nutrients, notably phosphorus (Table 2).

TABLE 2.

EXTRACTABLE PHOSPHORUS (PPM) (MORGAN'S EXTRACT) AND COPPER (PPM) (0.05 *M* EDTA SOLUBLE) PRESENT IN SOIL AFTER APPLICATION OF SLURRY AT 20 m^3 ha^{-1} ON 6 OCCASIONS

Nutrient	Soil horizon (cm)	Site 3		Site 1a	
		Control	Slurry	Control	Slurry
		Extractable Nutrient (ppm)			
Phosphorus	0-5	6	41	10	24
	5-10	3	18	5	33
Copper	0-5	4.1	8.0	1.0	2.9
	5-10	3.5	5.7	0.9	3.9

(4) Soil/mat-plant: In pots, different grass species which received combined slurry applications equivalent to 750 m^3 ha^{-1} had their copper contents significantly increased, especially the coarser grasses. The copper content of clovers was not affected. Destruction of the sward with paraquat, followed by incorporation of the surface mat into the top 10 cm of soil, likewise increased the concentration of copper in herbage by the following amounts: *L. perenne* (S23) 5.4 ppm, (Vigor) 5.6 ppm, (Reveille) 3.8 ppm; *L. multiflorum* (Lemtal) 3.8 ppm; *Festuca arundinacea* (Alta) 8.0 ppm; *Dactylis glomerata* (Rano) 9.0 ppm; *Phleum pratense* (Climax) 9.6 ppm; *Trifolium repens* (Blanca) 1.2 ppm, (Kent wild white) 0 ppm; and *Trifolium pratense* (Hungaropoly) 4.0 ppm. Thus contrasting results were obtained for fine grasses and clovers on the one hand and for coarse grasses on the other; greater elevation of copper by contamination with slurry occurred with the former and greater uptake of copper occurred with the latter.

Results from other experiments indicated that the uptake of copper initially was at least as great from slurry-derived copper as from copper salts with 100 - 200 ppm copper added to the soil. Soil texture or CEC had no significant effect on copper uptake. Growth of *L. perenne* seedlings was retarded when copper from slurry in soil was greater than 220 ppm. With this, and greater concentrations of copper in soil from slurry or copper salts, herbage attained a maximum of about 30 ppm copper. Present results indicate that uptake of copper from slurry, even more than from copper salts, decreased with successive cuts and was not significant after 12 months.

Large concentrations of copper in herbage were found at the end of the season, a time when copper uptake tends to be greatest (Fleming, 1973). Copper concentration in herbage tended to increase most on the lightest (driest) soil even with least slurry applied and increases generally exceeded 10 ppm. Contamination of herbage with slurry (4a) may have contributed to this effect in the field. Copper concentration in herbage tended to increase more under grazing than under cutting regimes at similar rates of slurry application, again reflecting mat disturbance.

(5) Mat-soil: The factors leading to destruction of the mat have not been examined. However, where large repeated applications of pig slurry were made to pasture, sward damage and/or weed growth tended to increase, which may eventually necessitate cultivation and reseeding.

(6) Plant-animal (grazing experiments): Table 3 gives results from the grazing trial, showing that copper concentrations in pasture were increased, on average, by approximately 30 ppm over each of two seasons. Estimation of faecal copper confirmed that much of this copper was ingested by the grazing sheep, and Table 4 gives liver copper concentrations, confirming that copper had been deposited in the liver. In neither year did copper reach potentially toxic levels.

(6, 7) Mat/plant-animal (feeding trials): Values in Table 5 from feeding a pelleted grass meal supplemented with slurry to housed sheep indicated copper absorption and liver storage. Increases in copper content were similar to those in the grazing experiments. In two separate trials (I and II) copper as the sulphate was shown to be more available than copper in slurry: in one trial (I) copper as sulphide was less available than copper in slurry. Experimental inclusion of the mineral components of slurry in trial III decreased copper absorption but did not account for the full difference between copper sulphate and copper in slurry.

(8) Soil-animal: The addition of soil to the feed in two trials (II and III) depressed copper uptake by sheep, but probably not to the same extent as that measured previously by Suttle et al. (1975) using a different technique.

In the grazing experiments conducted on two sites in two years, ingested soil was found to amount to about 2.5 percent of dry matter intake for much of the time, but at the end of

TABLE 3

AVERAGE COPPER CONTENT (ppm) OF HERBAGE, SAMPLED ON 2 GRAZING SITES IN 1977 AND 1978

Year	Site 3		Site 1a	
	Control	Slurry	Control	Slurry
	Copper content of herbage (ppm)			
1977	10.0	31.0	5.9	38.6
1978	10.6	42.1	4.9	33.1

TABLE 4

COPPER CONTENTS OF SHEEP LIVERS (ppm DRY WEIGHT BASIS) ON 2 GRAZING SITES IN 1977 AND 1978

Date	Site 3		Site 1a	
	Control	Slurry	Control	Slurry
	Copper content of sheep livers (ppm)			
1977				
April 18[1]	58	63	48	41
November 15[2]	82	231***	31	341***
1978				
April 18[1]	40	98	50	68
July 21[2]	149	274	39	134**
November 22[3]	114	251**	56	364***

[1] Samples were obtained by liver biopsy

[2] Samples were obtained following slaughter

[3] Value was significantly different from control at P = 0.05 (*), 0.01 (**) or 0.001 (***).

the grazing season rose to about 10 percent, which is similar to extreme values found elsewhere (Healy, 1973). Soil ingestion at the latter level of soil containing 200 ppm added copper would of itself raise the copper content of the diet by 20 ppm.

TABLE 5

INCREASE IN COPPER CONTENT OF LIVER ABOVE CONTROL FOR TRIALS I AND II AND ABOVE SLURRY ONLY TREATMENT FOR TRIAL III IN THREE EXPERIMENTS

Trial	Treatment	Liver copper difference (ppm)	LSD (P = 0.05)
I	Copper sulphide	68	94
	Slurry	121	
	Copper sulphate	361	
II	Soil 3 (2.5%)	-18	76
	Slurry	119	
	Slurry + soil 3 (2.5%)	80	
	Slurry + soil 3 (10.0%)	59	
III	Slurry + soil 1a (2.5%)	-48	243
	Slurry + soil 3 (2.5%)	-54	
	Slurry + soil 4 (2.5%)	-48	
	Copper sulphate	427	
	Copper sulphate + slurry components	237	

CONCLUSIONS

These studies have shown that there is a risk of copper accumulation in susceptible animals arising from the spreading of copper rich pig slurry. Its magnitude will depend on (a) type of animal (b) prior copper status (c) interaction of copper with other nutrients and with soil and (d) management factors including extent of exposure of animals to slurry copper. For this reason every situation where susceptible species, particularly sheep or young cattle, are exposed to amounts of copper in excess of those required for adequate nutrition should be considered with care. Nevertheless, slurry application to short grass as a fertiliser at a rate of 20 m^3 ha^{-1} yr^{-1} would be most unlikely to present a hazard to grazing sheep or young cattle. However, heavily contaminated pasture whether grazed or conserved does subject susceptible species to the risk of copper toxicity.

Therefore, in the short term, the responsible use of copper-enriched pig slurry does not appear to impose any unusual restriction. Indiscriminate disposal of slurry will, in the long term lead to the irreversible accumulation of copper in addition to zinc and phosphorus. The major effects are likely then to be on plant growth and composition, rather than directly through soil ingestion, to animal nutrition.

In Ireland, where animal copper status is often low, the need for a further restriction in the use of copper-enriched pig slurry as a fertiliser on grassland is not indicated.

ACKNOWLEDGEMENTS.

The authors wish to thank Mr. R.J. McCormack, Mr. J. Sinnott, and Mr. W. Moore for skilled technical assistance, the staffs of the Analytical Laboratories, Johnstown Castle and Dunsinea for analysis, Mr. J. Dwyer for technical assistance with pot experiments and Mr. M. O'Keeffe for processing data.

REFERENCES

Batey, T., Berryman, C. and Line, C., 1972. The disposal of copper-enriched pig-manure slurry on grassland. J. Br. Grassld. Soc. 27: 139-143.

Dalgarno, A.C. and Mills, C.F., 1975. Retention by sheep of copper from aerobic digests of pig faecal slurry. J. agric. Sci., Camb. 85: 11-18.

Feenstra, P. and van Ulsen, F.W., 1973. Hay as a cause of copper poisoning in sheep. Tijdschr. Diergeneesk. 98: 632-633.

Fleming, G.A., 1973. Mineral composition of herbage. In: Chemistry and Biochemistry of Herbage. Vol. I. Ed. G.W. Butler and R.W. Bailey Academic Press, London and New York, pp 529-566.

Gracey, H.I., Stewart, T.A., Woodside, J.D. and Thompson, R.H., 1976. The effect of disposing high rates of copper-rich pig slurry on grassland on the health of grazing sheep. J. agric. Sci., Camb. 87: 617-623.

Healy, W.B., 1973. Nutritional aspects of soil ingestion by grazing animals. In: Chemistry and Biochemistry of Herbage. Vol. I. Ed. G.W. Butler and R.W. Bailey. Academic Press, London and New York, pp 567-588.

Suttle, N.F., Alloway, B.J. and Thornton, I., 1975. An effect of soil ingestion on the utilisation of dietary copper by sheep. J. agric. Sci., Camb. 84: 249-254.

Suttle, N.F. and Price, J., 1976. The potential toxicity of copper-rich animal excreta to sheep. Anim. Prod. 23: 233-241.

DISCUSSION

J.H. Voorburg *(The Netherlands)*

The Animal Health Service in the Netherlands investigated copper toxicity on farms some years ago, and found about 20 or 30 ppm in the soil. However, two interesting points were that it occurred mostly on pig producing farms; and that the worst cases were in February when some concentrates were being fed. The hypothesis was made that sheep grazing on grassland manured with pig slurry were already receiving a potentially dangerous level of copper and that the concentrates supplied extra copper to the amount already ingested.

D. McGrath *(Ireland)*

That is quite possible. Perhaps I should have stressed that sheep are susceptible to copper toxicity and anything that increases the copper level in the herbage can put them at extra risk, for example during the compounding of rations when trace elements may be added to the diet.

D.B.R. Poole *(Ireland)*

The availability of copper to the animal from conserved forage, such as hay and cereal feeds and possibly even silage, is much higher than that from grass with an equivalent copper level. In a general way this may be the answer.

SIMULATION OF ENVIRONMENTAL POLLUTION BY LANDSPREADING OF MANURE

H. Laudelout and R. Lambert

Departement Science du Sol, Place Croix du Sud 2, Université de Louvain, B-1343 Louvain-la-Neuve, Belgium

The quantitative description of processes of importance in the agricultural environment has received increased attention in the last few years; the level of complexity of the description has varied widely as well as the degree of simplification of the mechanisms involved. Most of the scepticism about the usefulness of mathematical models comes from the failure to identify the relationship between the structure of the model and its intended purpose. For example, the purpose may be limited to a sufficiently accurate description of salinity effects following slurry application. The results will be adequate for practical purposes but the model will not give further information and important processes will be entirely omitted. At the other extreme, the model may try to encompass too many features and may become so unwieldy that study of the real system is easier.

This paper summarises the attempts that have been made to develop a model for treating quantitatively the nitrogen pollution from land application of manure. The processes involved include the quantitative description of the mineral-isation of organic matter, the utilisation of mineral nitrogen, the death and lysis processes of biomass and the nitrification of ammmonium nitrogen.

Similar modelling approaches have been used recently, some being much more comprehensive, others being less realistic in their treatment of the processes.

We limit ourselves to the process indicated above with the emphasis on the nature of the rate laws used.

1. ELEMENTARY RATE PROCESSES

Rate processes in soil biology and biochemistry refer either to the rate of change of concentrations of catalysts or to the kinetics of reactions; these processes occur within a convection-diffusion flow of the soil solution through the profile. Chemical engineers face similar problems when studying problems of reactors. Chemical kinetics further differ in that the rate laws which may be used represent grosser formulation, such as the use of the hyperbolic relationship for expressing rate of either substrate disappearance or the increase in bacterial numbers.

For instance, the rate of increase in bacterial numbers, n, will be given by

$$\frac{dn}{dt} = \frac{k_o nc}{C_1 + C} \qquad (1)$$

where C is the substrate concentration, C_1 and k_o constants and t is the time.

On the other hand, the rate of substrate disappearance is related in a similar way to the bacterial numbers and substrate concentration by

$$-\frac{dC}{dt} = \frac{AnC}{K_m + C} \qquad (2)$$

where A is the specific activity of the biomass for a substrate which has a saturation constant K_m.

The two differential equations are not independent, the molar growth yield R being defined by

$$-\ R = (dn/dt)/(dC/dt) \qquad (3)$$

may be constant in which case exponential growth occurs until most of the substrate is transformed; if enzymatic transformation of the substrate only occurs, then R is zero.

One of the most important relationships which can be used for those equations is that which obtains between the specific growth rate, the specific activity and the molar growth yield: knowledge of any two of them allows the calculation of the third.

Combining equations (1) (2) and (3) gives

$$\frac{k_o}{R} = \frac{1 + C_1/C}{1 + K_m/C} \qquad (4)$$

if, (i) C_1 differs little from K_m and (ii) C is comparable to either C_1 or K_m, then

$$k_o = AR$$

Knowledge of the relationship allows solution of the equation from the choice of whichever two of the parameters k, R and A are most easily determined, and the validity of equation (5) is justified by the following examples.

TABLE 1

RELATIONSHIP BETWEEN MOLAR GROWTH YIELD, SPECIFIC ACTIVITY AND GROWTH RATE FROM EXPERIMENTAL RESULTS

	1 Soils	2 Pure	3 culture
Growth rate (day^{-1})	.22	1.11	1.1
Specific activity (pM $cell^{-1} day^{-1}$)	3.8	.27	0.1 to 0.2
Growth yield (cell pM^{-1})	.06	5.2	4 to 10
Growth yield calculated from equat. (5)	.06	4.0	5.5 to 11

Columns (1) and (2) from Rennie and Schmidt (1977). Column (3) from Laudelout et al (1968), Dessers et al (1970).

Dommergue et al. (1979) have reported on the basis of data published by Rennie and Schmidt (1977) values of the parameters given in Table 1 which we have completed by data from this laboratory. Clearly the agreement between the calculated value for the molar growth yield and those observed is satisfactory and does not need experimental determination. The agreement observed is worth stressing since accurate determinations of yield in soils are difficult to obtain.

If two substrates are involved in a process, the combined effect of their concentration C' and C" can feasibly be modelled, by considering the reaction as approximating to a two substrate reaction, by stating that:

$$-\frac{dC'}{dt} = \frac{AnC'C''}{(K'_m + C')\,(K''_m + C'')}$$

If the concentrtion of one substrate is constant, this rate equation is identical with that described by equation (2).

The effect of the hydrogen ion concentration $[H^+]$ on the specific activity A is expressed by dividing the latter by a function of $[H^+]$ of the Michaelis type:

$$A = A_o/(1 + [H^+]/K_1 + [H^+]).$$

These rate relationships are illustrated in the following section.

II. APPLICATION TO MODELLING

The problem of predicting environmental contamination by effluents has many aspects, two of which seem particularly amenable to a quantitative description. The first is the calculation of the effect of nitrification on the nitrogen load of the effluent, the second is the more complex pattern of the fluxes of nitrogen between a number of compartments, such as living or dead biomass, soil organic matter, added organic material, stable humus and mineral nitrogen.

A. Nitrification

Since internationally acceptable concentrations of nitrogen in water are fairly low, being 0.5 mg NH_4-N l^{-1} and 11.3 mg NO_3-N l^{-1}, the role of nitrification on the fate of nitrogen is understandably important. Therefore the first phase of this programme devoted most attention to modelling biological oxidation of ammonium.

The factors included were - substrate concentration, initial population of nitrifiers, their rate of growth, specific activity, molar growth yield, oxygen partial pressure, temperature, pH and substrate inhibition.

Using the method of formulation outlined above, this involved writing the following six differential equations, in which the values of 16 parameters and 7 initial conditions have to be supplied. The first is:

$$\frac{dn}{dt} = \frac{k_o \; n \; C \; f_{O_2}}{(C_1 + C)(1 + [H]/k_a)} \qquad (6)$$

describing the growth of *Nitrosomonas* in numbers ml^{-1} (n) as determined by its substrate concentration, hydrogen ion concentration and the oxygen partial pressure occurring in the expression f_{O_2} defined by

$$f_{O_2} = \frac{[O_2]}{k_m^{O_2} + [O_2]} \qquad (7)$$

Similary we have for the rate of ammonia disappearance

$$-\frac{dC}{dt} = \frac{A n \; C \; f_{O_2}}{(K_m + C)(1 + H/K_b)}$$

where A is the specific activity defined above (equation (2)).

Identical equations may be written for *Nitrobacter* with appropriate symbols for the parameters characteristic of this organism.

The rate of change of oxygen concentration is given by

$$\frac{d\,[O_2]}{dt} = K_{1a}\ \{[O_2]_o - [O_2]\} + \frac{1}{2}\frac{dC'}{dt} + \frac{3}{2}\frac{dC}{dt}$$

where k_{1a} is the transfer rate of oxygen to the solution which has a temperature-dependent saturation value of $[O_2]_o$; the last two terms refer to the oxygen demands for the oxidation of nitrite or ammonium respectively.

Finally, the rate of production of hydrogen ions is represented by

$$\frac{d\,[H]}{dt} = -\,2\,\frac{dC}{dt}$$

No empirical adjustment of the 16 parameters can possibly be carried out since this would preclude any objective validation; because for example, the number of parameters has to be increased to include the temperature characteristics in order to have the model operational at any temperature differing from the optimum for the process.

Collecting, determining or checking the numerical values of some 30 parameters is a lengthy process. However, it offers the advantage of producing a model which is fairly general and quite versatile.

Since the validation of this model has been published elsewhere, there is no need to repeat it here. However one point is worth emphasising, within the general theme of our programme.

The combination of experimental results and simulation has shown that the transient accumulation of nitrite was directly related to the temperature of ammonium oxidation. The initial population of ammonium and nitrite oxidisers had, within the limits of their range in natural conditions, little influence on the existence of this transient concentration.

B. Mineralisation of organic nitrogen

In order to describe the mineralisation of organic nitrogen quantitatively, two problems have to be solved. The first problem is conceptual; if parameters are known for the rate processes, how are the various compartments containing nitrogen connected so that the overall behaviour of the model simulates as closely as possible the natural system. The second problem is experimental: although most of the processes involved in nitrogen mineralisation have been studied for many years, quantitative data are notably lacking. Alexander (1965) stated: "In many papers, and in more languages than any mortal agronomist should be expected to comprehend, it has been stated, restated and again stated that increasing temperature will favour oxidation" but trying to find data on the numerical values of rate constants in the literature, even without their temperature characteristics, is an unrewarding task.

Nevertheless, sufficient data on the quantitative behaviour of the microbial population of the soil seem to exist so that a model can be checked firstly against these observations. Figure 1 illustrates the scheme chosen.

Total microbial growth from the various substrates introduced into the soil or synthesised within it by the microflora can be represented by a sum of terms characteristic of each substrate, taking account of its N content and C/N ratio. Furthermore the growth from a given substrate will be considered formally as a two substrate reaction, namely:

$$\frac{dn}{dt} = \frac{k_o \, n \, C\rho}{C_{1c} + C\rho} \left\{ \frac{C}{C_{1N} + C} + \frac{C_{NH_4}}{C_{2N} + C_{NH_4}} \right\}$$

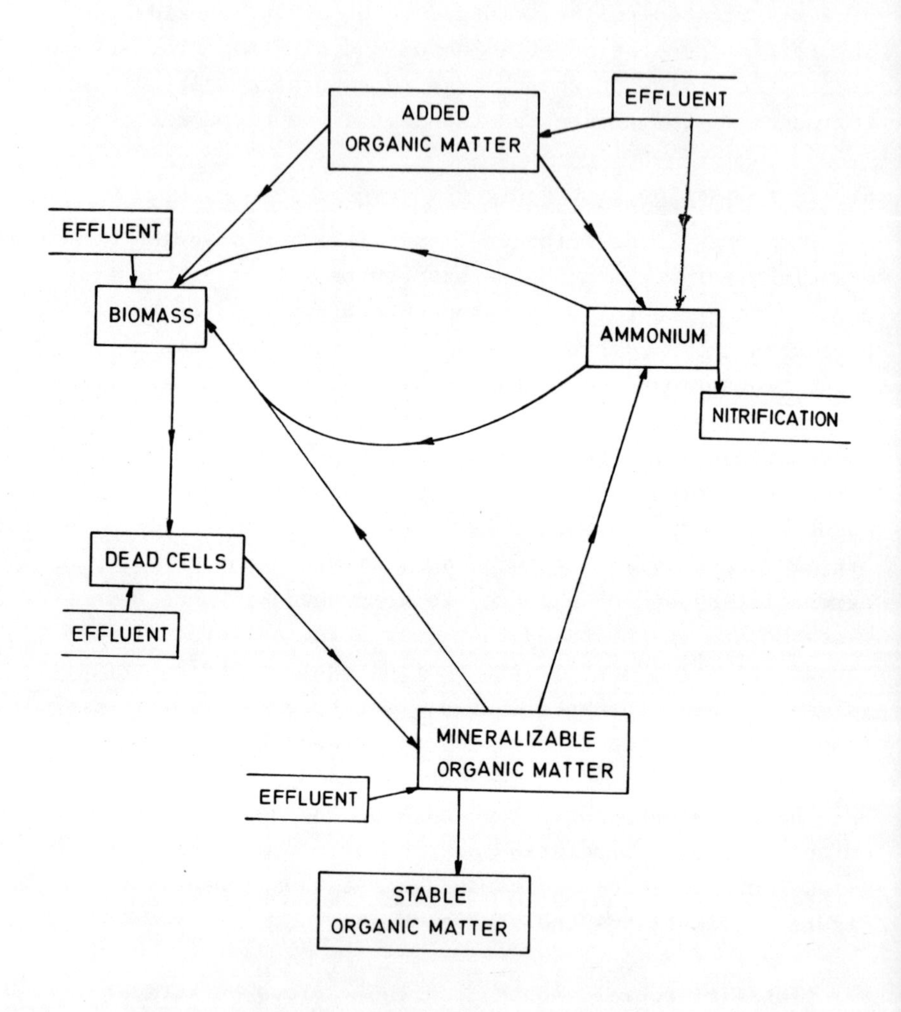

Fig. 1. Simplified scheme for nitrogen fluxes in the soil.

in which C is the N concentration corresponding to the substrate; C_{NH_4} that of the NH_4-N; and ρ the C/N ratio characteristic of the substrate: C_{1c}, C_{1N}, C_{2N} are the saturation constants corresponding to substrate C, substrate N and NH_4-N respectively; n is a quantitative expression of biomass concentration.

For all substrates occurring in the soil:

$$\frac{1}{n}\frac{dn}{dt} = \Sigma \frac{k_o^{(i)}\, C^{(i)}\, \rho^{(i)}}{C_{1c}^{(i)} + C^{(i)}\, \rho^{(i)}} \left\{ \frac{C^{(i)}}{C_{1N}^{(i)} + C^{(i)}} + \frac{C_{NH_4}}{C_{2N} + C_{NH_4}} \right\}$$

A constant term $-K_L$ may be added to this expression in order to take account of the biomass decrease or what amounts to the same formallly, of the maintenance consumption of the biomass.

This results from the relationship:

$$\frac{dn}{dt} = kn - MYn$$

in which Y is the molar growth yield for a given substrate in the absence of maintenance energy expenditure and M is the maintenance energy expended (unit of time)$^{-1}$ and (unit amount of biomass)$^{-1}$ in the absence of growth.

Formally, the product MY or the lethality coefficient k_L are identical, it is thus superfluous to try to determine separately the maintenance energy coefficient, which is difficult to measure even in pure culture.

The fate of dead cells also needs to be considered. The process of lysis is supposed to be first order with respect to n_D.

$$\frac{dn_D}{dt} = k_L\, n - k_1\, n_D$$

The nitrogen set free by the lysis of cells will increase the amount of N contained in the soil N compartment which as shown in Figure 1 may enter the biomass or the mineral N compartment or stable organic compounds according to values of the fluxes the sum of which is

$$\frac{dN}{dt} = k_1 n_D - \frac{k_3 nN}{k_m^{(s)} + k_s N}$$

where k_3 represents the specific activity of the biomass for mineralising soil N, k_s represents the first order constant for the transformation of soil N into stable compounds.

For the mineralisation of organic material added to the soil at a concentration V one would similarly have

$$\frac{dV}{dt} = - \frac{k_3^V \; nV}{K_m^{(v)} + V}$$

The balance between mineralisation and immobilisation of N may thus be expressed:

$$\frac{dCNH_4}{dt} = - \frac{dV}{dt} + \frac{k_3 nN}{K_m^N + N} - \frac{dn}{dt} - k_L n - \frac{k_n n_s CNH_4}{K_m^{(ns)} + CNH_4}$$

The last term corresponds to the disappearance of ammonium through the action of the ammonium oxidisers, where population density is n_s, specific activity k_1 and saturation constant $k_m^{(ns)}$.

Operation at temperatures other than the reference temperature is obtained by calculating the values of the parameters by the Arrhenius formula which is written

$$k_T = k_{ref} \exp \frac{\mu}{RT_{ref}^2} \{(T - T_{ref})\}$$

where k_{ref} is the value of the parameter at the reference temperature, T_{ref}, R is the gas constant and μ is the temperature characteristic.

Many simulation experiments may be carried out with a model such as the one which has just been described. We have chosen for the purpose of illustration in Figure 2, the influence of temperature on the rate at which biomass accumulates in the soil and the maximum amount of N which can be immobilised temporarily in the biomass. Immobilisation of added N clearly occurs more slowly at lower temperatures but most is also immobilised at these temperatures. Models are useful for identifying the lack of actual experimental data and are a sufficient reason for trying to improve our quantitative understanding of pollution processes.

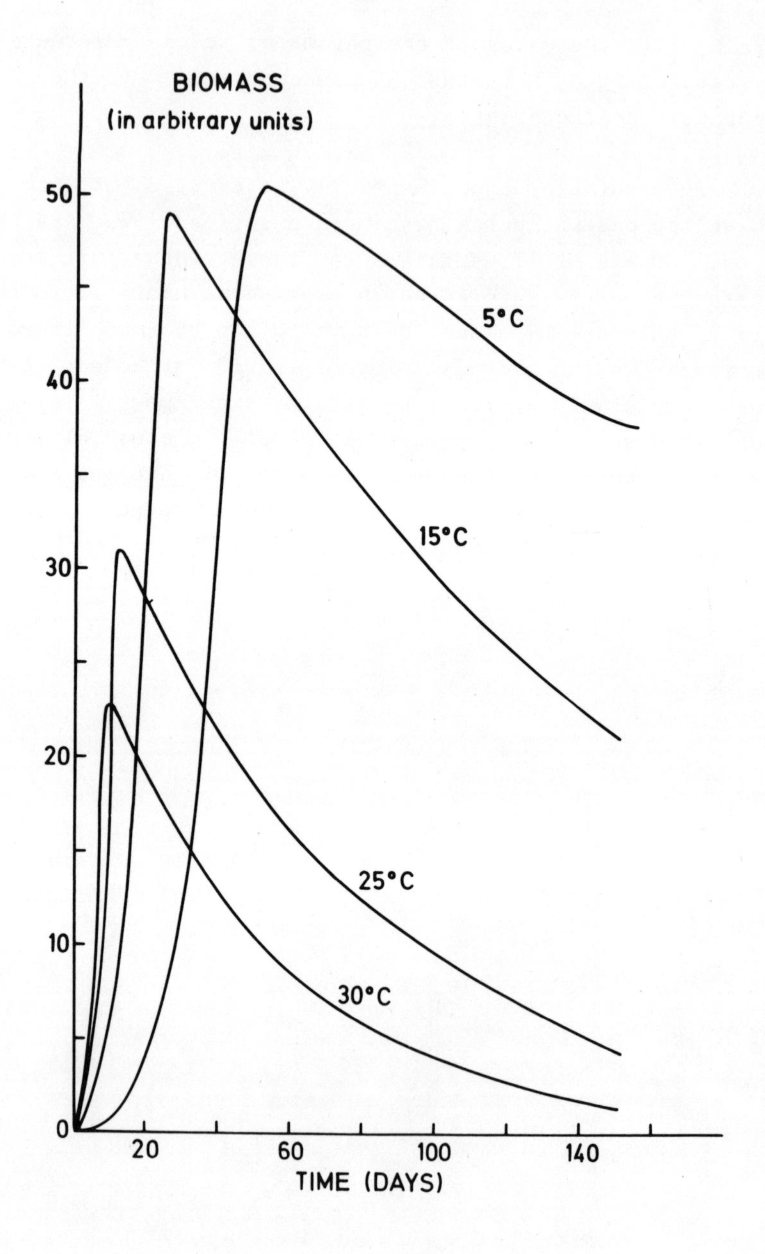

Fig. 2. The influence of temperature on change in biomass nitrogen after a single addition of organic matter.

REFERENCES

Laudelout, H., Lambert, R., Fripiat, J.L. and Pham, M.L., 1974. Effet de la température sur la vitesse d'oxydation de l'ammonium en nitrate par des cultures mixtes de nitrifiants. Ann. Microbiol., 125B, 75-84.

Laudelout, H., Lambert, R. and Pham, M.L., 1974. Influence du pH et de la pression partielle d'oxygene sur la nitrification. Ann. Microbiol., 127A, 367-382.

Laudelout, H., Lambert, R. and Pham, M.L., 1978. Variation saisonnière de la population microbienne du sol. Ecol. Biol., 15(2), 157-168.

Laudelout, H., 1978. Mathematical modelling of biological ammonium oxidation. Microbiology, 384-386.

OTHER RELEVANT LITERATURE

Alexander, M., 1965. 'Nitrification' In: Soil Nitrogen, W.V. Bartholomew and Clark, F.E. Ed. American Society of Agronomy. Monographs, 309-346.

Dessers, A., Chiang, C. and Laudelout, H., 1970. Calorimetric determination of free energy efficiency in *Nitrobacter*. Journal of General Microbiology, 64, 71-76.

Dommergues, Y.R., Belser, L.W. and Schmidt, E.L., 1978. Limiting factors for microbial growth and activity in soil. Advances in Microbial Ecology, M. Alexander Ed. Plenum Press, 49-104.

Laudelot, H., Simonart, P.C. and van Drogenbroeck, R., 1968. Calorimetric measurements of free energy utilisation by *Nitrosomonas* and *Nitrobacter*. Archiv fur Mikrobiologie, 63, 265-277.

Rennie, R.J. and Schmidt, E.L., 1977. Autecological and kinetic analysis of competition between strains of *Nitrobacter* in soils. Ecol. Bull. (Stockholm) 25.

THE USE OF MATHEMATICAL MODELS IN STUDYING THE LANDSPREADING OF ANIMAL MANURES

P.D. Herlihy
Economics and Rural Welfare Research Centre,
The Agricultural Institute, Dublin, 4., Ireland

INTRODUCTION

This paper considers some models which have been developed to study the application of animal manures to agricultural land. These models have been so formulated that they can be readily handled using available mathematical techniques, in particular the technique of Linear Programming (Danzig, 1963). However the emphasis in this paper is on the concepts underlying these models. Results obtained using the models are given for some examples.

THE CENTRAL PROBLEM

Animal manures are the link between crop production and animal production. Since the manures from livestock are returned to the local crops the balance between crops and livestock is an important consideration at farm and regional level. This balance is usually studied by looking at nutrient cycles for each nutrient separately (Cooke, 1969). That is the total available quantity of each nutrient taken separately is compared to the requirements of the crops. While this method is widely used it is based on an untenable assumption i.e. that in studying manure the nutrients can be considered separately. The key fact about manures is that the nutrients occur in fixed proportions to each other and they cannot be easily separated. So any realistic model of manure use must deal with all the nutrients simultaneously. The following simple example will illustrate the danger of taking each nutrient separately.

Consider a farm of 40 hectares with 172 cows. The crops grown are 20 hectares of fodder beet and 20 hectares of barley. The available nutrients in a cow's manure is given in Table 1 and the nutrient requirements of the crops are given in Table 2. Only P and K are considered.

TABLE 1

NUTRIENT IN MANURE

	Available nutrients in manure produced in 1 year in kg animal^{-1}	
	P	K
1 cow	7	37

TABLE 2

NUTRIENTS REQUIRED BY CROPS

	Nutrient required in kg hectare^{-1}	
	P	K
Barley	30	83
Fodder Beet	30	208

TABLE 3

TOTAL AVAILABLE AND REQUIRED NUTRIENTS

	P	K
Total nutrient available from manure	1204	6364
Total nutrient required by crop	1200	5820

In Table 3 we give the total available nutrients and the total requirements of the crops. If we consider the nutrient cycles for P and K separately it appears that the requirements of the crops can be met using the manures. However in practice this is not the case. To supply the P requirements of the two crops about half the manure must be applied to each crop. If this is done then the K requirement of the fodder beet is not supplied (less than 3 200 kg of K will be applied to the fodder beet which requires 4 160 kg). Thus considering the nutrients separately can be misleading.

This paper considers models for use with animal manures which consider all the nutrients at the same time.

AT THE REGIONAL LEVEL

An important problem at the regional level is to determine to what extent can the crops in a region cope with the amount and type of manures being produced in the region. A measure for a given region of this ability is required which takes account of the nutrients in the manures not one at a time but together. The measure proposed is donated by ISAP (Index of Structural Agricultural Pollution). It is defined as the ratio of the agricultural area of the region to A min, where A min is the minimum area with the same cropping pattern as the region on which all the manures produced in the region could be disposed without the build up of any nutrient. This minimum area will depend not only on the numbers of animals of the different types within the region but also on how well the manures produced match the crops. The problem of calculating the ISAP can be formulated as a Linear Programming (Appendix 1) problem and readily solved (International Business Machines).

As an example this approach was applied to various EEC countries (Appendix 2). The results are given in Figure 1. A value of the Index of less than 1 indicates that the manures in the country could be disposed of on the crops in the country (together with suitable chemical fertilisers) without the build

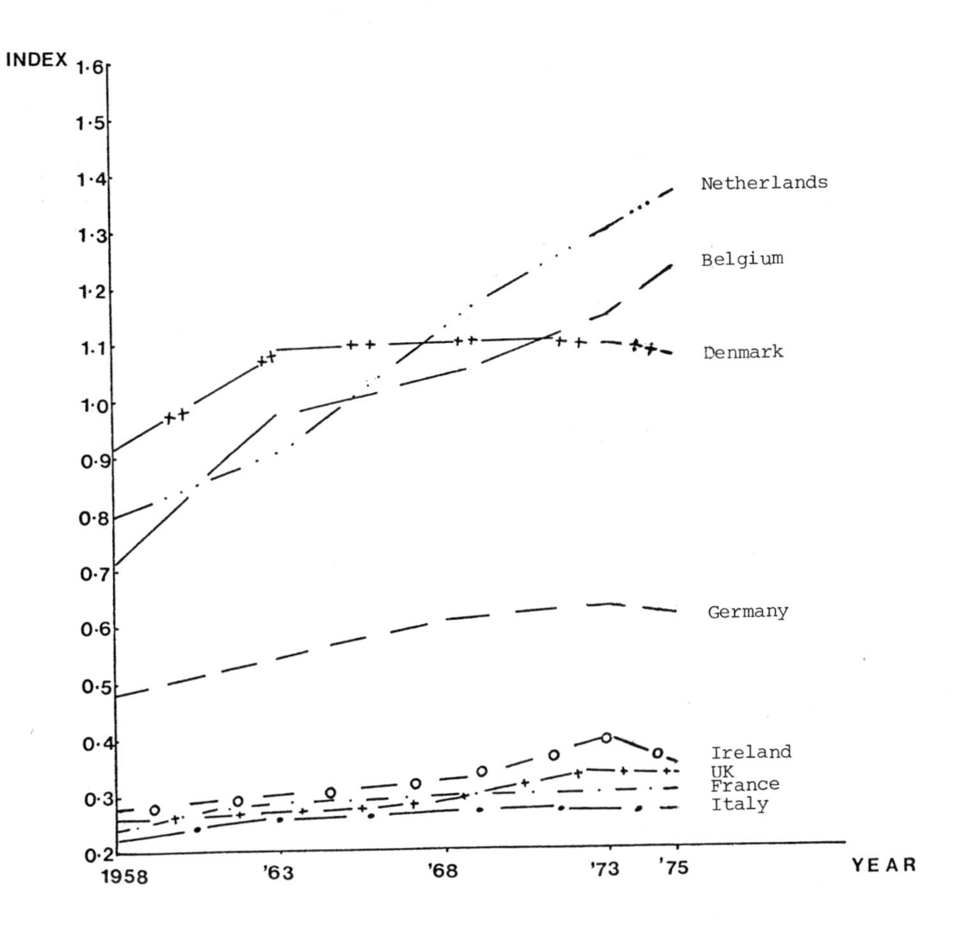

Fig. 1: Historical Trend of ISAP

up of nutrient. A value of the Index of greater than 1 means that the manures cannot be disposed of without the build up of some nutrient.

We see that the countries divide into two groups. France, Italy, Ireland and the United Kingdom would appear to have ample crops on which to spread their manure, (some smaller regions within these countries may have problems however). The Netherlands, Belgium and Denmark are at or beyond the stage where they can dispose of their manures without build up of at least some nutrient. More alarmingly the situation in the Netherlands and Belgium is rapidly getting worse. Germany falls between the two groups. An analysis of the results indicates that in all cases the nutrient which is causing the problem is P.

It must be stressed that these are based on the assumption that the optimum policy is followed in disposing of manures. In practice this is unlikely to be done and the build up of nutrient would be even more serious.

AT FARM LEVEL

The central fertilisation problem for a farmer is to determine, given a range of fertilisers (chemical fertilisers and manures) and their associated costs what is the cheapest way to satisfy the requirements of his crop. This can be formulated as a Linear Programming problem (Dodd et al., 1975). In the following example which illustrates the type of results which can be obtained from models of this type, attention is restricted to chemical fertilisers. Even in this simple case the results are interesting.

In this example, given the cost of a range of fertilisers, the combination of fertilisers which provides at minimum cost, a fertiliser with nutrients in any required proportion to each other is determined.

Table 4 gives the prices for a range of fertilisers at a location in Ireland at a given time. It also shows the cost of the combination which would have the same contents at minimum cost.

In the case of a number of the compound fertilisers it is possible to produce a combination with the same constituents at lower cost using the other fertilisers available. Table 5

TABLE 4

POSSIBLE SAVING USING LEAST COST COMBINATIONS BASED ON PRICES PREVAILING AT A CERTAIN LOCATION AT A CERTAIN TIME. (EXAMPLE USED MONAGHAN AT END OF FEBRUARY, 1978)

Type of Fertiliser	Retail price £ $tonne^{-1}$ (Farm Bulletin, Dept. of Agriculture)	Cost of least cost combination	Saving £ $tonne^{-1}$	% Saved
Calcium Ammonia Nitrate (27.5% N)	86.00	86.00	0.00	0.0
Superphosphate - 16% P	100.00	100.00	0.00	0.0
Muriate of potash - 50% K	74.00	74.00	0.00	0.0
0-10-20	100.00	92.10	7.90	7.9
10-10-20	110.00	110.00	0.00	0.0
14-7-14	N.A.	98.57	-	-
0-7-30	97.00	88.15	8.85	9.0
18-6-12	105.00	102.98	2.02	1.9
27-2½-5	103.00	103.00	0.00	0.0

TABLE 5

THE COMBINATIONS CORRESPONDING TO THE LEAST COST FIGURES GIVEN IN TABLE 4

Fertiliser required	Least cost combination: Fraction of a tonne $tonne^{-1}$				
	C.A.N. 27 ½% N	Super phosphate 16% P	Muriate of Potash 50% K	10-10-20	27-2½-5
C.A.N. (27½% N)	1.000				
Superphosphate (16% P)		1.000			
Muriate of potash (50% K)			1.000		
0-10-20		0.625	0.400		
10-10-20				1.000	
14-7-14				0.629	0.286
0-7-30		0.438	0.600		
18-6-12				0.478	0.490
27-2½-5					1.000

shows the combinations which achieve minimum costs.

The application of these techniques to prices at a number of locations at a number of different times indicates that the results above are typical in that substantial savings (≈5%) are usually possible in at least some of the fertilisers on offer. This is even true when average prices are taken at a number of locations.

This example shows how models of this type can determine optimal-fertiliser purchase policies which are superior to more routine methods of fertiliser purchase. Whether a farmer will use these policies will depend not only on the size of possible savings but also on the difficulty of implementing the implied spreading programme at farm level.

It is possible to take account of many individual features of a particular farm in models of the above type. In particular it is possible to allow for

(a) the supply of animal manures, their chemical constituents and the availability of these chemicals to crops,
(b) the purchase cost, if any, and the cost of applying animal manures,
(c) the costs of the available artificial fertilisers and the costs of applying them,
(d) the acreages of different crops to be grown, their yields and their requirements for different nutrients,
(e) the time of application of fertilisers,
(f) the type of soil.

THE COMPLETE SYSTEM

Many of the models used at present in animal production are inadequate in their treatment of animal manures. These models (Swanson, 1955; Kearney, 1971) take no account of the value of nutrients in manures for fertilisation purposes or the effect of diet on the level of nutrients in the manures. In particular in pig and poultry production animals are fed 'least cost' rations. These are designed to select what ingredients should be included in the animal's rations to give certain minimum nutritional standards at least cost. No allowance is made for the fertilisation value of the manure produced. It is possible to take account of the fertilisation value of different manures in 'least cost' rations by an analagous method to that proposed for conventional diets (Evans, 1960; Committee of Ministry of Agriculture, 1946). The prices of the ingredients can be adjusted for their manurial value and an 'adjusted least cost' ration determined using these adjusted costs. The effect of doing this will vary depending on the prevailing prices of the different ingredients. In general it will lead to rations being fed which, while slightly more expensive than the original 'least cost', will give manures richer in nutrients.

The importance of this adjustment to current practice is not only in the consequences which it can have on the type of rations being fed but the emphasis it places on manures as the link between animal production and crop production. As animal production and crop production systems become more specialised their common interest in manures being produced can be forgotten. These manures have to be disposed of on the crops and this should be emphasised at all times.

CONCLUSIONS

1. Techniques for studying the application of animal manures to crops which consider each of the nutrients in the manure separately are inadequate.

2. Models can be developed which allow all nutrients to be considered simultaneously.

3. The use of such models shows that for some countries within the European Economic Community the animal numbers are such that all the chemicals in the manures being produced cannot be taken up by the crops in these countries. In the long term this will lead to a build up of some chemicals, particularly phosphorus, in the soil. The situation has been rapidly deteriorating in some countries.

4. It is possible to determine least cost fertilisation programmes which are agronomically sound using models of the above type. These fertilisation programmes not only make the best use of available manures but can also exploit, to the considerable advantage of the farmer, the prevailing market for artificial fertilisers.

5. Animal rations which are determined using the least cost principle and which place no value on the nutrients in the manure produced can lead to manures which are less rich in nutrients than would otherwise be the case. This in turn can lower the fertilisation value of the manure being produced.

REFERENCES

Committee of Ministry of Agriculture, 1946. Agriculture 53, 169.

Cooke, G.W., 1969. Plant Nutrient Cycles Transition from extensive to intensive Agricultural with Fertilisers. International Potash Institute.

Danzig, B.G., 1963. Linear Programming and Extensions, Princeton U.P.

Dodd, V.A., Lyons, D.F. and Herlihy, P.D., 1975. A system of optimising the use of Animal Manures on a Grassland Farm. J. Agric. Engng. Res. 20. pp. 391-403.

Evans, R.E., 1960. Rations for Livestock. British Ministry for Agriculture, Fisheries and Food.

International Business Machines, MPS360, Users Guide. International Business Machines.

Kearney, B., 1971. Linear programming and feed mix formulation. Ir. J. Agric. Econ. and Rur. Sociol., 3, pp. 145-155.

Swanson, E.R., 1955. Solving minimum cost feed mix problems. J. Fm. Econ. 37, 135.

APPENDIX 1

Let X_{ik} be the number of animals of type k whose manure is applied to crop i.

Let P_i be the proportion of the region devoted to crop i.

Let A_{ij} be the number of kilograms of nutrient j required by crop i.

Let B_{jk} be the number of kilograms of nutrient j supplied by each animal of type k each year.

Let Q_k be the number of animals of type K.

Then A_{min} is got by

Minimising Z

Subject to: $\quad {}_{k} \; B_{jk} X_{ik} - A_{ij} P_i Z \quad 0$

for all ij

and $\quad {}_{i} \; X_{ik} = Q_k$

for all k.

APPENDIX 2

In this example 3 animal types were considered i.e. pigs, hens and cattle. Table 1 gives the chemical contents used for their manure.

TABLE 1

KG CHEMICALS PRODUCED BY ANIMAL

	Chemical produced/annum in kg		
Animal type	N	P	K
Pig	8.1	2.7	3.1
Hen	0.8	0.3	0.2
Cattle	44.4	7.0	54.5

Eight types of crop were considered and the nutrient removed per hectare are given in Table 2.

TABLE 2

NUTRIENTS ON AVERAGE CROP YIELDS

	Nutrients removed/hectare in kg		
Crop	N	P	K
Barley	70	12	30
Other Cereals	80	12	40
Fodder and Sugar Beet	230	20	220
Other root crops	180	25	250
Grassland	250	30	250
Clover	150	10	100
Lucerne	280	30	180
Maize	360	52	230

A MODEL FOR THE SIMULATION OF THE FATE OF NITROGEN IN FARM WASTES ON LAND APPLICATION

K.K.S. Bhat, T.H. Flowers and J.R. O'Callaghan
School of Agriculture,
University of Newcastle upon Tyne, UK.

ABSTRACT

A mathematical model is presented for predicting the various transformations undergone in the soil by nitrogen applied in farm wastes, in response to variations in soil and climatic factors. The soil is divided into layers and a simple mass balance model is used to describe the movement and redistribution of water within the soil profile, as a function of rainfall, evapotranspiration and soil moisture characteristics.

Mineralisation of native and applied organic N is assumed to follow the first-order reaction kinetics and nitrification and denitrification are treated as zero-order reactions. The rates or rate constants for all three reactions were related separately to soil temperature, moisture content and pH.

Uptake of nitrogen by a grass crop during a growing season was assumed to follow a quadratic relationship with time, to provide a sink-term for the N transformations model.

INTRODUCTION

The principal considerations in devising a schedule of application of animal manures on cultivated land are a) the availability of nutrients to the crop and b) the possibility of pollution of ground water. In order to assess both plant availability and pollution potential, a detailed understanding of the fate of plant nutrients applied in animal manures is essential. The lack of consistency in the results of experiments to evaluate the effectiveness of animal manures applied to grassland (see Whitehead, 1970 for a review) clearly demonstrates the need for such detailed knowledge.

The criteria on which a model describing the pathways of plant nutrients in animal manures on land application should be based have been described by O'Callaghan et al. (1978). They suggested that in the short-term such a model should be based on the transformations of nitrogen in the soil, whereas long-term models should take into account elements other than nitrogen which tend to accumulate in the soil since they may set limits to the annual applications of wastes.

In this paper an attempt is made to bring together a series of quantitative relationships expressing the various ways nitrogen applied as farm wastes is transformed in the soil as a function of environmental variables. This model attempts to find how the plant available forms of nitrogen accumulate in the soil or disappear from the soil in response to variations in the climate (temperature and rainfall), water and nutrient demands by the crop and losses through leaching. Being a simple practical model, it is specifically designed to accept as input variables the weather data published by the Meteorological Office and soil data usually available to the Agricultural Development and Advisory Service.

THE MODEL

Nitrogen is present in animal manures in the form of NH_4^+-N and in organic combination, the relative proportions of the two forms varying according to conditions of handling and storage. On land application, all the NH_4^+-N should be adsorbed by the soil colloids and retained in the surface layer of soil. Also, in the absence of surface run-off, most of the organic-N being insoluble, should remain in the top soil. Moreover, about four-fifths of the root systems of grass crops is confined to the top 20 cm of field soil. Hence, in this model, the soil is divided into two layers: 0 - 20 cm and 20 cm to the lower boundary of the root zone. The transformations of nitrogen, changes in soil moisture and plant uptake are considered separately in the two soil layers.

The model consists of three sub-models:

I. a water model: to describe the movement, redistribution and availability to plants of water within the soil profile as a function of rainfall, evapotranspiration and soil water characteristics;

II. a nitrogen model: that considers the various transformations undergone by the native and applied nitrogen in the soil profile in response to changes in the environmental conditions; and

III. a nitrogen uptake model: to account for the removal of N from the soil by a grass crop over a growing season.

I. THE WATER MODEL

The water model was based on a simple water mass balance for each soil layer. The water content of each layer at the end of a time step (at present the balance is calculated weekly) was permitted to vary between moisture contents of a characteristic field capacity (0.05 bar tension) and permanent wilting point (15 bar). Only water uptake by plants and downward leaching of water were considered in the model.

Plant water uptake

Actual evapotranspiration was calculated from potential evapotranspiration (which was supplied as data to the programme) depending on percentage of crop cover and the available water content of the soil to rooting depth. The relationship between actual evapotranspiration (AE) and potential evapotranspiration (PE), depending on the available water content of the soil, was taken from Baier (1969). The ratio AE/PE was assumed to be 1.0 and independent of soil available water content until it was less than 70 percent of its value at field capacity. From this point a linear decrease to zero at permanent wilting point was used.

The partition of the calculated actual evapotranspiration between the two soil layers contributing to plant water uptake depended on the crop and stage of growth. Root distribution data were used for calculating the contribution of the two layers. For a grass sward 90 percent of the actual evapotranspiration was taken from the top soil layer (AE1) and 10 percent from the second layer (AE2), provided there was sufficient available water present in each layer to meet this demand. If one soil layer was unable to meet its full demand the deficit was made up by the other layer.

Water balance

The whole of the rainfall for one time step was added to the top layer, the calculated AE1 subtracted and any excess over the field capacity moisture content calculated. This excess was added to layer two and the calculations repeated.

Average moisture content

An average moisture content for each time step was required for use in the moisture relationships within the Nitrogen Transformations model. A water model with a long time step was not well suited to supply such a value. Hence, a daily water budget to obtain a daily average moisture content was used for this purpose.

Leaching

Only movement of nitrate was considered. Complete mixing of incoming water and nitrate with the water and nitrate already present in the soil layer was assumed. The calculations are made in sequence down the profile starting with the top layer.

II. THE NITROGEN TRANSFORMATIONS MODEL

A detailed scheme for the pathways of N in farm wastes on land application is given by O'Callaghan et al. (1978). However, in the absence of sufficient quantitative information on all the steps involved, some of them were simplified, as illustrated in Figure 1.

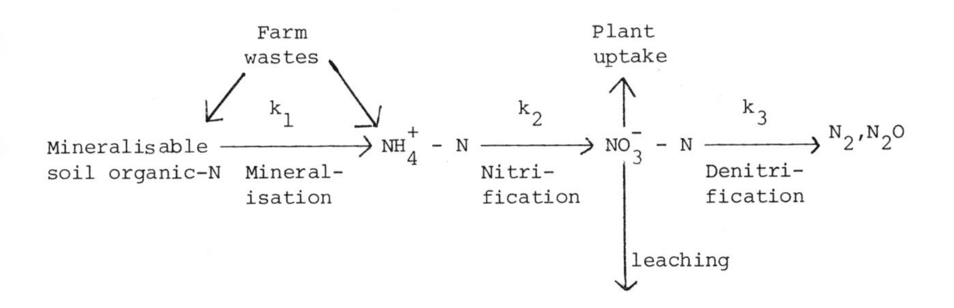

Fig. 1 Simplified flow diagram for nitrogen

A. Mineralisation

The potentially mineralisable organic-N in the wastes is assumed to undergo mineralisation twice as fast as the native soil mineralisable organic-N. Similarly, the NH_4^+-N applied as farm wastes is simply added to the nitrifiable NH_4^+- pool of the soil. The immobilisation of mineral-N was not considered separately and the term 'mineralisation' as used here refers to the net production of nitrifiable NH_4^+-N in the soil as a result of both these reactions occurring simultaneously as well as adsorption, desorption, fixation and release in the soil. Volatilisation losses of NH_3 are not considered separately.

Following earlier workers (Stanford and Smith, 1972; Campbell et al., 1974; Cameron and Kowalenko, 1976), the mineralisation process was assumed to obey the first-order reaction kinetics, whereby the rate of mineralisation is proportional to the amount of potentially mineralisable-N present in the soil according to the equation

$$\frac{dN}{dt} = - k_1 N \qquad (1)$$

and

the N mineralised in time $t = N_o \left(1 - \exp(-k_1 t)\right)$ (2)

where k_1 is the mineralisation rate constant in day -1.

The initial value N_o for the mineralisable soil-N was calculated as:

$$\underset{(ppm)}{N_o} = 53.5 + 113.2 \underset{(\%)}{N_{total}} \qquad (3)$$

as suggested by Hagin and Amberger (1974). The value of N_o was updated at each slurry application, assuming 80% of the total organic-N in the manure is mineralisable (Pratt et al., 1976; Sluijmans and Kolenbrander, 1977).

Effect of environmental factors

(1) Soil temperature:

The value of k_1 was calculated first, for a soil at optimum moisture content and pH, as a function of temperature. For temperatures above 10^oC, this was done according to a modified Arrhenius equation as suggested by Stanford et al. (1973):

$$\log_{10} k_1 = 7.71 - \frac{2758}{273.15 + {}^oC} \qquad (4)$$

Between 0° and 10^oC, the value of k_1 varied linearly with soil temperature as:

$$k_1 = 0.0009218 \times {}^oC \qquad (5)$$

(2) Soil pH:

From earlier data (Stanford and Smith, 1972; Dancer et al., 1973; Cornfield, 1959) it appears that at pH values between 4.5 and 8.0 the rate of mineralisation is insensitive to variations in soil pH. Figure 2 shows that the lower limit at which mineralisation ceases was assumed to be pH 3.5 and the k_1 values calculated according to equations (4) and (5) were modified by multiplying by a pH factor obtained as:

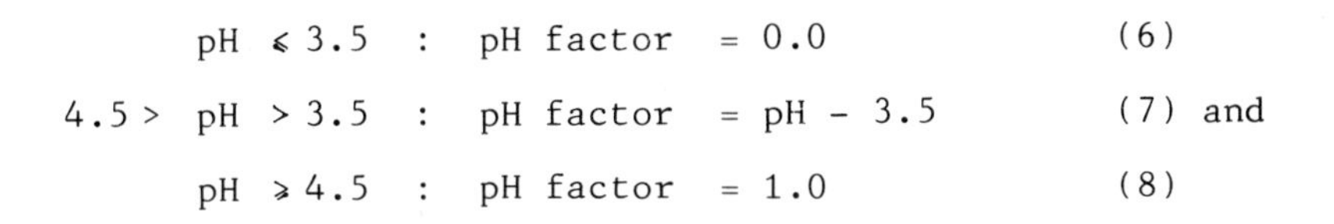

$$\text{pH} \leqslant 3.5 \quad : \quad \text{pH factor} = 0.0 \qquad (6)$$

$$4.5 > \text{pH} > 3.5 \quad : \quad \text{pH factor} = \text{pH} - 3.5 \qquad (7) \text{ and}$$

$$\text{pH} \geqslant 4.5 \quad : \quad \text{pH factor} = 1.0 \qquad (8)$$

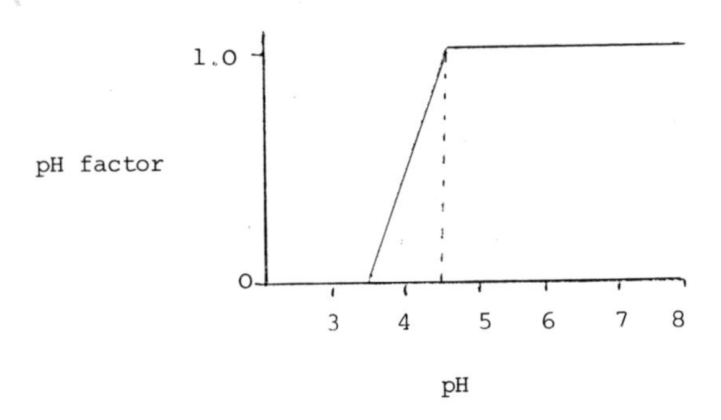

Fig. 2.

(3) Soil moisture content:

The amount of NH_4^+-N produced during incubation has been shown to increase linearly with volumetric moisture content up to a moisture content corresponding to 0.33 bar tension, above which N mineralisation is unaffected by variations in moisture content (Stanford and Epstein, 1974; Sindhu and Cornfield, 1967; Reichman et al., 1966). Accordingly, the NH_4^+-N produced calculated from equations (2) to (8) above was then multiplied by a moisture correction factor as follows:

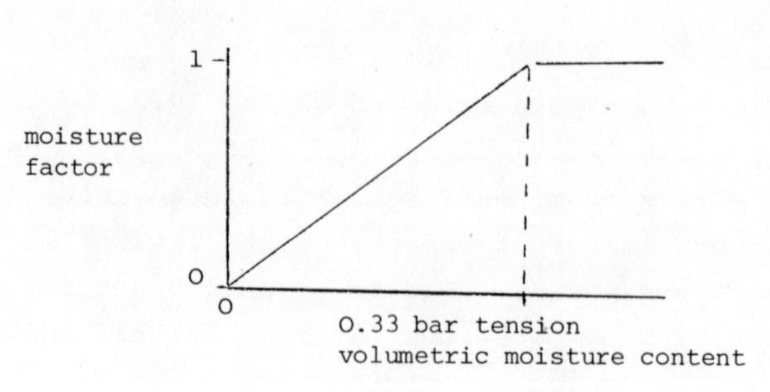

Fig. 3.

Volumetric moisture content $\geq$ value at 0.33 bar tension:

$$\text{moisture factor} = 1.0 \tag{9}$$

Volumetric moisture content $<$ value at 0.33 bar tension:

$$\text{moisture factor} = \frac{\text{moisture content}}{\text{moisture content at 0.33 bar tension}} \tag{10}$$

B. Nitrification

The conversion of NH_4^+ to NO_3^- was treated as a single-step zero-order reaction since sufficient quantitative data could not be found to account for the effect of environmental variables on the production and further oxidation of the intermediate ion, NO_2^-. The maximum value of k_2 (at 25^oC and optimum conditions of moisture and pH) was set at 100 ppm NO_3^--N week^{-1}, which appeared to be reasonable from data reported in the literature (Sabey et al., 1956; Sabey, 1969; Frederick, 1956).

Effect of environmental variables:

(1) Soil temperature:

Between 10^oC and 25^oC, an Arrhenius relationship with a Q_{10} of 2.0 was assumed (from an approximation of the data of Sabey et al., 1956; Frederick, 1956) so that

$$\underset{(\text{ppm week}^{-1})}{\ell n\, k_2} = 24.5782 - \frac{5954.96}{273.15 + {}^oC} \tag{11}$$

k_2 varied linearly between 0^oC and 10^oC as $k_2 = 3.4712 \times {}^oC$ (12)

(2) Soil moisture content:

Several studies have shown that the nitrification rate varies linearly with the volumetric moisture content between values corresponding to 15 bar tension and the optimum. However, the optimum value itself is quoted as being at 0.1 bar (Sabey, 1969), 0.2 bar (Reichman et al., 1966) and as at between 0.15 and 0.5 bar (Miller and Johnson, 1964) tensions. We chose the moisture content corresponding to 0.33 bar tension as the point at which nitrification rate is at its maximum. The k_2 calculated from (11) and (12) above was multiplied by the moisture factor as follows, and as shown in Figure 4.

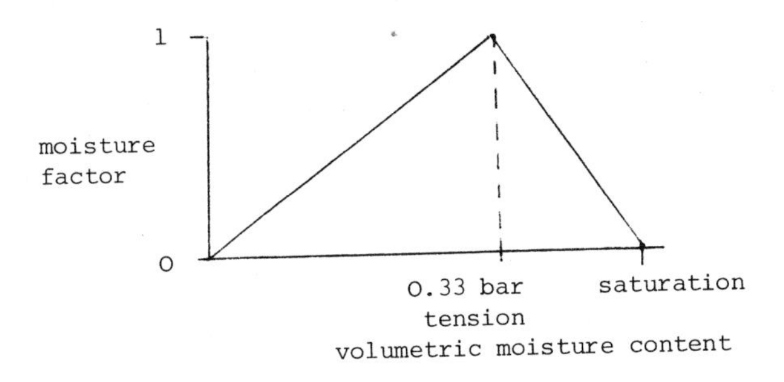

Fig. 4.

Volumetric moisture content	moisture factor	
≤ vmc at 0.33 bar tension	$\frac{\text{moisture content}}{\text{moisture content at 0.33 bar tension}}$	(13)
> vmc at 0.33 bar tension:	$1 - \frac{\text{vmc-vmc at 0.33 bar tension}}{\text{saturation - vmc at 0.33 bar tension water holding capacity}}$	(14)

(3) Soil pH:

Following Dancer et al. (1973) and Morrill and Dawson (1967), k_2 calculated as above was multiplied by a pH factor which increased linearly with pH from 0 at pH 4.5 to 1.0 at pH 6.5 with no change thereafter, as shown in Figure 5.

$$pH \leq 4.5: \quad \text{pH factor} = 0.0 \tag{15}$$

$$4.5 < pH < 6.5: \quad \text{pH factor} = \frac{pH - 4.5}{2.0} \tag{16}$$

$$pH \geq 6.5: \quad \text{pH factor} = 1.0 \tag{17}$$

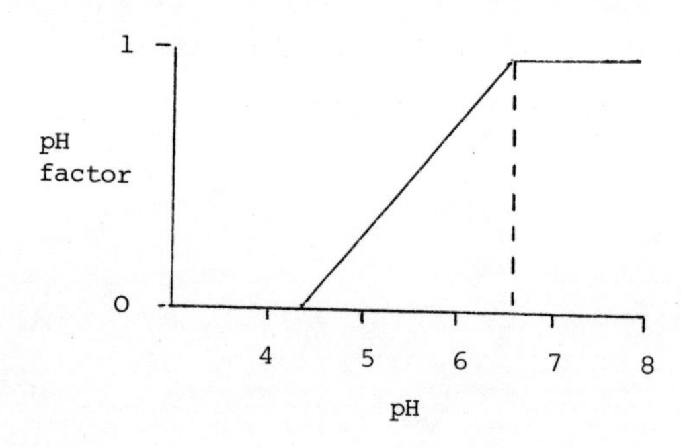

Fig. 5.

C. Denitrification

The loss of nitrate by denitrification was considered as a single-step zero-order reaction, dependent on the same environmental factors as above. However, since denitrification is known to be accentuated by the application of farm wastes because of the increased oxygen demand imposed (Burford, 1976; Burford and Greenland, 1976), the BOD_5 of slurry and the soluble organic C content of the soil were regarded as the factors of overwhelming importance in determining the value of k_3.

From their experiments on large numbers of soils, Burford and Bremner (1975) and Stanford et al. (1975b) have derived relationships between denitrification rates and total, soluble or mineralisable organic-C content of soil. We chose to use the relationships given by Burford and Bremner (1975), modified to take into acount the absence of denitrification beyond 5 days of incubation in their experiments:

The value of k_3 due to soil organic-C was first calculated as:

$$k_{3soil}(\text{ppm day}^{-1}) = 10.0 \times \% \text{ total soil organic C} + 1.24 \quad (18)$$

The oxygen demand of slurry was assumed to be entirely due to mineralisable organic-C and at each application of slurry, the value of k_3 was updated by adding k_3 due to slurry calculated as:

$$k_{3slurry} \; (\text{ppm day}^{-1}) = 0.1712 \times (\text{slurry BOD}_5 \text{ ppm soil}) \times \frac{12}{32} - 4.62 \quad (19)$$

The slurry BOD_5 was allowed to decay in the soil in the same way as the slurry mineralisable organic-N.

Effect of environmental variables:

The relationships (18) and (19) were obtained at 20°C in total absence of oxygen. So, the values of k_3 calculated as above were then corrected for variations in temperature, pH and moisture content as follows:

(1) Soil temperature:

The temperature effects on denitrification rates were expressed by an Arrhenius equation by Focht (1974). The results of Stanford et al. (1975a) and Cooper and Smith (1968) suggested a Q_{10} of 2.0, and based on these observations we used the following relationships to obtain the temperature correction factor:

$$T \leq 10^{\circ}C : \text{temperature factor} = 0.05 \times \text{temp } ^{\circ}C \quad (20)$$

$$T > 10^{\circ}C : \text{temperature factor} = \exp 19.6265 - \frac{5753.5}{273.15 + {}^{\circ}C} \quad (21)$$

(2) Soil moisture content:

At moisture contents above the critical required for denitrification to occur (75 percent of water-holding-capacity according to Bremner and Shaw, 1958), the main effect of moisture content is in governing the air-filled porosity and hence the level of oxygen supply. The critical values of air-filled porosity observed by Pilot and Patrick (1972) varied between 11 and 14 percent, depending on the soil texture. Focht (1974) showed a linear decrease in denitrification rate from maximum to zero when the percentage of air-filled pores increased from 0 to 22. We assumed a linear increase in the rate of denitrification from zero at field capacity (corresponding to 0.05 bar tension) to maximum at saturation water-holding-capacity, as shown in Figure 6.

$$\text{vmc} \leq \text{FC}: \quad \text{moisture factor} = 0 \tag{22}$$

$$\text{vmc} \geq \text{WHC}: \quad \text{moisture factor} = 1 \tag{23}$$

$$\text{FC} < \text{vmc} < \text{WHC}: \quad \text{moisture factor} = \frac{\text{vmc} - \text{FC}}{\text{WHC} - \text{FC}} \tag{24}$$

(3) Soil pH:

Based on the observations of Bremner and Shaw (1958) and Alexander (1961), we assumed that the denitrification rate increased linearly from zero at pH 5.0 to maximum at pH 7.5 so that:

$$\text{pH} \leq 5.0: \quad \text{pH factor} = 0 \tag{25}$$

$$7.5 > \text{pH} > 5.0: \quad \text{pH factor} = \frac{\text{pH} - 5.0}{2.5} \tag{26}$$

$$\text{pH} \geq 7.5: \quad \text{pH factor} = 1.0 \tag{27}$$

and as is shown in Figure 7.

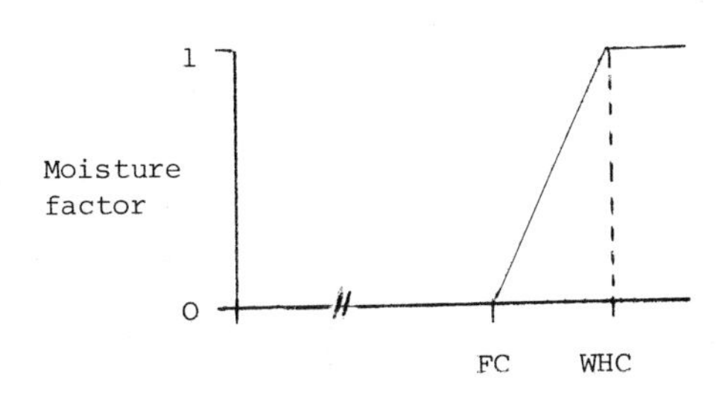

Fig. 6.

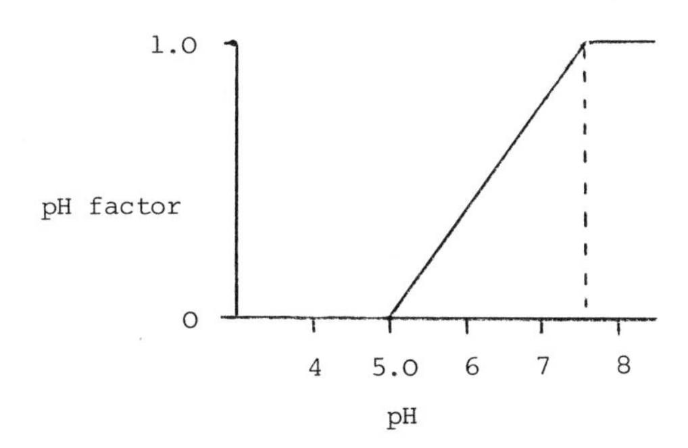

Fig. 7.

III. PLANT UPTAKE MODEL

As mentioned previously, the main objective of the current model is to simulate the transformations of N in the soil from meteorological data and a simple water model rather than predicting plant response. So the plant uptake model used at this stage serves only to provide a sink term to the N transformations model. N is assumed to be taken up only in the form of NO_3^-.

Experiments at Hurley (see Whitehead, 1970; or Spedding, 1971) have shown a very rapid uptake of N by the grass crop from the beginning of the growth period to the point of 50 per cent ear emergence and practically no uptake thereafter, in the case of an uninterrupted growth of a sward. Hunt (1973) also observed that during the primary growth of ryegrass, all

N uptake takes place within the initial month of the growth period (April 16 - May 15). However, considerable N uptake during the regrowth after cutting has been reported by Wilman et al. (1976) and Hunt (1974).

We assumed a maximum total removal of 500 kg N ha^{-1} (Reid, 1970; Lee et al., 1977) over a 7-month period of uptake (April to October inclusive). Uptake was calculated from the equation:

$$\text{N uptake (kg ha}^{-1}) = 34.4127t_{(weeks)} - 0.5913t^{2}_{(weeks)} \quad (28)$$

according to which, if soil supply of N is not limiting, 57 percent of the total uptake of 500 kg ha^{-1} occurs during the first 10 weeks of growth and the remaining 43 per cent over the last 18 weeks.

Eighty percent of the total uptake in any week is assumed to be removed from the top 20 cm soil and 20 percent from below, if these amounts are present in the two layers. If however, the demands on the first layer cannot be met, this is compensated for by a larger uptake from the lower layer, provided amounts in excess of 20 percent of the total demand are available in this layer. Further, when uptake can be limited due to lack of moisture in a dry summer, the uptake is apportioned between the two layers in the same ratio as water uptake.

PREDICTIONS

The computer programme is written in FORTRAN. The programme was run (i) over a single year to predict closely, week-by-week, the transformations, plant uptake and leaching of N; and (ii) over a period of 10 years, to examine the changes in the overall balance sheet for N on an annual basis, resulting from 10 years of application of pig slurry.

Weekly rainfall, soil temperature and potential evapotranspiration data averaged over the 30-year period 1941 - 1970

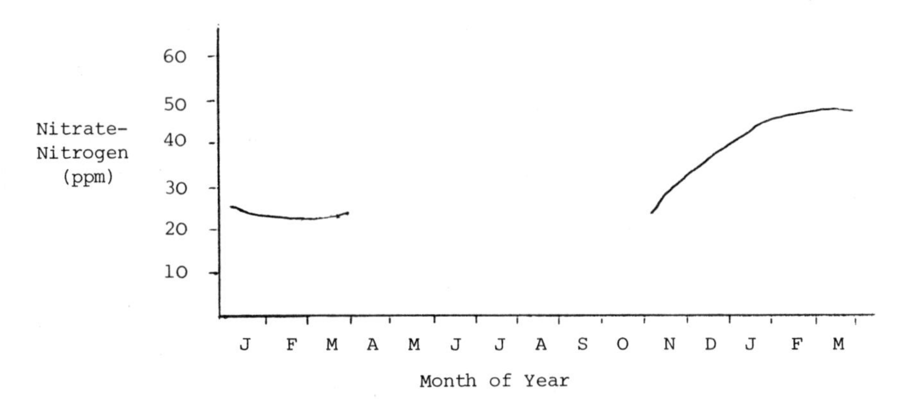

Fig. 8. Model prediction of Nitrate concentration of Leachate.

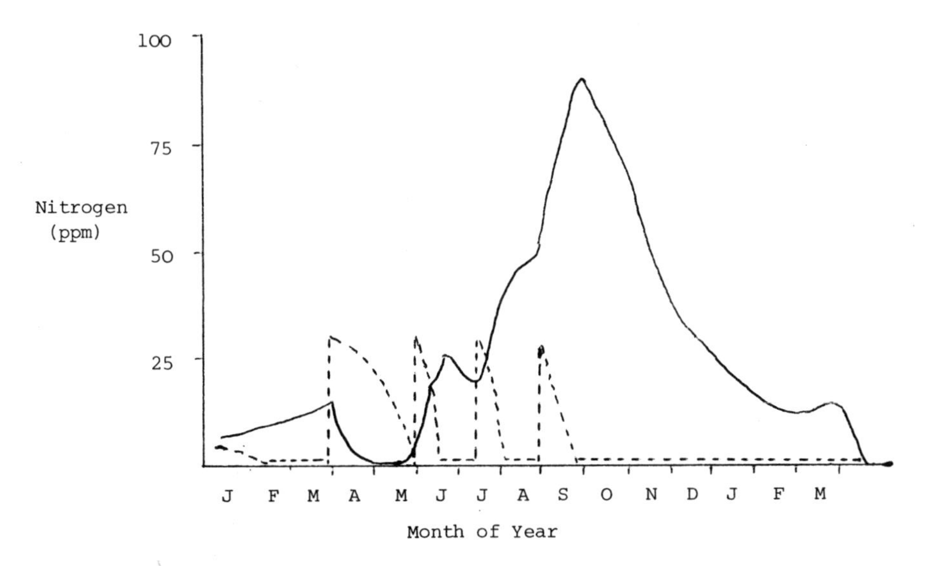

Fig. 9. Model predictions of Ammonium and Nitrate Nitrogen contents of Soil layer 1.

(NO_3 - N ———— , NH_4 - N --------)

for England and Wales (calculated from the monthly figures reported by MAFF, 1975) were used as inputs for environmental variables. Predictions were made for a sandy-loam soil (pH in water 5.3, total organic-C 4.2%, total-N 0.36%), which was used in lysimeter experiments. The effect of four annual applications (1 April, 1 June, 15 July and 1 September), each of 25 tonnes ha^{-1} of pig slurry containing 0.3% of org-N and 0.3% of NH_4^+-N and having a BOD_5 value of 25 000 ppm was tried.

A complete account of the results and a comparison of the predictions with values obtained in the lysimeter experiments over a period of two years will be given in a separate publication, but some of the trends are summarised in Figures 8 and 9.

REFERENCES

Alexander, M., 1961. Introduction to Soil Microbiology. John Wiley and Sons, London and New York.

Baier, W., 1969. Concepts of soil moisture availability and their effect on soil moisture estimates from a meteorological budget. Agr. Meteorol. 6, 165-178.

Bremner, J.M. and Shaw, K., 1958. Denitrification in soil II. Factors affecting denitrification. J. Agric. Sci., Camb. 51, 40-52.

Burford, J.R., 1976. Effect of the application of cow slurry to grassland on the composition of the soil atmosphere. J. Sci. Fd. Agric. 27, 115-126.

Burford, J.R. and Bremner, J.M., 1975. Relationships between the denitrification capacities of soils and total, water-soluble and readily decomposable soil organic matter. Soil Biol. Biochem. 7, 389-394.

Burford, J.R. and Greenland, D.J., 1976. Effects of heavy dressings of slurry and inorganic fertilisers applied to grassland on the composition of drainage waters and the soil atmosphere. Proc. ADAS/ARC Conf. on Agriculture and Water Quality. MAFF Tech-Bull. 32, 432-443.

Cameron, D.R. and Kowalenko, C.G., 1976. Modelling nitrogen processes in soil: mathematical development and relationships. Can. J. Soil Sci. 56, 71-78.

Campbell, C.A., Stewart, D.W., Nicholaichuk, W. and Biederbeck, V.O., 1974. Effects of growing season soil temperature, moisture and NH_4^+-N on soil nitrogen. Can. J. Soil Sci. 54, 403-412.

Cooper, G.S. and Smith, R.L., 1963. Sequence of products formed during denitrification in some diverse western soils. Soil Sci. Soc. Am. Proc. 27, 659-662.

Cornfield, A.H., 1959. Mineralisation, during incubation, of the organic nitrogen compounds in soils as related to soil pH. J. Sci. Fd. Agric. 10, 27-28.

Dancer, W.S., Peterson, L.A. and Chesters, G., 1973. Ammonification and nitrification of N as influenced by soil pH and previous N treatments. Soil Sci. Soc. Amer. Proc. 37, 67-69.

Focht, D.D., 1974. The effect of temperature, pH and aeration on the production of nitrous oxide and gaseous nitrogen - a zero-order kinetic model. Soil Sci. 118, 173-179.

Frederick, L.R., 1956. The formation of nitrate from ammonium nitrogen in soils. I. Effect of temperature. Soil Sci. Soc. Amer. Proc. 20, 496-500.

Hagin, J. and Amberger, A., 1974. Contribution of fertilisers and manures to the N- and P- load of waters. A computer simulation model. Rep. Deutsche Forschungs Gemeinschaft.

Hunt, I.V., 1973. Studies of response to nitrogen fertiliser 3. The development of response to fertiliser nitrogen in primary growth of ryegrass. J. Br. Grassld. Soc. 28, 109-118.

Hunt, I.V., 1974. Studies of response to nitrogen fertiliser 6. Residual responses as nitrogen uptake. J. Br. Grassld. Soc. 29, 69-73.

Lee, G.R., Davies, L.H., Armitage, E.R. and Hood, A.E.M., 1977. The effects of rates of nitrogen application on seven perennial ryegrass varieties. J. Br. Grassld. Soc. 32, 83-87.

MAFF, 1975. Agricultural Climate of England and Wales. Tech. Bull. 35. HMSO, London.

Miller, R.D. and Johnson, D.D., 1964. The effect of soil moisture tension on carbon dioxide evolution, nitrification, and nitrogen mineralisation. Soil Sci. Soc. Amer. Proc. 28, 644-646.

Morrill, L.G. and Dawson, J.E., 1967. Patterns observed for the oxidation of ammonium to nitrate by soil organisms. Soil Sci. Soc. Amer. Proc. 31, 757-760.

O'Callaghan, J.R., Bhat, K.K.S., Flowers, T.H. and Parkes, M.E., 1978. Modelling the pathways of plant nutrient elements in livestock wastes on land application. Proceedings, EEC Seminar on Engineering Problems with Effluents from Livestock, Cambridge 63-78, EUR 6249 EN.

Pilot, L. and Patrick, W.H. Jr., 1972. Nitrate reduction in soils: effect of soil moisture tension. Soil Sci. 114, 312-316.

Pratt, P.F., Davis, S. and Sharpless, R.G., 1976. A four-year field trial with animal manures II. Mineralisation of nitrogen. Hilgardia 44, 113-125.

Reichman, G.A., Grunes, D.L. and Viets, F.G. Jr., 1966. Effect of soil moisture on ammonification and nitrification in two northern plains soils. Soil Sci. Soc. Amer. Proc. 30, 363-366.

Reid, D., 1970. The effects of a wide range of nitrogen application rates on the yields from a perennial ryegrass sward with and without white clover. J. Agric. Sci. Camb. 74, 227-240.

Sabey, B.R., 1969. Influence of soil moisture tension on nitrate accumulation in soils. Soil Sci. Soc. Amer. Proc. 33, 263-278.

Sabey, B.R., Bartholomew, W.V, Shaw, R. and Pesek, J., 1956. Influence of temperature on nitrification in soils. Soil Sci. Soc. Amer. Proc. 20, 357-360.

Sindhu, M.A. and Cornfield, A.H., 1967. Effect of sodium chloride and moisture content on ammonification and nitrification in incubated soil. J. Sci. Fd. Agric. 18, 505-506.

Sluijmans, C.M.J. and Kolenbrander, G.J., 1977. The significance of animal manure as a source of nitrogen in soils. Proc. Seminar on Soil Environment and Fertility Management. Tokyo. 403-411.

Spedding, C.R.W., 1971. Grassland Ecology. Oxford University Press, London.

Stanford, G. and Epstein, E., 1974. Nitrogen mineralisation - water relations in soils. Soil Sci. Soc. Amer. Proc. 38, 103-106.

Stanford, G., Frere, M.H. and Schwaninger, D.H., 1973. Temperature coefficient of soil nitrogen mineralisation. Soil Sci. 115, 321-323.

Stanford, G. and Smith, S.J., 1972. Nitrogen mineralisation potentials of soils. Soil Sci. Soc. Amer. Proc. 36, 465-472.

Stanford, G., Dzienia, S. and Vander Pol, R.A., 1975a. Effect of temperature on denitrification rate in soils. Soil Sci. Soc. Amer. Proc. 39, 867-870.

Stanford, G., Vander Pol, R.A. and Dzienia, S., 1975b. Denitrification rates in relation to total and extractable soil carbon. Soil Sci. Soc. Amer. Proc. 39, 284-289.

Whitehead, D.C., 1970. The Role of Nitrogen in Grassland Productivity. Commonwealth Agric. Bureau, Farnham Royal.

Wilman, D., Ojuederie, B.M. and Asare, E.O., 1976. Nitrogen and Italian ryegrass 3. Growth up to 14 weeks: yields, proportions, digestibilities and nitrogen contents of crop fractions and tiller populations. J. Br. Grassld. Soc. 31, 73-79.

DISCUSSION

J.K.R. Gasser *(UK)*

We have had three presentations, do the authors have any comments on the other papers?

H. Laudelout *(Belgium)*

The difference between what Dr. Herlihy presented and what Professor O'Callaghan and I presented, is essentially that Dr. Herlihy is more concerned with economics. Farmers work for money, and economic constraints have to be taken into account. If a more comprehensive and more scientific approach, such as the one presented by Professor O'Callaghan, is being used as a base for these econometric models, one can superimpose a linear programme for optimising financial yields for farmers.

J.R. O'Callaghan *(UK)*

I would like to reinforce what Professor Laudelout has said. I think a problem is transposing research results for the people who want to use them. In scientific research, experiments are done under controlled conditions and one part of a process is taken and experimentation is made on that, but the farmer has got to transpose the results into his farming system, in which many factors interact including temperature, soil condition and its composition. Modelling enables the pieces of information to be assembled and show the overall shape of the pattern. If some of the information is inadequate, then the model itself will be less good, but sometimes we find that compensating errors enter into calculations of processes, such as zero order reactions, which give you an order of result sufficiently accurate to be able to advise the farmer. There is much information in the literature on work that has been done over the years, and we do need to synthesise it in some way. I think putting it into a quantitative form as a model is an exacting discipline, and is a useful tool for the researcher.

F.A.M. de Haan *(The Netherlands)*

I am not familiar with the models that Professor O'Callaghan and Professor Laudelout are using, and the introduction of a number of biological processes complicates the model as they do in the nitrogen cycle. For denitrification an easily consumable carbon source is needed for the micro--organisms. Are your models sufficiently advanced to include this parameter? Very closely related to this question is the problem of whether the models are sensitive to the alternation of aerobic and anaerobic conditions in the soils, because some parts of the soil are known to remain anaerobic all the time and other parts remain aerobic all the time. This is a very general question but it is a matter of interest to know how advanced your models are at this stage.

J.R. O'Callaghan

First of all I would say that my attitude to modelling is to try to make it simple, as we have done. A good model predicts the main results and conclusions required and reflects practice, without needing difficult computations, because then it becomes very expensive. We spent much time trying to avoid the micro-biological component, although we have had to introduce a micro-biological component related to the C : N ratio.

H. Vetter *(West Germany)*

I would expect that the aim of modelling would be to answer the three questions which were put yesterday. I doubt if we have enough information yet to answer these questions.

J.K. Grundey

My concern and interest, from the extension point of view, is on behalf of farming. Legislation on manure disposal is likely sometime in the future. The danger is that this question may be proposed and discussed in relation to single elements. However, slurries contain the three main nutrients and may contain copper, and the standard for disposal should reflect the need to avoid pollution in different areas, so that unnecessary costs are not incurred by the farmer.

We need a much better method of deciding the framework than those seeming to be used at the moment. Therefore, the idea is suggested that if modelling brings greater discipline, then this is beneficial.

TRANSFER OF OXYGEN IN PIGGERY WASTES

A. Heduit

Centre technique du Genie Rural des Eaux et Forêts,
Ministere de l'Agriculture, Division Qualité des Eaux,
Pêche et Pisciculture, 14, Avenue de Saint-Mandé,
75012 Paris, France.

Our research has been conducted in order to improve knowledge of factors determining the size of aeration devices for livestock effluents. Tests have been conducted in clean water, activated sludge and aerated slurry. Some of the experiments have been carried out on a laboratory scale, others in actual plants.

I. PRINCIPLE OF OXYGEN TRANSFER MEASUREMENTS IN CLEAN WATER AND IN ACTIVATED SLUDGE

In studying oxygenation, three parameters can be selected:

- the overall transfer coefficient K_La (h^{-1})
- the oxygenation capacity O.C (kg O_2 h^{-1})
- the aeration efficiency N (kg O_2 kWh^{-1})

Those parameters are determined from non steady state tests (Lewis and Withman, 1924; Kayser, 1978; Heduit, 1979), as follows:

<u>In clean water:</u> the water is first deoxygenated by dissolving in it an excess of sodium sulphite in the presence of cobalt ions as a catalyst. When all the sodium sulphite has been used, the build-up of dissolved oxygen with time is monitored by means of dissolved oxygen probes. From the analysis of the resulting reoxygenation curves, each of the parameters can be derived.

In activated sludge: the feed and return-sludge flows are interrupted and the aeration is continued until the oxygen uptake rate becomes constant. The aeration is then stopped and the oxygen concentration is allowed to decrease. When aeration is restarted, the oxygen concentration increases in a similar way to that shown by the reaeration curves derived from the clean water test. The analysis of the reaeration curves is similar with clean water and with sludge.

II. LABORATORY SCALE EXPERIMENTS

Aeration tank:

Transparent PVC with a square base and rectangular sides

base: 30 x 30 cm

height: 50 cm

Aerators:

Diffused air aeration is provided by blowing air through a porous diffuser in the bottom of the tank. Air flow is monitored by means of a rotameter and slow-speed stirring is provided (Figure 1).

Surface aeration:

Two surface aerators (8 cm and 4 cm in diameter) have been developed (Figure 2) and two oxygen meters are provided to monitor the dissolved oxygen level in the contents of the tank. A vibrating blade stirrer provides water circulation in the vicinity of the sensing membrane of each probe.

III. CLEAN WATER EXPERIMENTS IN THE FIELD

The influence of the water depth on oxygen transfer has been studied in a tank equipped with a sub-surface aerator and Table 1 summarises the results obtained.

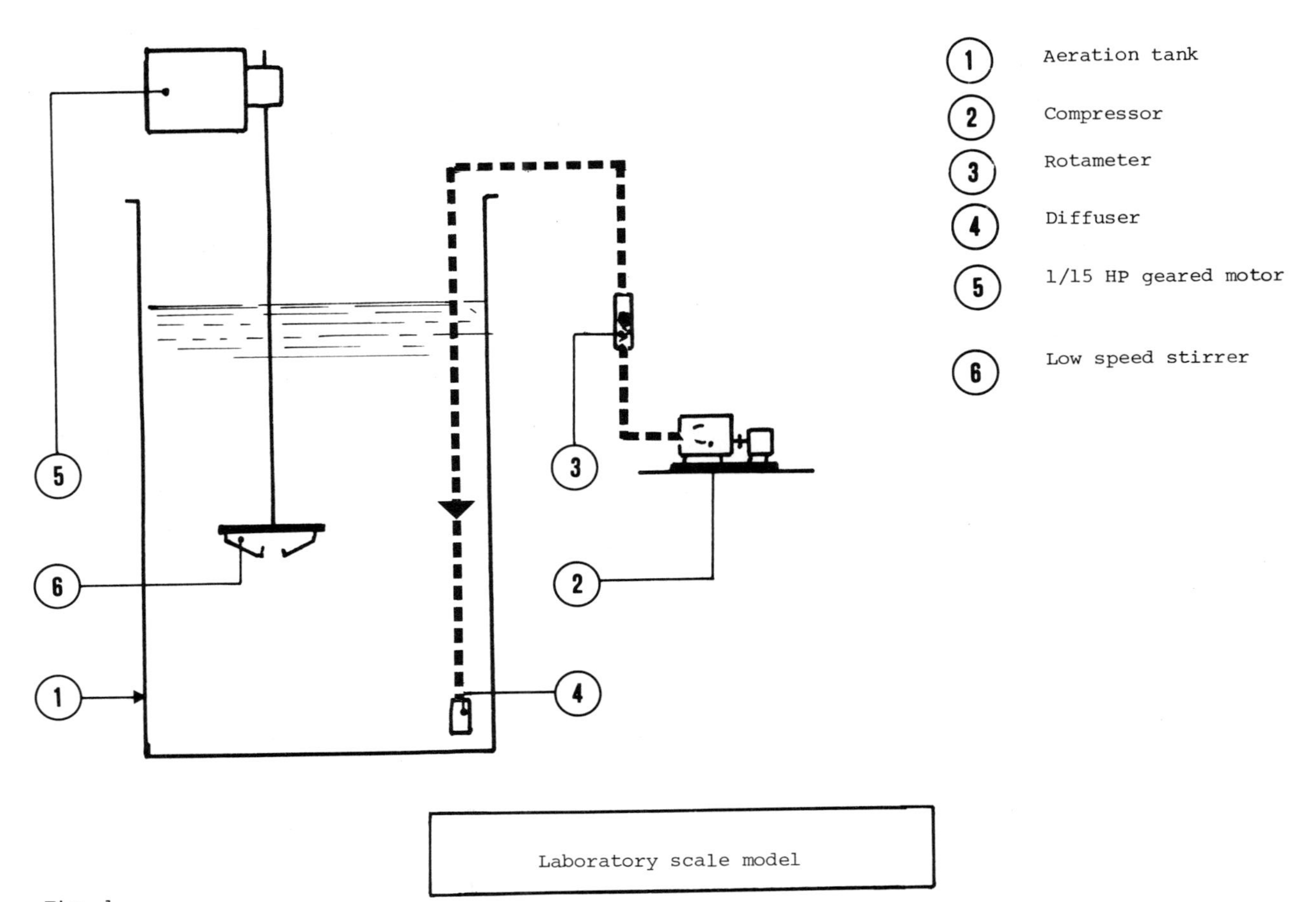
5
6
1
3
2
4
1 Aeration tank
2 Compressor
3 Rotameter
4 Diffuser
5 1/15 HP geared motor
6 Low speed stirrer
Laboratory scale model

Fig. 1.

1. Aeration tank
2. Axle 8 mm diameter stainless steel
3. 1/15 HP gearted motor
4. Surface aerator

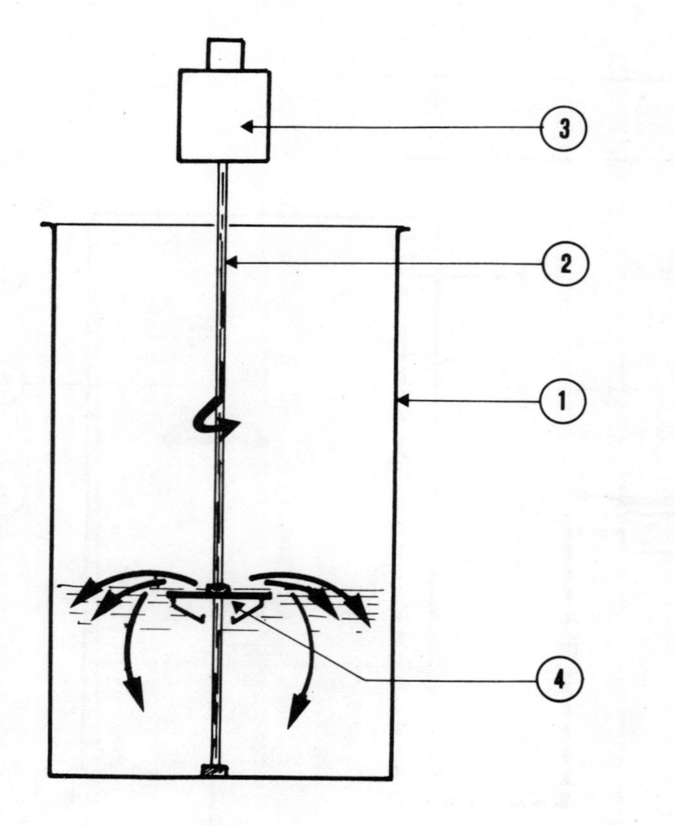

Fig. 2. Laboratory scale model

Oxygenation capacity and aeration efficiency did not depend on the water depth in the range 1.1 - 2.5 m, under the conditions of the experiment (square base tank; base sides: 5 m). Similar experiments carried out with a high speed mechanical surface aerator gave similar results in the range of 1.55 - 2.8 m deep (Table 2).

IV. CLEAN WATER TESTS IN THE LABORATORY

Tests were carried out to determine the influence of conductivity and pH on oxygen transfer. They were performed with the fine bubble diffusion system.

Conductivity effect

The results obtained are given in Table 3.

Oxygen transfer proved to be not significantly affected by variation of conductivity in the range 700 - 1 600 μMho cm^{-1} (the range of values usually encountered in clean water aerator tests in the field). Nevertheless, we observed a slight increase of transfer coefficient with salt content. This finding, in agreement with some bibliographical data (Benedek, 1971; Roustan, 1979), shows that oxygen transfer in a highly saline solution may be higher than in clean water.

pH effect

Variations in pH were obtained by adding sulphuric acid and sodium hydroxide; Table 4 summarises the results obtained.

The pH of a solution has no significant effect on the value of K_La when measurements are made with fast response probes. Low values obtained for a K_La beyond 10.5 can be ascribed to the presence of cobalt hydroxyde flakes in the solution.

TABLE 1

Test number	Water depth (m)	Power (kw)	Specific power ($w\ m^{-3}$)	Oxygenation capacity ($kg\ O_2\ h^{-1}$)	Aeration efficiency ($kg\ O_2\ kWh^{-1}$)
1	2.5	7.20	117	4.35	0.6
2	1.8	7.45	168	4.20	0.55
3	1.1	7.45	278	4	0.55

TABLE 2

Test number	Water depth (m)	Power (kw)	Specific power ($w\ m^{-3}$)	Oxygenation capacity ($kg\ O_2\ h^{-1}$)	Aeration efficiency ($kg\ O_2\ kWh^{-1}$)
1	2.78	2.26	32.75	2.18	0.97
2	2.26	2.29	41.1	2.17	0.95
3	1.55	2.3	60.6	2.17	0.95

TABLE 3

Test number	1	2	3	4	5
Conductivity ($\mu Mho\ cm^{-1}$)	660	850	1 000	1 400	1 600
K_La (h^{-1})	4.45	4.50	4.45	4.55	4.60

TABLE 4

Test number	1	2	3	4	5	6	7
pH	3	5.2	7.9	7.9	8.3	10.8	11
K_La (h^{-1})	3.7	3.8	3.5	3.7	3.6	3.3	3.3

V. ACTIVATED SLUDGE EXPERIMENTS IN TREATMENT PLANTS

Oxygenation tests were performed in the treatment plant of the Experimental Pig Unit of Institut Technique du Porc (South of France). We measured the performances of an aeration system consisting of two rotors (bladed), 3 m wide and driven by 5.6 kW motors in a 560 m^3 aeration ditch. Three tests in activated sludge (suspended solid concentration 14.8 g l^{-1}, 35 percent mineral matter) and one in clean water were carried out (Table 5). A mean correction coefficient α of 0.83 was thus established for oxygen transfer by comparing the results of the tests in activated sludge with those of the similar test made in tap water under the same conditions.

TABLE 5

	Activated sludge			Clean water
	1	2	3	1
$(K_La)_{20}$ (h^{-1})	2.25	2.4	2.3	2.8
Oxygenation capacity ($kg\ O_2\ h^{-1}$)	11.2	11.9	11.4	14.5
Power (kW)	11.3	11.3	11.3	10.5
Oxygen uptake rate $mg\ l^{-1}\ h^{-1}$	9.4	10	7.6	-

VI. ACTIVATED SLUDGE LABORATORY EXPERIMENTS

The influence of suspended solid concentration on oxygen transfer has been studied in the laboratory apparatus equipped with air diffusers and slow speed stirring. The results obtained are given in Table 6.

The α factor is a descending function of suspended solid concentration. The higher transfer in activated sludge than in clean water in this case may be explained by the presence of surface active agents and by a high level of mixing.

The influence of the aeration device on the value of α for a given activated sludge has been tested (Table 7). The Table shows that the α factor could vary with the aeration device and with the rotation speed of a surface aerator.

TABLE 6

Suspended solid concentration ($g\ l^{-1}$)	0.4	3.4	4.6	5.6	9.9	11.7
$\alpha = \dfrac{(K_L a)_T \text{ (activated sludge)}}{(K_L a)_T \text{ (clean water)}}$	1.4	1.3	1.37	1.32	1.21	1.18

TABLE 7

Type of aerator	Turbine dia. 8 cm	Turbine dia. 4 cm	Turbine dia. 4 cm	Air diffusion
Rotation speed ($rev\ min^{-1}$)	270	680	370	-
Peripheral speed ($m\ s^{-1}$)	1.13	1.42	0.77	-
Suspended solid concentration ($g\ l^{-1}$)	9.4	8.8	8.5	9.2
Mean α measured	.88	.91	.65	1

VII. EXPERIMENTS ON AERATED SLURRY UNDER PRACTICAL CONDITIONS

The coefficient α in the aerated slurry cannot be determined by the reoxygenation method because of the low dissolved oxygen levels found in these liquids. Such a determination should be performed in an airtight scale model by monitoring the oxygen level of the air used to oxidise the slurry.

The results obtained from scale models are not applicable to full scale plants and such experiments seem to be of limited interest, since the Suspended Solid Concentration fluctuations during the treatment will probably lead to large variations of the α coefficient. Then, the aim of the full scale experiments is the verification of the rule usually applied to determine the size of the aeration device for the aerobic storage of slurry:- the supply of 1 kg of O_2 (under standard conditions) kg^{-1} of BOD plus more than 20 W m^{-3} for mixing.

a) Stages in the experiment

Determination of the aerator performance (under standard conditions) in clean water in the storage tank.

After emptying the tank, refilling with an initial amount of slurry.

During seven weeks aeration, and periodical additions of given loads of slurry.

Every week, measurement of the load in the tank and determination of the amounts of COD and BOD eliminated.

The eliminated loads can be linked to the operating time of the aerator and consequently to the amount of oxygen that it would introduce under standard conditions. The efficiency of the COD and BOD removal allow the determination of the oxygenation capacity available for the load applied.

TABLE 8

Week number	1	2	3	4	5	6	7
Aeration time (h)	32.7	39.1	39.1	26.8	33	31	32
Oxygen supplied (kg) (standard conditions)	134	160	160	110	135	127	131

TABLE 9

COD

Week number	Number of additions	Initial load kg	Added load kg	Initial + added load kg	Final load kg	Eliminated load kg	Efficiency %	Volume loss (m^3)	Temperature ^{o}C
1	5	1 115	352	1 467	806	661	45	2.5	12 → 28
2	4	806	223	1 029	414	615	60	1.65	30
3	6	414	335	749	607	142	19	0.85	30
4	3	607	137	744	628	116	16	0.8	30
5	5	628	298	926	630	296	32	0.75	30
6	5	630	326	956	655	301	32	0.75	30
7	4	655	445	1 100	865	235	22	0	30

TABLE 10

BOD

Week number	Number of additions	Initial load kg	Added load kg	Initial + added load kg	Final load kg	Eliminated load kg	Efficiency %	Volume loss (m^3)	Temperature ^oC
1	5	397	147	544	195	349	64	2.5	12 → 28
2	4	195	115	310	99	211	68	1.65	30
3	6	99	133	232	119	113	49	0.85	30
4	3	119	58	117	149	28	16	0.8	30
5	5	149	65	214	143	71	33	0.75	30
6	5	143	129	272	157	115	42	0.75	30
7	4	157	88	245	194	51	21	0	30

b) Results

The weekly amounts of oxygen which would have been supplied under standard conditions over the operating times of the aerator are summarised in Table 8. The load balances are given in Tables 9 and 10.

The values for the elimination efficiency vary with time and those variations have no significance since the weekly additions were constant neither in quantity nor quality, and the initial week was under different conditions (high load and temperature lower than 30°C). The weekly ratios of oxygen supplied under standard conditions per unit of load eliminated are reported in Table 11.

TABLE 11

Week number	1	2	3	4	5	6	7
O_2 supplied COD^{-1} eliminated	0.2	0.26	1.13	0.95	0.46	0.42	0.56
O_2 supplied BOD^{-1} eliminated	0.38	0.76	1.42	3.93	1.9	1.10	2.57

The ratio O_2/COD eliminated and O_2/BOD eliminated were related to the percentage COD and BOD eliminated (Figure 3).

The ratios increased as the percentage COD and BOD eliminated decreased due to the stabilisation of the slurry.

At high elimination percentages (COD 50 to 60 percent; BOD 60 to 70 percent) during the first weeks of aeration, the ratios O_2/COD eliminated and O_2/BOD eliminated can reach 0.2 and 0.4 respectively.

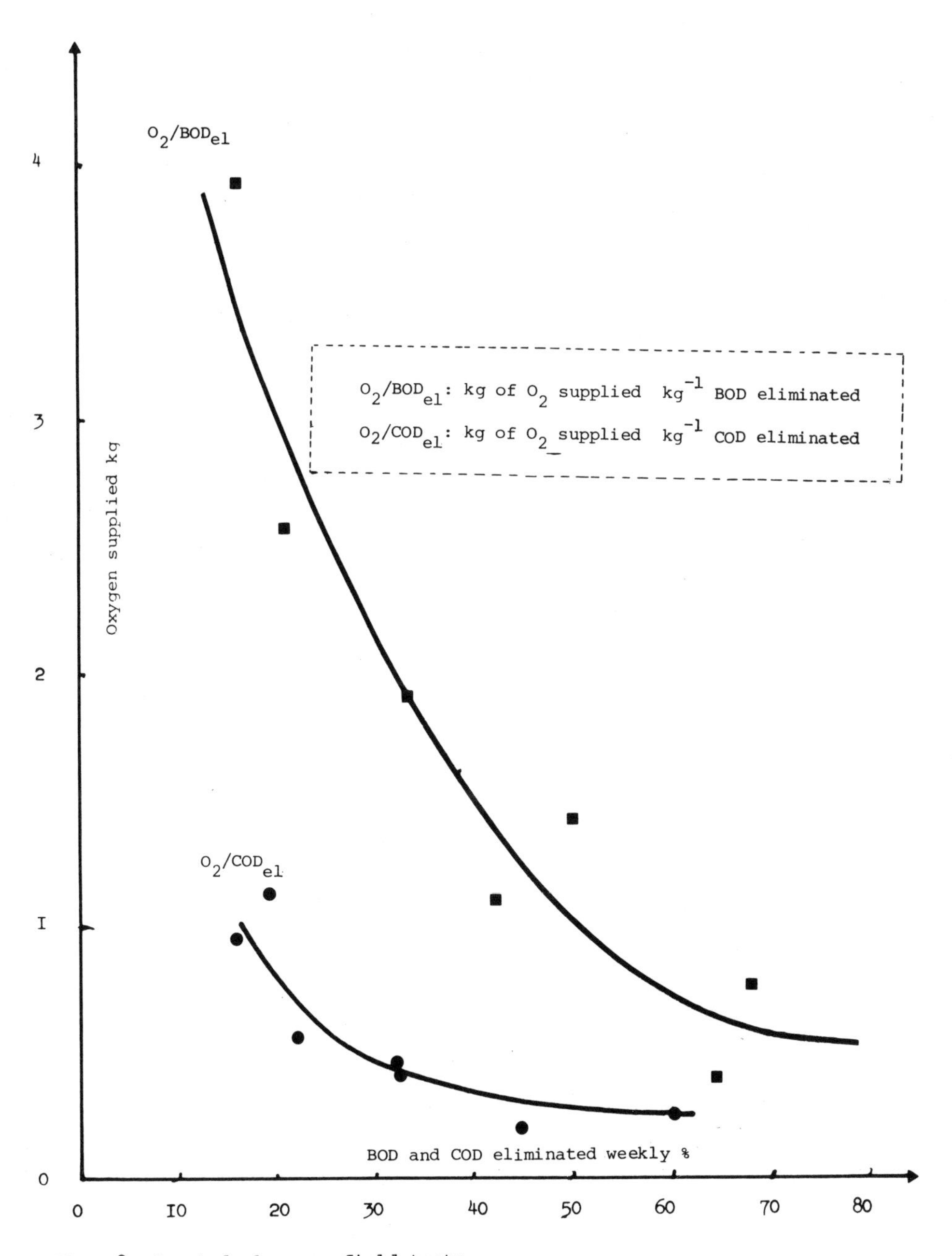

Fig. 3. Aerated slurry - field tests.

c) Conclusion

With available oxygen capacities of:

0.3 kg O_2 kg^{-1} of COD load

0.85 kg of O_2 kg^{-1} of BOD load

we achieved in 47 days with a temperature of the mixed liquor close to $30^{o}C$, a non-odorous slurry, and the following efficiencies:

BOD	80 percent
COD	70 percent
Kjeldahl N	35 percent
Suspended solids	55 percent

The rule usually applied to determine the size of the aeration device - Supply of 1 kg O_2 kg^{-1} of BOD added - seems to lead to an excess of oxygen in favour of a good treatment when mixing level and retention time are adequate.

VIII. CONCLUSION

In order to improve knowledge of factors determining the size of aeration devices, it is essential to perform measurements on full scale plants.

We cannot rely on the correction coefficients established in laboratory scale plants because of their dependence on the aeration system used.

The concentration of suspended solids affects the oxygen transfer, but the performances of an aeration device may be independent of the liquid depth in the tank over a range yet to be accurately determined.

The classical calculation for slurry stabilisation - supply of 1 kg O_2 kg^{-1} of BOD added - proves to be sufficient when mixing level and retention times are adequate.

Economics in energy input should be sought by adjusting the aeration time according to the load.

REFERENCES

Lewis, W.K. and Withman, W.G. 1924. Principles of gas absorption. Ind. Eng. Chem. 14 - 1216.

Kayser, R. 1978. Measurements of oxygen transfer in clean water and under process conditions. AIRPE Conférence sur l'aération (Amsterdam).

Heduit, A. 1979. Mesure des performances d'aérateurs. Méthode et résultats. CEBEDEAU 32 ème journées internationales (Liege).

Benedek, A., 1971. Problems with the use of sodium sulphite in aerator evaluation. Engineering Bulletin of Purdue University, 947-955.

Roustan, M., 1979. Fondements théoriques des transferts de matière gaz-liquide. INSA Toulouse journées d'étude gaz-liquide, 1-13.

BALANCE AND EVOLUTION OF NITROGEN COMPOUNDS DURING THE TREATMENT OF SLURRY

C. Besnard
Centre technique du Genie Rural des Eaux et Forêts,
Ministère de l'Agriculture, Division Qualité des Eaux,
Pêche et Pisciculture, 14, Avenue de Saint-Mandé,
75012 Paris, France.

Aerobic purification or treatment to limit the smell of slurry usually leads to the formation of oxidised products.

The transformation of nitrogen during aerobic treatment has a special interest; because the concentration of NH_3-N in the slurry limits the purposes for which the treated slurry can be used.

The objects of this study were to draw up a balance sheet of the different forms of N for several types of aerobic treatments and to examine variations of reduced and oxidised forms of N in the slurry during treatment, as well as the influence of the treatment modifications on the forms of N.

EXPERIMENTAL

The experiments were done in three areas:

i) After a review of analytical methods, a laboratory study was made using 4 series of 16 small scale units each containing 1 l of aerated liquid to study nitrification in slurry.

ii) Two detailed studies lasting for several months, were carried out on two plants of a different type:

an aerated storage tank working in an experimental piggery of the Pork Institute and,

an aerated lagoon at a piggery.

iii) Meanwhile, measurements made on large scale treatment plants, enabled the experimental results to be verified.

RESULTS AND DISCUSSION

1. In the studies with small scale units, the two independent parameters varied were the daily nitrogen load (NH_3-N and organic-N) and the organic load (solid loading, Cm, from 0 to 0.26 g BOD g VSM^{-1} day^{-1}).

For a temperature of about 20^oC, an 0_2 concentration from 5 to 8 mg 1^{-1}, and a retention time of 5 days, the percentage of nitrification $\frac{[NO_3^-]}{[NH_4^+ + NO_3^-]}$ x 100 in the effluent varied from 9 to 92 percent.

Figure 1 shows that the production of nitrate did not depend on the N load (0.002 to 0.05 g N g VSM^{-1} day^{-1} and Figure 2 shows that the percentage of nitrification decreased with increasing nitrogen load at all organic loadings.

Figures 3 and 4 show that for a given nitrogen load, NO_3-N produced and nitrification percentage, increase with the organic load applied at all N loadings.

These statements have to be linked to the pH changes of the aeration units which decreased from pH 6.6 to pH 4.7 during the experiment, because Figure 5 shows that the largest nitrification ratios were found at the highest pH. The optimum pH range for nitrification varies from pH 6.5 to pH 8.5 (Wong Chony and Loehr, 1975); the pH conditions achieved in this experiment were therefore far from this optimum. The high organic loads seemed to buffer the medium, and therefore in their presence acidification was less. Moreover, the conditions associated with a high organic load seemed to favour nitrification because, at similar pH (pH 6.4), nitrification was more complete with a large load of solids.

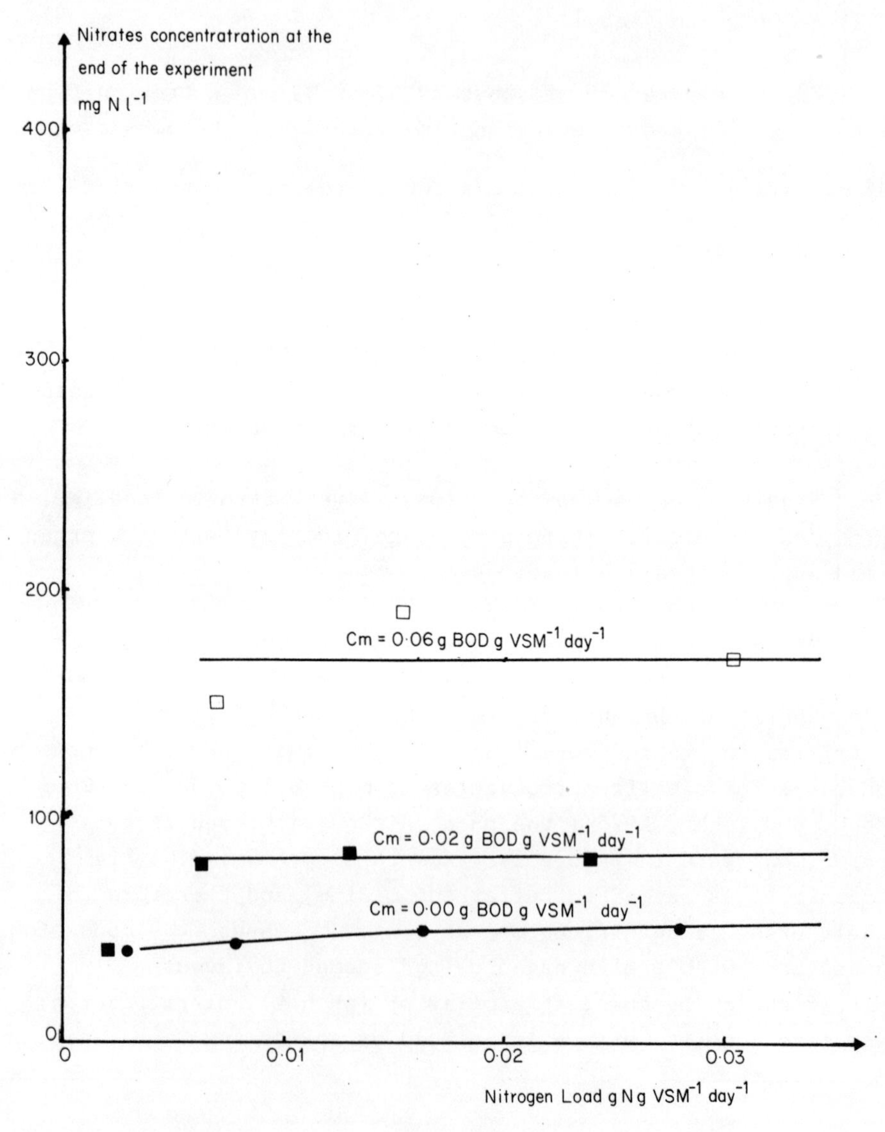

Fig. 1. Effect of the applied nitrogen load on the final nitrate concentration.

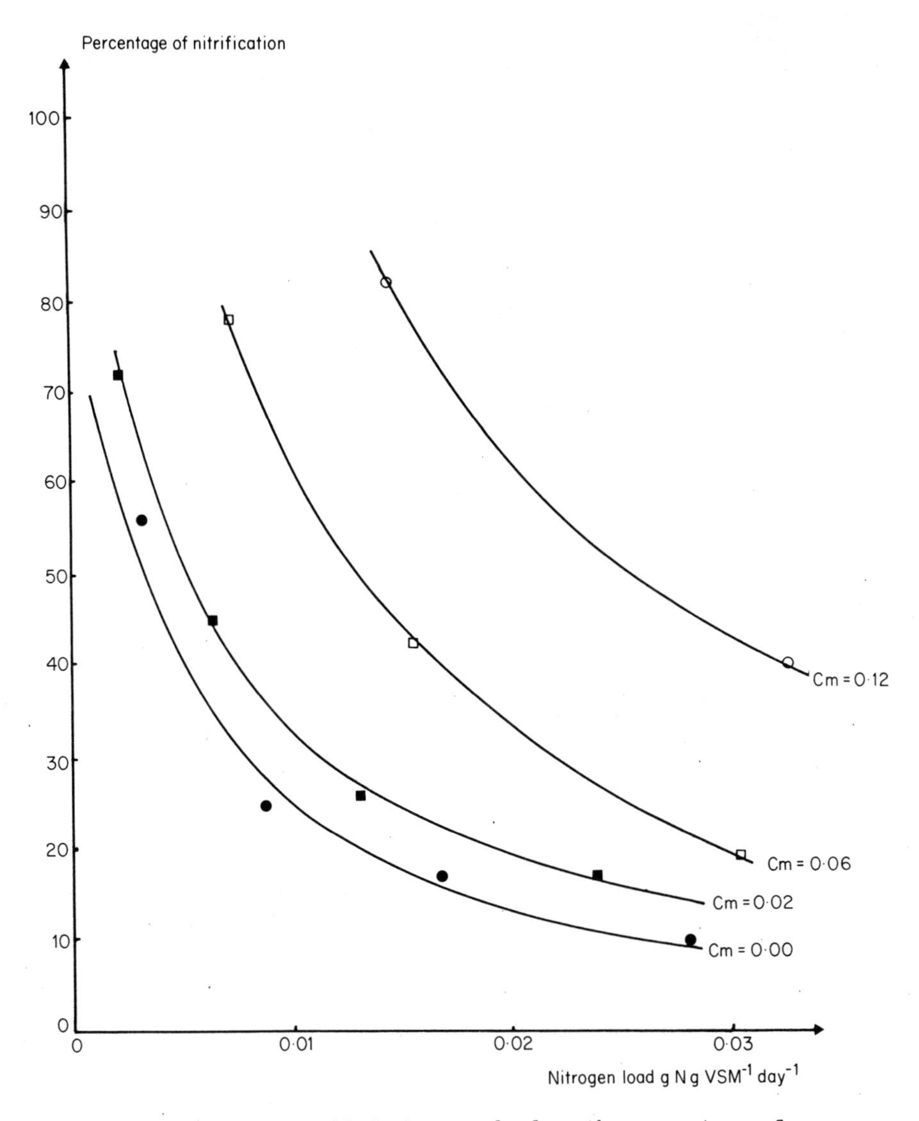

Fig. 2. Effect of the applied nitrogen load on the percentage of nitrification.

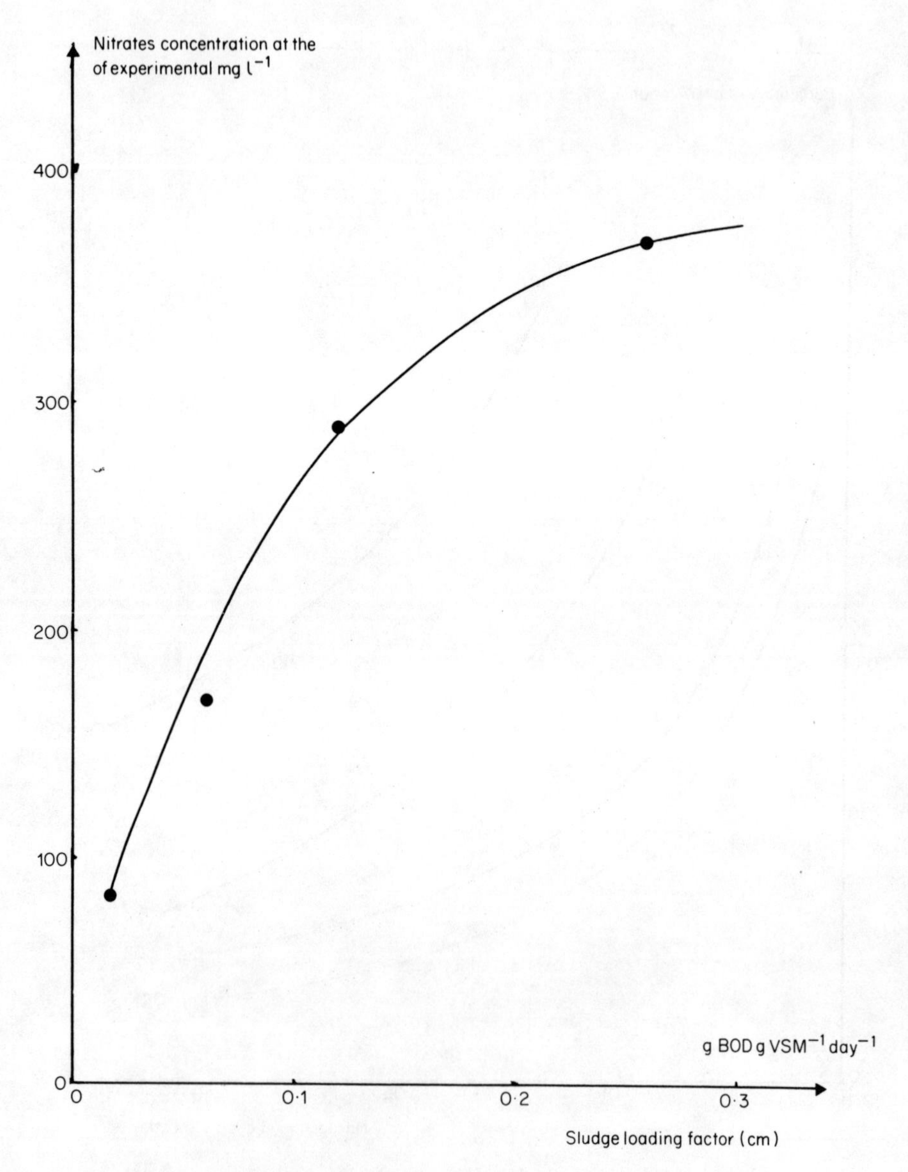

Fig. 3. Effect of the sludge loading factor on the final nitrate concentration.

This observation agrees with those of other workers, especially by Loehr et al. (1973), who have observed nitrification at pH 4.9 in a very concentrated effluent, with a solid load of 0.8 g COD g VSM^{-1}.

Also, Hockenbury et al. (1977) have noted that the organic loads did not inhibit the autotrophic nitrifiers.

2. The study was made on slurry stored and directly aerated in the oxidation ditch. The changes in the various forms of N were followed for two periods of fattening pigs lasting for 26 weeks.

The aeration periods were varied with the changes in the organic load. Regular observations have also been made on the variation of odour inside the piggery.

For the minimum aeration required to prevent anaerobic conditions, the ammonia concentration was markedly increased up to 3.85 g l^{-1} with pH 9 and a temperature of 27 - 30°C. Under these conditions gaseous ammonia is lost to the atmosphere, and the amounts can be evaluated (Loehr et al., 1973).

The concentrations of dissolved ammonia, estimated at 1.2 g NH_3-N l^{-1}, for the last month prevented the development of nitrifying bacteria because *Nitrosomonas* is completely inhibited by 150 mg NH_3-N l^{-1} and *Nitrobacter* by 10 mg NH_3-N l^{-1} (Anthonisen et al., 1976).

On the other hand, when much higher concentrations of dissolved oxygen were maintained in the liquid (4 to 6 mg l^{-1}), at first both nitrates and nitrites were formed; later the formation of nitrate ceased while the NO_2 concentration continued to increase. The cessation of nitrate formation seemed to be caused by the considerable concentration of non--dissociated nitrous acid (2.5 g l^{-1}) arising from the high initial concentration of ammonia (2.6. g l^{-1}) and total-N in

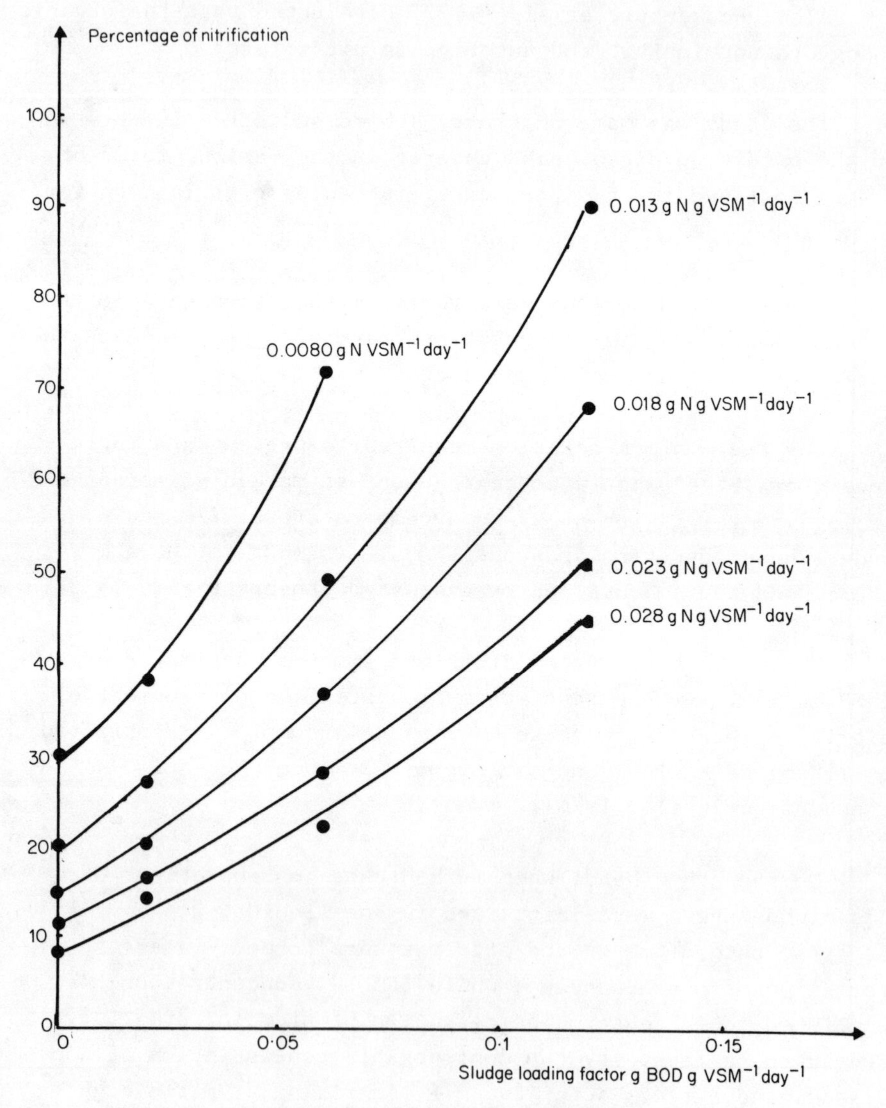

Fig. 4. Effect of the sludge loading factor (Cm) on the percentage of nitrification.

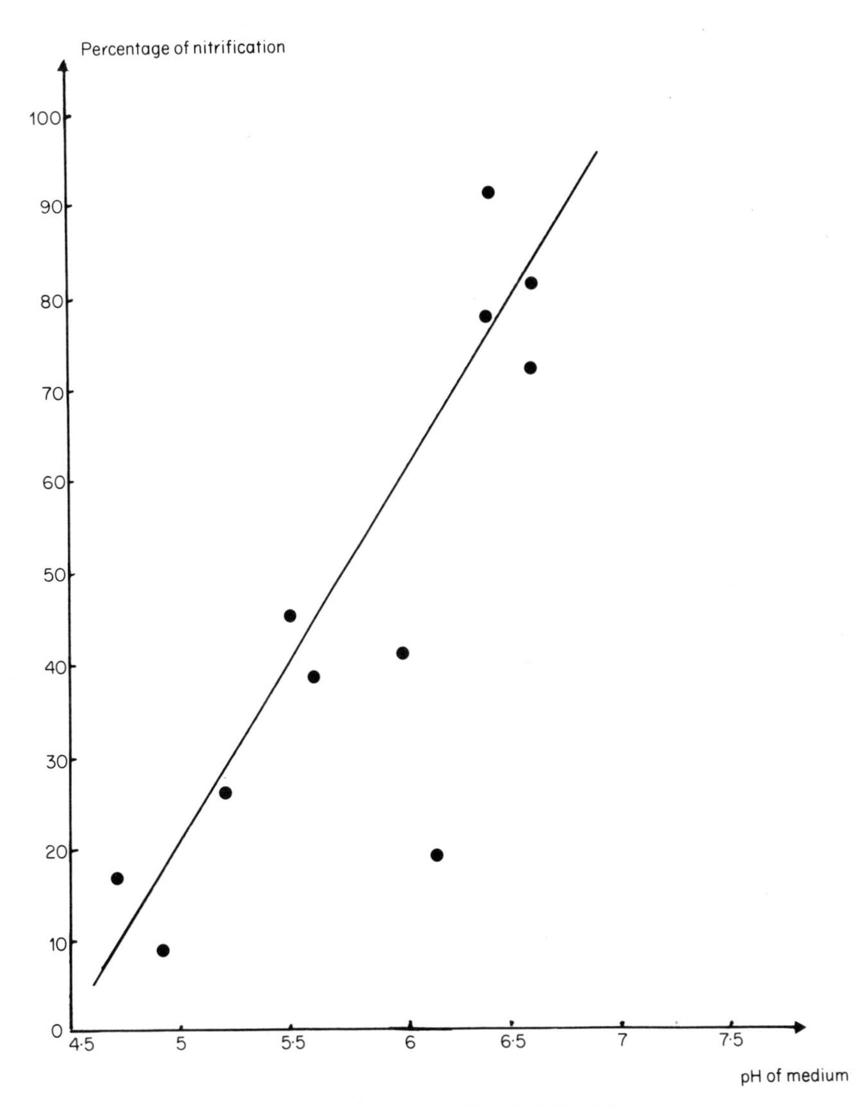

Fig. 5. Effect of pH on percentage of nitrification.

the aerated slurry at pH 6.5. *Nitrobacter* is inhibited by HNO_2 concentrations varying from 0.2 to 2.8 mg l^{-1} (Anthonisen et al., 1976). Although the pH remained stable at pH 6.6, when the formation of nitrate ceased, it could also be the cause of this stoppage. The loss of nitrogen in this case, essentially occurred by denitrification of the nitrite.

3. Observations were made repeatedly, in the course of several series of measurements during treatment by aeration of slurry in a lagoon.

The efficiency of nitrogen elimination from the liquid effluent varied much especially when compared with the efficiency of transfer of BOD to COD in the sludge (See Table 1). The percentage loss of the N compounds as a whole can be increased, for example up to 94 percent was lost during the measurements made in 1972. In the aeration tank, proportionately more NH_4-N was always lost than organic-N.

TABLE 1

ELIMINATION OF NITROGEN IN AN AERATED LAGOON

		March 1977 %	April 1977 %	July 1977 %
N lost in the aeration tank	Organic-N	74	21	56
	NH_4-N	85	38	68
N lost in the settling pond	Organic-N	68	43	78
	NH_4-N	37	-	70
Total efficiency	Total-N	89	55	90
Total efficiency	change to COD	94	84	94

In alkaline conditions, pH more than pH 8, and for large concentrations of ammonium salts, much N can be eliminated by gaseous loss, which was confirmed by the very strong ammoniacal smell emanating from the tank in these conditions.

At relatively low pH, nitrite and nitrate were noted in the effluent so that there was partial nitrification of the ammonium.

Nitrogen transformed into an organic form becomes part of the sludge and is eliminated with this. The cycle of aeration and removal of the sludge has an important influence upon the efficiency of processes affecting N compounds.

CONCLUSIONS

Although incomplete, the observations on aerated storage plants lead to the following conclusions:

1) When slurry is aerated at a minimal level for preventing unacceptable odour emission, nitrogen is mainly in the form of ammonia, with no oxidised forms. Under these conditions N is essentially lost by volatilisation of ammonia, and the amount varies from 30 to 40 percent of that present.

2) If the concentration of 0_2 dissolved in the aerated liquid is sufficient to measure (about 1 mg l^{-1}) nitrites and nitrates are formed, and in the treated liquid, nitrification and denitrification take place simultaneously. Hardly any odour is emitted from the liquid and more than 50 percent of the N is lost.

REFERENCES

Anthonisen, A.C. et al., 1976. Inhibition of nitrification by ammonia and nitrous acid. Jour. Water Poll. Control Fed., 48, 835-852.

Hockenbury et al., 1977. Factors affecting Nitrification, Journal of the Environmental Engineering Division, ASCE, 103, EE1, p. 9-19.

Loehr, R.C. et al., 1978. Development and Demonstration of Nutrient Removal from Animal Wastes. EPA. R2 73 - 095 EP Technol. Ser.

Wong Chony, G.M. and Loehr, R.C., 1975. The kinetics of microbial nitrification, Water Research, 9, 1099-1106.

DISCUSSION

J.H. Voorburg *(The Netherlands)*

I would like to comment because Dr. Besnard has some experience of the treatment of slurry from veal calves. Our experience is that, on a farm scale, nitrogen content can be decreased by 90 percent, by nitrification/denitrification. This plant, with a capacity for slurry from 7 000 veal calves is managed by a farm worker. He only has two types of test papers, one for nitrite and the other for ammonium. If the nitrite concentration is increasing the amount of aeration must be decreased and if the ammonium concentration is increasing, more oxygen is needed. We did an experiment on removing phosphate by adding lime to give pH 8.5 during aeration, which not only removed more than 90 percent of the phosphates but also positively influenced nitrification.

J.C. Hawkins *(UK)*

I think the work on oxygen transfer is particularly important because aerating slurry is unprofitable. Energy is put in, nitrogen is lost and money is spent. If control of odour is the objective of aerating, knowledge of how little energy is needed to reach the degree of control required becomes important. This was stressed in the first paper.

One other factor can perhaps contribute to economy. If the liquid fraction of a separated slurry is aerated, much less energy will probably be used because much of the power in aeration is needed to stir the slurry, not to introduce oxygen.

A. Heduit *(France)*

I think we had good mixing in our experiments and this could possibly be decreased. However, the low-speed mechanical surface aerator and the high-speed system such as we used, differ in efficiency; variation of the depth in the tank is also a problem.

SUMMARY OF SESSION 3

J.K.R. Gasser

1. COPPER IN SLURRY

There is much concern about the effects on the environment of copper in slurry from pigs which have received supplementary copper in their diet. This copper can have both short and long term effects. In the short term, copper in slurry can be toxic to livestock particularly sheep and calves grazing recently treated grassland. Continued addition of copper can result in very large concentrations. In grassland, it will be concentrated in the surface layer and in arable soils, throughout the cultivated layer. In extreme cases, the soil fauna are killed, only copper tolerant plants will grow and the soil is unproductive. However, the reverse is true, some soils are deficient in copper, and the application of slurry containing copper may be beneficial. Copper will also interact with other trace elements with occasional undesirable effects on animal health.

Copper in the soil-plant-animal system is important and particularly the effects of adding copper enriched slurry. This area of work requires much more investigation before satisfactory guidelines on slurry disposal can be formulated. More information is needed on the changes in the chemical forms of copper in the soil after adding it in slurry.

2. MODELLING

Modelling has been widely discussed during the recent programme. It received particular attention at the Seminar at Chimay in October 1978 and the more recent meeting in Brussels on surface runoff. Both confirmed the need for two types of model. One is the scientific model, describing processes and giving quantitative measures of their effects. The other model required is simpler being a practical econometric solution to the farmers' problems and enabling them to devise a satisfactory farming system.

3. ENGINEERING

The successful handling, storage and spreading of slurry requires much engineering and the needs were reviewed at a Seminar at Cambridge in September 1978. The treatment of slurry may be aerobic or anaerobic using whole slurry or slurry separated into solid and liquid fractions. Whatever treatment is used it should allow the smell to be kept to an acceptable level. Cost is important, and for aerobic systems, efficient transfer of oxygen is necessary to minimise expense. With anaerobic treatment, methane can be produced and recovered for use as a fuel but there is still some cost. Any treatment will still leave a material containing plant nutrients which should be used most profitably. The treatment itself only makes the animal effluent more acceptable in terms of its smell.

GENERAL DISCUSSION

The discussion following the Chairman's summary and conclusions was devoted entirely to the problems of copper in the soil-plant-animal system. The following points were made:

1. Soil ingested by ruminants may absorb copper and decrease the toxic effects of the ingestion of herbage polluted with slurry enriched with copper.
2. The use of fungicidal sprays based on copper (Bordeaux mixture) for many years in vineyards, hop gardens and orchards had enriched the surface soil so much that problems of copper toxicity were encountered when these perennial crops had to be re-planted.
3. Information is needed on the extent of these problems and the way they may be overcome.
4. Work is required to find how to prevent the problem of copper accumulation to toxic levels in soils.
5. There is a complete lack of information on the effects of adding slurry enriched with copper to soils known to be deficient in copper, particularly on animal health.

SESSION IV

ODOUR CHARACTERISATION AND MEASUREMENT

Chairman: J.H. Voorburg

DEVELOPMENT OF INSTRUMENTAL METHODS FOR MEASURING ODOUR LEVELS IN INTENSIVE LIVESTOCK BUILDINGS

J. Schaefer
Central Institute for Nutrition and Food Research TNO,
Zeist, The Netherlands.

ABSTRACT

An investigation was carried out to identify the volatile compounds responsible for the odour of laying hen houses, based on the assumption that hen manure was the main cause of the odour.

The volatile compounds present in high concentrations in liquid hen manure appeared to be the same as those in liquid pig manure, namely phenols, indoles and aliphatic carboxylic acids. In addition, very low concentrations of a number of sulphur compounds were observed, which a panel judging odour found to play an important role in the odour of the hen manure. Because of the low concentrations, even in 50 kg of manure, it was not possible to obtain sufficient material for identification by a gas chromatograph - mass spectrometer computer (GC-MS-COMP).

In a model system consisting of an aqueous solution of aldehydes, H_2S, methylmercaptan and ammonia at pH 5 in a concentration ten times as high as that in chicken manure, several dozen sulphur compounds had been formed after several days. The greater part of these compounds showed the same gas chromatographic properties as those present in chicken manure. So far, five compounds have been identified with certainty, three of which also appeared to be present in liquid chicken manure in an estimated concentration of 0.1 - 1.0 $\mu g\ kg^{-1}$.

One of these compounds - 2,6-dimethylthi-3-ine-carbonaldehyde - was detected in the ventilating air of laying hen houses. A significant correlation was observed between the odour concentration of the ventilating air and the relative concentration of this sulphur compound. It may be possible to develop an instrumental method suitable for measuring the odour level of laying hen houses with the aid of this compound.

1. INTRODUCTION

In 1973 an investigation was started by the Central Institute for Nutrition and Food Research TNO (CIVIO-TNO) in co-operation with the Central Technical Institute TNO (CTI-TNO) and the Institute of Agricultural Engineering (IMAG), in order to develop an instrumental method of measuring odours emitted from intensive livestock buildings.

In the first instance, the ventilating air from piggeries was examined, because quantitatively these buildings present the biggest problem in the Netherlands. It appeared that phenols, indoles and aliphatic carboxylic acids played an important role in the odour (Schaefer et al., 1974; Schaefer, 1973).

In the summer of 1975 an investigation comprising a number of instrumental and sensory measurements of odour concentration was carried out (Logtenberg and Stork, 1976; Schaefer, 1977). The results showed p-cresol concentration to have the highest correlation with odour concentration. The regression equation was calculated to be:

odour concentration in $ou^{*} m^{-3}$ = 20.0 + 2.4 x p-cresol concentration in $\mu g\ m^{-3}$.

The p-cresol measurement is being used at present for estimating the odour reducing capacity of biological air scrubbers that are used in the odour control of the ventilating air of piggeries. For this purpose p-cresol determinations are carried out in the air going in and coming out of the scrubbers.

The odour concentration of the ventilating air of piggeries can be determined by first measuring the p-cresol concentrations over one day. With the aid of the mean p-cresol concentration for that day, the mean value of the odour concentration can be calculated by the regression equation.

*ou = odour units

The study of the odour emitted from laying hen houses described in this paper was started in 1976. As in the case of the pig buildings, the manure was considered to be the main cause of the odour and was therefore used as basic material. The investigation took place in laying hen houses with a liquid manure system. For the investigation of the volatile compounds responsible for the odour, the following procedures were applied:

1. Isolation, concentration and analysis of the odorous compounds from liquid manure.
2. Analysis of a model system, consisting of an aqueous solution of volatile carbonyl compounds, H_2S and methylmercaptan, all of which had been identified in liquid manure before (Merkel et al., 1969).

2. EXPERIMENTAL

2.1. Manure extract

A manure extract was prepared from units of 5 kg of liquid manure which were put through a coarse sieve and continuously extracted in a 'Kutscher-Steudel' extractor with a mixture of pentane and diethylether (1 + 1).

Apart from the volatile odour compounds, the crude extract contained a lot of non-volatile compounds, which were removed by high vacuum distillation. The volatile compounds as well as the solvent were collected in the distillate, while the non-volatiles remained in the residue. Then the extracting solvent was carefully evaporated until a volume of 0.5 ml was left. This manure concentrate had a very strong and characteristic odour. For identification purposes, 50 kg of manures were extracted and the concentrate reduced to 5 ml, which were finally evaporated to a volume of 1 ml.

2.2. Gas chromatography

2.2.1. Aromagram

The manure concentrate was separated on a gas chromatographic column. Part of the separated compounds was led via a splitter to a flame ionisation detctor (FID) and part outside the gas chromatograph, where the smell was judged by a panel. Such a combination of detector signal and panel judgement is called an aromagram.

2.2.2. Sulphur compounds

The gas chromatographic analysis of the sulphur compounds was performed with a flame photometer detector (FPD), which is very sensitive to sulphur compounds and hardly responds to other organic compounds.

2.2.3. Collection

For identification purposes, some of the gas chromatographic fractions that were led outside the gas chromatograph were condensed in a glass capillary, cooled with a stream of air drawn through liquid N_2. Immediately after collection the capillary was sealed at both ends with a gas burner. Figure 1 shows a schematic drawing of the apparatus used.

2.3. Identification

Identification was carried out with a gas chromatograph-mass-spectrometer-computer (GC-MS-COMP) with two types of samples:

a. a total extract of manure or an extract of the model solution (see 2.5.) to obtain information about the main compounds,

b. the collected fractions selected by the panel or via gas chromatographic analysis with the FPD.

The samples were separated in the gas chromatograph, while the eluted compounds were directly led into the mass spectrometer. Every two seconds a mass spectrum was recorded on tape and given a number. The total ion stream chromatogram, looking like an ordinary gas chromatogram with the numbers of the recorded mass spectra on the time axis, was recorded at the same time. The spectra were recorded and plotted while a relevant peak in the gas chromatogram was visible, and were compared with those of a reference system. If the unknown spectrum was identical with that of a reference spectrum, the relevant compound was considered to be tentatively identified. If the gas chromatographic properties of the unknown and reference compounds were also the same, the identification was considered to be positive.

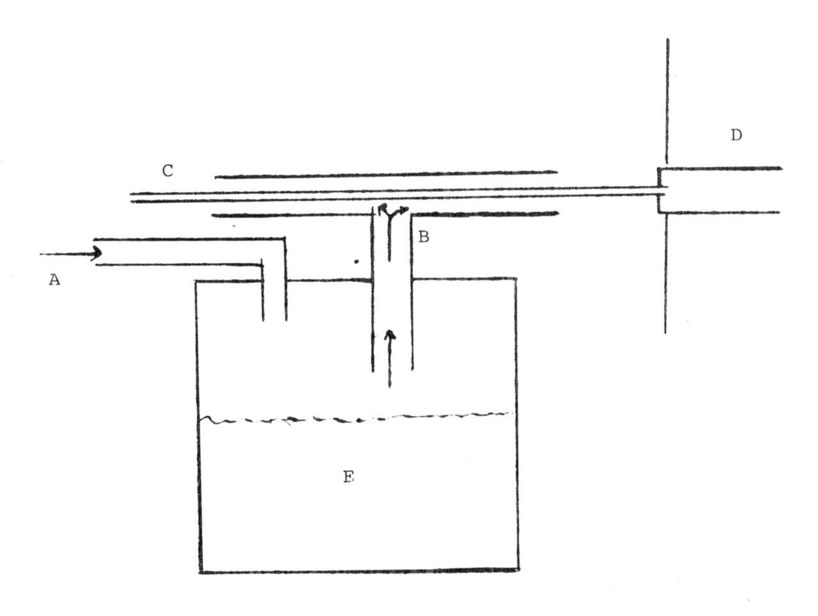

Fig. 1. Schematic drawing of apparatus for collecting fractions separated by gas chromatography.

A. air inlet
B. T-piece
C. cooling capillary
D. gas chromatograph
E. dewar flask with liquid N_2.

When no identical reference spectrum was available, a number of possible structures for the unknown compound was deduced from the break-down pattern. Accordingly, several compounds were synthesised and their mass spectra as well as gas chromatographic properties were compared with those of the unknown compounds.

If sufficient material of the unknown compound could be collected, a nuclear magnetic response spectrum (NMR) was recorded, which gave extra information about the structure of the molecule and thus considerably reduced the number of possibilities.

2.4. Quantitative measurements in manure

2.4.1. Aldehydes

Saturated aldehydes with 2 - 5 C-atoms were converted into semicarbazones by reaction with semicarbazide. The semicarbazones could then be separated; subsequently the aldehydes were liberated again after the addition of acid. The aldehydes were quantitatively determined by gas chromatographic analysis of the air over the solution.

Unsaturated aldehydes, i.e. acrolein and crotonaldehyde were quantitatively determined by gas chromatographic analysis of the air over liquid manure.

2.4.2. Sulphur compounds

H_2S and methylmercaptan were determined by gas chromatographic analysis with the FPD.

2.4.3. Ammonia

Ammonia was distilled from the manure at pH 7.8 and then determined by titration with potassium biiodate.

2.4.4. Acids, phenols, indoles

These compounds were determined by gas chromatographic analysis of the manure extract.

2.5 Model system

Aqueous solutions were prepared in 1 l units. In 1 l water at pH 5 all or some of the following compounds were mixed:

a. saturated aldehydes with 2 - 5 C-atoms

b. unsaturated aldehydes: acrolein and crotonaldehyde

c. sulphur compounds: H_2S and methylmercaptan

d. ammonia

e. acids, phenols, indoles.

From all solutions prepared, the following three were used for the investigation, i.e.:

I. a, b, c, d and e in concentrations as found in chicken manure

II. a, b, c, d in concentrations as found in chicken manure

III. a, b, c, d in 10-fold concentrations as found in chicken manure.

The reaction period was fixed at 7 days; every day extra H_2S and methylmercaptan were added to compensate for evaporation losses.

After the reaction period, an extract was prepared as described in 2.1; high vacuum distillation was not necessary, because no non-volatile compounds were present.

2.6. Synthesis

2.6.1. 2,6-dimethylthi-4-ine-3-carbonaldehyde (I)

This compound was prepared according to Kleipool et al. (1976), in which method crotonaldehyde and H_2S are brought together in a basic solution.

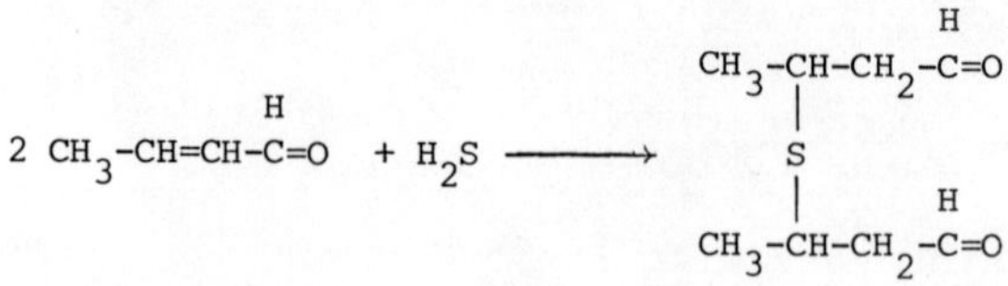

$$2\ CH_3\text{-}CH{=}CH\text{-}\overset{H}{C}{=}O + H_2S \longrightarrow \begin{array}{c} CH_3\text{-}CH\text{-}CH_2\text{-}\overset{H}{C}{=}O \\ | \\ S \\ | \\ CH_3\text{-}CH\text{-}CH_2\text{-}\overset{H}{C}{=}O \end{array}$$

H_3C — (ring with S) — CH_3, $C{=}O$ OH, $\overset{H}{C}{=}O$, H_3C — (ring with S) — CH_3

(I)

The identity was confirmed with GC-MS-COMP.

2.6.2. <u>3,5-dimethyl-1,2,4-trithiolane</u> (II)

This compound was prepared according to Dubs and Joho (1978), starting with α-chloroethylsulphurenylchloride and sodium sulphide.

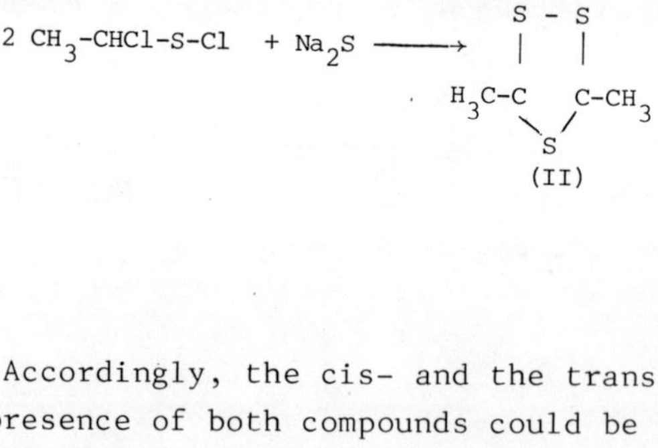

$$2\ CH_3\text{-}CHCl\text{-}S\text{-}Cl + Na_2S \longrightarrow \begin{array}{c} S - S \\ |\quad\ \ | \\ H_3C\text{-}C\quad C\text{-}CH_3 \\ \diagdown S \diagup \end{array}$$

(II)

Accordingly, the cis- and the trans-form are produced; the presence of both compounds could be confirmed by GC-MS-COMP. A small amount of 3.6.-dimethyltetrathiane as a by-product was found.

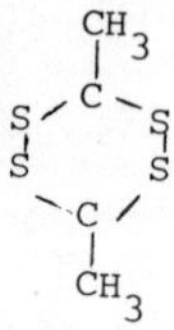

$$\begin{array}{c} CH_3 \\ | \\ C \\ S \quad S \\ S \quad S \\ C \\ | \\ CH_3 \end{array}$$

2.6.3. 3,5.-dialkyl-1,2,4-trithiolanes

Starting with a mixture of

α-chloroethylsulphenylchloride,
α-chlorobutylsulphenylchloride and sodium sulphide,
a mixture of 3,5-dimethyl-(II), 3-methyl-5-propyl-(III)
and 3,5.-dipropyl-1,2,4-trithiolane (IV) was prepared
according to Dubs and Joho (1978).

CH_3-CHCl-S-Cl

CH_3-CH_2-CH_2-CHCl-S-Cl

+ Na_2S ⟶

H_3C-C(-S-S-)(-S-)C-CH_3 (II)

CH_3-CH_2-CH_2-C(-S-S-)(-S-)C-CH_3 (III)

CH_3-CH_2-CH_2-C(-S-S-)(-S-)C-CH_2-CH_2-CH_3 (IV)

The presence of these compounds in the cis- as well as in the trans-form could be confirmed by GC-MS-COMP.

3. RESULTS AND DICUSSSION

Figure 2 shows the aromagram of the chicken manure concentrate as well as the fractions collected for identification with GC-MS-COMP.

A gas chromatogram of the manure concentrate recorded with the FPD is given in Figure 3; a similar gas chromatogram of an extract of the model system is shown in Figure 4.

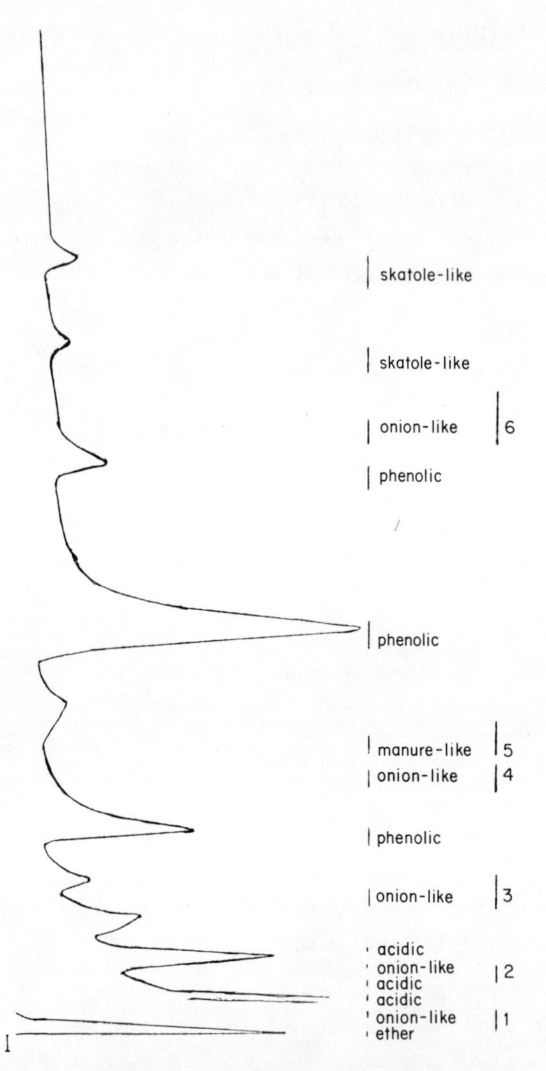

Fig. 2. Aromagram of chicken manure concentrate

Column: glass; 1m x 3mm coiled
Phase: 10 percent SE 30 on Chromosorb W-AW, DMCS
Temperature oven: isothermal 15 min. 70°C, subsequently programmed to 150°C with 4°C $min.^{-1}$
detector: 200°C
injection point: 200°C
Splitter ratio detector/heated outlet= 1 : 50
Detector: FID
Flow: N_2 30 ml $min.^{-1}$; N_2 ml $min.^{-1}$; air 500 ml $min.^{-1}$.

The figure shows the joint panel judgements, together with the fractions collected and analysed by GC-MS-COMP.

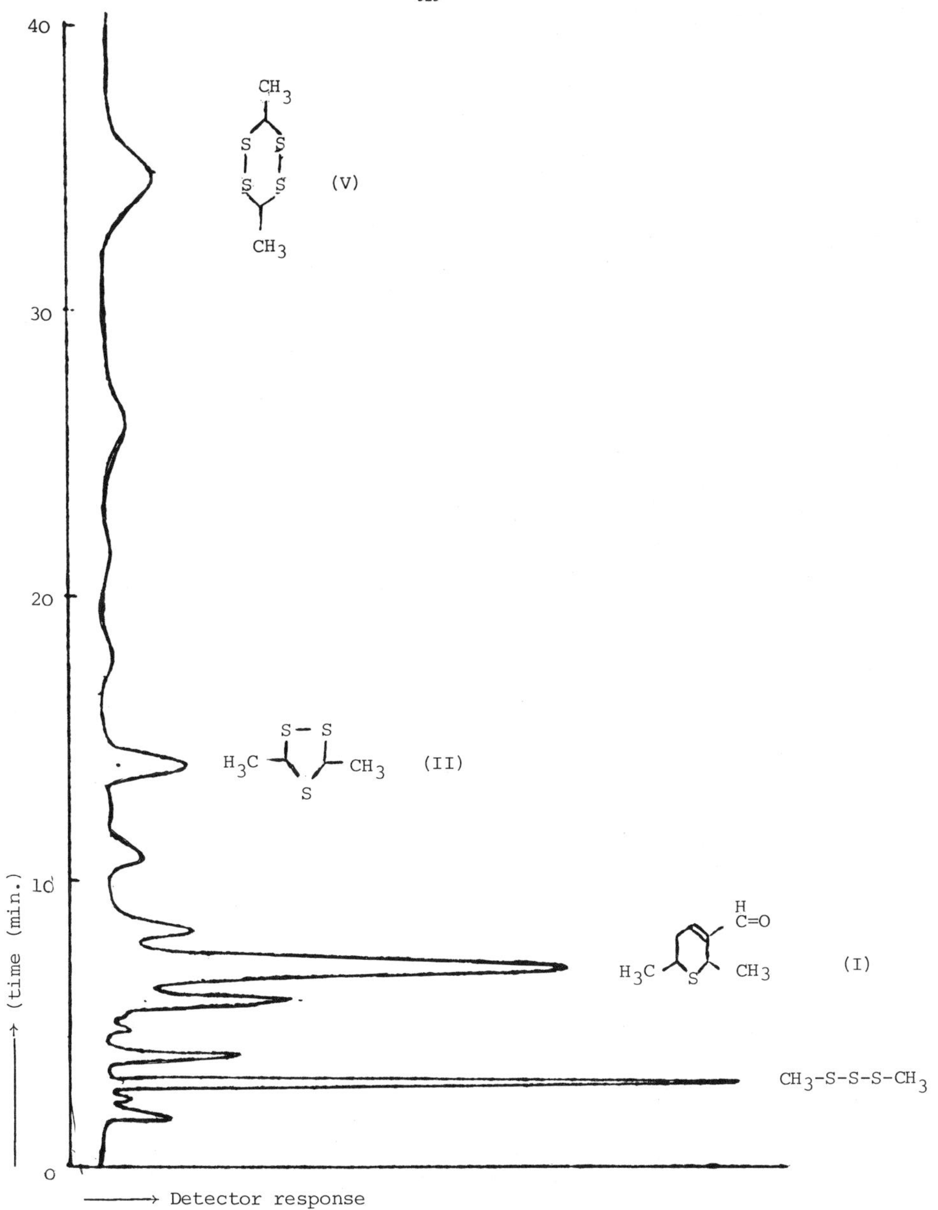

Fig. 3. Gas chromatogram of chicken manure concentrate.

Column: glass 2m x 3mm i.d. filled with 10 percent SE 30 on Chrom. W. 60 - 70 mesh
Temperature oven: 110°C
Carrier gas : N_2 30 ml min.$^{-1}$
Detector : FPD H_2 80 ml min.$^{-1}$
air 100 ml min.$^{-1}$

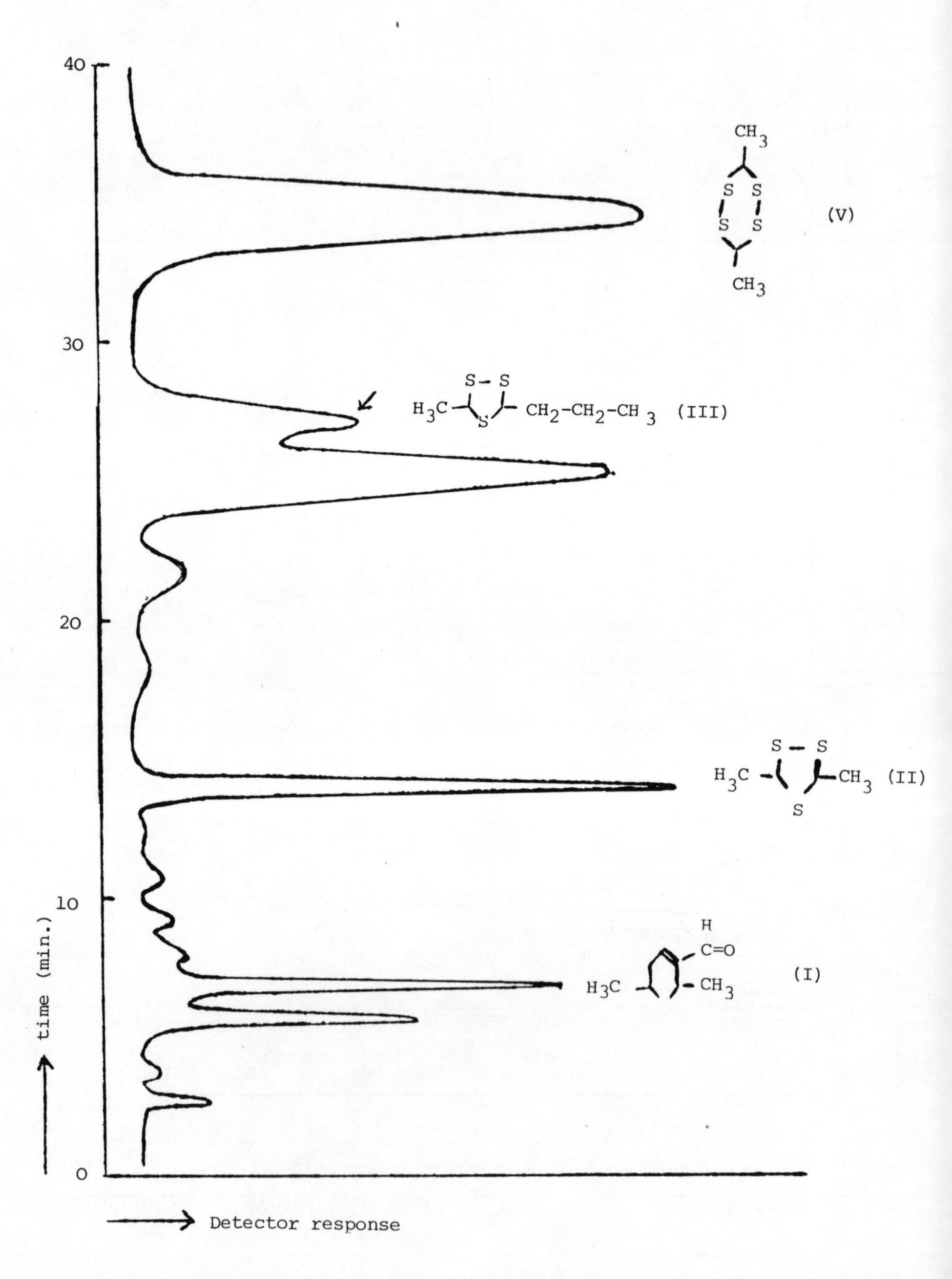

Fig. 4. Gas chromatogram of extract of model system (conditions as in Figure 3.

The volatile compounds that are identified in the manure concentrate with GC-MS-COMP are summarised in Table 1. The results of the quantitative determination in liquid chicken manure are given in Table 2. Table 3 shows the sulphur compounds that are identified in the model system with GC-MS-COMP.

TABLE 1

COMPOUNDS IN CHICKEN MANURE, IDENTIFIED BY GC-MS-COMP

Compound
phenol
p-cresol
ethylphenol
indole
skatole
acetic acid
propionic acid
i-butyric acid
n-butyric acid
i-valeric acid
n-valeric acid
tetrachloroethene
2-methylcyclopentanone
hex-3-ene-1-ol
hexanol
heptanol
octanone
limonene
benzylalcohol
α-methylbenzylalcohol
dimethyltrisulphide
dodecane

TABLE 2

CONCENTRATIONS OF SOME VOLATILE COMPOUNDS IN LIQUID CHICKEN MANURE

Name	Concentration
acetaldehyde	20 μg kg^{-1}
i-butyraldehyde	125 μg kg^{-1}
n-butyraldehyde	15 μg kg^{-1}
i-valeraldehyde	10 μg kg^{-1}
n-valeraldehyde	40 μg kg^{-1}
acrolein	165 μg kg^{-1}
crotonaldehyde	480 μg kg^{-1}
hydrogensulphide	2 mg kg^{-1}
methylmercaptan	590 μg kg^{-1}
ammonia	6.3 g kg^{-1}
phenol	23 mg kg^{-1}
p-cresol	80 mg kg^{-1}
indole	2 mg kg^{-1}
skatole	10 mg kg^{-1}
acetic acid	3.5 mg kg^{-1}
propionic acid	3.8 mg kg^{-1}
i-butyric acid	2.0 mg kg^{-1}
n-butyric acid	6.6 mg kg^{-1}
i-valeric acid	5.3 mg kg^{-1}
n-valeric acid	1.4 mg kg^{-1}

TABLE 3

POSSIBILITIES CONCERNING THE IDENTITY OF SULPHUR COMPOUNDS IN THE MODEL SYSTEM, OBTAINED BY GC-MS-COMP.

Basis*	
r	3-methyl-1,2,4-trithiolane
i	3-propyl-1,2,4-trithiolane
r	3,5-dimethyl-1,2,4-trithiolane
i	3,5-dipropyl-1,2,4-trithiolane
i	3-methyl-5-propyl-1,2,4-trithiolane
i	3-propyl-1,2,4-trithiane
i	3,5- or 3,6-diethyl-1,2,4-trithiane
t	3,6-dimethyltetrathiane
i	2-methyl-1,3-dithiol-4-ene
r	2- or 6-methylthi-3-iene-3-carbonaldehyde
i	2,6-dimethylthi-3-iene-3-carbonaldehyde
i	oxane-3-carbonaldehyde-4-thiol

* r = reference spectrum present
i = interpretation via spectrum of homologous compound
t = 'tentative' possibility based on examining the breakdown pattern in spectrum.

3.1. Chicken manure concentrate

3.1.1. Aromagram

Four odour types were observed in the aromagram (Figure 2), three of which, acidic, phenolic and skatole-like, at a time when a peak was clearly visible. In addition, an odour that could be described as onion-like and manure-like was detectable six times; this odour type is mostly caused by the presence of sulphur compounds. However, the fact that in all six instances no peak was visible indicated the presence of very low concentrations of certain compounds in the manure.

3.1.2. Identification

During the analysis of the total manure concentrate by GC-MS-COMP the main volatile compounds of chicken manure could be identified (Table 1). Several of these, causing the highest peaks in chromatograms, are the same as the ones present in high concentrations in liquid pig manure (Schaefer et al., 1974), i.e.:

phenol
p-cresol
ethylphenol
indole
skatole
acetic acid
propionic acid
i-butyric acid
n-butyric acid
i-valeric acid
n-valeric acid

The retention times of these compounds coincide with those at which the odour descriptions phenolic, skatole-like and acidic were given in the aromagram.

In the gas chromatogram recorded with the FPD dimethyltrisulphide was visible as one of the highest peaks. The estimated concentration in manure amounts to 1 - 10 μg kg^{-1} manure.

The six fractions that, according to the panel, suggested the presence of sulphur compounds were collected and analysed by GC-MS-COMP. However, the amount of material that could be obtained in this way was not sufficient to give a mass spectrum which could be interpreted positively. This means that in manure these compounds are present in concentrations lower than 1 μg kg^{-1}.

3.2. Model system

3.2.1. Preparation

It is well-known (Badings et al., 1975) that in an aqueous solution of aldehydes, H_2S and methylmercaptan sulphur compounds are easily formed as a result of, for example, addition and condensation reactions. These sulphur compounds often have intensive odours and very low odour threshold values. The basic materials are all present in chicken manure (Merkel, 1969), while the pH 5 - 7 of the manure is conducive to such reactions.

Aqueous solutions (para 2.5. I) were prepared of the above mentioned volatile compounds plus ammonia, acids, phenols and indoles, all in concentrations as found in chicken manure. Part of the solution had pH 5 and part pH 7. The extracts of the solutions at pH 5 were found to have a strong manure odour.

Similar aqueous solutions were prepared without acids, phenols and indoles (II). Gas chromatographic analysis of the extracts with the FPD after a reaction period of 7 days at pH 5 showed that in solutions I and II the same sulphur compounds were formed.

Aqueous solutions of 10-fold concentrations of aldehydes, H_2S, methylmercaptan and ammonia (III) provided extracts with an odour strongly suggesting sulphur compounds. Gas chromatographic analysis with the FPD showed the extracts of solutions III to contain the same sulphur compounds as the extracts of solutions I and II, although in much higher concentrations.

Gas chromatographic analysis of the chicken manure concentrate with the FDP showed for the greater part peaks with the same retention times as the extracts of the aqueous solutions. The peaks in the chromatograms of the latter were higher, especially those of the extracts of solution III.

Based on these observations it was concluded that an aqueous solution with a 10-fold concentration of aldehydes, H_2S, methylmercaptan and ammonia was suitable as basic material for the identification of a number of sulphur compounds in chicken manure.

The extract of such an aqueous solution was, after a reaction period of 7 days at pH 5, used for the identification with GC-MS-COMP.

3.2.2. Identification

Analysis with GC-MS-COMP suggested the identity of a number of sulphur compounds (Table 3), based on the following criteria:

r: The mass spectrum of the unknown compound is identical with a reference spectrum.

i: The mass spectrum of the unknown compound is very similar to a reference spectrum, which suggests the presence of a homologous compound.

t: After studying the breakdown-pattern of the mass spectrum a possible identity is suggested.

Since no reference compounds were available it was decided first to synthesise compounds having a gas chromatographic behaviour similar to the relevant fractions in the chromatogram.

In this way the presence of the following compounds could be confirmed in the model system:

3,5-dimethyl-1,2,4-trithiolane
3-methyl-5-propyl-1,2,4-trithiolane
3,5-dipropyl-1,2,4-trithiolane
2,6-dimethylthi-3-ine-3-carbonaldehyde
3,6-dimethyltetrathiane.

In chicken manure, in which the concentration of sulphur compounds was much lower, the presence of only the following compounds could be confirmed:

3,5-dimethyl-1,2,4-trithiolane
2,6-dimethylthi-3-ine-3-carbonaldehyde
3,6-dimethyltetrathiane.

The estimated concentration of these compounds in chicken manure is 0.1 - 1.0 $\mu g\ kg^{-1}$; their odour threshold values are not known. Of some heterocyclic sulphur compounds this value is known. For thiazoles it is 100 - 1 000 times lower than the above mentioned concentration range and for thiophenes it is within that range. Based on these figures it can be concluded that the identified sulphur compounds do play a role in the odour of chicken manure.

Figures 3 and 4 show the peaks representing the identified compounds. Since dipropyltrithiolane has a retention time of about 60 min., it is not visible in these chromatograms, which only record retention times of up to 40 min.

3.2.3. Formation of sulphur compounds in chicken manure

The identified sulphur compounds are probably formed by the following reactions in the manure.

3,5-dimethyl-1,2,4-trithiolane (I)

$$2\ CH_3\text{-}\overset{H}{C}\text{=}O + 3\ H_2S \longrightarrow CH_3\text{-}\underset{SH}{CH}\text{-}S\text{-}\underset{SH}{CH}\text{-}CH_3$$

$$\downarrow O_2$$

```
      S —— S
      |    |
  H3C-C    C-CH3
       \  /
        S
       (I)
```

The propyl derivatives are likely to be formed from n-butyraldehyde and H_2S.

3,6-dimethyltetrathiane (II)

$$2\ CH_3\text{-}\overset{H}{C}\text{=O} + 2\ H_2S \longrightarrow CH_3\text{-}CH(SH)_2 \quad (HS)_2CH\text{-}CH_3$$

$$\downarrow O_2$$

$$CH_3\text{-}C \underset{S - S}{\overset{S - S}{\ }} C\text{-}CH_3$$

(II)

The formation of 2,6-dimethylthi-3-ine-3-carbonaldehyde is likely to take place according to the reaction equations given in 2.6.1.

4. INSTRUMENTAL MEASURING METHOD

The purpose of the investigation was to develop an instrumental method for measuring the odour level of the ventilating air of laying hen houses.

In this connection, measurements of the ventilating air of three laying hen houses (Jansen, 1979) were carried out in co-operation with IMAG and the Central Technical Institute TNO.

The relevant procedure was as follows. Of ventilating air that was led through adsorption traps filled with 50 mg activated carbon, samples of 1 500 l were collected during an eight-hour period. Gas chromatographic analysis with the FPD showed all air samples of the three laying hen houses to contain seven sulphur compounds. One of these compounds appeared to be 2,6-dimethylthi-3-ine-3-carbonaldehyde, the relative concentration of which was found to have a significant positive correlation with the odour concentration. (The correlation was calculated by means of a non-parametric statistical method, the rank-correlation of Spearman.)

This investigation led to the conclusion that, with the aid of this compound, it will in principle be possible to develop an instrumental method for measuring the odour concentration. However, since the concentrations of sulphur compounds in the ventilating air of laying hen houses were found to be very low, large quantities of such air samples have to be collected over a period of at least 7 - 8 hours in order to get an estimate of the mean odour concentration.

5. REFERENCES

Badings, H.T., Maarse, H., Kleipool, R.J.C., Tas, A.C., Neeter, R. and Ten Noever de Brauw, M.C., 1975. Formation of odorous compounds form hydrogen sulphide amd methanethiol, and unsaturated carbonyls. Proc. Int. Symp. Aroma Res., Zeist, The Netherlands, 1975, eds. Maarse, H. and Groenen, P.J., PUDOC, Wageningen, p. 63-73.

Dubs, P. and Joho, M., 1978. A short synthesis of 3,5-dimethyl-1,2,4-trithiolane. Helv. Chim. Acta 61, 1404-1406.

Jansen, C.M.A., 1979. Instrumenteel-sensorisch onderzoek aan de ventilatielucht van leghennenstallen. MT-TNO Rapport Nr. 8707-4013.

Kleipool, R.J.C., Tas, A.C., Maarse, H., Neeter, R and Bading, H.T., 1976. Reaction of hydrogen sulphide with 2-alkenals. Z. Lebensm. Unters. Forsch. 161, 231-238.

Logtenberg, M.Th. and Stork, B., 1976. Het ontwikkelen van meetmethoden voor het bepalen van de stank van ventilatielucht van mestvarkensstallen. CTI-Rapport Nr. 01-4-40130.

Merkel, J.A., Hazen, T.E. and Miner, J.R., 1969. Identification of gases in a confinement swine building atmosphere. Trans. Am. Soc. Agric. Eng. 12, 310-315.

Schaefer, J., 1973. Onderzoek naar de voor de stank van varkensmesterijen verantwoordelijke componenten. CIVO-Rapport Nr. R. 4265.

Schaefer, J., 1977. Sampling, characterisation and analysis of malodours. Agric. Environm. 3, 121-127.

Schaefer, J., Bemelmans, J.M.H. and ten Noever de Brauw, M.C., 1974. Onderzoek naar de voor de stank van varkensmesterijen verantwoordelijke componenten. Landbouwk. Tijdschr. 86, 228-232.

DISCUSSION

A. Maton *(Belgium)*

Have you made comparisons between the concentrations you were showing us on the one hand, and the reactions of the human panel, on the other hand. If so are you sufficiently far advanced to be able to make comparisons or establish correlations between concentrations of components and odour?

J. Schaefer *(The Netherlands)*

No, the aromagram was just a qualitative judgement of the odours. However, odour concentration measurements made on the ventilation air from some buildings were compared with the concentrations of the sulphur compounds in the air, when a statistically significant correlation was found.

J.H. Voorburg *(The Netherlands)*

However, it was found not to be sufficiently well correlated for use.

J.K. Grundey *(UK)*

Can this test be done on the farm, Dr. Schaefer, or must it be done in the laboratory?

J. Schaefer

The sensory measurement is done at the farm. Air samples for analysis are transported to the laboratory, because they cannot be analysed in the field.

TENTATIVE MEASUREMENTS OF ANIMAL MANURE ODOUR

J.L. Roustan[1], A. Aumaitre[1] and C. Bernard[2]
[1]INRA, 78350 Jouy-en-Josas, France.
[2]UCAAB, 02400 Chateau Thierry, France.

INTRODUCTION

The experimental data from chemical analyses of pig manure were presented at the Ghent Seminar. The main results from the storage of samples on commercial farms were:

1) the content of volatile fatty acids (VFA) increased; the relative percentages of the different acids remained constant and appeared to be independent of the sample unless a biological treatment had been used.

2) ammonia was the principal volatile nitrogen compound in pig slurry; amines (methyl and dimethyl-amine) seemed to represent only a very small fraction (1 percent) of the volatile nitrogen.

3) ammonia and hydrogen sulphide appeared very rapidly at the beginning of storage and reached a final concentration proportional to the dry matter (DM) content of the slurry.

During the research programme, chemical analyses for some other compounds were made in addition to an attempt to determine odour by means of sensory methods. The effect of some biological treatments on odour from slurry was also studied.

Attempts to perform experiments on the laboratory scale were unsuccessful. For example, Figure 1 shows that manure samples collected in metabolic crates and then stored at laboratory temperature tended to give a 'false pattern' of phenolic compounds, with a larger concentration of phenol than

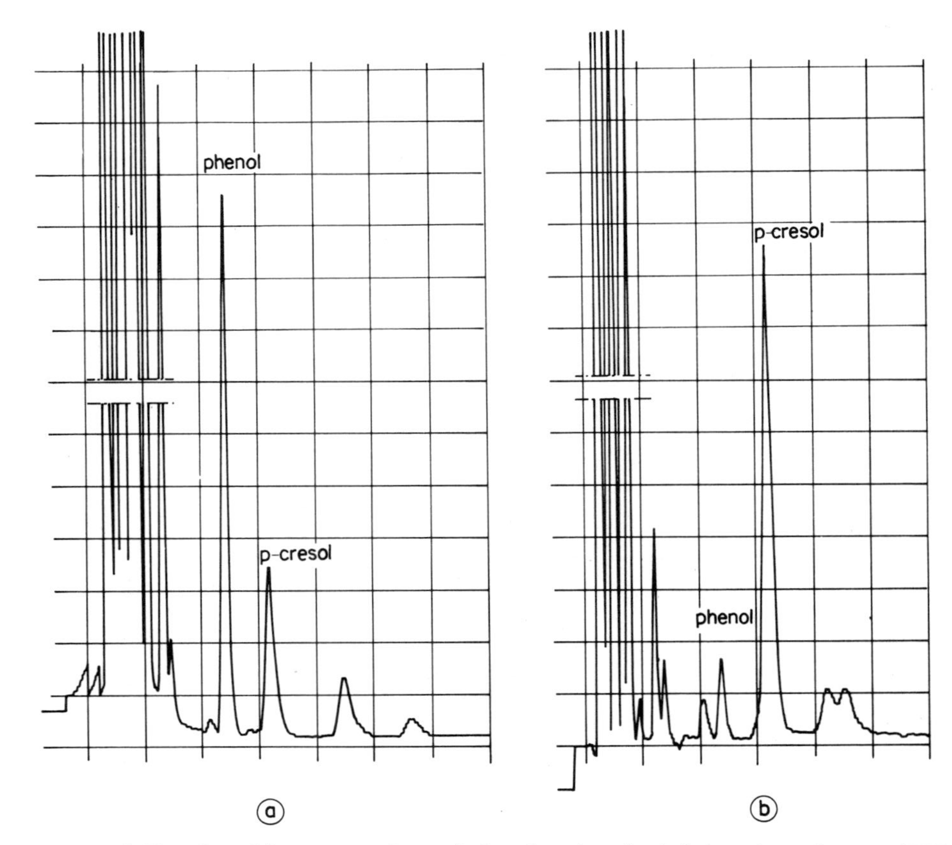

Fig. 1. Chromatogram of the phenolic compounds. a) Sample stored at laboratory temperature; b) sample taken from a conventional storage pit.

of p-cresol. The temperature of storage or composition of the gas phase over the sample during storage seemed also to promote this unusual composition. A similar effect was shown on the VFA composition by McGill and Jackson (1977), but Spoelstra (1977) studying the metabolism of phenolic compounds in manure found a normal pattern although the slurry was incubated anaerobically in 100 ml flasks.

Our results were obtained (i) on a commercial farm equipped with a liquid composting system fitted with a pilot plant for deodorising slurry from the UCAAB or (ii) six experimental pilot scale anaerobic digesters at the Institute.

ANALYTICAL INVESTIGATIONS

Experimental facilities

At the UCAAB plant

A 300 m^3 pit was used to treat 100 m^3 fresh liquid manure by a batch method with forced aeration by means of a floating aerator (centrirator from Alfa-Laval) with a power input of 3.5 kWh (see Figure 2).

A high rate biological filter with a volume of 4.5 m^3 built from bricks was developed in collaboration with CNEEMA (Mr. Vasseur). It has been used for treatment of the liquid phase of slurry separated by means of a 500 μ screen.

At the Institute

Figure 3 shows the flow diagram of the plant on the commercial farm. The anaerobic digesters with a capacity of 3.5 m^3 were insulated by means of 15 cm thick polyurethane lagging with electrical heating to a constant temperature of 35°C or 40°C. The contents were mixed continuously by paddles. The fresh manure transferred to the digester was obtained from the primary storage tank of the commercial farm (see Figure 2).

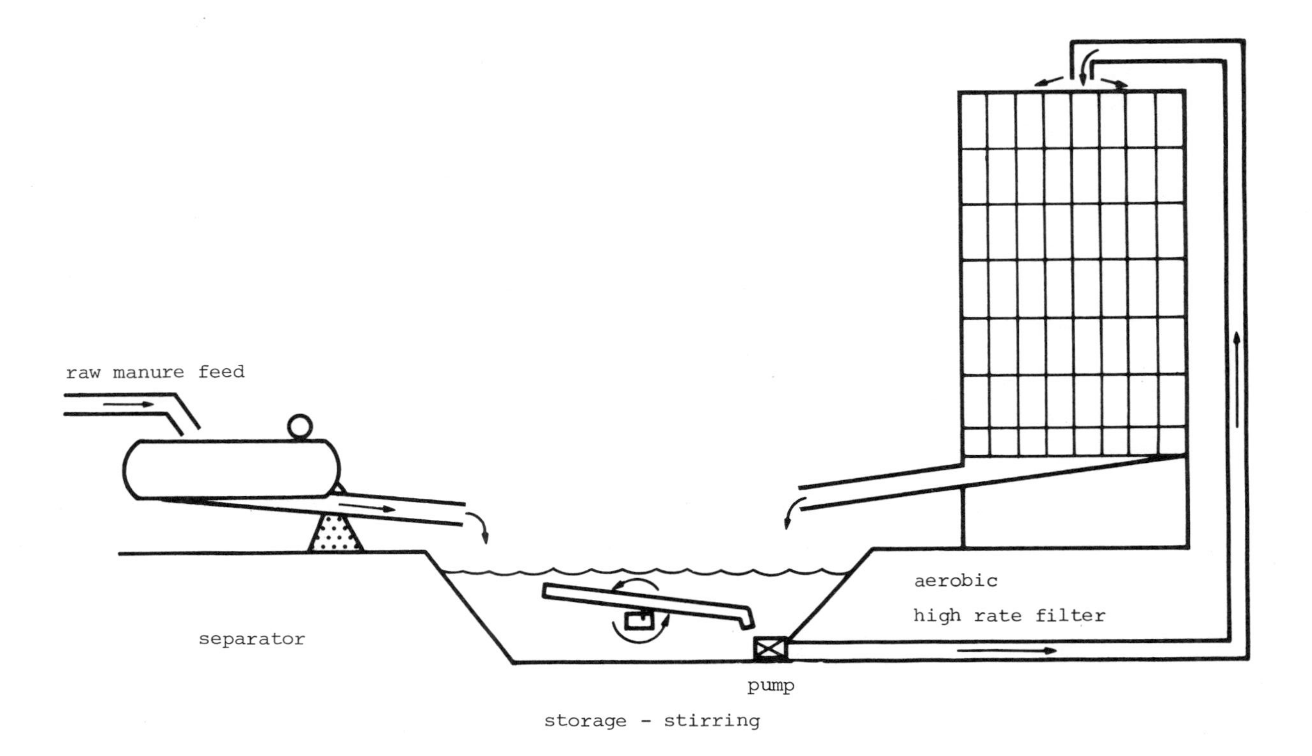

Fig. 2. Device for deodorisation of the slurry: high rate aerobic filter (UCAAB - CNEEMA).

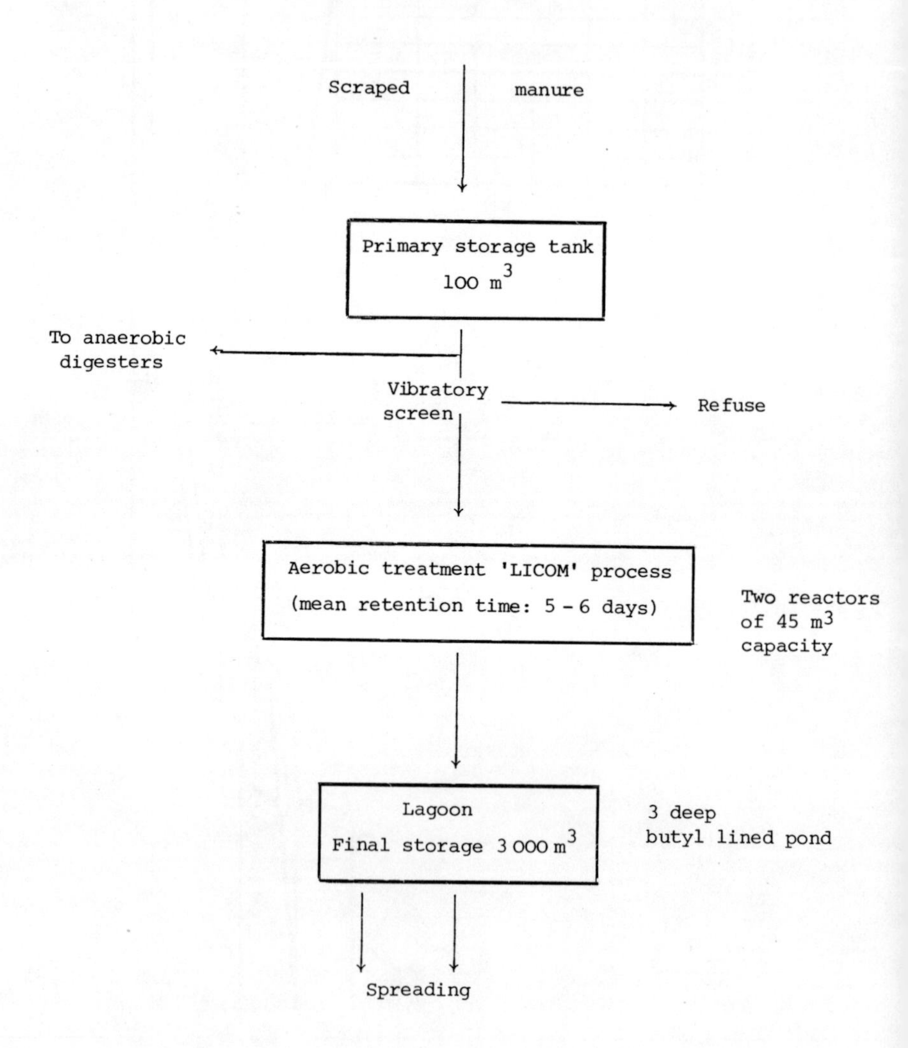

Fig. 3. Flow diagram: commercial farm (180 sows and progeny)

Results

High rate aerobic filter

Table 1 (Experiment 1a) summarises results obtained during a four months' winter period resulting in low biological activity on the filter. In this trial, a single load of 120 m^3 screened slurry was transferred into the storage pit. During treatment, the liquid was recirculated continuously through the filter at the rate of 2 m^3 h^{-1}. The results show that although very little organic matter was mineralised during the treatment, nitrogen was lost mainly as ammonia. These losses represented 45 percent of the total-N and 70 percent of the NH_3-N.

TABLE 1

SOME CHARACTERISTICS OF THE SLURRY BEFORE AND AFTER TREATMENT, WITH A HIGH RATE FILTRATION TOWER

	Experiment			
	1a		1b	
	Batch treatment for 120 days in winter		Continuous addition (2 m^3 day^{-1}) for 60 days, then batch treatment for 30 days	
Chemical components	Before	After	Before	After
Dry matter %	1.5	0.7	2.9	1.4
Ash (% DM)	36.7	38.6	34.4	38.8
Total-N [(1)] %	0.1	0.0	0.2	0.1
NH_3-N %	0.1	0.0	0.1	0.0
Total COD g O_2 l^{-1} [(2)]	21.2	9.4	33.8	15.0

(1) Total Kjeldahl nitrogen expressed as N

(2) Chemical oxygen demand

The slurry was effectively deodorised after this 4 months period and had an average rating of less than 1 for manure odour estimation (see odour tests for explanation of this value); thus the treatment capacity was about 0.9 m^3 day^{-1}.

Table 1 (Experiment 1b) shows the results of a trial, in which manure was added continuously to the storage ditch at a rate of 2 m^3 day^{-1}. After a period of 60 days, during which slurry was added, a further 30 days was necessary to obtain a deodorised manure. As in the previous case the losses of ammonia and total-N were important, the latter being 60 percent. Over the 90 days of the trial, 116 m^3 manure were put into the system; its treatment capacity was thus 1.3 m^3 day^{-1}.

Floating aerator

At the start of aeration, a layer of foam (1 m deep) was on the liquid surface, providing efficient thermal insulation. Because of the microbial activity, the temperature in the pit rose and the average values obtained during the treatment were 36°C in the winter and 45°C in the summer. Another consequence of the rise in temperature was the increase in evaporative losses to about 20 percent of the volume treated. Deodorisation was obtained within 20 days in the winter and 50 days in the summer, during which some problems occurred with the mixing. The treatment capacity ranged between 2 and 5 m^3 day^{-1}.

Table 2 compares the results. Mineralisation was larger than with the high rate filter, and the losses of ammonia ranged similarly from 60 to 80 percent. At the end of the run, the solids from the mixed liquor tended to settle very rapidly and, for example, in the winter trial sludges with a DM content of 12 percent, in which the organic matter (OM) represented 40 percent, occupied 6 percent of the total volume.

Another beneficial effect of this treatment was that it improved the microflora of the manure as is shown by bacterial analysis before and after treatment (Table 3).

TABLE 2

INFLUENCE OF SURFACE AERATION (FLOATING AERATOR) OF THE SLURRY ON ITS CHEMICAL COMPOSITION

	Experiment			
	2a		2b	
	Aeration for 50 days			
	Winter		Summer	
Chemical components	Before	After	Before	After
Dry matter %	4.0	2.1	2.8	1
Ash (% DM)	31.6	39.9	30.5	51.6
Total-N %	0.3	0.1	0.2	0.0
NH_3-N %	0.1	0.0	0.1	0.0
Total COD g O_2 l^{-1}	44	7.3	32	5

TABLE 3

BACTERIOLOGICAL ANALYSIS OF THE SLURRY BEFORE AND AFTER SURFACE AERATION (10^3 COUNTS ml^{-1} OF SLURRY)

	Experiment			
	2a		2b	
	Surface aeration for 50 days			
	Winter		Summer	
Bacteriological analysis	Before treatment	After treatment	Before treatment	After treatment
Aerobic mesophilic flora	9 900	4 400	5 600	3 100
Coliforms	1 000	0.01	1 000	0.01
E. coli	100	0	1 000	0.01
Anaerobic sulfito-reductors (thermo-resistant species)	30	1.1	10	10
Faecal streptococci	100	0.1	100	0.1
Enterobacteria	350	0.16	490	3.5

Similar trends were observed on the commercial farm, where the same treatment in the mesophilic temperature range was applied continuously with an average retention time of 6.5 days.

Table 4 gives an example of single values measured before and after treatment, from samples taken in the primary storage tank and in the lagoon before spreading respectively. Sampling from the lagoon was difficult and the value reported was that from continuing action of the biological treatment and the effect of long storage in the lagoon. There was also a significant activity in the lagoon, because during at least two periods of the year the lagoon turned black. Part of the black sludge fermented again and floated on the lagoon surface. A sample taken at that period showed increased percentage of OM in the DM and larger concentration of hydrogen sulphide. Thus, the odour of the liquid was as offensive as that of the fresh manure.

Anaerobic fermentation

Pig manure ferments rapidly under anaerobic conditions at temperatures from 35 to 40°C. Fermentation develops after a lag phase or more rapidly by using seeding material. Since anaerobic digestion allows the recovery of methane, as a fossil fuel substitute, its use as a pretreatment should be considered.

Table 5 gives an example of how some of the properties of slurry changed during fermentation compared with values of the sample taken from the lagoon. Anaerobic digestion preserved the fertilising value of the substrate, particularly N. The P contents of the two treated manures differed because, in the aerobic treatment, P was transferred to the sludge deposited at the bottom of the lagoon.

Figure 4 shows the changes in concentration of the phenolic and VFA compounds during fermentation. In every cycle phenolic compounds only seemed to be degraded after disappearance of the VFA. A long retention time was required and in some experiments the phenol and p-cresol concentration increased during the first period of the cycle.

TABLE 4

CHANGES IN SOME PROPERTIES DURING TREATMENT OF PIG SLURRY (COMMERCIAL FARM AERATION PLANT, SEE FIGURE 3)

Treatments	Control before treatment	After aerobic treatment
Chemical components		
Dry matter %	4.27	0.94
Ash % DM	25.7	68
Total-N %	0.4	0.2
NH_3-N %	0.3	0.1
K %	0.2	0.1

TABLE 5

CHANGES IN SOME PROPERTIES OF MANURE DURING ANAEROBIC TREATMENT

Treatments	Control before treatment	After anaerobic treatment 5 weeks
Chemical components		
Dry matter %	5.3	3.2
Ash % DM	22	35
Total-N %	0.4	0.4
NH_3-N %	0.2	0.3
P %	0.1	0.1
K %	0.1	0.1
Soluble COD $g\ l^{-1}$	20	10

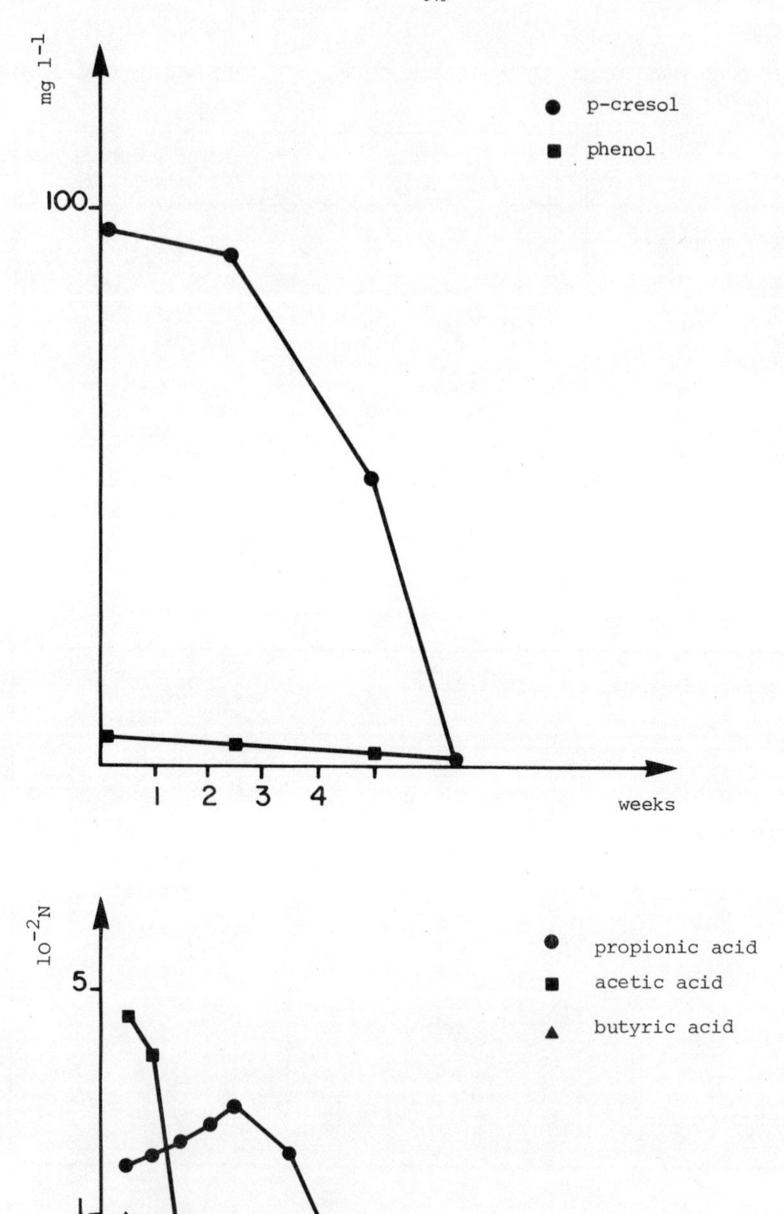

Fig. 4. Evolution of some chemical components during anaerobic fermentation.

The total sulphide content was the same at the beginning and at the end of the fermentation period. Attempts to evaluate the total sulphur content showed that the ratio $\frac{\text{total sulphur}}{\text{total sulphide}}$ ranged from 3.6 to 3.7. This suggested a balance similar to that of the ratio observed for the nitrogenous compounds in which the NH_3-N/total-N ratio was constant. Hydrogen sulphide being the last step in the anaerobic metabolism of sulphur, the anaerobic treatment would not be able to remove this product. Nevertheless, samples of fermented slurry were found with an odour intensity less than the minimum detectable by the laboratory olfactometer using 20 times dilution.

Conclusions

1) Among the aerobic treatments 'high rate filtration' appears to be the most sensitive to seasonal variations and requires the use of a solid-liquid separator thus increasing the cost of the plant. However, the filtration tower was simply constructed and could be made with cheap material. For a filter volume of 1 m^3 a pit storage volume of 20 m^3 is necessary and effluent from 100 pigs requires 2 m^3 of filter capacity.

2) Direct aeration in the storage tank by a floating aerator requires careful estimation of the energy input of the system: settlement must be avoided and the aeration device must be able to mix the contents of the tank without an excessive energy consumption.

3) Oxidation as an odour control system causes a loss of at least 50 percent of the N in the slurry mainly due to a stripping effect. Long term stabilisation of the effluent in the continuous process does not appear to be possible in all cases, since new fermentation sometimes takes place during the storage.

4) Anaerobic fermentation would deodorise pig manure, but the potential of such a system working on a continuous basis has to be accurately defined because batch fermentation of the liquid manure is not economical.

ODOUR TESTS

Table 6 summarises information from the literature about odour measurements of animal slurry, exhaust air from animal buildings or air in the environment of sewage treatment plants, except for references from the works co-ordinated by EEC (Stork-Hilliger-Dorling).

Odour measurements of slurries are often made using a rating scale recommended by Sobel (1969, 1972). This technique is relatively simple to perform but requires a panel of more than ten people. Presentation of the sample is rather difficult and can influence the panellist because of the colour of the slurry, for example. The sample was diluted with odour-free water by Burnett et al. (1969), simultaneous presentation of blanks and diluted samples in a triangular test procedure, in order to obtain a threshold odour number (TON), takes rather a long time to perform. In addition, the odour of the slurry can be modified, as the dilution alters the pH and therefore destroys the equilibrium of some components between gas and liquid phase (Barth and Hill, 1972).

Dilution of the air over the slurry overcomes the problem of sample presentation by using some techniques recommended for measurement of air pollution. The procedure becomes similar for example to that used for estimating the odour of exhaust air from animal buildings.

Figure 5 shows schematically the device used in our study. The test can be made simultaneously from ten ports, seven with dilutions of the sample and three control samples randomly presented to the panellists, who are selected from the staff of the laboratory. They were asked to smell the

TABLE 6

SOME LITERATURE REFERENCES TO MANURE ODOURS

Substrate	Authors		Methods of measurement	Scope
Poultry manure	Burnett	1969	Liquid dilution	Changes during storage
Poultry manure	Sobel	1969	Liquid and vapour dilution	Comparison of the two methods
Cattle slurry	Ifeadi	1972	Vapour dilution	Tentative to evaluate the contribution of various components to the total odour
Synthetic mixture	Hill	1976	Syringe dilution (ASTM)	Theoretical study of the Threshold odour number of a NH_3 - H_2S CH_3 - NH_2 mixture
Cattle faeces	Kellems	1976	Rating scale 0 → 10	Effect of feed additives
Cattle feedlots	Miner Sweeten	1975 1977	Scentometer	Effect of chemical treatment on feed additives
Liquid dairy manure	Stalling	1978	Rating scale 0 → 10	Effect of oxidising agents
Poultry manure	Sobel	1972	Rating scale 0 → 10	Effect of DM (odour increase with manure content)
Pig manure	Cole	1976	Rating scale 0 → 10	Effect of oxidising agents
Pig manure	Lindvall Noren-Lindvall	1974 1977	Vapour dilution magnitude matching with H_2S	Odour during spreading. Effect of various treatments
Pig manure	Welsh	1977	Rating scale (Sobel method)	Effect of an aerobic digestion
Dairy manure	Barth and Hill	1972	Liquid dilution	Correlation OII (Odour Intensity Index) on odorous component
Sewage sludge	Molton	1978	Liquid dilution	Correlation OII and chromatographic trace
Sewage sludge	Huang	1979	Vapour dilution	Activated carbon adsorption for level odour control

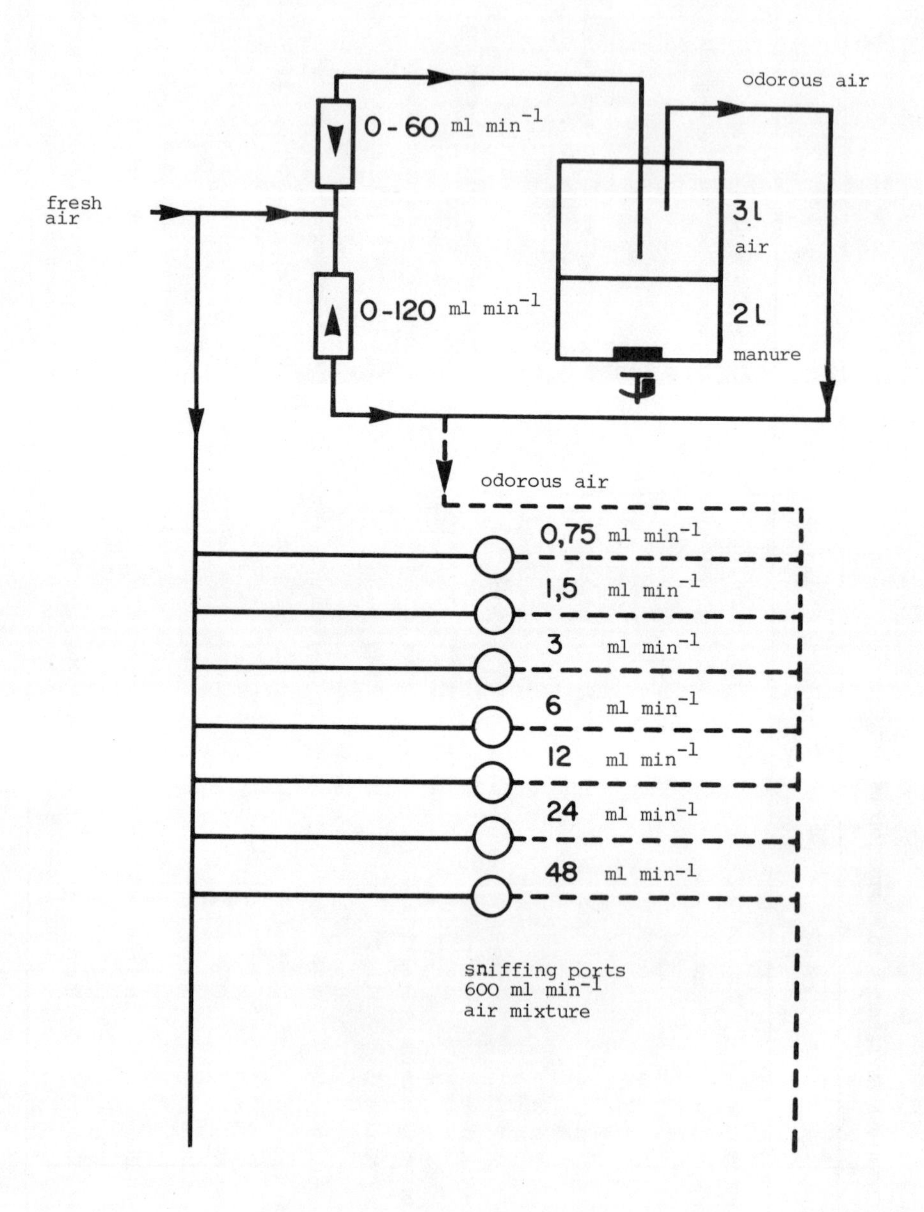

Fig. 5. Scheme of the Olfactometer.

different dilutions presented in increasing order of concentration. The first sniffing port was flushed with pure air.

The strength of the test odour is presented as a geometric progression with intervals of two. The gas was diluted through capillary tubes (liquid chromatography teflon tubing) with an internal diameter ranging from 0.3 to 0.8 mm giving air flow rates from 0.75 to 50 ml min^{-1} with a pressure of 7 cm water, and this flow rate is made up to 600 ml min^{-1} with dilution air and draught to sniffing ports. The dilution ratios vary from 20 to 800. Additional dilution could be obtained by diluting the carrier air before entering the sample vessel by means of two flowmeters (see Figure 5). All connections were made with teflonor low odour silicone tubes. The dilution air was compressed air passing through a filter. The gas flow-rate used in the presentation of a sample was low in order not to disturb the equilibrium between the gas and the liquid phase inside the sample vessel. The manure sample at a constant temperature was equilibrated for one hour with the vapour phase before starting the measurement, following a rapid screening to select the mean dilution to ensure that a correct dilution range had been chosen.

Each panel member chose the dilution at which odour was detected indicating in this way its detection threshold and not its recognition threshold. The analytical interpretation of the gross values obtained was similar to that described by Dravnieks and Prokop (1975). Table 7 gives an example and Figure 6 shows the results obtained with samples from a commercial farm: line 1 represents the raw manure corresponding to the calculation presented in Table 7b; line 2 the manure after treatment, and line 3 the effluent after treatment and storage in a lagoon. The odour unit expressed as a 'dilution to threshold' varied from 5 760 before treatment to 80 after treatment and to 40 after storage.

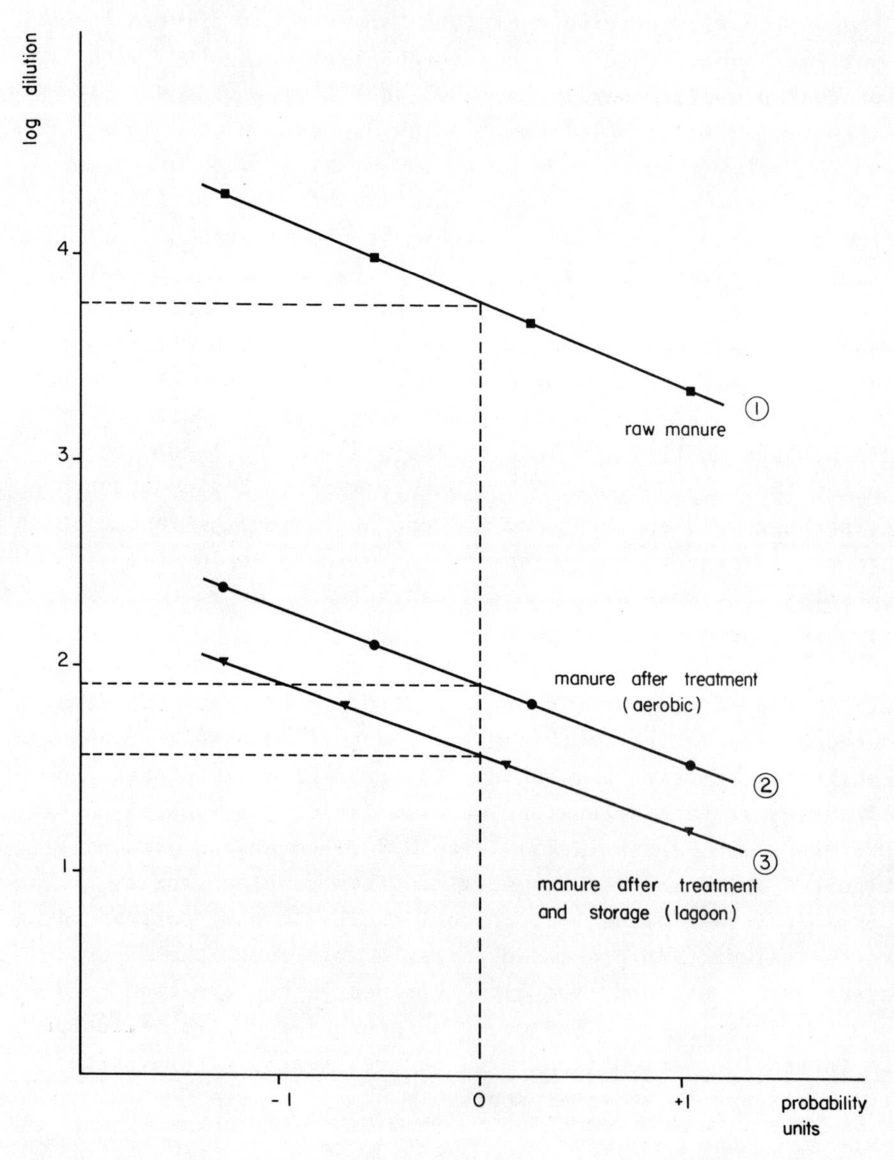

Fig. 6. Example of the analytical interpretation of the organoleptic tests.

TABLE 7

DETERMINATION OF THE ODOUR THRESHOLD OF FRESH PIG MANURE BY 9 PANNELLISTS

Dilution factor	13650	6500	3080	1530	820
Panellist number					
1	-	-	+	+	+
2	-	-	+	+	+
3	+	+	+	+	+
4	-	+	+	+	+
5	-	-	-	+	+
6	-	-	+	+	+
7	-	+	+	+	+
8	-	+	+	+	+
9	-	-	-	+	+
Frequency to threshold	1	3	3	2	

b) Calculation of the probability units (see Dravnieks and Prokop, 1975)

Dilution	Log dilution	Log tolerance level	Frequency to threshold	Average rank	Probability units
13650	4.135	4.2864	1	1	- 1.28
6500	3.8129	3.9743	3	3	- 0.52
3080	3.488	3.6507	3	6	+ 0.25
1530	3.1847	3.3368	2	8.5	+ 1.04
820	2.912				

The method has only been applied to some samples in the laboratory giving indications of the advantages and limitations of such an apparatus. Generally the required dilution is around 5 000 for untreated samples (typical values were 2 700, 7 100, 9 900, 5 700). Treated samples show threshold values less than 1 000.

The dry matter content of the sample presented might affect the final score, and this trend has to be ascertained if such a method should be generally used.

Another simple technique of measurement by a 4 point rating scale was used at the UCAAB Laboratory. The 0 value was attributed to the blank (fresh water) and the value of 4 to a raw manure sample. Half dilution of the various samples were presented to eight to twelve panel members. An arithmetic mean of the scores was used as a measure of odour intensity. Figure 7 gives examples of results obtained.

CONCLUSIONS

According to Molton and Cash (1978), 'Two methods are better than one', which is the reason why an attempt was made during the research programme to perform simultaneously an analytical and a sensory measurement in the study of pig manure odours. The duration of the programme did not allow us to draw a definitive conclusion from our results.

The analytical method seems to be the simplest one but it requires more and more sophisticated techniques due to the need to identify trace amounts of components leading to an increase in the numbers involved in identification. The great difficulty with this approach is to create storage conditions in the laboratory similar to those prevailing in a larger tank or pit. Nevertheless, the method is sufficiently good to provide average values for the main chemical compounds.

The sensory method has to be adapted to the type of research to be performed. For the air pollution problem, official standard methods are not available at present for determining acceptable levels of 'air pollution standard', since only results obtained by the same technique (apparatus, presentation of the sample, and mathematical treatment of the scores), can be compared, emphasis should be laid at the present time on the need for more co-operation between the participants

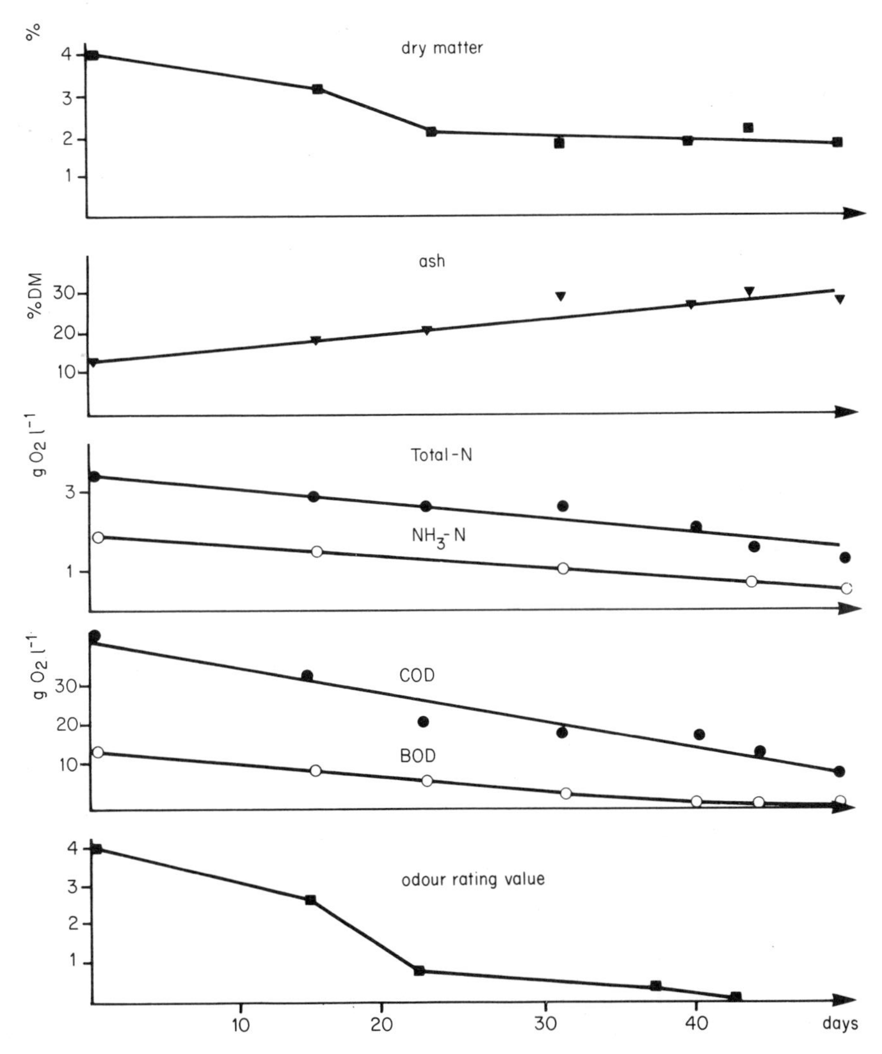

Fig. 7. Evolution during the treatment period of some pig manure components (aerobic high rate filter)

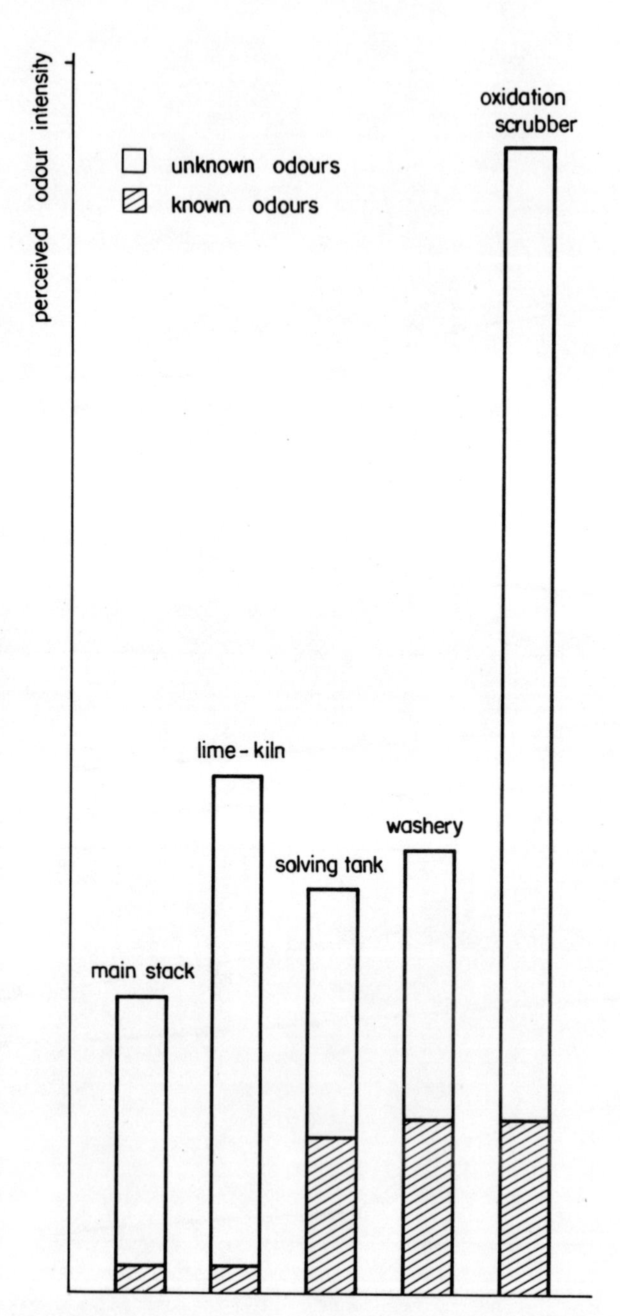

Fig. 8. Perceived odour intensity of five pulp mill effluents (from Berglund)

in order to test simultaneously some typical situations with the same method.

Correlations found between the level of some particular compounds in the slurry and the odour intensity could not be generalised for all situations. Thus Berglund (1977), measuring the odour intensity of gaseous effluents from pulp mills, mentioned that using vector sum of the individual contribution of each constituent, not more than 50 percent of the perceived odour intensity could be accounted for (Figure 7). Such results should prompt other workers to be careful in their generalisation of individual results.

At the beginning of this programme, spreading of pig slurry was considered to be a source of smell nuisance by the pig breeders. At present, because of the socio-economic conditions this nuisance must be turned into profit by recovering the energy by using an anaerobic fermentation process to lessen the odour.

REFERENCES

Barth, C.L., Hill, D.T., et al., 1972. Correlating OII and odorous components in stored dairy manure. ASAE paper 72-950.

Berglund, B., 1977. Quantitative approaches in environmental studies. International journal of Psychology, 12, 111-123.

Burnett, W.E. et al., 1969. Microbial and chemical changes in poultry manure in Animal waste management. Cornell University conference on agricultural waste management. 411 p.

Cole, C.A. et al., 1976. Efficacy of certain chemical and biological compounds for control of odour from anaerobic liquid swine manure. J. Anim. Sci., 42, 1-7.

Dravnieks, A. and Prokop, W., 1975. Source emission odour measurement by dynamic forced choice triangle olfactometer. Journal of the Air Pollution Control Association 1975. 25, 28-35.

Huang, J.V.C. et al., 1979. Evaluation of activated carbon adsorption for sewer odour control. Journal Water Pollution Control Federation 1979. 51, 1054-1062.

Hill, D.T. and Barth, C.L., 1976. Quantitative prediction of odour intensity. Transaction of the ASAE. 19, 939-944.

Ifeadi, C., 1972. Quantitative measurement and sensory evaluation of dairy waste odour. Ohio State University, Ph. D. Thesis. 185 p.

Kellems, R., 1976. The effect of ration formulation on the subsequent generation of volatile gases on odours from Oregon State University. Ph. D. Thesis. 59 p.

Lindvall, T., Noren, O. and Thyselius, L., 1974. Odour reduction for liquid manure systems. Transaction of the ASAE, 17, 509-512.

MacGill, A.E. and Jackson, N., 1977. Changes in the short-chain carboxylic acid content and chemical oxygen demand of stored pig slurry. Journal Science Food and Agriculture. 28, 424-430.

Miner, J.R., 1975. Evaluation of alternative approaches to control of odours from animal feedlots. Idaho Research Foundation Inc. 83 p.

Molton, P.M. and Cash, D., 1978. Novel combination method assesses sewage odours. Water and Wastes Engineering. February 78. 47-52.

Noren, O. and Lindvall, T., 1977. In: Animal Wastes. Applied Science publisher. 423 p.

Sobel, A.T., 1969. Measurement of the odour strength of animal manures in: Animal wastes management. Cornell University conference on agricultural waste management. 414 p.

Sobel, A.T., 1972. Olfactory measurement of animal manure odour. Transaction of the ASAE., 15, 696-703.

Spoelstra, S.F., 1977. Simple phenols and indoles in anaerobically stored piggery wastes. Journal Science Food and Agriculture. 28, 415-423.

Stalling, C.C. et al., 1978. Chemical treatment for odour abatement measured organoleptically in liquid dairy manure. Journal of Dairy Science. 61, 1509-1516.

Sweeten, J.M. et al., 1977. Odour intensities at cattle feedlots. Transaction of the ASAE. 20, 502-508.

Welsh, F.W. et al. 1977. The effect of anaerobic digestion upon swine manure odours. Canadian Agricultural Engineering. 19 122-126.

DISCUSSION

J.K. Grundey *(UK)*

You compared aerobic and anaerobic treatment of slurry. Have you tried to apply the results to farm systems in order to compare costs?

A. Aumaitre *(France)*

When farms are equipped with aerobic systems, these aerators are not always operated. For instance, if two 'Licom' aerators are installed only one is used, to convince the neighbours that it is still working. Much equipment is only used at about half capacity, because of the cost of running the system. Aeration is expensive for the farmer, with estimates of between £2 and £4 for each fattening pig produced.

A. Maton *(Belgium)*

Anaerobic treatment with the production of methane only appears economically justified when the gas produced can be used immediately and continuously. If the gas has to be stored, which is usual, is anaerobic treatment economically justified?

A. Aumaitre

I agree. Methane production may be one of the primary aims of the system or a by-product which will help to balance the energy utilisation, this is not known. If the methane produced is used to heat the anaerobic digester, we would be very satisfied, because the control of malodour would not require extra energy.

ODOUR CHARACTERISATION IN ANIMAL HOUSES BY GAS CHROMATOGRAPHIC ANALYSIS ON THE BASIS OF LOW TEMPERATURE SORPTION

J. Hartung and H.G. Hilliger
Institut für Tierhygiene der Tierärztlichen Hochschule
Hannover, Bünteweg 17 P, D 3000 Hannover 71,
West Germany.

ABSTRACT

Work under the project was done in five parts:

A. Development of methods.

B. The total integral values of the chromatograms.

C. The characteristics of the gas chromatograms using flame ionisation detection (FID).

D. The characteristics of the gas chromatograms using flame photometric detection (FPD) for sulphur compounds.

E. The gas chromatographic analysis of phenolic compounds.

In addition to gas chromatographic analysis measurements were made of the climatic and husbandry conditions in the animal houses. The odour intensity should be characterised sensorily by the paper strip test. The results of the total investigations are summarised as follows:

1. *The temperature gradient tube (TGT) allows the trace gases in the air of animal houses to be sampled quantitatively for analyses by gas-liquid chromatography (GLC).*
2. *A general purpose column (OV17) with flame ionisation detection (FID) gave chromatograms from the air samples with many incompletely resolved peaks. Single peaks are difficult to estimate and identify. Indicator compounds for odour evaluation could not be determined by this method. Therefore in our opinion this summary description of an air sample is unsuitable for characterising the odour intensity in animal houses.*

3. *Using the flame photometric detector (FPD) with a special 394 nm filter for sulphur containing compounds, the chromatogram normally shows five well resolved peaks. The largest peak is dimethylsulphide; hydrogen sulphide is only found in small amounts. Part of the chemically labile mercaptans might have changed into sulphides and were contained in the sulphide peaks. Our results do not indicate either that one of the sulphur containing peaks, or that some or all of them, correlate with the odour intensity in animal houses or with the housing systems.*

4. *Short glass tubes filled with Tenax-GC as adsorption material are suitable for sampling phenolic components from the air. The quantitative and qualitative results correspond throughout with the results of the Dutch investigators (Logtenberg and Stork, 1976), but we could not find a satisfactory correlation between the amount of phenolic compounds detected and the odour intensity either in hen or in pig houses.*

5. *Our simple paper strip test is suitable for an approximate estimation of the odour in pig houses, but is less suitable in hen houses.*

6. *No correlations were found between the GC-results and the climatic and husbandry conditions in the houses, except for the tendency for decreasing air volumes (m^3 animal^{-1}) corresponding with increasing total integral values of the FID.*

7. *The future work should try to standardise the sensory methods used in the different research groups. The research for main components of the odour should be intensified, especially by employing specific detectors as FPD, N-FID or ECD.*

Malodours emitted from animal husbandry represent one of the major environmental problems confronting livestock producers in populated areas, but there is at present no generally accepted procedure for measuring odours (Miner, 1977). Between 1975 and 1978, investigations were carried out in our institute to characterise the odour in animal houses especially by gas chromatographic means. In addition to the gas chromatographic analysis, measurements were made of relative humidity, temperature and velocity of air, contents of CO_2, NH_3 and H_2S, inside and outside the buildings; the housing system, waste removal and numbers and weight of animals were also recorded.

After the development of special methods for the sampling procedure and analysis, the investigations were carried out in three main series:

Series I: 12 pig fattening houses during 6 months with 151 samplings

Series II: 9 pig fattening houses during 9 months with 144 samplings

Series III: 9 hen houses during 6 months with 94 samplings

Each sampling included duplicate air samples resulting in chromatograms. At each sampling inside the animal houses, the procedure was completed by also sampling the ambient air.

A. DEVELOPMENT OF METHODS

The project first required the development of a procedure allowing air samples to be taken both in animal houses and in the ambient air. The sampling procedure should enrich all natural gaseous compounds occurring in small concentrations in the air, to provide later on in the analysis as much information as possible about the trace gases in the air. On the basis of the sampling tube described by Kaiser (1973) our temperature gradient tube (TGT) was developed. It is shown schematically in Figure 1. For the analysis a gas-chromatograph (GC) was used which was equipped with both flame ionisation detection (FID) and flame photometric detection (FPD). The procedure for sampling, transporting and analysing was arranged to be optimal (Hartung, 1977; Hartung and Hilliger, 1977). The following results were obtained:

1. A mobile sampling technique was developed suitable for use in both animal houses and the ambient air, by using Tenax-GC as sorption material and liquid nitrogen for cooling. Each TGT usually required a sampling volume of 2 - 3 l and a sampling time of 4 - 6 min.

2. The TGT was modified to provide a reproducible temperatur gradient from $-160^{o}C$ to $+10^{o}C$ along the tube.

3. The transportation of the charged TGT from the sampling place to the laboratory can be done at normal temperatures without losses if, after sampling, both ends of the tube are well closed and if Tenax-GC is used as absorbing material.

4. The sample should be transferred (injected) from the TGT to the GC by heat-desorption at a temperature of $+210^{o}C$ and by means of a specially developed injection technique.

5. The temperature of the GC-column should be kept at $-75^{o}C$ during the transfer to prevent clogging of the column by freezing carbon dioxide.

6. When using the phosphorus specific filter of the FPD for GC-analysis of air from animal houses, no peaks were recorded. With the sulphur specific filter of the FPD a few well resolved peaks were obtained.

7. Because the air samples contain sensitive sulphur containing compounds, all-glass-systems should be used for sampling, transfer and analysis.

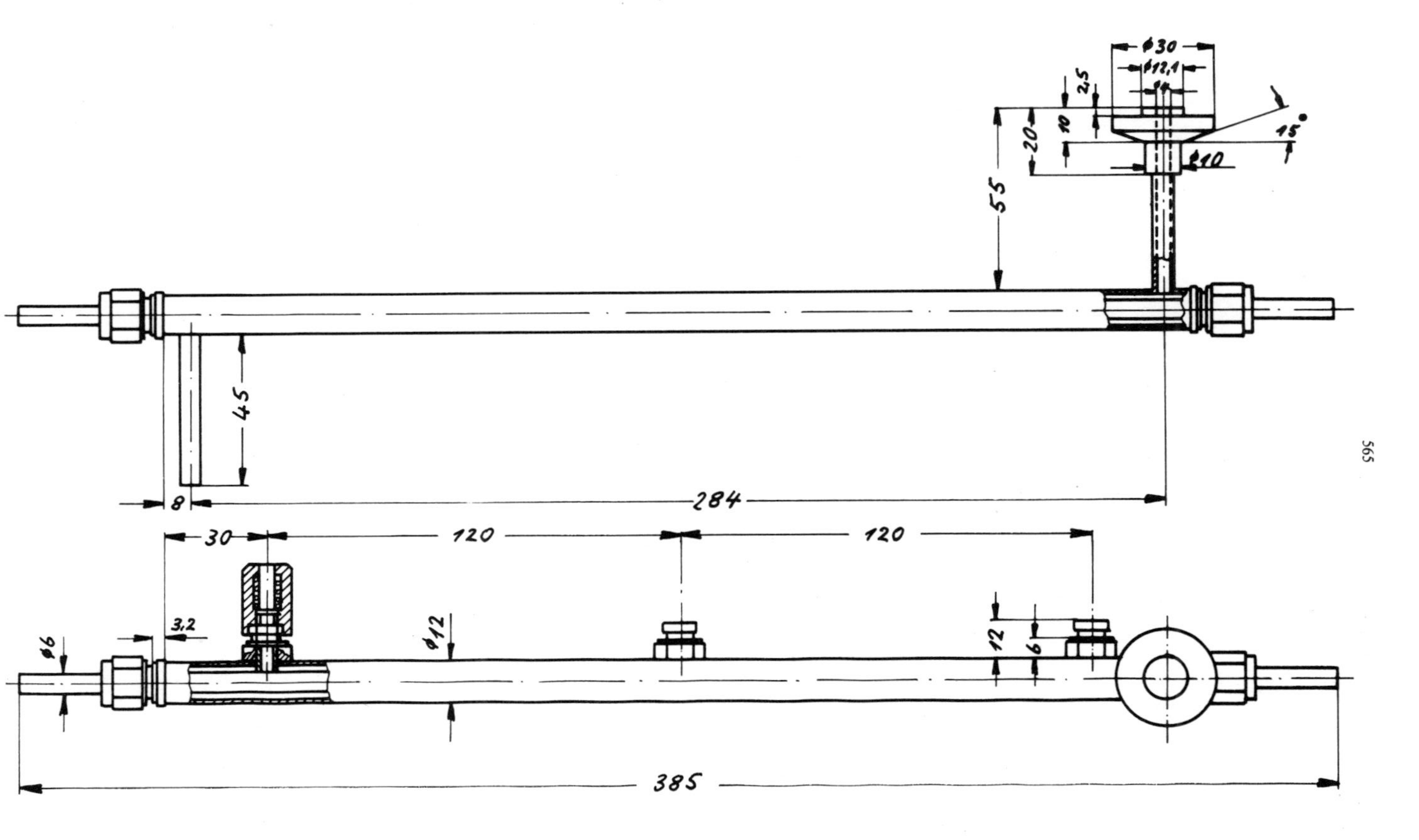

Fig. 1. Schematic diagram of the temperature gradient tube (TGT) for sampling air for odour analysis in a gas chromatograph (Hartung, 1977; Hartung and Hilliger, 1977)

B. THE TOTAL INTEGRAL VALUES OF THE CHROMATOGRAMS

After the development of a suitable sampling method for trace gases (Part A), an investigation was started in 12 pig fattening houses with different housing systems, most without litter. The following experimental work was done at two week intervals in each animal house: sampling and analysis of trace gases from the air of animal houses and from the ambient air by the TGT and the GC, subjective sensory smell test and the measurement of the concentrations of NH_3, H_2S and CO_2 in the houses; temperature and humidity measured inside and outside the houses. The aim of these investigations was to find out, if the total integral of the FID and FPD chromatograms of the inside air would relate to corresponding integrals in the ambient air and to values of the other measurements inside and outside the animal houses. The reproducibility of the results was also investigated and an attempt was made to demonstrate differences between the different housing systems in this way (Hilliger and Hartung, 1977, 1978a). The following results were obtained:

1. The total integral values of the FID chromatograms from the pig houses of air were always larger than from the ambient air and varied widely in value.

2. The total integral values of the FID chromatograms, from the ambient air, had relative values between 16 percent and 68 percent of those in the houses. The variation in values of the ambient air depended more on the location of sampling than on the housing system.

3. The total integral values of the FPD chromatograms varied widely too and no recognisable relationship was found between the total integral of the FID results and the other measurements either in the air of the animal houses or in the ambient air (Figure 2).

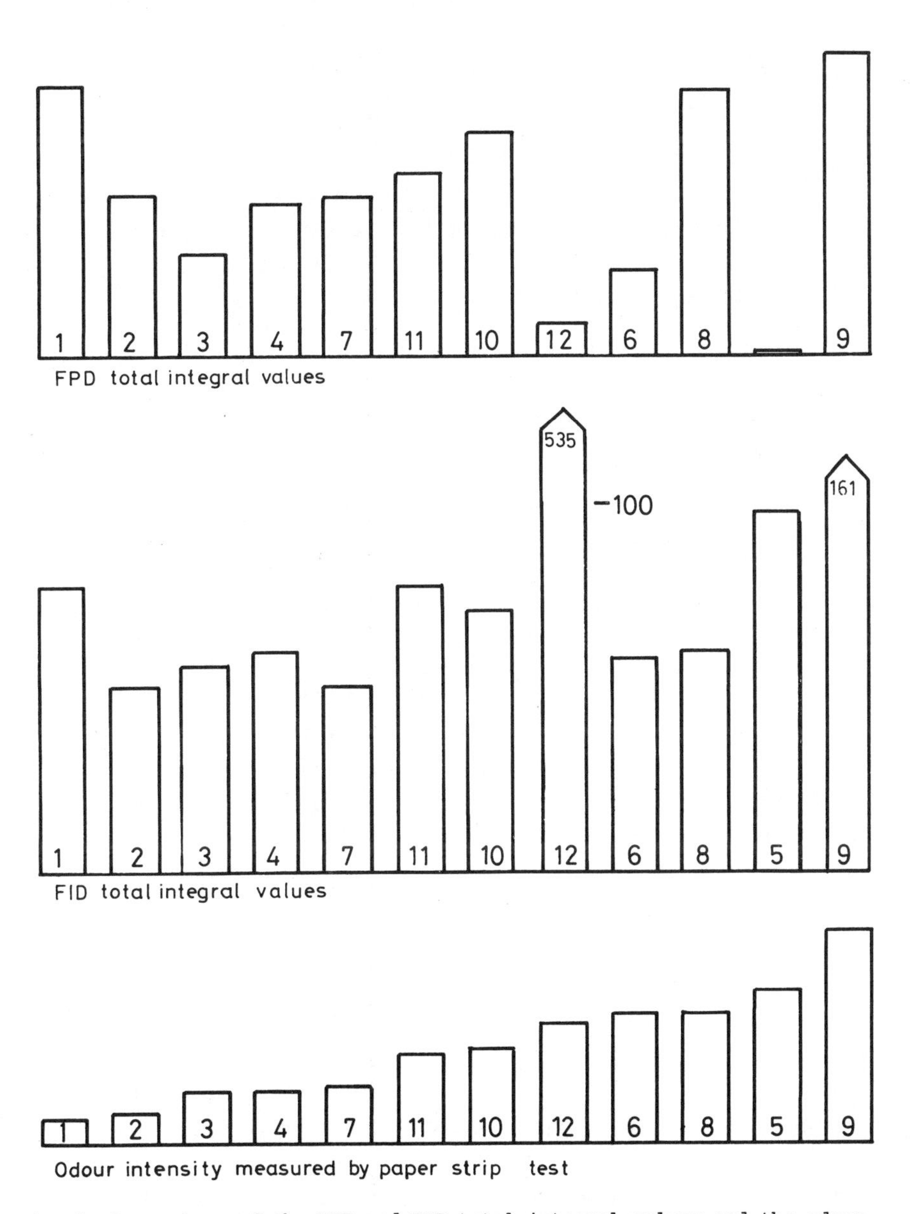

Fig. 2. Comparison of the FID and FPD total integral values and the odour intensity in 12 piggeries. The piggeries are listed from left to right with increasing odour intensity. The numbers in the columns stand for the animal house numbers. The height of the column indicates the amount of the integral values obtained on average of ten measurements (Hilliger and Hartung, 1977).

4. In pig fattening houses, the simple paper strip test is more suitable as a sensory method for odour characterisation and more reliable than direct sensory evaluation by smell (Hartung, 1977; Hilliger and Hartung, 1978a).

5. The paper strip test did not correlate with the total integral of the FID and FPD results nor to the other characteristics of the inside air including CO_2, H_2S and NH_3 contents.

6. The conditions in the animal houses and in the ambient air changed from sampling to sampling, so that the results by all methods used showed large variations.

7. There was no correlation between a single or several measured factors and the housing systems of the animal houses.

C. THE GAS CHROMATOGRAPHIC CHARACTERISTICS OF THE FID RESPONSE

The FID regularly gives chromatograms with at least 25 peaks which are more or less incompletely resolved or fused (Figure 3). The chromatograms should be able to be evaluated by the characteristics of their corresponding total integral values. The electronic equipment registers each peak by dropping a vertical from the low point between two peaks to the baseline. The integrator registers the retention time at the top of the peaks and integrates the area under the curves. Many small peaks are considered to be superimposed on the large ones and the small peaks could not be registered. Our analytical system did not allow individual peaks to be identified. Therefore the qualitative and quantitative results were only investigated to find differences between animal species or between the housing systems. For this purpose the findings of the series II (9 pig fattening houses with 242 chromatograms) and series III (9 hen houses with 181 chromatograms) were used. The following results were obtained:

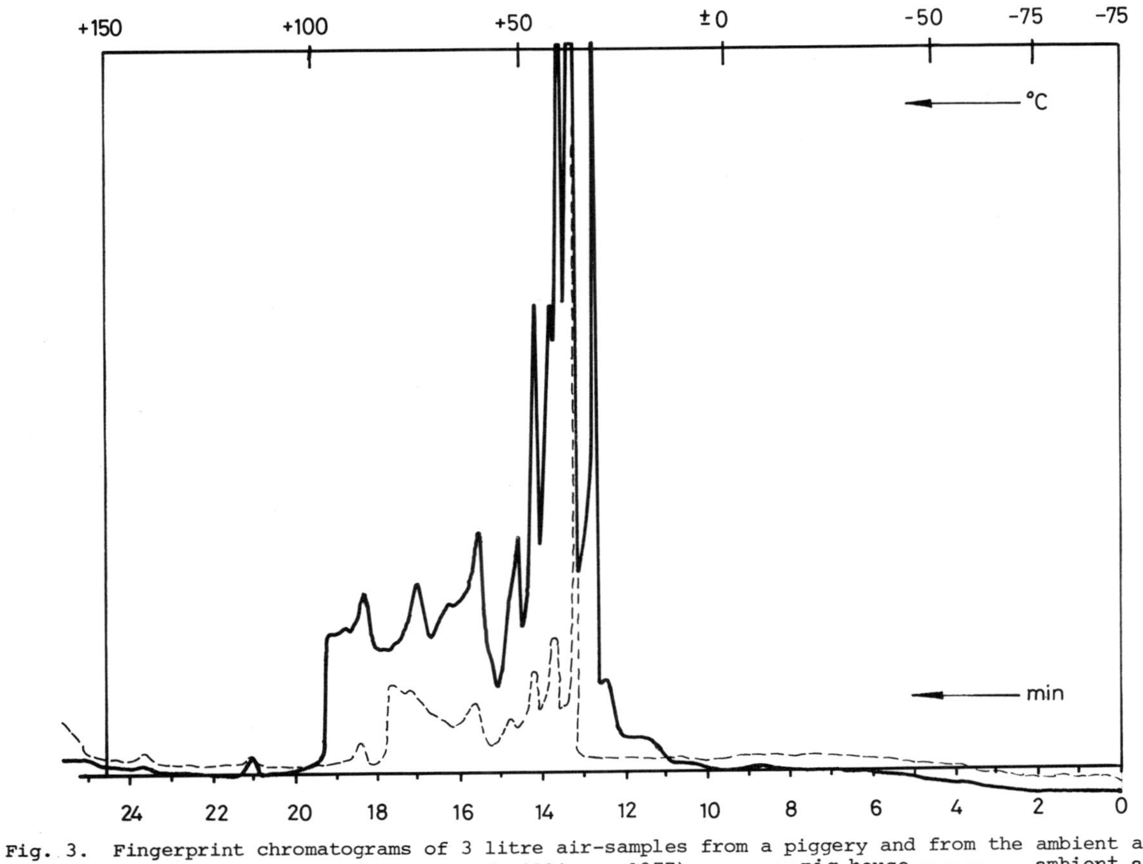

Fig. 3. Fingerprint chromatograms of 3 litre air-samples from a piggery and from the ambient air by means of the FID (Hartung and Hilliger, 1977) ——— pig house - - - - ambient air

1. On average, the total integral values of the pig houses were 3.6 times as large as those of the hen houses. Based on the gas chromatographic results, the air in the pig fattening houses seemed to be distinctly more contaminated than in hen houses. This finding corresponds to the subjective experience and odour testing by panels.
2. In the pig house, 30 peaks were registered, and in the hen houses only 22 peaks. The peak integral varied widely not only in one animal house between the periodical samples but also between the different animal houses.
3. In all chromatograms of the pig houses 21 peaks appeared with a frequency of less than 50 percent and only three peaks had a frequency of more than 70 percent. In the chromatograms of the hen houses, 10 peaks appeared with a frequency of less than 50 percent and 8 peaks had a frequency of more than 70 percent.
4. Table 1 shows the relative average proportions of the total integral values for the nine peaks with a frequency of more than 50 percent for pigs and 65 percent for hens. Peaks with the same number are not necessarily the same compound.
5. No correlation was found between the appearance and the value of single peaks and the housing system, or between the pig houses and the hen houses.
6. No correlations were found in either pig or hen houses between the appearance or the value of single peaks and the odour intensity or the content of NH_3, CO_2 and H_2S in the air.

TABLE 1

RELATIVE FREQUENCY OF THE NINE LARGEST SINGLE PEAKS AND THEIR RELATIVE PROPORTIONS OF TOTAL INTEGRAL VALUE IN PIGHOUSES AND HENHOUSES

	PIGS			HENS		
Range of frequency	Peak number	Relative frequency of single peak	Relative proportion of total integral value with coeff. of variation	Peak number	Relative frequency of single peak	Relative proportion of total integral value with coeff. of variation
1	4	100%	46.7% ±22%	4	100%	25.8% ±53%
2	19	80%	2.0% ±40%	12	94%	13.9% ±45%
3	11	78%	2.6% ± 9%	22	94%	2.0% ±55%
4	29	68%	0.9% ±36%	8	91%	5.6% ±49%
5	15	67%	2.9% ±23%	9	88%	4.4% ±63%
6	16	63%	2.7% ±81%	19	87%	2.7% ±66%
7	10	61%	5.1% ±119%	16	85%	3.2% ±50%
8	18	58%	1.5% ±38%	15	77%	3.2% ±75%
9	20	55%	1.6% ±55%	6	66%	6.0% ±50%

D. THE GAS CHROMATOGRAPHIC CHARACTERISTICS OF THE SULPHUR FPD RESPONSE

The gas chromatographic analysis of the air in the animal houses using the sulphur specific FPD regularly gave from three to five completely resolved peaks (Figure 4), however even in a single animal house the findings were not consistent (Hartung and Hilliger, 1978). Therefore the peaks should be considered both qualitatively and quantitatively to try to discover differences with respect to the species and to the housing systems. For this purpose the findings of the series II (9 pig fattening houses with 242 chromatograms) and the series III (9 hen houses with 181 chromatograms) were evaluated. The following results were obtained:

1. Three of the sulphur containing peaks were identified: peak 1 was hydrogen sulphide; peak 3 was dimethylsulphide; and peak 4 was dimethyldisulphide. The other two peaks were not identified. However, the sulphide-peaks may also include mercaptan-compounds.

2. In general most dimethylsulphide was produced while little hydrogen sulphide was detected and often not at all, especially in hen houses. After the further evaluation of the sulphur findings, the preliminary results, reported by Hilliger and Hartung (1977), have to be corrected using the proportions of S-containing compounds shown in Table 2.

3. The results within a single animal house varied considerably for both pigs and hens; values also varied between animal houses.

4. The total quantity of gases differed between pig houses and hen houses, the total integral of the sulphur containing compounds in the pig houses being from 1.5 to three times as great compared with the value in the hen houses. Air from pig houses contained about 0.5 mg m^{-3} of sulphur-containing compounds.

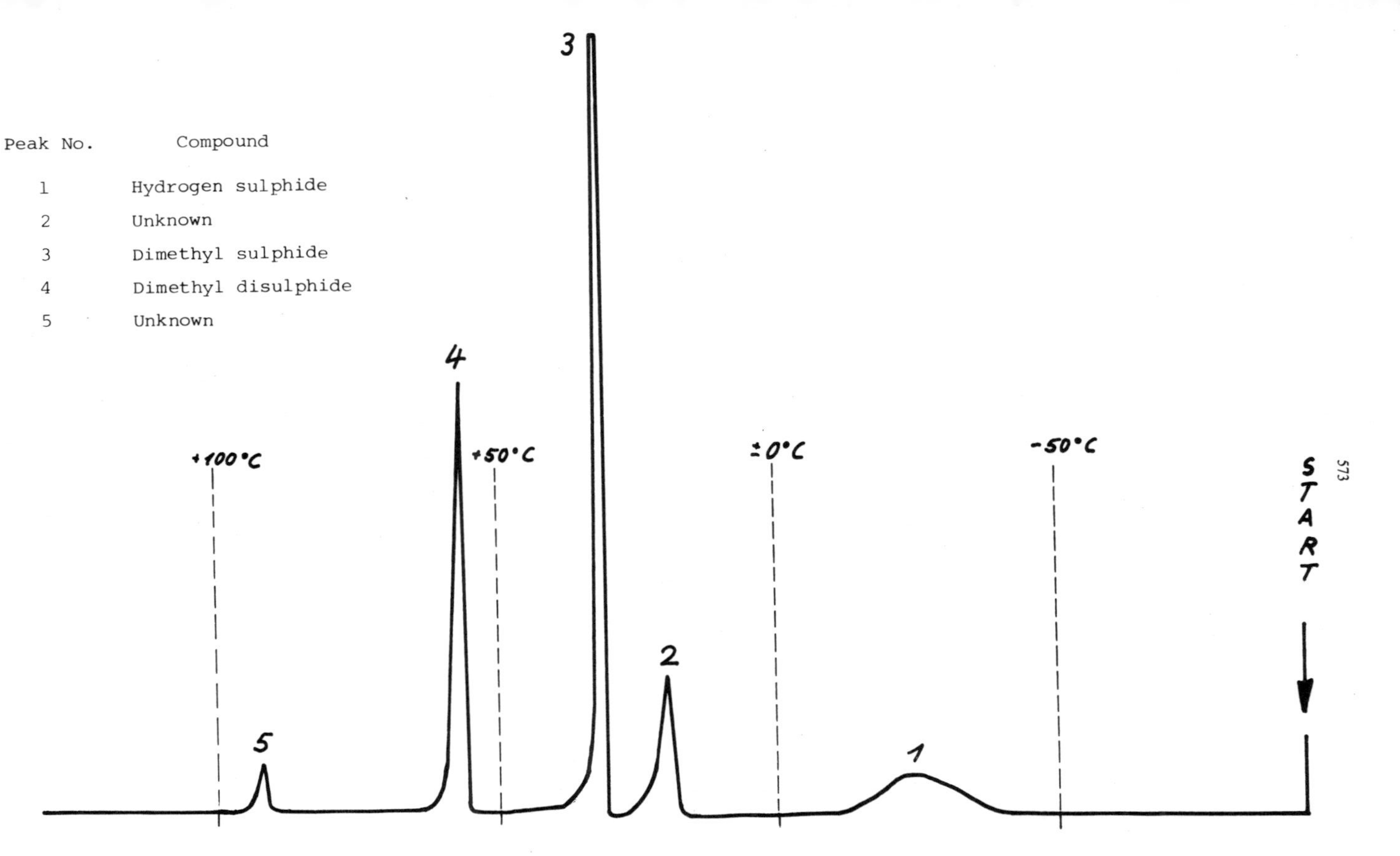

Fig. 4. Fingerprint chromatogram of a 3 litre air-sample from a piggery by means of the FPD in the sulphur specific mode (Hartung and Hilliger, 1978)

TABLE 2

COMPARISON OF THE OCCURRENCE OF SULPHUR-CONTAINING COMPOUNDS IN THE AIR OF 9 HEN HOUSES AND 12 PIGGERIES

Peak number	ø frequency of appearance		ø proportion of the total integral		Coeff. of variation of the proportion		Identified peaks
	Pig	Hen	Pig	Hen	Pig	Hen	
1	57%	31%	0.3%	0.1%	± 237.8%	± 127%	Hydrogen sulphide
2	56%	31%	3.5%	7.9%	± 72.8%	± 104%	Unknown
3	99%	99%	69.6%	60.1%	± 21.4%	± 33%	Dimethyl sulphide
4	98%	94%	21.1%	26.6%	± 58.1%	± 74%	Dimethyl disulphide
5	96%	70%	5.5%	5.3%	± 66.9%	± 74%	Unknown

5. There was only a slight tendency for the total amount of sulphur-containing compounds detected to increase with decreasing intervals of removal of waste.

6. No correlation existed between the quantitative and qualitative composition of the sulphur-containing compounds and the housing system whether for pigs or hens.

7. No correlation was found between the total amount of sulphur in all samples or single sulphur peaks and the odour intensity or the content of NH_3 and CO_2 in the air of the animal houses of both species.

E. THE GAS CHROMATOGRAPHIC ANALYSIS OF PHENOLIC COMPOUNDS

Logtenberg and Stork (1976) stressed the possible importance of some phenolic compounds as indicators for odour intensity. Therefore our programme included the analysis of phenolic compounds, using a slightly modified gas chromatographic procedure. We directed our attention especially to m+p-cresol, indole and skatole (Hilliger and Hartung, 1978b). This additional analysis was carried out on some of the samples from series II (pigs) and on all samples from series III (hens). The following results were obtained:

1. On average, the total amount of phenolic compounds detected was about 40 $\mu g\ m^{-3}$ in the air from the pig houses and 4.6 $\mu g\ m^{-3}$ in the air from the hen houses, giving a ratio of 8.5 : 1.

2. The values varied widely both within a single animal house and between different animal houses.

3. The frequency of appearance of m+p-cresol, indole and skatole is given in Table 3 together with their average relative proportions referred to the total phenolic amount. After the further evaluation of the cresol

findings the preliminary results reported by Hilliger and Hartung (1977) have to be corrected using the proportions of phenols shown in Table 3.

4. Our results suggest that the amount of phenolic compounds detected and the odour intensity are not well correlated.

5. No correlation was found between the quantitative and qualitative composition of the phenolic compounds and the housing system either for pigs or hens.

TABLE 3

COMPARISON OF THE FREQUENCY OF APPEARANCE OF M+P-CRESOL, INDOLE AND SKATOLE IN THE AIR OF HEN AND PIG HOUSES AND THEIR RELATIVE PROPORTIONS AS A PERCENTAGE OF THE TOTAL PHENOLS PRESENT

Compound	Pig		Hen	
	Frequency	Relative proportion	Frequency	Relative proportion
m+p cresol	100%	58.0%	17%	1.1%
Indole	95%	35.5%	56%	24.1%
Skatole	89%	6.5%	60%	74.8%

CONCLUSIONS

Our results suggest that future work in the field of odour characterisation from rural areas should include:

1. The sensory methods used for odour characterisation and measurement should be made reliable and, as far as possible, standardised to allow better comparisons to be made of the sensory results obtained by the different research groups.

2. The research on main components of the odour should be intensified using both sensory and chemical-analytical methods.

3. For the gas chromatographic characterisation of odorous air from animal houses containing large numbers of different chemical components, selective sampling methods and specific detectors should be employed.

ACKNOWLEDGEMENT

This work has been supported by the Commission of the European Communities (CEC), Brussels, Treaty No.220, as part of the research programme, 'Effluents from Livestock'; also by the Umweltbundesamt (Department of the Environment), Berlin.

REFERENCES

Hartung, J.,1977. Ergänzung und Anpassung eines Probenahmeverfahrens zur Erfassung von Fremdgasen aus Stalluft mit der Gaschromatographie. Diss. Vet. med. Tierärztliche Hochschule Hannover.

Hartung, J. and Hilliger, H.G., 1977. A Method for Sampling Air in Animal Houses to Analyse Trace Gases including Odorants with the Gas-chromatograph. Agric. Environ., 3, 139-145

Hartung, J. and Hilliger, H.G., 1978. Schwefelhaltige Spurengase in der Luft von Mastschweineställen. Fortschritte Vet. med. Heft 28 (12. Kongressbericht) 54-58.

Hilliger, H.G. and Hartung, J., 1977. Progress Report of the EG Project 220 for 1976 presented at the EG meeting in Gent, March, 1977.

Hilliger, H.G. and Hartung, J., 1978a. Geruhsbewertung im Stall durch Verknüpfung von sensorischen und analytischen Methoden. In Aurand, K., H. Hässelbarth u. Mitarb, (Hsg):
Organische Verunreinigungen in der Umwelt - Erkennen, Bewerten, Vermindern.
Erich Schmidt Verlag Berlin (West) S.475-482.

Hilliger, H.G. and Hartung, J., 1978b. Zur Chemie der Stalluft Wiener tierärtzl. Mosch. 65, 341-343.

Kaiser, R., 1973. Enriching volatile compounds by a temperature gradient tube. Analytical Chemistry 45, 965-967.

Logtenberg, M.Th. and Stork, B., 1976. Het ontwikkelen van meetmethoden voor het bepalen van de stank van ventilatielucht van mestvarken-stallen. Rapport de Central Technisch Instituut, Nr.01-4-40130, Zeist, Holland.

Miner, J.R., 1977. Characterization of odours and other volatile emissions. Agric. Environ., 3, 129-137.

DISCUSSION

A. Maton *(Belgium)*

Would you agree that collecting good samples of air from buildings still requires improvement?

J. Hartung *(Denmark)*

Yes, I agree. We are trying to improve the methods.

INVESTIGATIONS INTO THE TREATMENT OF ANIMAL EXCREMENTS BY AERATION TO REDUCE SMELL, AID DISINFECTION AND REDUCE VOLUME

R. Vetter and W. Rüprich
Institute for Agricultural Engineering Technology,
University of Hohenheim,
Stuttgart-Hohenheim, West Germany.

Industry offers various processes for the treatment of animal excreta, which are critically assessed by considering their technical and economic fields of application under varying conditions.

The systems examined are those which reduce the smell of the exhaust air and those working at thermophilic temperatures to achieve disinfection.

In Table 1 the most important design figures of four aerobic aeration systems for treating liquid manure are compiled. The farms applied these processes for several years. Some of the results are from investigations over several years, others from shorter periods.

FARM A: OXIDATION DITCH

The aeration of the slurry under a slatted floor in a piggery is done by two helix-aerators of 2.2. kW each, for each oxidation ditch (Figures 1, 2 and 3).

The installed specific aerator capacity is between 45 and 53 W m^{-3}. In the piggery for weaner pigs there is one 2.2 kW helix-aerator under a part-slatted floor (Figures 4 and 5).

Up to one-third of the ditch contents is drained into the settling tank (60 m^3) twice a week from the piggery for weaner pigs and once a week from the piggery for fattening pigs (Figure 6).

TABLE 1

AEROBIC TREATMENT OF LIQUID MANURE. DESIGN FIGURES

	Unit	Farm				
		A	B	C1	C2	D
1. Species		pig	pig	pig	pig	laying hens
2. Quantity	head	760	480	336	336	90 000
3. Quantity (livestock unit)*	GV	91.2	57.6	40.3	40.3	91.2
4. Type of the aeration basin		Oxidation ditch in the piggery	Storage tank	Storage tank	Mixing pit	Storage tank
5. Livestock units per aeration basin	GV $unit^{-1}$	30.4	57.6	40.3	40.3	45.6
6. Capacity per aeration basin	m^3 $unit^{-1}$	90	32 - 200	25 - 186	14	400
7. Aerator		Helix-	Surface- + flushing	Surface- + flushing	Submerged- + flushing	Rotary compressor + turbo-mixer
8. Installed energy	kW	2 x 2.2	2.2	2.2	1.1 + 0.35	2.2 + 11
9. Installed energy	W m^{-3}	49	68 - 11	88 - 12	79 + 25	33 (5.5 + 27.5)
10. Energy consumption	kWh m^{-3}	46.5	19.6	23.3	17.4	7.5

(Table 1 Cont.)

* 1 Livestock unit = 500 kg liveweight

TABLE 1 (Cont.)

	Unit	Farm				
		A	B	C1	C2	D
11. Objectives						
aa) Odour reduction of the exhaust air		+	+	+	+	-
ab) Odour reduction during spreading		+	+	+	+	+
b) Disinfection		-	-	-	+	+
12. Temperature	^{o}C	25 - 35	5 - 20	5 - 20	35 - 52	15 - 48
13. Capital requirement specific to the process	DM $place^{-1}$	38	20	46		0.23
14. Costs specific to the process	DM m^{-3}	10.50	4.72	7.22		1.72

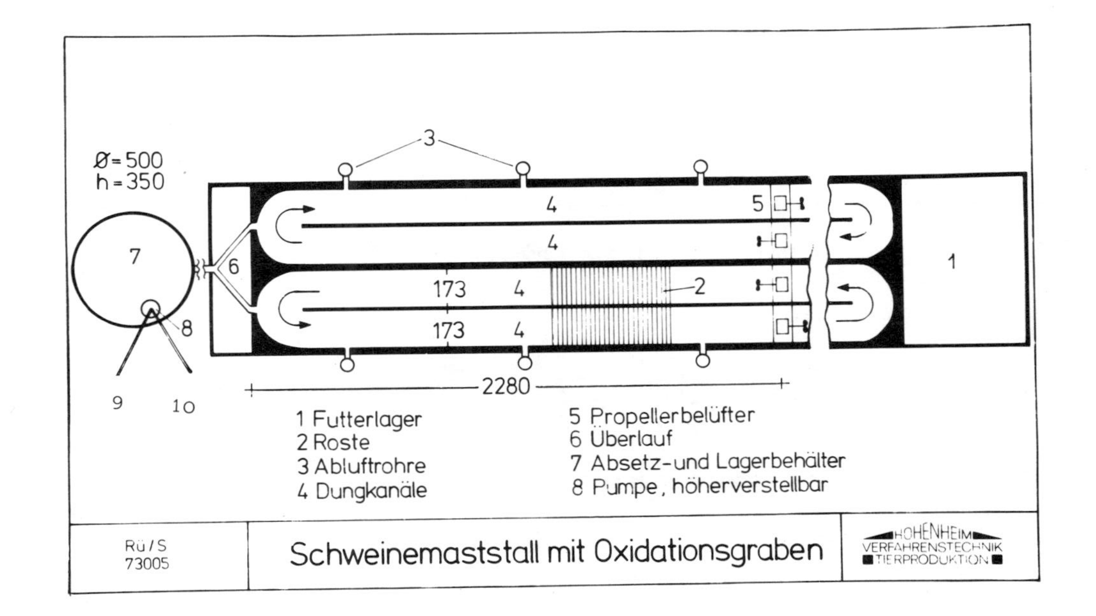

Fig. 1. Piggery with an oxidation-ditch under slatted floor

1. Feed store
2. Slatted floor
3. Air exhaust
4. Slurry channels
5. Helix-aerator
6. Overflow
7. Settling tank
8. Pump, vertically adjustable
9. Composting bed or storage tank
10. Tanker feeder

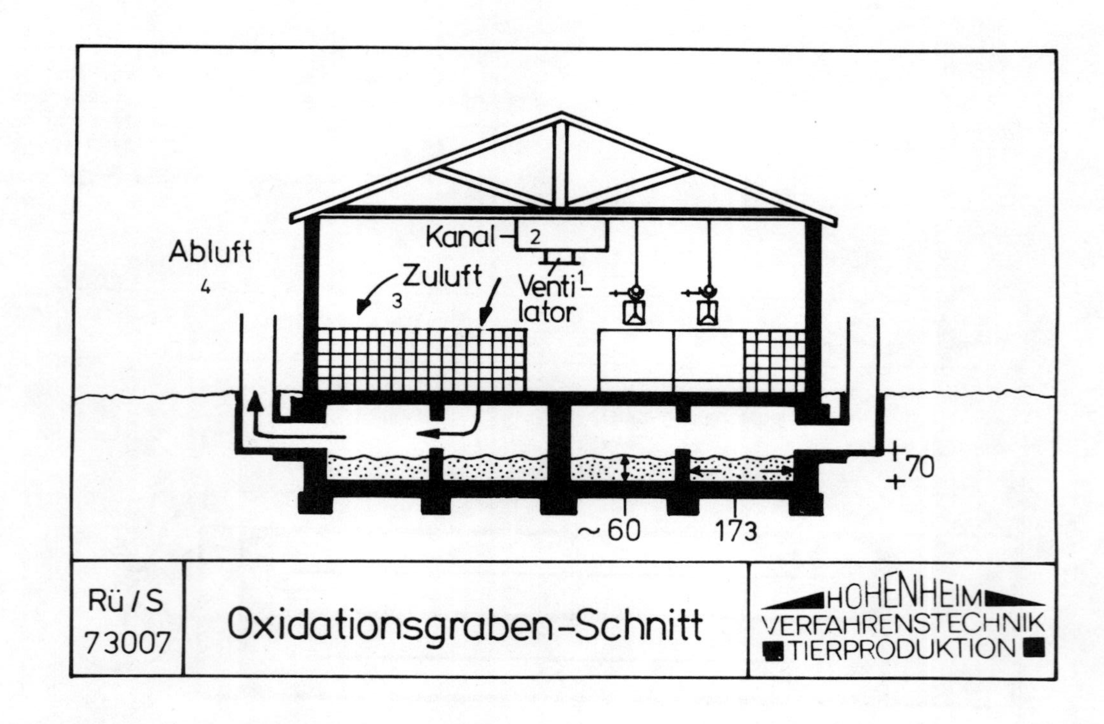

Fig. 2. Piggery with an oxidation ditch under slatted floor - cross section

1. Fan
2. Air flue
3. Fresh air
4. Exhaust air.

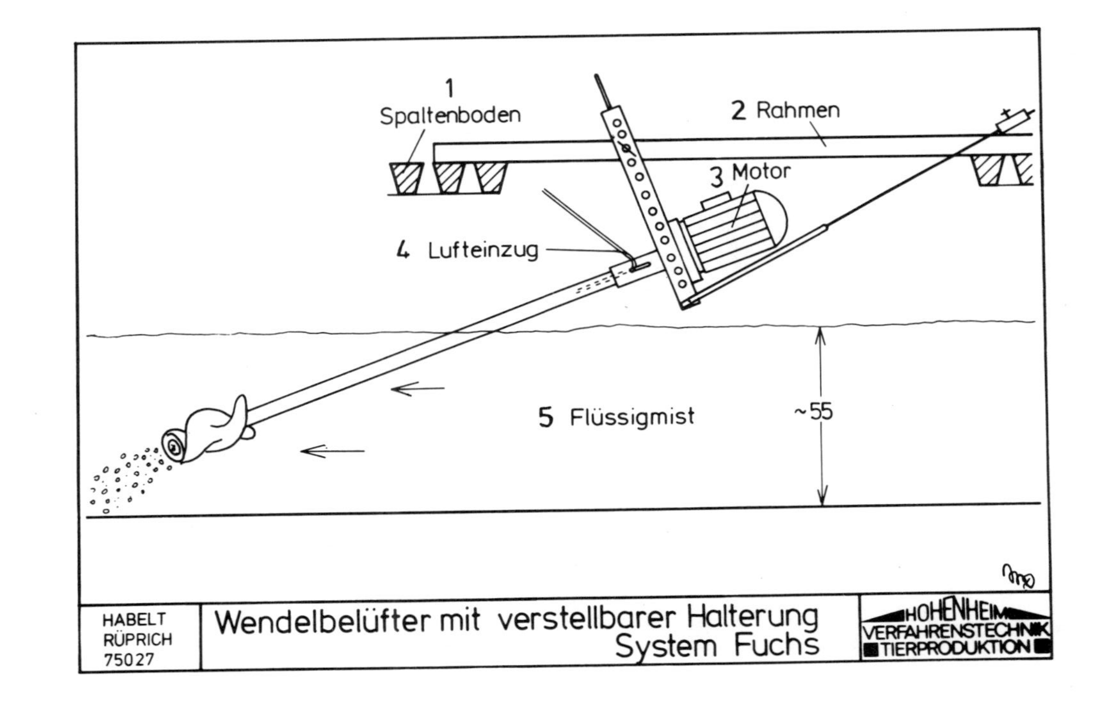

Fig. 3. Helix-aerator with adjustable fixing (System FUCHS)

1. Slatted floor
2. Frame
3. Motor
4. Air intake
5. Slurry

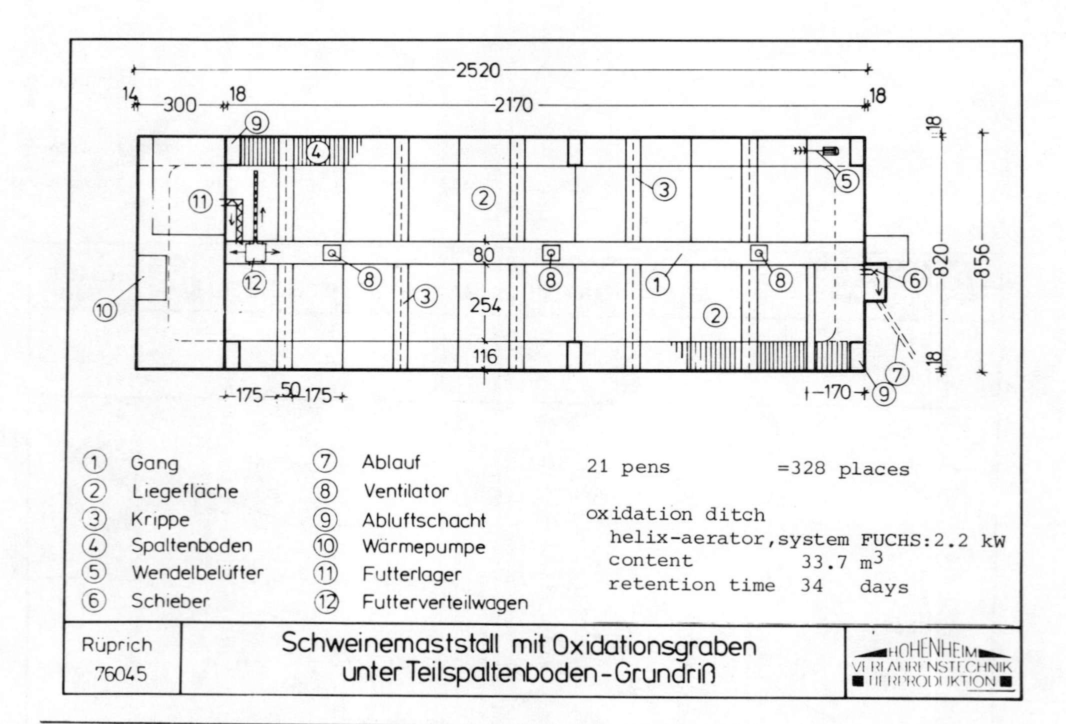

Fig. 4. Piggery with an oxidation ditch under partially slatted floor

1. Corridor
2. Bed
3. Crib
4. Slatted floor
5. Helix-aerator
6. Slide valve
7. Drain
8. Fan
9. Chimney
10. Heat pump
11. Feed store
12. Feed distributor

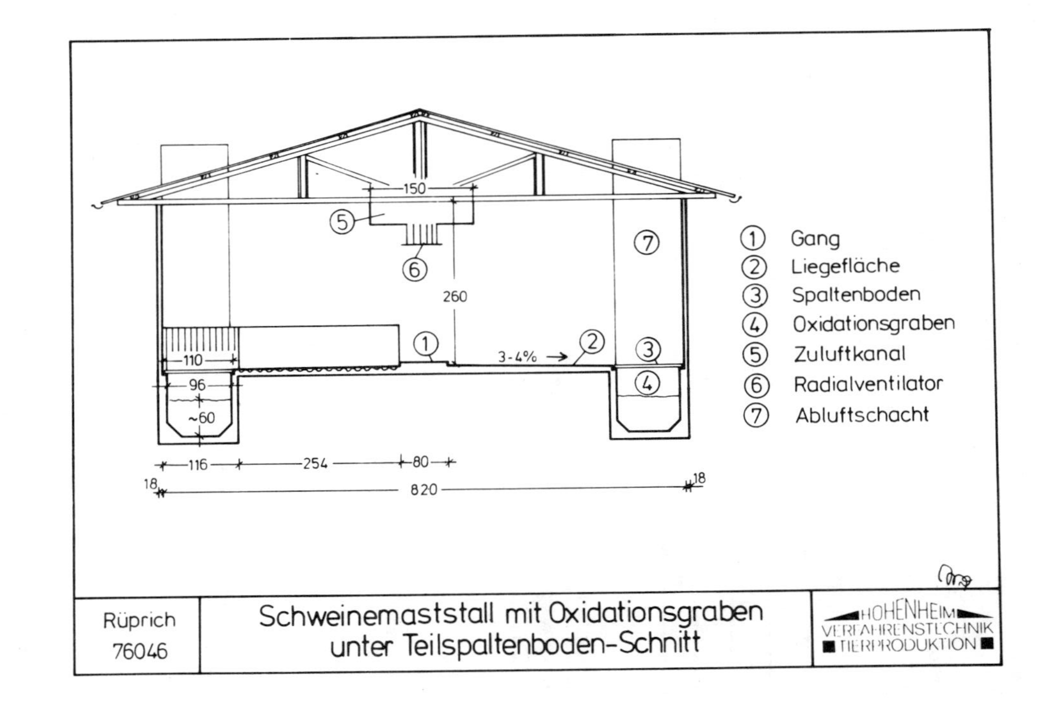

Fig. 5. Piggery with an oxidation ditch under partially slatted floor - cross section

1. Corridor
2. Bed
3. Slatted floor
4. Oxidation ditch
5. Air flue
6. Radial fan
7. Chimney

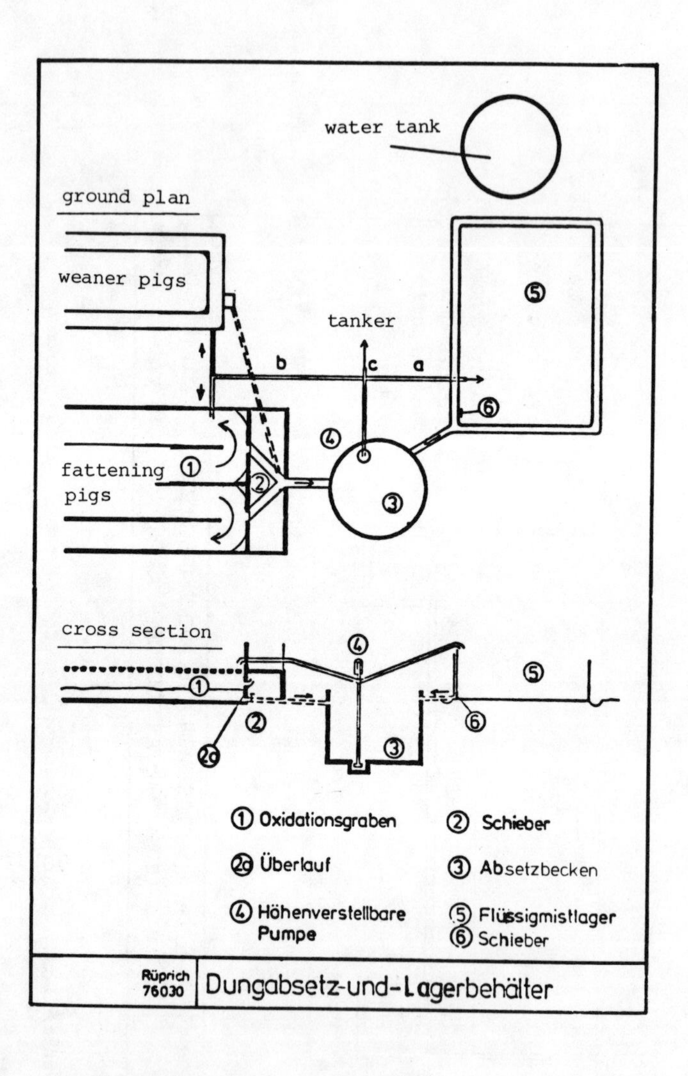

Fig. 6. Settling and storage tank

1. Oxidation ditch
2. Slide valve

2a. Overflow

3. Settling tank
4. Pump, vertically adjustable
5. Storage tank
6. Slide valve

After sedimentation, the relatively clear liquid is pumped back into the ditches. This is done by a pump which is adjustable in height. The solid fractions are removed from the settling tank and either spread directly or conserved in the storage tank before spreading. Only liquid manure which has not been stored longer than three weeks after draining out of the oxidation ditches, is spread on the fields in the neighbourhood of dwelling-houses.

As the volume in the oxidation ditch increased, the current velocity decreased and the energy consumption of the aerators increased. In the ditch the current velocity should not be less than 0.2 m s^{-1} in order to avoid sedimentation.

The reduction of total nitrogen reached 36 percent, the dry matter content decreased to 59 percent. Altogether there was a volume reduction of between 20 and 25 percent.

FARM B: AUTOMATIC-FLUSHING-SYSTEM

In 1974 the automatic-flushing-system was installed at farm B (Figures 7 and 8).

The liquid manure is treated by a surface aerator in the subterranean storage tank (Figure 9).

The electricity consumption is between 68 and 11 W m^{-3}, depending on the volume of the manure in the storage tank.

From the storage tank, aerated liquid is taken and used as flushing liquid. Several times per day flushing of the channels is carried out consecutively, controlled by a time switch. A dam at the end of the channels retains aerated liquid to a depth of about 8 cm under the slatted floor in the piggery. Because the excreta always fall into treated liquid, the smell substances are widely neutralised, and there is a reduction of the smell in the piggery and in the exhaust air.

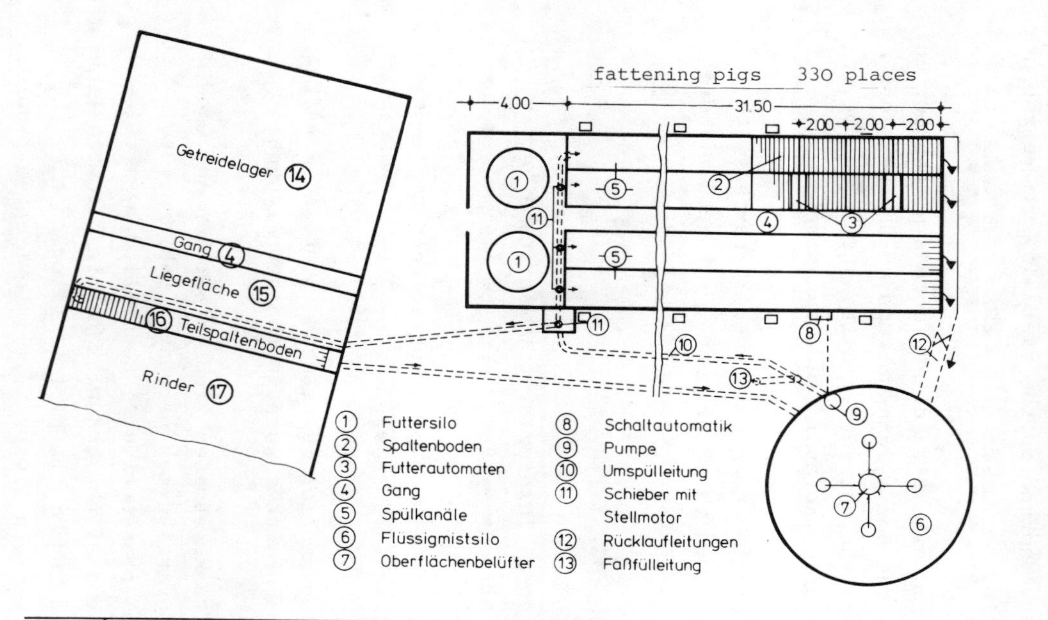

Rüprich 77042 | Mastschweinestallanlage mit automatischem Stau-Umspül-Belüftungssystem | HOHENHEIM VERFAHRENSTECHNIK TIERPRODUKTION

Fig. 7. Piggery with automatic-flushing-system

1. Feed silo
2. Slatted floor
3. Automatic feeder
4. Corridor
5. Flushing channels
6. Storage tank
7. Surface aerator
8. Control unit
9. Pump
10. Flushing pipe
11. Slide valve with control motor
12. Return pipe
13. Tanker feeder
14. Grain store
15. Bed
16. Partially slatted floor
17. Cattle

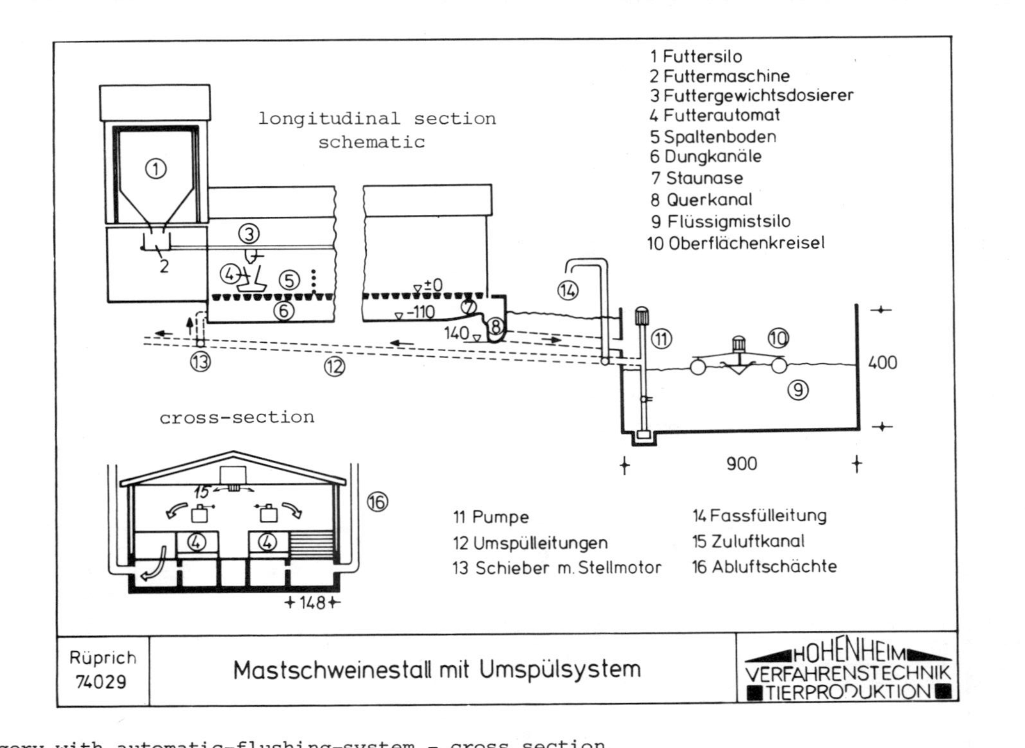

Fig. 8. Piggery with automatic-flushing-system - cross section

1. Feed silo
2. Feeding machine
3. Feeding metering unit (by weight)
4. Automatic feeder
5. Slatted floor
6. Slurry channels
7. Dam
8. Cross channel
9. Storage tank = aeration basin
10. Surface aerator
11. Pump
12. Flush pipes
13. Slide valve with control unit
14. Tanker feeder
15. Air flue
16. Chimney

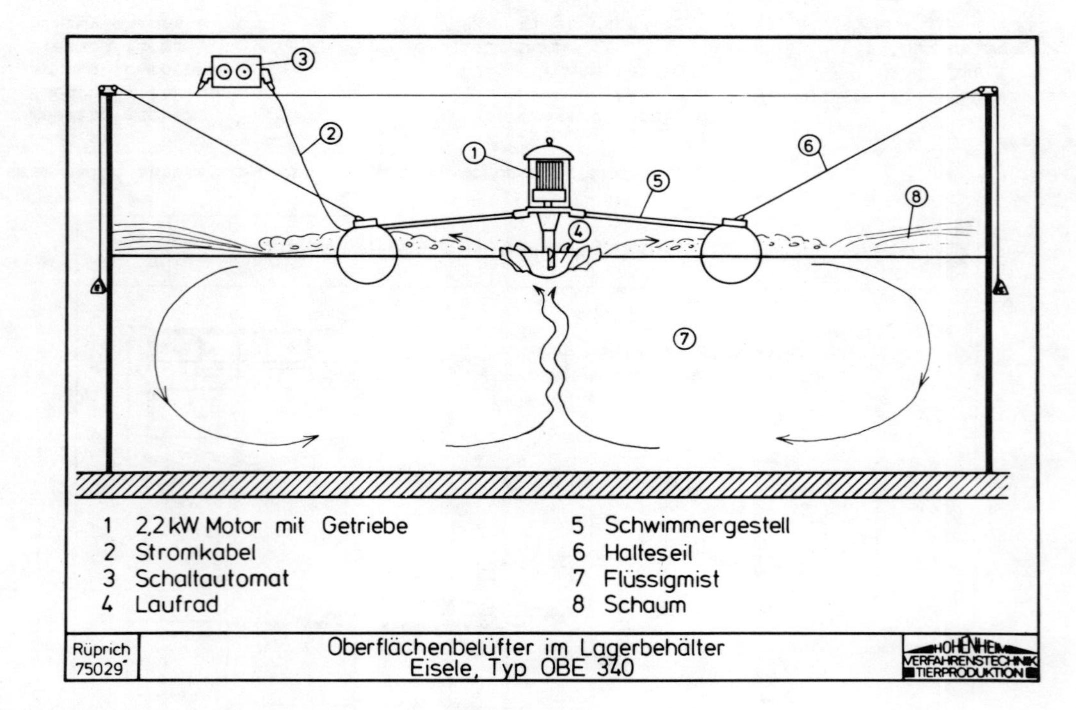

Fig. 9. Surface aerator in the storage tank (EISELE, type OBE 340)

1. 2.2. kW-motor with transmission
2. Power cable
3. Control unit
4. Paddle
5. Floats
6. Cable
7. Slurry
8. Foam

During flushing the slurry flows from the piggeries into the storage tank. Under certain conditions during flushing smell may occur near the storage tank.

During summer, at some days, all the liquid in the slurry produced can evaporate, with the result that it must be replaced by water.

According to the investigations carried out by the Institute of Veterinary Medicine of the Hohenheim University the automatic-flushing-system combined with a surface aerator does not produce disinfection. This could not be expected at temperatures between 5 and 20°C in the storage tank.

The costs specific to the process of the automatic-flushing-system at farm B with 480 places for fattening pigs are 10.23 DM place^{-1} year^{-1} or 47.20 DM t^{-1} produced.

FARM C: AUTOMATIC-FLUSHING-SYSTEM

Farm C runs the automatic-flushing-system in a piggery with part slatted floor (Figures 10 and 11).

An important difference compared to farm B, is the aeration of the liquid manure in a storage tank above ground, so that a mixing pit and a second pump are necessary. The flushing takes place from the storage tank.

The costs specific to the process are higher for the 336 fattening places. They amount to 15.83 DM place^{-1} year^{-1} or 72.20 DM t^{-1} produced.

After extensive experiments on the treatment of slurry at thermophilic temperatures in a pilot plant (Figure 12), a submerged aerator (System FUCHS) was installed in the uninsulated mixing pit on Farm C (Figure 13). At a volume of 14 m^{3} maximum, the installed energy capacity was 79 W m^{-3} for the aerator and 25 W m^{-3} for the foam cutter.

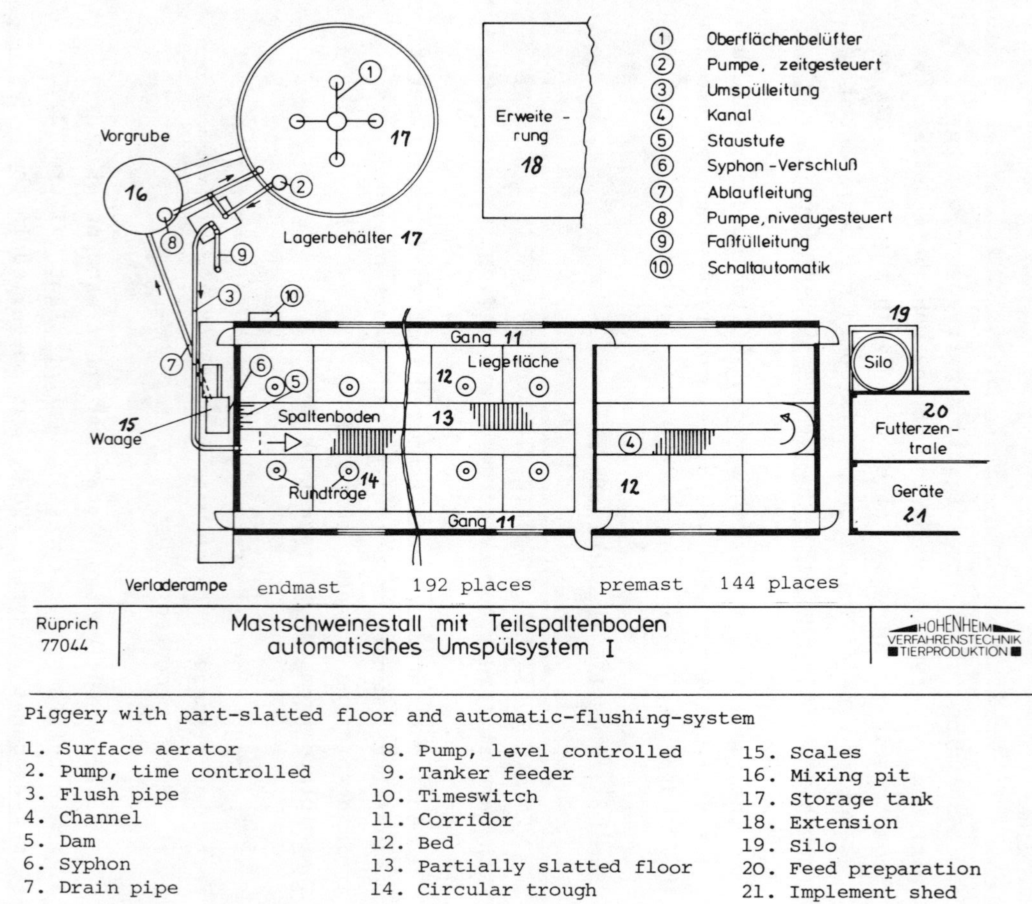

Fig. 10. Piggery with part-slatted floor and automatic-flushing-system

1. Surface aerator
2. Pump, time controlled
3. Flush pipe
4. Channel
5. Dam
6. Syphon
7. Drain pipe
8. Pump, level controlled
9. Tanker feeder
10. Timeswitch
11. Corridor
12. Bed
13. Partially slatted floor
14. Circular trough
15. Scales
16. Mixing pit
17. Storage tank
18. Extension
19. Silo
20. Feed preparation
21. Implement shed

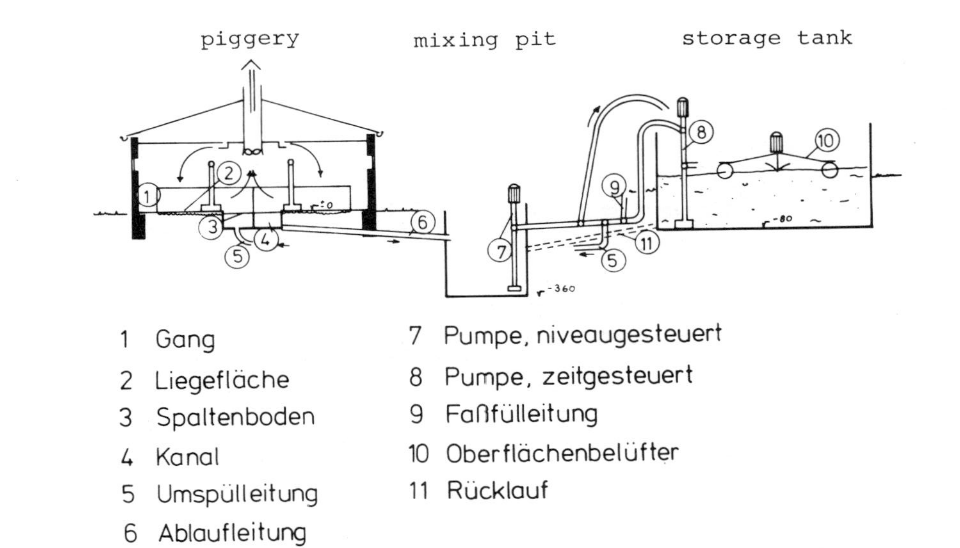

Fig. 11. Piggery with partially slatted floor and automatic-flushing-system - cross section

1. Piggery
2. Bed
3. Slatted floor
4. Channel
5. Flush pipe
6. Drain pipe
7. Pump, level controlled
8. Pump, time controlled
9. Tanker feeder
10. Surface aerator
11. Return pipe

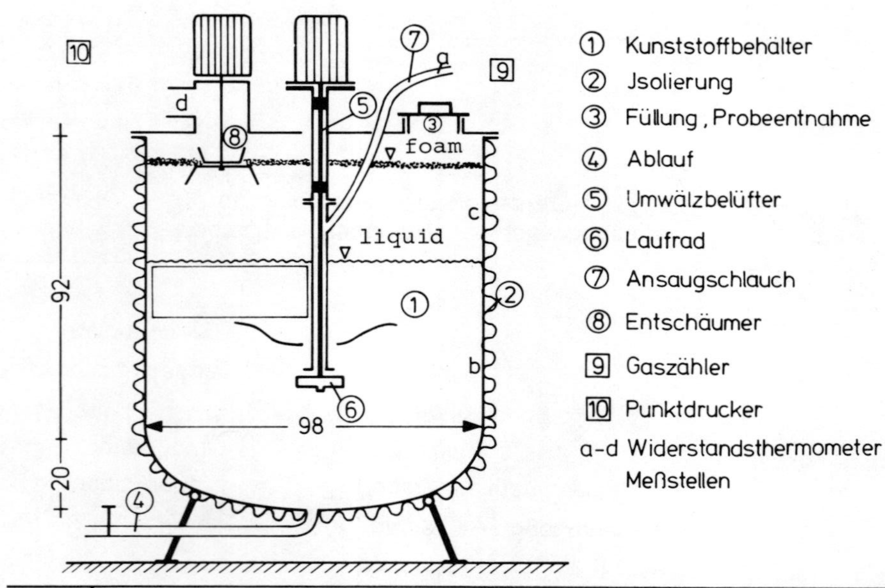

Rü/Sr
73037

Flüssigkompostierung-System Fuchs
Halbtechnische Versuchsanlage

Fig. 12. Pilot plant for the treatment of slurry at thermophilic temperatures (System FUCHS)

1. Fibre glass tank
2. Insulation
3. Filling, sampling
4. Outlet pipe
5. Submerged aerator
6. Impeller
7. Intake pipe
8. Foam cutter
9. Gas meter
10. Printer

a-d resistance thermometer points

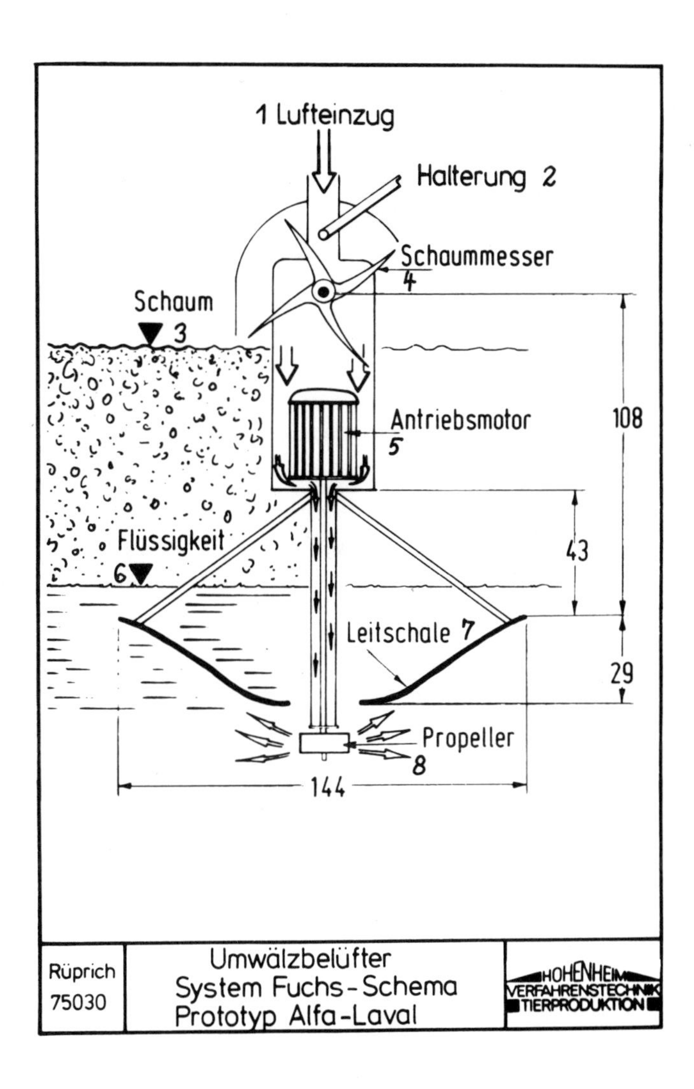

Fig. 13. Submerged aerator (System FUCHS)

1. Air intake
2. Suspension
3. Foam
4. Foam cutter
5. Motor
6. Liquid
7. Eddy guide
8. Impeller

The temperatures reached in the pit were between 35 and 52°C depending on the quantity of slurry and rate of flushing.

The experiments made by the Institute of Veterinary Medicine at the Hohenheim University showed a destruction of pathogenic organisms such as salmonella.

During flushing, carried out from the pit, the temperature inside the piggery increased only slightly, as did the concentrations of ammonia.

The energy consumption of the submerged aerator and the foam cutter amounted to an average of 29 kWh day^{-1}. Compared with the surface aerator the energy consumption was reduced to about 69 percent. The costs specific to the process cannot be calculated exactly because this aerator is not offered commercially.

The results of measurements of odour on these three farms were less than the minimum distances from dwelling-houses with the maximum of 100 points as required in the 'VDI-Richtlinie Nr. 3471' (Figure 14). The systems are therefore able to reduce odour to a satisfactory level.

FARM D: TURBO-MIXER AND ROTARY COMPRESSOR

The test farm keeps 90 000 laying hens in tiered cages in four houses. The droppings are pushed by means of rams to the ends of the hen houses, whence they are carried by worm conveyors into a collecting pit. The storage tanks which are installed at a higher level are filled by a pump (Figures 15 and 16).

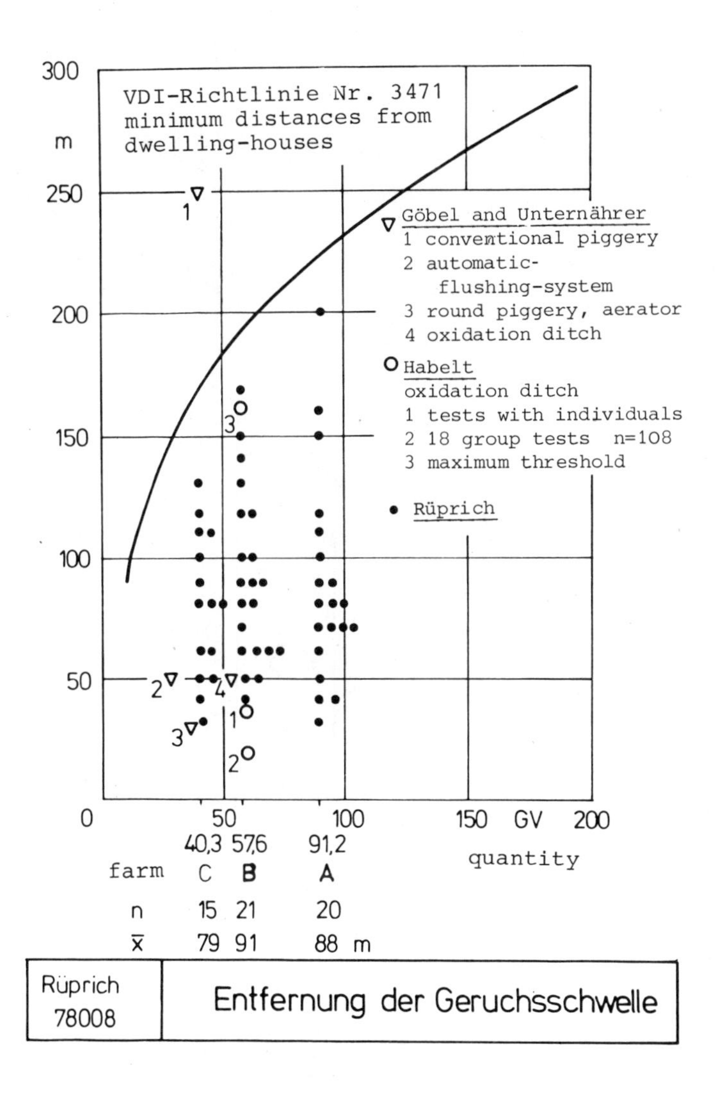

Fig. 14. Threshold of the odour

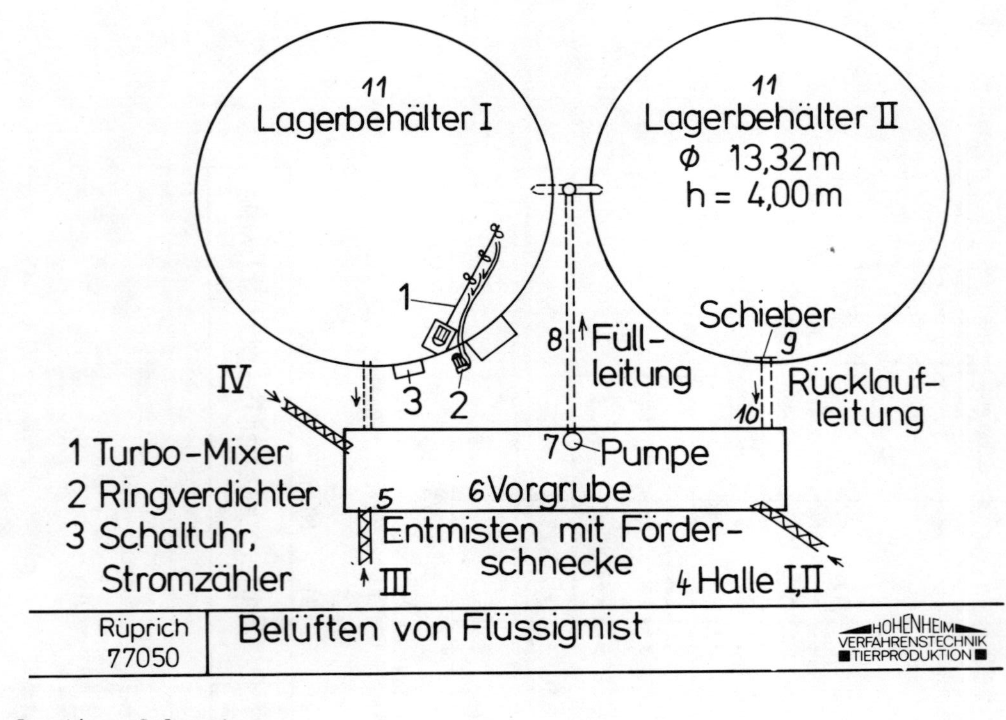

Fig. 15. Aeration of droppings

1. Turbo-mixer
2. Rotary compressor
3. Timeswitch, electric meter
4. Henhouses
5. Manure removal with auger
6. Mixing pit
7. Pump
8. Feed pipe
9. Slide valve
10. Return pipe
11. Storage tank

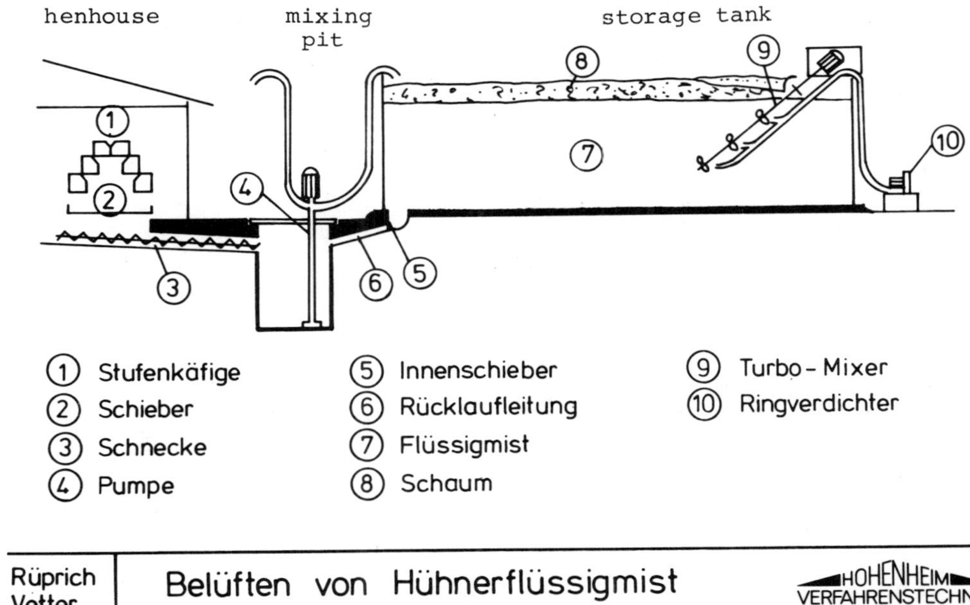

Fig. 16. Aeration of droppings - functional layout

1. Tiered cages
2. Scraper
3. Auger
4. Pump
5. Slide valve
6. Return pipe
7. Liquid manure
8. Foam
9. Turbo-mixer
10. Rotary compressor

The aeration system comprises a combination of a turbo-mixer and a rotary compressor. The 3-paddle turbo-mixer is powered by an 11 kW motor, in addition there is the 2.2 kW of the rotary compressor giving an installed capacity of 33 W m^{-3}. The volume of the air supplied was between 180 and 230 m^3 $hour^{-1}$. As a result of supplying oxygen, bacterial activity brings about heating of the liquid manure to reach thermophilic temperatures. The rise of temperature is well correlated with the daily aeration time and the supply of air (Figure 17). Measurements showed that the tank contents were well mixed.

The agitated slurry was distributed within two or three days. When the liquid manure was being spread, the odour threshold was at 100 to 300 metres distance from the field; when unaerated manure was spread, the odour limits were found to be at 1 600 to 2 200 metres.

Calculations of capital requirements and of costs specific to the process are based on data gathered on the test farm. The calculations are based on an aeration period of a fortnight with 20.6 hours aeration per day. The specific power consumption was 7.5 kWh m^{-3} of liquid manure. Based on a price of 0.10 DM kWh^{-1} for electricity and 10.00 DM man^{-1} $hour^{-1}$ for labour, the total costs come to 12 575 DM $year^{-1}$ or 1.72 DM m^{-3} of liquid manure.

CONCLUSIONS

Both oxidation ditch and automatic-flushing-system reduce odour from the stables and from the exhaust air.

The highest costs among the investigated systems were for the oxidation ditch. Here the temperatures were between 25 and 35°C and recovery of heat should be possible.

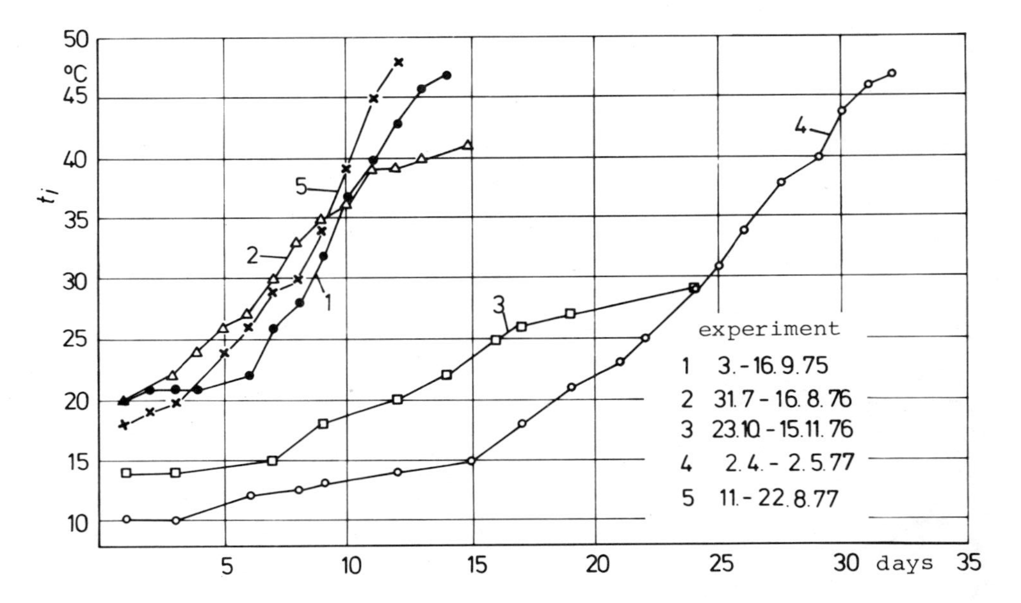

Fig. 17. Temperature changes during the aeration of droppings with a turbo-mixer and rotary compressor.

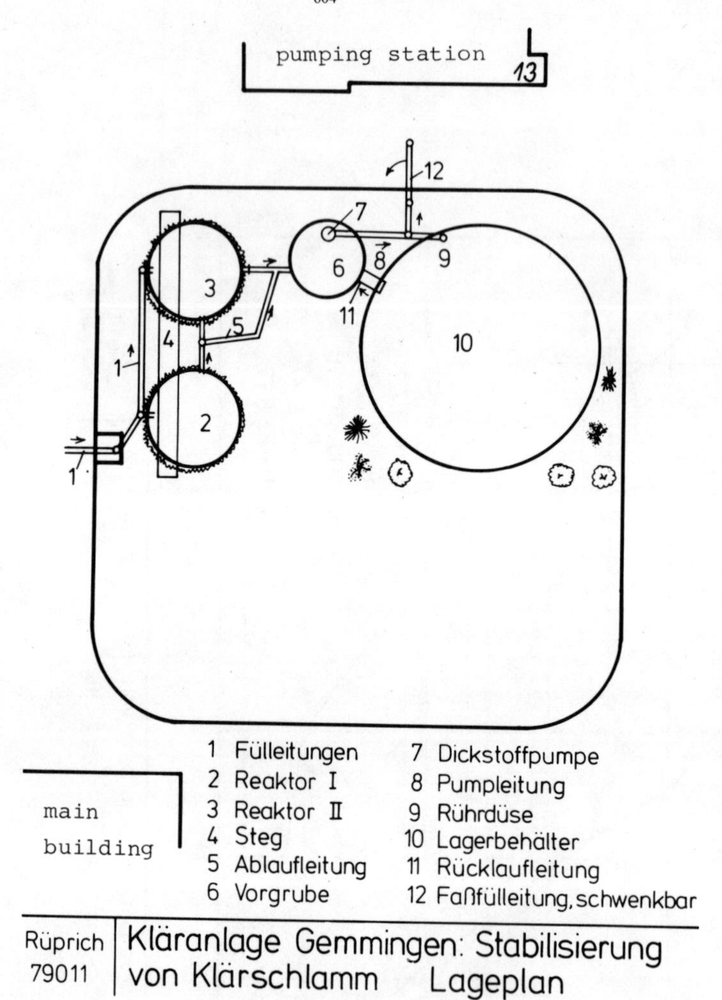

Fig. 18. Sewage plant: aerobic-thermophilic stabilisation of human sludge

1. Feeding pipe
2. Reactor I
3. Reactor II
4. Bridge
5. Drain pipe
6. Mixing pit
7. Sludge pump
8. Pump line
9. Stirring nozzle
10. Storage tank
11. Return pipe
12. Pivot tanker feeder

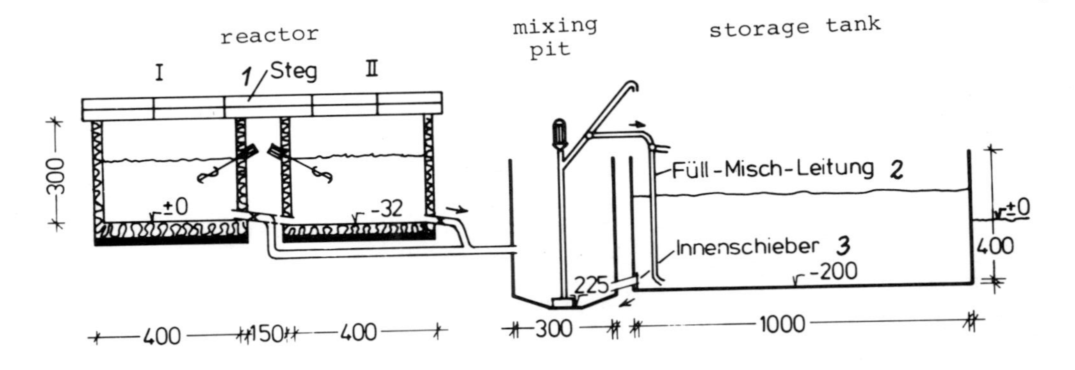

Schematischer Schnitt

Rüprich
79010

Kläranlage Gemmingen: Aerob - thermophile
Stabilisierung von Klärschlamm

Fig. 19. Sewage plant: aerobic-thermophilic stabilisation of human sludge

1. Bridge
2. Filling and mixing pipe
3. Slide valve

By the aeration of liquid manure with a submerged aerator, temperatures up to 52°C have been reached. Such temperatures guarantee the destruction of pathogenic organisms. Parallel hygienic investigations at farm C were carried out by the Institute of Veterinary Medicine of the Hohenheim University.

The aeration of liquid manure from laying hens needs a high input of energy. At the beginning of the aeration, the dry matter content was about 20 percent.

Based on the positive results obtained by the treatment of liquid manure in the thermophilic temperatures range a plant was developed for an aerobic-thermophilic stabilisation of human sludge (Figures 18 and 19).

This plant is under construction, so that after 1980 investigations on processes and hygienic effects can be undertaken.

REFERENCES

Habelt, J., 1977. Verfahrenstechnische Untersuchung des Oxidationsgrabens im Mastschweinestall unter Ganzspaltenboden. Dissertation Hohenheim.

Rüprich, W., 1973. Application of Agitator/Aerator Unit in the processing of liquid manure. (Einsatz des Umwälzbelüfters für die Flüssigmistaufbereitung). Scientific Information Section, National Institute of Agricultural Engineering, Wrest Park, Silsoe, Bedford. Translation No. 343.

Rüprich, W., 1974. Der Mastschweinestall mit Oxidationsgraben Bauen auf dem Lande, 25. Jg. H.5, S.142-148.

Rüprich, W., 1974. Aufbereitung von tierischen Exkrementen mit Umwälzbelüfter und Oberflächenkreisel. In: Landtechnik 29 H.3, S.62-67.

Rüprich, W., 1975. Mastschweinestall mit Stau-Umspülsystem Landtechnik, Heft 11, November 1975. S.484-487.

Rüprich, W. and Strauch, D., 1976. Verfahrenstechnische und hygienische Probleme bei der Aufbereitung von Flüssigmist. Zuchtungskunde, Bd. 48, H.3, S.242-249.

DISCUSSION

A. Heduit *(France)*

Submerged aerators should provide good treatment with a low energy cost because of the increased temperature and microbiological activity compared to those obtained with surface aerators. Surface aerators are known to be less efficient in clean water in terms of kg O_2 kWh^{-1}. An investigation could be carried out in order to select better submersible devices for aerating slurry.

R. Vetter *(West Germany)*

Several types of aerators were tested and the best one was used with an efficiency of about 0.6 to 0.8 kg O_2 kWh^{-1}.

J.C. Hawkins *(UK)*

The efficiency of oxygen transfer does not use all the energy, much is used for mixing the slurry.

R. Vetter

I agree, this system used about 90 percent of the energy for oxygen transfer; the rest was used for mixing.

MEASUREMENT OF ODOUR EMISSIONS AND IMMISSIONS

H.H. Kowalewsky, R. Scheu and H. Vetter
Landwirtschaftliche Untersuchungs- und Forschungsanstalt,
Mars-la-Tour-Str. 4, 2900 Oldenburg,
West Germany.

ABSTRACT

Some odour components and the sensory detectable odour levels were determined close to animal houses, slurry tanks and fields with slurry applied, to try to answer the question, whether the determination of main components allows a reasonable estimation of the odour level. The results can be summarised as follows:

- *determination of the small concentrations of odour components requires enrichment during sampling,*
- *to absorb the odiferous substances different solutions are needed such as NaOH, H_2SO_4 and $CdSO_4$,*
- *photometric and gas chromatographic methods are suitable for chemical analyses,*
- *the sensory odour classification of odour levels differed little between individuals,*
- *NH_3 is suitable for use as a main component because its concentration correlated well with the sensory odour levels,*
- *the propionic acid concentration was potentially suitable for odour characterisation,*
- *the suitability of other components must be examined,*
- *the determination of the main components at different distances from animal houses, slurry tanks and fields with slurry applied showed different rates of odour dilution in the immediate area of these sources of odour emission.*

These results suggest that the determination of odour components may be a practicable way for an objective measurement of odour emission and immission.

INTRODUCTION

Pig and poultry farms have been criticised for the strength of their odour emissions. At present no objective method for the measurement of odour intensity is available. Therefore, new methods of decreasing the odour emission and immission cannot be objectively assessed.

Starting with the idea that odour is caused by different substances, the detection of odorous compounds and the measurement of their concentration seemed a useful approach paying particular attention to the areas of emission and immission. The aim of our work was to find substances whose concentrations correlated with the sensory evaluation of odour intensity. After identification of these compounds we wanted to discover their origin and mode of spreading.

In the last two years thirteen compounds were chosen from the large numbers present in odorous emissions which appeared suitable as main components of smell. A main component is defined as a compound whose concentration correlates with the sensory evaluation of odour intensity. After preliminary screening, only eight of these thirteen compounds were found to be suitable. These were ammonia, acetic acid, propionic acid, butyric acid, phenol, p-cresol and hydrogen sulphide and, with some restrictions, the content of the inflammable gases. In the following paper, we summarise the methods of air sampling used, the chemical analyses and our first results.

MATERIAL AND METHODS

About two thousand air samples have been taken at different distances from animal houses, open slurry tanks and fields treated with slurry. Initially some different methods for sampling air were tested, such as:

the use of special glass containers for transporting the air-samples;

the value of solid substances such as activated charcoal and Tenax for absorbing gases;

the absorption of odour components in the different liquids; and

the direct determination of offensive odours with a portable detector.

Each method had its advantages and disadvantages. After extensive tests, absorption in liquids and determination by a portable detector were chosen. However, this detector can only be used for measuring high concentrations of inflammable gases, such as occur in the areas of emission.

Air-sampling

For absorbing the odour components a gas train was used, comprising a standard wash bottle, a gasmeter and a pump. Figure 1 shows such an arrangement. The exhaust air was sucked through the wash bottle and gasmeter by the suction pump. In the wash bottle the exhaust air diffuses throughout the solution, so that there was good absorption of the odour components. Depending on the absorbing solution and the individual odour component, absorption varied between 94 and 97 percent.

As well as a high rate of absorption, this method has the following advantages:

- large volumes of exhaust air are quickly sucked through the absorbing solutions
- different substances can be absorbed by different solutions
- when absorbed in solution the odour components are stable

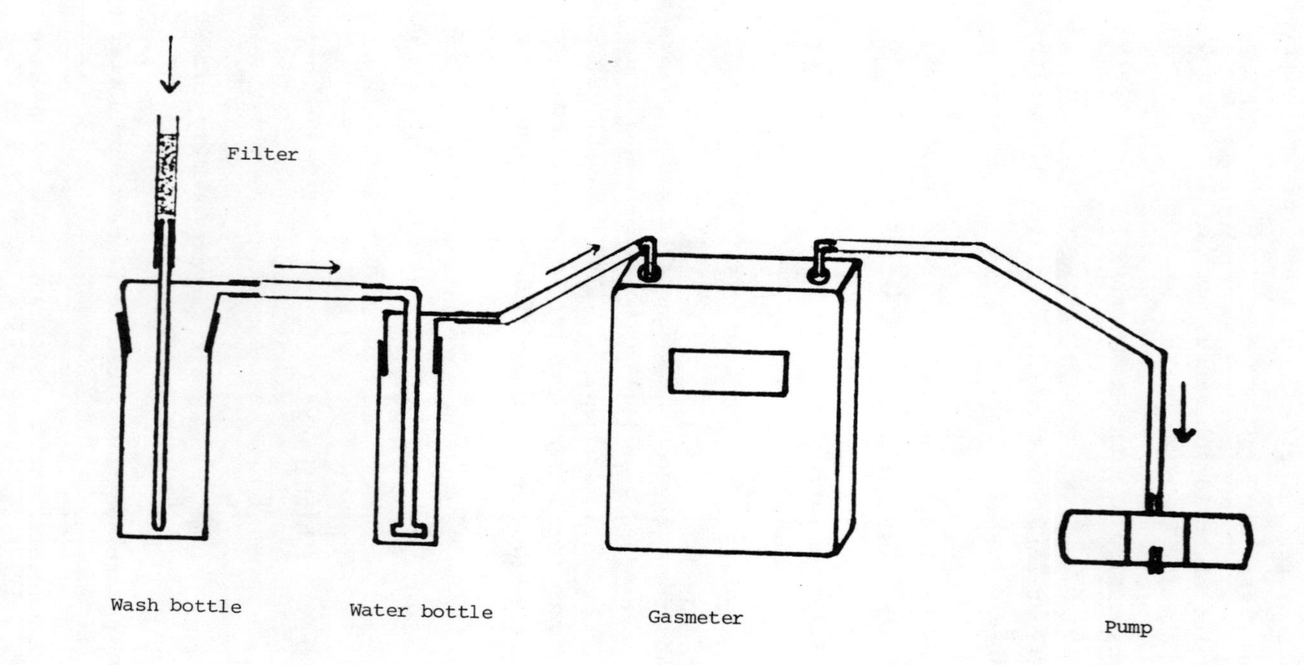

Fig. 1. Equipment for air sampling.

- the moisture content of the air does not influence the rate of absorption
- variations in the concentrations of the main components and changes in wind direction do not influence the results so much as with a short time of air sampling.

Table 1 gives a summary of the absorbed substances, the solutions required, the time for air sampling and the volume of exhaust air.

TABLE 1

SUMMARY OF THE ESSENTIAL CONDITIONS OF AIR-SAMPLING

Odour component	Absorbing solution	Sampling time	Volume of air required
Ammonia	100 ml 0.1 N H_2SO_4	30 min	1 m^3
Acetic acid Propionic acid Butyric acid Phenol p - cresol	100 ml 0.1 N NaOH	45 min	1.5 m^3
Hydrogen sulphide	50 ml 0.1 N NaOH + 50 ml $CdSO_4$ sol	45 min	1.5 m^3

Figure 2 shows how the eight air samplers were placed at different distances on the lee side of the sources of emission. Changes in wind direction required the re-establishment of the air samplers; on one occasion, it was necessary seven times.

During air sampling the following parameters were measured - the velocity and direction of the wind, the moisture and the temperature of the air and exhaust air, the height and the breadth of the sources of emission, the number of animals in the houses and the level of slurry in the tanks.

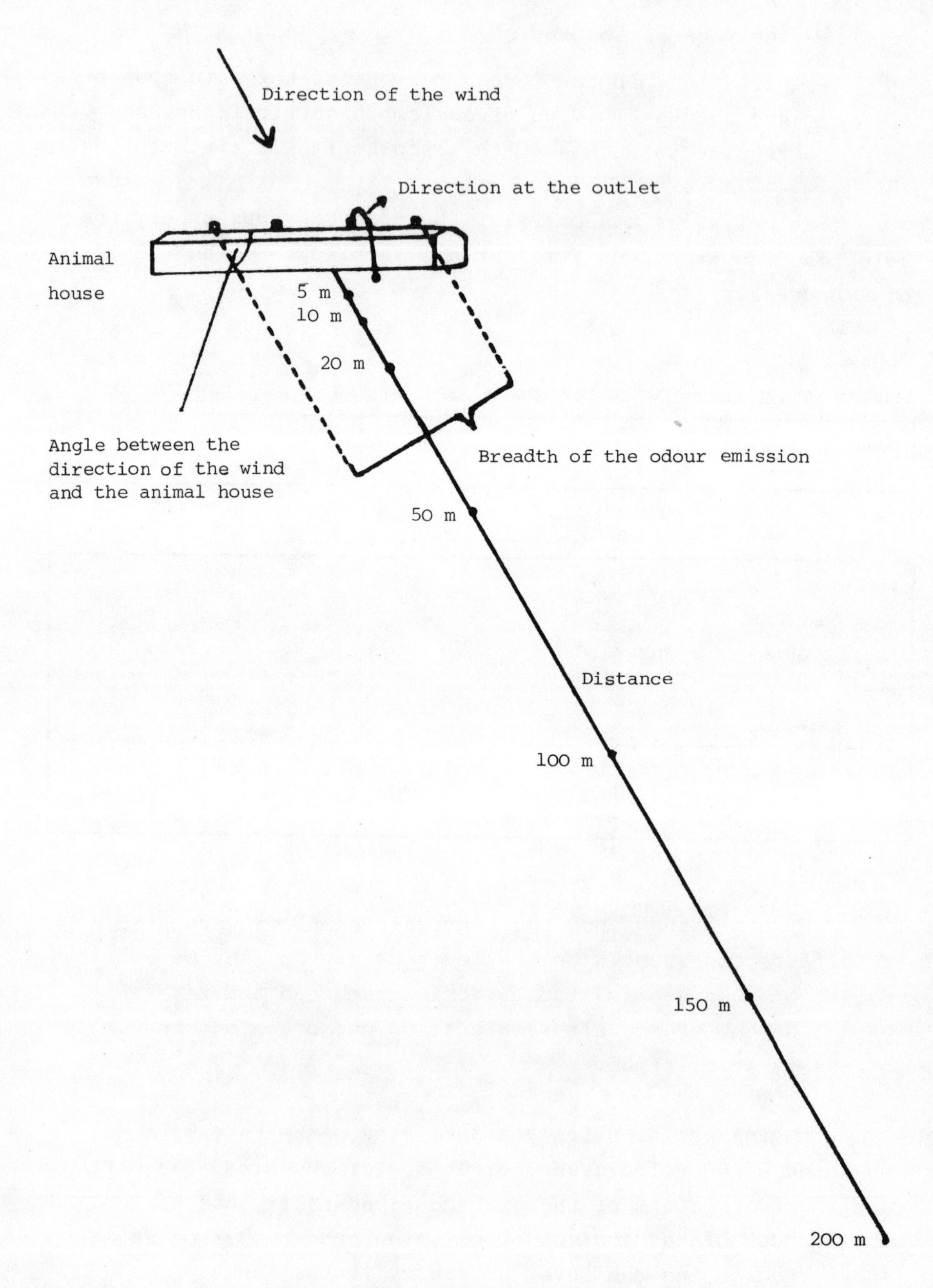

Fig. 2. Location of the air samplers from the source of emission.

After sampling the air, the solutions were placed in plastic bottles for transport to the laboratory where storage was necessary, because more time was needed for chemical analyses than for sampling the air.

Figure 2 also shows how the breadth of the odour flag was measured, by calculating the distance between the exterior ventilators and the angle between the direction of the wind and the length of the animal house.

During air sampling the odour level was assessed by different persons. The level was placed in one of four classes:

Odour level	
I	no odour detectable
II	faint odour detectable
III	distinct odour detectable
IV	strong odour detectable

A finer classification did not seem to be useful because the human nose can only detect three or four odour levels. For an exact classification, members of the panel have to stay at each distance for a longer time. Farmers, chemists, clerks, pensioners and students were members of the smell panel.

Later we recognised that sometimes a definite classification into one of the levels could not be made. In these cases, when such indefinite classifications occurred, intermediate odour levels were used.

Chemical analyses

Some of the methods for the determination of different odour components were taken from VDI-directions and some we developed. Table 2 summarises the methods used.

TABLE 2

CHEMICAL ANALYSES FOR DIFFERENT ODOUR COMPONENTS

Odour component	Method	Minimum detectable amounts
Ammonia	Photometrically, following VDI-direction 3496	0.1 $\mu g\ m^{-3}$
Acetic acid Propionic acid Butyric acid	Gaschromatographically, method developed	1 $\mu g\ m^{-3}$
Phenol p-cresol	Gaschromatographically, method developed	10 $ng\ m^{-3}$
Hydrogen sulphide	Photometrically, following VDI-direction 2454	0.1 $\mu g\ m^{-3}$
Inflammable gases	Portable detector	1 ppm

The photometric methods for the determination of ammonia and hydrogen sulphide were improved to detect smaller concentrations, also the determination of hydrogen sulphide had to be simplified considerably. We are trying to determine H_2S gas chromatographically with greater precision.

For the gas chromatographic methods, three different kinds of detectors were needed. The organic acids were determined with a flame ionisation detector; an electron capture detector was used to determine phenol and p-cresol; and a flame photometric detector was used for the determination of hydrogen sulphide.

The development of the air sampling methods and the chemical analyses took much time, so that less analyses were done than had been planned, but the exact determination of the odour components seemed to be the most important work.

RESULTS

Correlation between the concentrations of different main components and the sensory determined odour-level

The sensory determination of the odour level during air sampling allows the calculation of the correlation with the concentration of different main components, except for the concentration of the inflammable gases when the detector was only used for measurements directly at the source of emission. We report, therefore, the correlations between odour intensity and the concentrations of ammonia, acetic acid, propionic acid and butyric acid.

NH_3-concentration and odour intensity

The NH_3-concentration in more than 1 000 air samples was determined. Resulting from improvements in air sampling and chemical analyses the sensitivity of this method was increased considerably, the errors were decreased and the correlation with the odour level was improved. Figure 3 shows the correlation of the NH_3-contents with the sensory determined odour level. Note that the NH_3-concentration is represented on a logarithmic scale.

The curvilinear regression clearly shows how the ammonia concentration decreases markedly with decreasing odour level. The differences between the odour levels were significant.

The correlation coefficient of $r = 0.85$ shows that the individual results lie close to the calculated curvilinear regression line and it is possible to assess the odour level from the ammonia concentration. Only four NH_3 determinations are necessary to determine the odour level with a confidence of 90 percent.

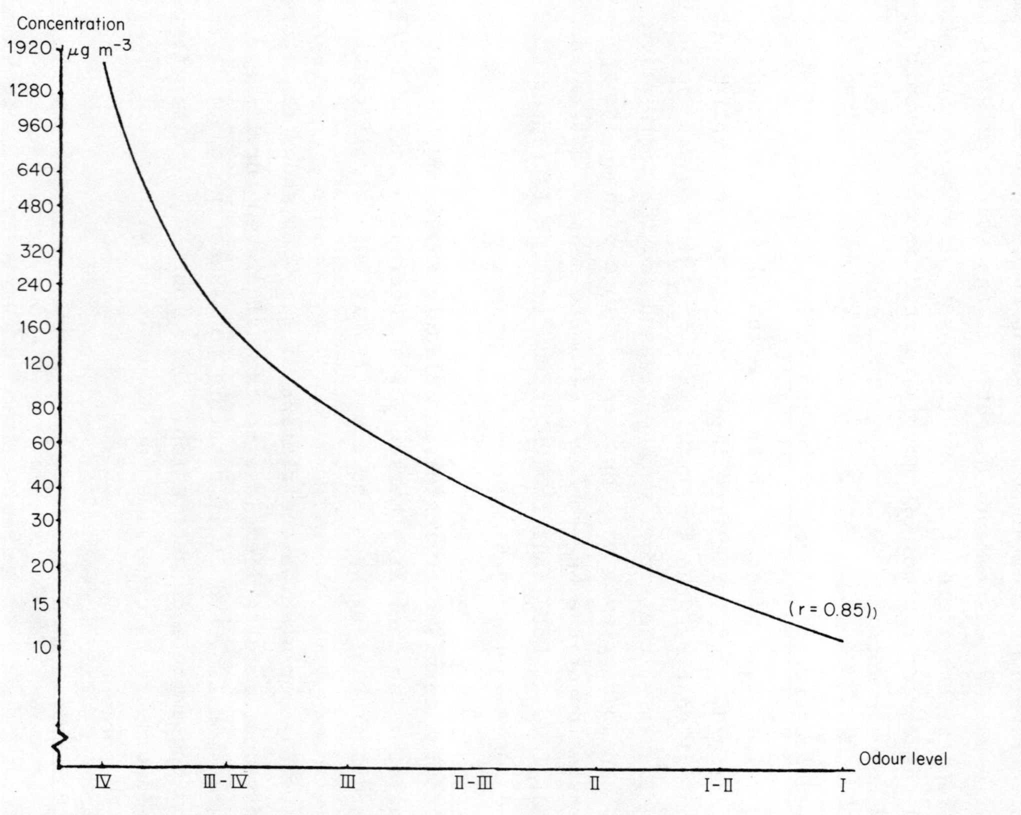

Fig. 3. Correlation between the NH_3-concentration and the sensory determined odour level.

Organic acid concentration and odour-units

The organic acids were only determined during the last two months. This correlation was calculated from about 210 results for every acid and Figure 4 shows the correlations.

The concentration of all three organic acids decreases with the lessening of the odour level. Acetic acid concentration decreases much, but the correlation coefficient, r = 0.50, is low. Acetic acid cannot therefore be used as a main component. The same is true for butyric acid.

Changes in the concentrations of propionic acid were highly correlated with odour measurement, the correlation coefficient had a value, r = 0.80. However, propionic acid has limited value for odour characterisation because the minimum detectable concentration is too large. Better correlations may be obtained if the method of determination is improved. Such a method is being tested at present.

Concentration of the main components at different distances from animal houses, open slurry tanks and fields treated with slurry

Figure 2 shows the locations of the air samplers at different distances from the sources of emission. There were special requirements for the determination of the odour level from slurry tanks, because of the need for a sufficient distance between animal houses and open slurry tanks, so that the wind did not mix the odours from the two sources.

NH_3-concentration

Figure 5 shows the NH_3-concentrations on a logarithmic scale determined at different distances from animal houses, open slurry tanks and fields treated with slurry. With increasing distance the NH_3-concentration from the various sources decrease at different rates. The largest concentration was found at the point of emission from the animal houses.

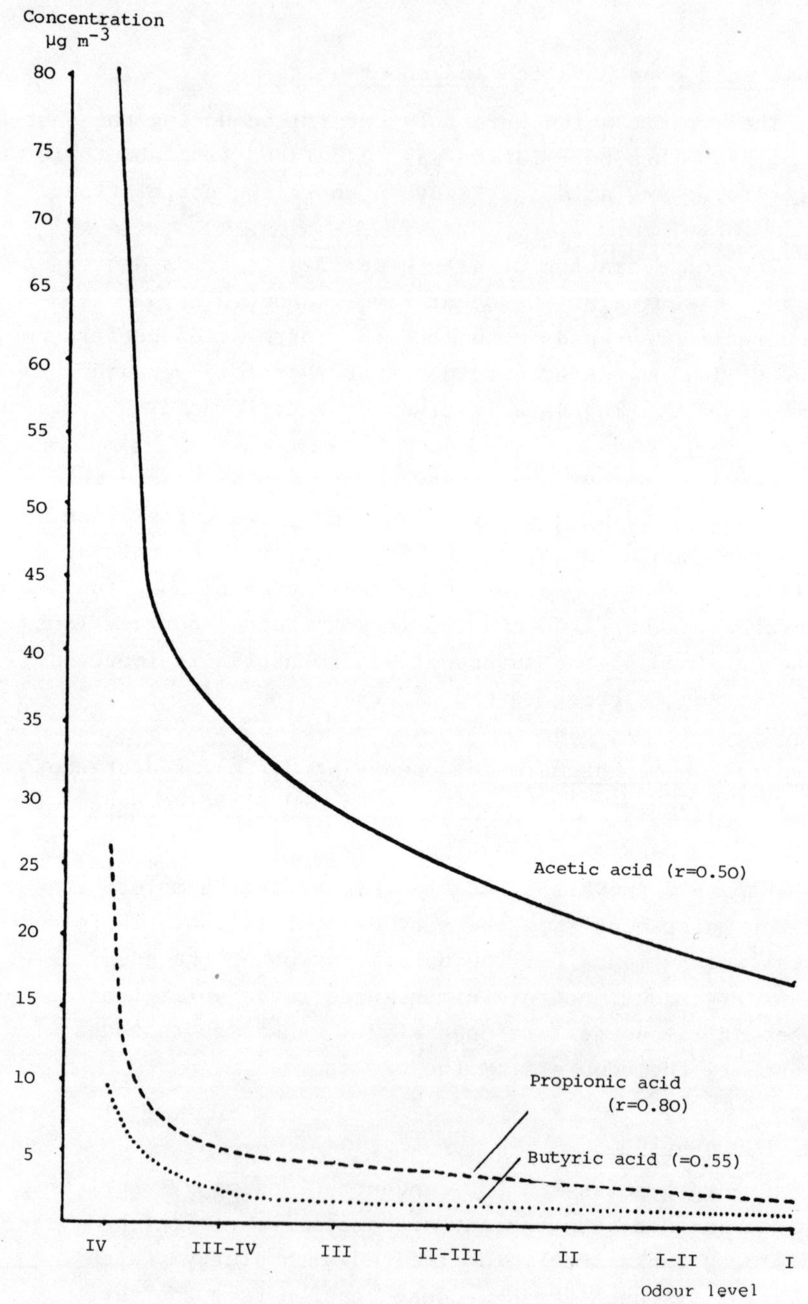

Fig. 4. Correlation between the acetic or propionic or butyric acid concentration and the sensory determined odour level

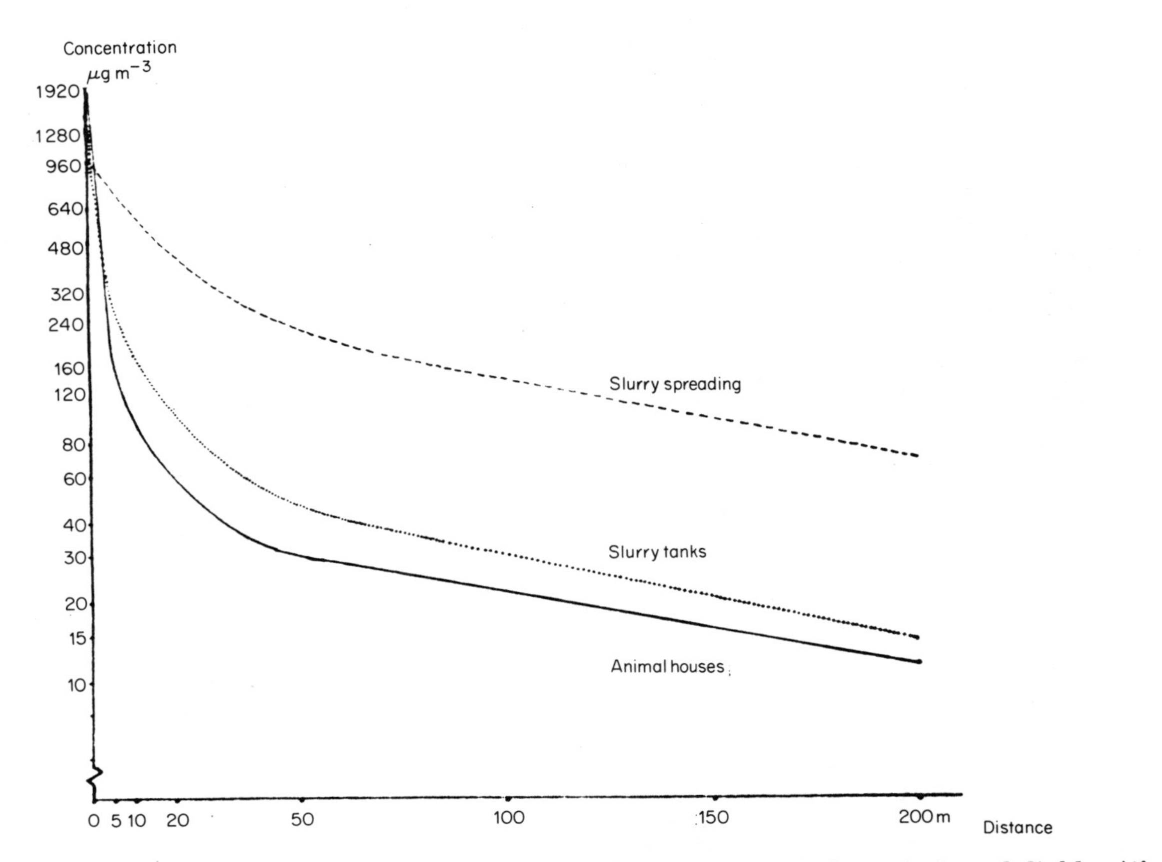

Fig. 5. NH_3-concentration at different distances from animal houses, open slurry tanks and fields with slurry.

The average concentration was 1 750 μg m^{-3}. At a distance of 50 m this concentration had decreased to 30 μg m^{-3}. From 50 to 200 m the concentration only decreased from 30 to 12 μg m^{-3}.

At the upper edge of five open slurry tanks the NH_3-concentration was 1 400 μg m^{-3}, which decreased slowly over the first 50 m, so that the NH_3-concentration at a distance of 50 m from the slurry tanks was greater than at the same distance from the animal houses. From 50 to 200 m the concentration of NH_3 decreases at the same rate as that from the animal houses.

The NH_3-concentration at the edge of fields treated with slurry was 980 μg m^{-3}, and decreased little with increasing distance. At 200 m the concentration was 72 μg m^{-3}, which is 5 - 6 times more, than at the same distance from animal houses and slurry tanks.

Propionic acid concentrations

Figure 6 shows the organic acid content of the air near animal houses, open slurry tanks and fields treated with slurry. At the point of emission from the animal houses, the concentration was 32 μg m^{-3}. Air at the upper edge of slurry tanks and directly at the edge of fields treated with slurry contained more propionic acid. With increasing distance, the propionic acid concentration from these three sources of emission decreased at different rates. The decline was rapid up to the distance of 20 m from animal houses and open slurry tanks. The concentration of propionic acid from fields treated with slurry decreased slowly, and at a distance of 200 m the content was 5 - 6 times more than in air at the same distance from animal houses and slurry tanks.

The dilution of the propionic acid concentration from different sources of emission differed from the dilution of the NH_3-concentration. p-cresol concentrations were similar to propionic acid.

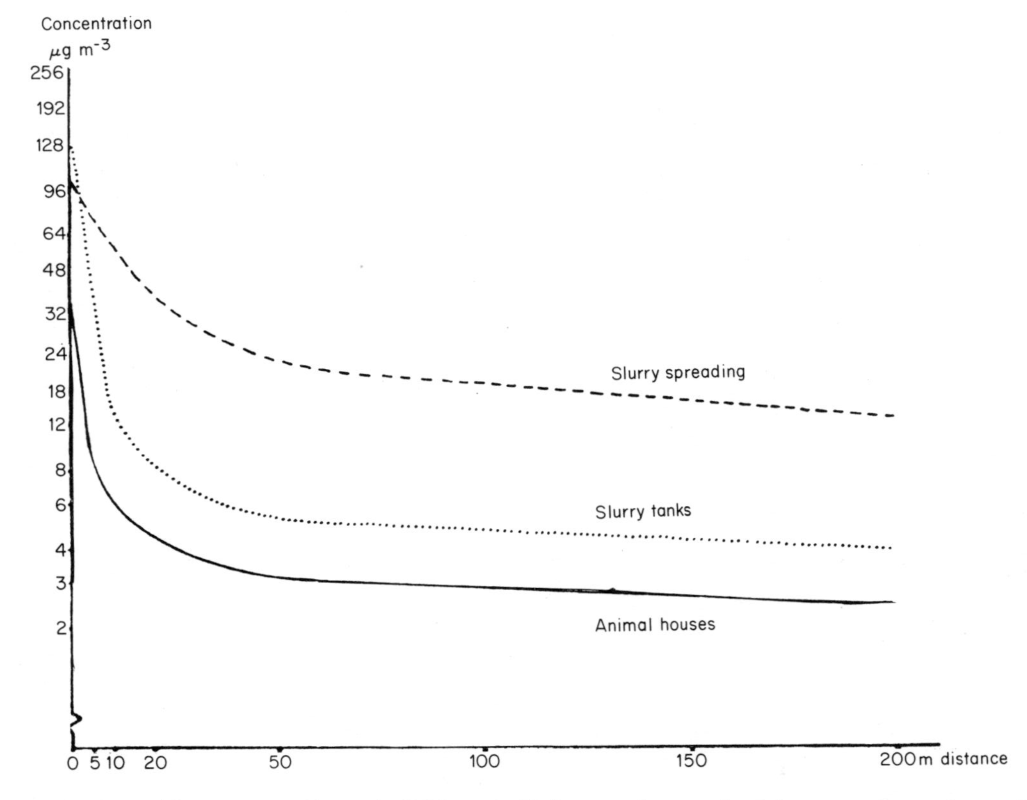

Fig. 6. Propionic acid concentration at different distances from animal houses, open slurry tanks and fields with slurry.

The differences between main component concentrations at different distances from animal houses, open slurry tanks and fields treated with slurry depend on both the differing breadths of the odour flag, and the different heights of the effective point of emission above the ground.

The extent of the odour flag depends on the breadth of the emission sources and the direction of the wind. A wide odour flag ensures a more exact air sample because changes in wind direction have less influence on concentrations of odorous compounds.

The height of the emission point from animal houses was 4.5 m and from slurry tanks 3.5 m above ground level. From fields treated with slurry, the smell emanated from the surface of the ground. The odour is diluted more rapidly, the higher the point of emission because the wind velocity increases with increasing height and because downward movement of odour is possible. At the animal houses the effective height of the odour emission depended on the height of the emission point and the velocity of the exhaust air. The average effective height of the source of emission was calculated as 6 m.

Because the concentrations of the main components and the sensory determined odour levels are well correlated, the odour immission during slurry spreading can be identified at greater distances than the odour immission from slurry tanks and animal houses.

DISCUSSION

Detection of the low concentration of the main components of odour in the area of immission, requires their concentration during sampling of the air. Experience with the analytical methods is also necessary, for example the determination of the NH_3-concentration was improved, resulting in an improvement of the minimum detectable amounts from 1 $\mu g\ m^{-3}$ to 0.1 $\mu g\ m^{-3}$

and in an improved value of the correlation coefficient between measurement and odour intensity from $r = 0.65$ to $r = 0.85$.

The sensory determination of odour by different people showed a surprisingly small spread of values. This may because of subdivision into only four odour levels and allowing the people to stay at different distances for a reasonably long time.

The results show that odour level can be measured by determining some of the main components, of which NH_3 has been found to be the best one in our investigations. Propionic acid, p-cresol and hydrogen sulphide are potentially also valuable as main components when the air sampling methods and the chemical analyses have been improved. More substances should be tested as main components, such as indole, skatole, mercaptans and amines.

DISCUSSION

J. Schaefer *(The Netherlands)*

When you measured the odour units by sensory methods at different distances, did you note the weather conditions?

H.H. Kowalewsky *(West Germany)*

We took the air samples in summer in varying weather conditions. Air samples cannot be taken in winter because the solutions will freeze.

J. Hartung *(Denmark)*

What amount of NH_3 was present in directions other than downwind?

H.H. Kowalewsky

At a distance of only 20 m away from the line of the wind there was no ammonia.

ODOUR MEASUREMENT RESEARCH IN INTENSIVE LIVESTOCK PRODUCTION UNITS

A.A. Jongebreur and J.V. Klarenbeek
Institute of Agricultural Engineering,
Wageningen, The Netherlands.

1. POULTRY OPERATIONS

1.1. Introdution

The emissions from three different poultry houses were investigated in co-operation with the Central Institute for Nutrition and Food Research, the Institute for Public Technology and the Institute for Poultry Research "Het Spelderholt". The poultry houses were selected using the manure handling system as the most important criterion. Ludington et al. (1971) suggested that drying poultry manure inside with forced ventilation and frequent removal of the droppings by belt-scraping results in a considerable reduction in odour strength compared to open inside storage of liquid manure.

1.2 Odour measurements

Chemical and sensory methods were used. The chemical analyses were directed towards the determination of the carboxylic acids C_2-C_5, phenol, p-cresol and some unidentified sulphur compounds. The sensory or psychophysical experiments served a double purpose:

i) the establishment of the threshold dilution number, by using a direct dilution method with a triangular presentation of the stimuli, the so-called "sniffcar" method of the Institute of Public Technology.

ii) comparison of the results obtained with two simple sensory methods. These were bag sampling in combination with a stationary dynamic olfactometer (e.g. Schoedder, 1977) and the paperstrip test (Sebek and Hilliger, 1973). The measurements were carried out in the three periods, April - May, June and August. Every section was measured on two days during each period.

Although the various reports are still being prepared, some conclusions can already be drawn.

1.3 Results

The effect of manure handling on the odour emission was clearly demonstrated. At the threshold level the ratio between liquid manure, dry manure and regular removal of manure was 100 : 15 : 18 (in odour units m^{-3} of ventilation air) using the sniffcar method. Although the threshold dilution numbers obtained using the bag sampling method with FEP-teflon bags differed by a factor 2 - 8, the trend was the same. The paper-strip test did not perform well with poultry house odours. The type of paper used in the paperstrip test is of importance. The exposure times, 0, 1, 5, 20 and 90 minutes, were not satisfactory and more research is needed on the processes of adsorption and desorption by the paper.

Using the bag sampling method, experiments have been carried out using more advanced psychophysical procedures such as signal detection theory to estimate the threshold, and magnitude estimation to investigate super-threshold sensation. The results of these experiments will be published later (Frijters et al., 1979).

Correlation analyses between threshold dilution numbers and concentrations of chemical components have not resulted in a workable formula. This agrees with views expressed by Nursten (1977). According to his classification, complex odours, for example, those emanating from intensive livestock operations, cannot be characterised by a single component or even a few. Therefore instrumental analyses of odour emissions are only considered to be suitable for screening purposes and not for a reliable odour determination. At present, the odour emission rates can only be established using sensory methods.

Until the sensory mechanism of man is better understood no progress in the application of instrumental analysis for reliable practical measurement can be expected in this field.

2. INSTRUMENTAL ANALYSES

Using gas-chromatographic methods to determine the concentration of the main odorants in the exhaust air of pig enclosures, two research programmes were recently completed.

2.1 Metabolism cages

The aim of this programme was to collect data on the contribution of the pigs to the total of the emitted concentration of odour from pig houses. For this purpose, exhaust air was sampled from pigs held in metabolism cages. The analyses showed that small quantities of carboxylic acids C_2 - C_5, phenol and p-cresol were detected, from which the conclusion was reached that the anaerobic degradation of stored pig manure, as described by Spoelstra (1978), is the main reason for the presence of malodours in the exhaust air of piggeries.

2.2 Measurements at farm piggeries

Knowing the origin of smells in the exhaust air of piggeries, the surface of the stored liquid manure becomes an important factor controlling the total amount of odours released to the atmosphere in the house. Since in Dutch livestock operations, the liquid manure is commonly stored directly underneath the slatted floors, this leads to a one to one relation existing between the manure surface and the slatted floor area. Taking the slatted floor area pen^{-1} as a prime variable, investigation has just been completed covering four different types of piggeries. Due to the fact that the results are still being interpreted no conclusions can be drawn yet.

3. FUTURE RESEARCH

Preparations have been made to start work into the odorous

emissions of various types of piggeries. The investigations will be made using the bag sampling method as a routine procedure. Occasionally the sniffcar method will be used to check whether the relationship between the two methods remains the same.

Furthermore, a ring test for olfactometers will be carried out with the aim of standardising the sensory method. Attention will also be paid to the rate of emission of odour during land-spreading of manure.

REFERENCES

Ludington, D.C., Sobel, A.T. and Gormel, B., 1971. Control of odours through manure management. Trans. ASAE 14 pp. 771-774.

Nursten, H.E., 1977. The important volatile flavour components of foods. In: Sensory properties of food, G.E. Birsch, J.G. Brennon and K.H. Parker ed. Applied Science Publishers, London.

Schoedder, F., 1977. Messen von Geruchstoffkonzentrationen Erfassen von Geruch, Grundl. Landtechn. 27. pp. 73-82.

Sebek, B. and Hilliger, H.G., 1973. Papierstreife als Hilfe der Geruchsbeurteilung, Bauen auf dem Landen. 73/6 pp. 153-155.

Spoelstra, S.F., 1978. Microbial aspects of the formulation of malodorous compounds in anaerobically stored piggery wastes. Dissertation, Agricultural University, Wageningen, Holland.

Frijters, J.E.R., Beumer, S.C.C., Klarenbeek, J.V. and Jongebreur, A.A., 1979. Psychophysical methodology in odour pollution research: the measurement of poultry house odour detectability and intensity. Chem. Sens. and Flav., vol.4, 1979 (in press).

DISCUSSION

D. McGrath *(Ireland)*

Could you tell me, in general terms, the principle of the paper strip method?

A.A. Jongebreur *(The Netherlands)*

The principle of the paper strip method is that a small piece of paper is exposed to the air inside the house for different times, we used 1, 5, 20 and 90 minutes. The paper strip is put in a small tube and corked. The sealed tube is transported to the laboratory where trained technicians compare the smells of the different paper strips. The differences, between the different exposure times may thus be established.

In response to comments from Dr. Vetter on the value of different methods of measuring odours Dr. Jongebreur agreed that with sensory methods, interpretation of the results obtained is very difficult. As the air of our pig houses and laying hen houses contains a complex mixture of odorous compounds the use of sensory methods appeared to be necessary.

SUMMARY OF SESSION 4

J.H. Voorburg

ODOUR MEASUREMENT

The analytical techniques for measuring the chemical components of contaminated air are well developed and small concentrations can be measured and components identified.

The lack of a sufficently good correlation between the concentration of one or more components and the odour intensity determined in a sensory way in the air from different stables is disappointing. However, analytical measurements can be applied to estimate the effect of odour-control techniques on odour reduction, although at present they are not suitable to make a reliable estimate of the odour intensity.

The research programme has resulted in the development of simplified sensory measurement techniques, which open the way for sensory measurement of odours at an acceptable cost.

ODOUR CONTROL

Research on odour control was only concerned with the aeration of slurry combined with a flushing system for removing slurry from the animal houses. This has given a maximum reduction of the odour emission from the animal houses and the fields.

The high energy costs make the system less attractive to the farmer.

PROPOSALS FOR FUTURE WORK

1. Present methods of assessing odour should be compared to provide improved simplified sensory or instrumental measurement techniques resulting in standardised methods.

2. The odour emitted from different housing and equipment systems should be measured in order to provide guidelines on livestock production with a minimum of odour emission.

3. The anaerobic digestion of slurry should be developed to provide a system with a low odour emission from the field and from the animal houses.

SESSION V

TRANSMISSION OF DISEASES AND PARASITES THROUGH SLURRY

Chairman: W.R. Kelly

A STUDY OF THE EFFECTS OF ANIMAL EFFLUENTS UTILISATION FOR GRASSLAND PRODUCTION ON THE LEVELS OF CERTAIN PATHOGENIC BACTERIA IN FARM ANIMALS AND THEIR CARCASES

W.R. Kelly

Faculty of Veterinary Medicine, University College Dublin, Dublin, Ireland.

INTRODUCTION

Because of the recognised importance of meat as a vehicle of infection with *Salmonella* species and *Escherichia coli* (Kelly et al., 1971; Linton, 1978; Hobbs and Gilbert, 1978), the influence of production conditions on the carrier rates of food animals was considered to require further investigation. Brophy et al. (1977) showed that cattle continued to excrete antibiotic-sensitive flora while grazing grass, up to the time of slaughter (Figure 1). However, grazing animals were known to become infected with salmonellae through grazing land which had been treated with infected slurry during the period immediately prior to the start of grazing (Kelly and Collins, 1978).

The present study was undertaken to determine the extent to which the landspreading of animal manures, under both commercial and experimental conditions, may influence the rate of infection with salmonellae and antibiotic-resistant *E. coli* of cattle and sheep grazing such land. Following slaughter under EEC-approved conditions for human consumption of the animals thus exposed, the transfer of such organisms onto the carcases was also investigated, as was the extent to which in vitro antibiotic resistance was transferred in these organisms during slurry storage.

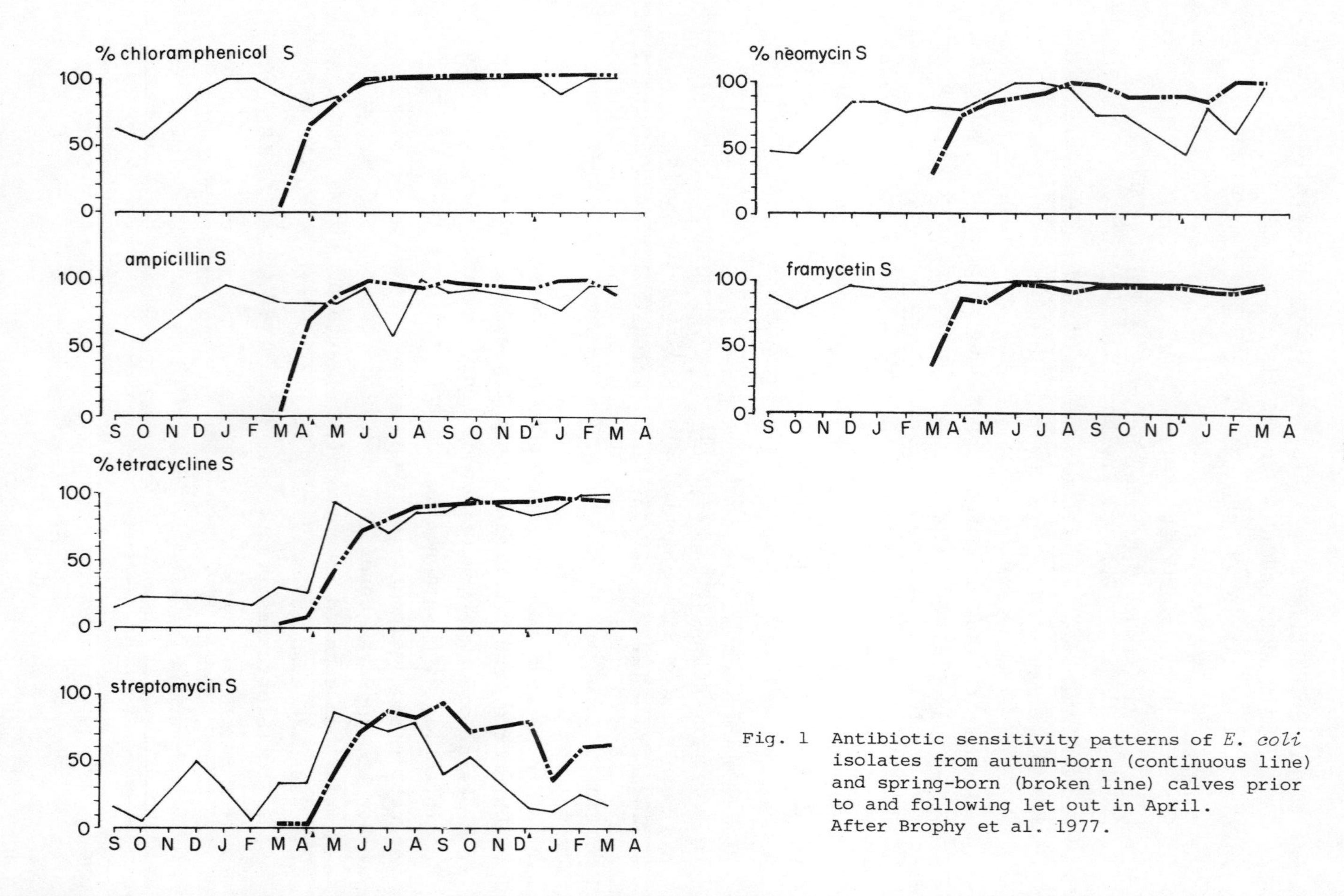

Fig. 1 Antibiotic sensitivity patterns of *E. coli* isolates from autumn-born (continuous line) and spring-born (broken line) calves prior to and following let out in April. After Brophy et al. 1977.

MATERIALS AND METHODS

Animals

1. Commercial calf-rearing unit

A total of 130 Friesian crossbred calves, from 1 300 purchased at 1 - 3 weeks of age and reared to slaughter, were examined (Table 1). The calves were fed a diet which included concentrate feed containing 100 ppm tetracycline during the first six months of life. They grazed grass for part of this period. These cattle were subsequently housed in slatted-floor houses. Studies were made on yearling heifers housed in the unit during the second year of the study and which had grazed pasture treated with slurry produced during the previous year in the calf-rearing unit.

TABLE 1

ANIMALS SUBJECTED TO EXPERIMENTAL INVESTIGATION

1. Commercial calf unit		
	1300 calves:	130 monitored
2. Commercial beef unit		
	4800 cattle:	3 x 30 monitored
3. Experimental Station		
	(i) 40 calves,	10 days - 2 years
	(ii) 70 lambs,	6 months

2. Commercial beef-fattening unit

Investigations were carried out at a fattening unit in which representative materials from some 4 800 one-to-two-year old cattle kept on slats were examined. The cattle, which were housed in groups of 30, were fed on chopped silage from grassland which had been dressed repeatedly during the preceding five years with slurry and effluent from the slatted-floored houses at the unit. The carcases and viscera of some 80 of these animals, which had been maintained at the unit, were also examined.

3. Experimental unit

A total of 40 Friesian and crossbred calves and about 70 crossbred sheep were examined during the period in which they were housed following purchase, and also in the course of two trials during which they grazed grassland which 10 days earlier had been treated with dairy-herd slurry inoculated with either (i) 2.0×10^6 *S. dublin* ml^{-1} or (ii) 10^6 *E. coli* 055B5 ml^{-1}, each at the rate of 25 000 l ha^{-1}. The *E. coli* 055B5 possessed multiple resistance to antibacterial substances. The carcases and viscera of these animals were also examined for the presence of the above organisms following commercial slaughter some 5 to 10 months after the grazing phase of the experiment.

Other Material

Slurry samples were collected at the commercial calf--rearing unit and the beef fattening unit at regular intervals during the study together with grass, soil, silage and water samples, and alginate swabs taken from the environs of the units.

METHODS

Alginate swabs taken from the recta of the above animals and the other materials were examined for the presence of *E. coli* and salmonellae, using methods described previously (Kelly, 1977) (Table 2). The in vitro sensitivity of the isolates to chloramphenicol, ampicillin, tetracycline, streptomycin and neomycin was also determined, and in the case of those isolates which displayed in vitro resistance to more than one of the reagents, each was examined for the presence of infectious drug resistance factors (R). Other measurements included (i) pH and temperature of slurry; (ii) water activity of slurry, soil and silage samples. The transfer of R-factors between *E. coli* strains and from multi-resistant *E. coli* isolates to *S. dublin,* using both sterilised and untreated slurries as the suspending medium, was also investigated according to an established procedure.

TABLE 2

METHODOLOGIES EMPLOYED AND SOURCES OF MATERIALS EXAMINED

1. *Escherichia coli* isolation
2. *Salmonella* isolation

   from - Rectal swabs (alginate)

   - Faeces
   - Slurry, in storage, during and after landspreading
   - Grass, soil, silage, water

3. In vitro sensitivity (plate techniques)
4. R-factor transfer

Also Slurry: pH, d.m., Aw., t^{o}

RESULTS AND DISCUSSION

Commercial calf-rearing unit

The feeding of 100 ppm tetracycline hydrochloride daily in concentrate feed to calves during the rearing period was associated with the development of in vitro resistance to ampicillin (50 percent), chloramphenicol (36 percent), tetracyclines (77 percent), neomycin (15 percent) and streptomycin (61 percent), in *Escherichia coli* recovered from rectal swabs taken from 130 calves (Table 3). The high prevalence of this resistance, which was usually multiple and of the transferable type (Table 4), persisted after let-out onto grassland which had been treated infrequently with farmyard manure during the previous ten years for as long as the calves continued to receive the medicated concentrates, but declined following withdrawal of the medication.

TABLE 3

ANTIBIOTIC RESISTANT *E. coli* IN CATTLE AT A CALF-REARING UNIT

Category	No. of isolates	% *E. coli* resistant to				
		Amp	C	Te	S	N
Housed calves	286	53.4	34.2	74.4	62.5	21.3
Calves, after let-out	155	58.0	40.6	92.2	79.3	4.5
Calves, at end of grazing	115	30.4	33.0	60.9	30.1	13.0
All sources:	556	50.0	35.8	76.6	60.6	14.9

TABLE 4

PREVALENCE OF MULTIPLE-RESISTANT *E. coli*. CALF UNIT

Category	% of *E. coli* resistant to 3 or more of Amp, C, Te and S
Housed calves	44.4
Calves, after let-out	50.3
Calves, at end of grazing	30.8
Heifers, grazing	59.0
Slurry (Calf Unit)	35.7
Slurry (Beef Unit)	40.0

Later, these cattle were transferred as yearlings to grassland which had been treated regularly with slurry from slatted-floored units used to house fattening cattle, and the excretion rate of multiple-resistant *E. coli* was not observed to increase. However, when these cattle were moved from grassland to slatted-floored houses, the excretion rate of multiple-resistant *E. coli* increased. The increase was attributed in part to direct or indirect contact within the unit with calves which had been introduced directly from the calf-rearing unit and which had continued to receive tetracycline medicated concentrates. Forty percent of the slurry samples taken from these houses yielded *E. coli* resistant in vitro to three or more of the antibiotics ampicillin, tetracycline, chloramphenicol and streptomycin. The resistance was of the transferable type.

In the second year of the study, yearling heifers were introduced into the calf-rearing unit and were grazed on pasture which had been treated with slurry produced during the previous year in the calf-rearing unit. About 36 percent of the isolates of *E. coli* recovered from this slurry were resistant in vitro to three or more of antibiotics (Table 4). The heifers, while grazing grass, yielded 59 percent of *E. coli* isolates resistant in vitro to three or more of the antibiotics ampicillin, tetracycline, chloramphenicol and streptomycin.

Commercial beef-fattening unit

Investigations carried out at a commercial unit in which some 4 800 cattle were kept demonstrated that antibiotic-resistant *E. coli* was not widespread in fattening cattle kept in slatted-floored houses and fed on concentrates mixed with chopped silage derived from grassland treated with slurry from the unit. However, the feeding of 9 g flavomycin daily to one group of 30 fattening cattle coincided with the emergence of *E. coli*, 9 percent of 32 such isolates being resistant in vitro to ampicillin, at a time when *E. coli* excreted by similar cattle receiving non-medicated feed displayed no in vitro resistance to any one of five antibacterial reagents (Table 5).

Experimental unit

Extensive studies carried out on 40 calves showed that the high prevalence of antibiotic resistance of the transferable type displayed by *E. coli* isolated during the period immediately following purchase and during the subsequent housing period declined after the calves were let out to grass. The low resistance levels were maintained during the subsequent housing period (Table 6).

TABLE 5

ANTIBIOTIC-RESISTANT *E. coli* IN CATTLE AT A BEEF-FATTENING UNIT

No. of Isolates	% of *E. coli* resistant to				
	Amp	C	Te	S	N
245	3.7	0	0.4	5.7	2.4
	No. isolates resistant to 3 or more of Amp, C, Te, S				

These cattle, as yearlings, together with 70 sheep, were used in two grazing trials on grassland treated with slurry from a dairy herd which had been inoculated with a multi-resistant strain of *E. coli* 055B5. A small number of sheep, but none of the cattle, were shown to acquire and excrete the multi-resistant *E. coli* 055B5 within one month following the start of grazing on the contaminated plot. In addition, a number of the sheep which, prior to grazing had excreted a sensitive flora, also excreted multi-resistant strains other than *E. coli* 055B5 during and following the grazing experiment.

Transmission studies involving the application of slurry containing *S. dublin* onto grassland as described under Methods and which was grazed by the experimental animals ten days after application, gave negative results.

Carcase findings

1. Commercial beef-fattening unit

In a further study involving 80 cattle, which were removed from the slatted-floored houses to untreated pasture prior to slaughter, these cattle were found to continue to excrete mainly

TABLE 6

PREVALENCE OF IN VITRO RESISTANCE TO ANTIBACTERIAL REAGENTS IN *E. coli* ISOLATED FROM CALVES KEPT AT THE EXPERIMENTAL UNIT

Date of sampling	No. of Samples	No. of *E. coli*	% of isolates resistant in vitro to:			
			Pn	C	Te	N
25/2/76	18	54	24.0	20.3	42.5	31.4
3/3/76	49	100	43.0	38.0	58.0	36.0
24/3/76	36	78	46.1	34.7	67.9	29.4
29/4/76	37	102	20.5	15.6	38.2	11.7
20/5/76	36	57	14.0	7.0	26.3	7.0
25/6/76	31	42	11.9	7.1	16.7	2.4
13/7/76	35	48	8.2	0	6.2	4.2
16/8/76	38	42	4.8	0	2.4	2.4
1/9/76	36	44	9.1	0	6.8	4.5
19/10/76	37	40	4.0	0	0	0
8/1/77	38	31	0	0	0	0
8/2/77	38	33	0	0	0	0
8/3/77	38	35	0	0	0	0

antibiotic-sensitive *E. coli* while at grass and immediately prior to slaughter, whereas a high proportion of *E. coli* isolated from their mesenteric lymph nodes were resistant in vitro to ampicillin (35 percent), chloramphenicol (8 percent), tetracycline (10 percent) and streptomycin (11 percent). Furthermore, 22 percent of *E. coli* isolates recovered from the carcase surfaces of these cattle were also resistant to ampicillin. In later studies on cattle slaughtered under similar conditions, *E. coli* isolated from the mesenteric lymph nodes displayed an in vitro sensitivity to antibacterial reagents which was similar to that shown by alimentary tract isolates.

Slurry samples collected from the slatted-floored houses yielded only one *E. coli* isolate which was resistant in vitro to two or more reagents. About 3 percent of *E. coli* recovered from grass, soil, silage and hay from the unit were resistant to three or more of the antibiotics ampicillin, chloramphenicol, tetracycline and streptomycin.

2. Experimental unit

These cattle and sheep were slaughtered at intervals from 5 - 10 months following the termination of the grazing trials. Neither *S. dublin* nor *E. coli* 055B5 was recovered from the carcases and/or viscera of 21 of the sheep, eight of which were potential sources of *E. coli* carrying infectious resistance factors (Tables 7 and 8). Included in these eight sheep were animals which had not grazed the plots treated with the slurry inoculated with *E. coli* 055B5, but these animals had been housed with sheep so exposed following the experimental period.

Slurry

In vitro studies demonstrated that multi-resistant *E. coli* did not lose resistance during storage in slurries of varying pH and water activity kept at ambient temperatures over a period of 27 days (Table 9). Infectious resistance was shown to be transferred both between *E. coli* strains and from *E. coli* to

TABLE 7

PREVALENCE OF ANTIBIOTIC-RESISTANT *E. coli* IN SHEEP AND CATTLE GRAZING PASTURE DRESSED WITH SLURRY CONTAINING R +ve *E. coli*

Species	Class/sample	No. of isolates	Amp	C	Te	S
Cattle	1. Pre-exposure	38	0	0	0	2.6
	2. Post-exposure					
	(a) intestinal	27	11.1	3.7	3.7	3.7
	(b) lymph node	7	0	0	0	0
	(c) carcase	11	9.9	0	0	0
Sheep	1. Pre-exposure	62	1.9	0	0	0
	2. Post-exposure					
	(a) intestinal	113	8.8	1.8	2.7	3.5
	(b) lymph node	69	13.0	0	0	1.4
	(c) carcase	192	0	0	0	0

Salmonella dublin in slurry kept at $2^{o}C$, ambient temperature and $37^{o}C$ (Table 10).

TABLE 8

PATTERNS OF ANTIBIOTIC RESISTANCE DISPLAYED BY *E. coli* RECOVERED FROM CATTLE AND SHEEP PRIOR TO AND FOLLOWING EXPOSURE TO PASTURE DRESSED WITH SLURRY CONTAINING R +ve *E. coli*

	Pre-Exposure	Post-Exposure
Sheep	Amp	Amp Te Amp, S Amp, C, Te, S
Cattle	-	Amp Amp, C, Te, S, N

TABLE 9

pH AND Aw VALUES OF LIQUID SLURRIES ON COLLECTION AT THE COMMERCIAL BEEF UNIT

	Mean ± s.d.	
	pH	Aw
Slatted floor house	8.51 ± 0.33	0.961 ± 0.014
Lagoon	7.31 ± 0.38	0.974 ± 0.008
Prior to landspreading	7.05 ± 0.21	0.957 ± 0.015

CONCLUSIONS

1. The level of contamination of slurry with multiple-resistant *E. coli* is directly related to the extent to which the animals producing the slurry are themselves infected (Figure 2). While antibiotics are being fed, these high contamination rates can be anticipated even when such animals are allowed to graze.

TABLE 10

R-FACTOR TRANSFER IN SLURRY

Donor: *E. coli* isolated from calves at commercial unit.
Recipient: *E. coli* E711

Slurry Type		@ 2°C			@ 12°C			@ 37°C		
	R-factor transferred for:	Amp	C	Te	Amp	C	Te	Amp	C	Te
Solid, sterile (2)		+	+	+	+	+	+	+	+	+
Solid, untreated (2)		+	+	+	+	+	+	+	+	+
Semi-solid, sterile (4)		+	+	+	+	+	+	+	+	+
Semi-soild, untreated (4)		+	+	+	+	+	+	+	+	+
Liquid, sterile (2)		+	+	+	+	+	+	+	+	+
Liquid, untreated (2)		+	+	+	+	+	+	+	+	+

Number of samples in brackets.

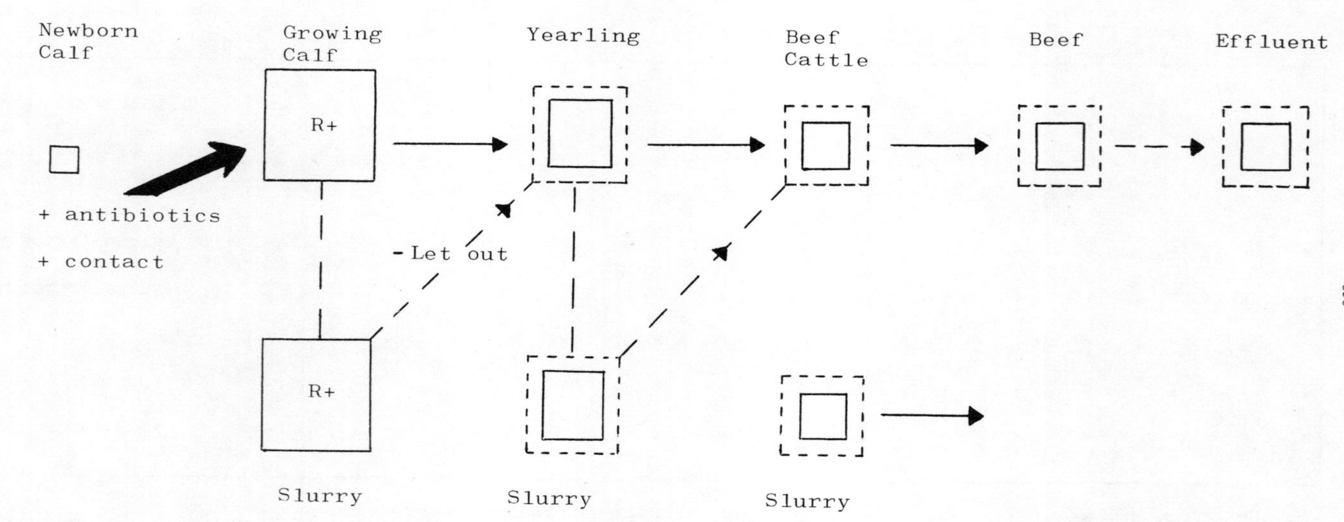

Fig. 2. Slurry, cattle and R+ve *E. coli*

2. Grazing cattle and sheep become infected from the application of slurry containing antibiotic-resistant *E. coli* to grassland under both commercial and experimental conditions.

3. Such infection can lead to the transfer of R-factors within the alimentary tract of the grazing animal.

4. In a few cases, carcases can become contaminated with such organisms under EEC-approved conditions of slaughter.

5. The extent of mesenteric lymph node contamination with multi-resistant *E. coli* in cattle and, more particularly, pigs, requires to be determined in view of the requirements for veterinary inspection, through incision of these nodes, under EEC law.

6. The effects of storage on the survival of multi-resistant bacteria, including *E. coli*, require to be examined further because resistant *E. coli* persist in slurry kept under experimental conditions and, more particularly, because of the demonstrated ability of such organisms to transfer resistance in slurry.

7. The transmission of *S. dublin* via slurry onto grazing land was uncertain.

REFERENCES

Brophy, P.O., Caffrey, P.J. and Collins, J.D., 1977. Sensitivity patterns of *Escherichia coli* isolated from calves during and following prophylactic chlortetracycline therapy. Brit. vet. J. 133: 340-345.

Hobbs, B.C. and Gilbert, R.J., 1978. In: 'Food Poisoning and Food Hygiene'. 4th Ed. Ed. Arnold, London.

Kelly, W.R., 1977. A study of the effects of animal effluent utilisation for grassland production on levels of certain pathogenic bacteria in food animals and their carcases. In: 'Utilisation of Manure by Land-spreading', pp. 381-390; Ed. J.H. Voorburg. Commission of the European Communities: Co-ordination of Agricultural Research Publication EUR 5672e.

Kelly, W.R. and Collins, J.D., 1978. The health significance of some infectious agents present in animal effluents. Vet. Sci. Comm., 2: 95.

Kelly, W.R., Collins, J.D., Hannan, J. and Conlon, P., 1971. The prevalence of contamination with resistant coliform bacteria and adulteration with antibacterial residues of the carcases of food animals. J. Irish med. Assoc. 65: 75-78.

Linton, A.H., 1978. In appendix to discussion on 'Identifying the priority contaminants'. In: 'Animal and human health hazards associated with the utilisation of animal effluents', pp. 86-95. Ed. W.R. Kelly. Commission of the European Communities: Co-ordination of Agricultural Research Publication EUR 6009en.

THE POSSIBLE ROLE OF ANIMAL MANURES IN THE DISSEMINATION OF LIVESTOCK PARASITES

N.E. Downey and J.F. Moore
The Agricultural Institute, Dunsinea, Castleknock, Co. Dublin, Ireland.

ABSTRACT

Farm surveys revealed widespread trichostrongyle contamination of slurry produced by cattle in their first winter (weanlings). The concentration of viable eggs, and thus the potential for disseminating parasites, was far greater in slurry stored under slatted-floor units than in material stored in slurry pits (dungsteads). This difference reflects the difference in storage duration: winter only for slatted-floor material and winter plus summer for pits. Also, low temperatures in winter favour egg survival whereas in summer the higher temperatures reduce survival.

Viable trichostrongyle contamination, though diminishing with time, can persist up to the end of May representing an accumulation and storage time of seven months.

Applications of trichostrongyle-contaminated slurry in the spring of one year increased over-wintering infection of pasture resulting in severe parasitic gastro-enteritis of calves in early June of the following year. Even when over-wintering infection was not markedly increased, contaminated slurry applied in two successive years so increased the level of parasitism as to reduce significantly the liveweight gain of calves in the second year. Slurry application to pasture is considered only to increase infection sufficiently to impair animal performance under some conditions. Of these the most important is grazing by susceptible calves following a spring application of contaminated slurry. When grazing was delayed until after cutting the sward for conservation, no increase in infection level was detected although further work on this aspect is needed.

Recommendations for minimising the occurrence of increased trichostrongyle infection by landspreading of cattle slurry are discussed.

INTRODUCTION

The eggs and other free-living forms of important livestock parasites are voided with the faeces of parasitised animals. In husbandry systems using bedding, faecal contamination during grazing was virtually the only source of parasitic infection on pasture, particularly in the case of ruminants. For a period after coming off pasture, the animals might continue to void eggs and larvae but these were destroyed by the composting effect of litter, such as straw, which was mixed with the excreta. With animal housing systems requiring little or no bedding material this effect is largely lacking. Thus, while most herbage infection still derives from the faeces of grazing animals, livestock effluent in the form of slurry could well contribute additional contamination when used as a fertiliser on grassland.

Bürger and Stoye (1978) described the biological factors concerned with the possible transmission of various parasites to their definitive hosts arising from the production and utilisation of animal excretions. They listed the parasites that may occur in stored manure derived from cattle, pigs and poultry. To these might be added the internal parasites of sheep (Anon, 1978). In the context of cattle slurry, the trichostrongyles are likely to present a prime hazard to animal hygiene since (i) they are very prevalent; (ii) certain of their free-living stages are rather resistant to external factors; (iii) cattle slurry is widely used to fertilise grassland. For these reasons and because trichostrongyles can cause serious economic losses in cattle production, they were investigated.

The trichostrongyles have a simple, direct life-cycle (Figure 1): the adult worms live in the digestive tract and the female lays eggs which pass out with the faeces. The eggs must hatch and the emergent larvae must develop to the third stage (l_3) for infection to occur in the ruminant host. The process from egg to l_3, which usually takes place on pasture, requires

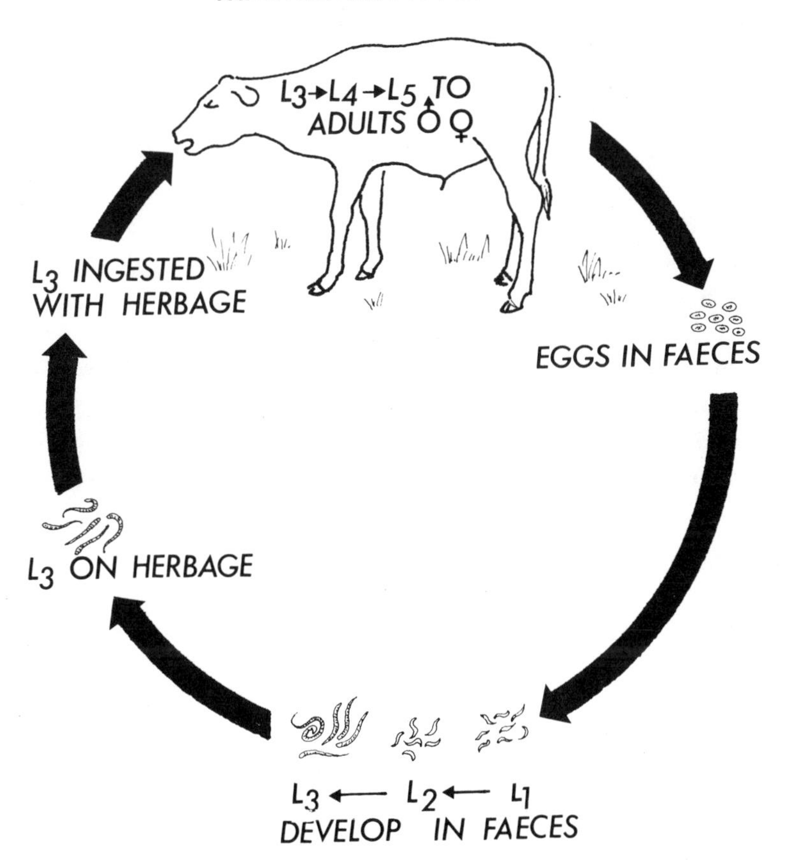

Fig. 1. The life cycle of trichostronglye worms.

from about two weeks to two or three months, depending on temperature. Inside the host the larvae undergo a further two or three weeks' development to reach the adult stage.

The two important species of trichostrongyles affecting cattle are *Ostertagia ostertagi* and *Cooperia oncophora*, the former being much more pathogenic. The epidemiology of parasitic gastro-enteritis (the infection caused by these worms) has been described in a number of publications (Michel, 1967, 1969; Downey, 1973). The infective larvae (l_3) are capable of surviving on the herbage over the winter, small numbers being present when calves begin to graze in April/May (that is, 'residual herbage infection'). The calves become infected and soon distribute eggs on the pasture. Between May and mid-July these eggs hatch, and the larvae develop to the infective stage at a rate which is dependent on temperature. As temperature increases during the early part of the grazing season, so also does the rate of larval development, resulting commonly in a big increase in pasture infection from about mid-July onwards, provided there is sufficient rainfall. The infection causes poor performance by the animals and possibly diarrhoea, and affects calves during their first season of grazing grass. In severe outbreaks, there will be actual loss of body-weight and sometimes mortality. The syndrome, occurring from late July to October, is known as type I ostertagiasis and its occurrence is widely recognised in Europe and elsewhere.

The impact of slurry application on the epidemiology of parasitism in cattle has received relatively little attention. Investigations by Persson (1974a, 1974b, 1974c, 1974d) in Sweden were mainly concerned with the occurrence of gastro-intestinal nematodes (*Ostertagia ostertagi* and *Cooperia oncophora*) in slurry produced by cattle in different regions and the survival of the parasites in slurry and on slurry-treated pasture. In addition, Persson (1974a) cites a personal communication from Nilsson who reported outbreaks of parasitic gastro-enteritis in herds grazing pasture which had not been contaminated by other cattle, but which had been fertilised with slurry.

Sampling on ten farms in Ireland indicated a high incidence of trichostrongyle contamination in slurry derived from young cattle (Downey et al., 1977), and this was later confirmed in wide-ranging investigations by Moore (1978). Downey and Moore (1977) in a preliminary report described the increased infection levels found on herbage and in calves as a result of fertilising a sward with contaminated slurry.

The present communication describes investigations into the incidence, amount and persistence of trichostrongyle contamination in slurry and the effects on levels of infection arising from the application of slurry to pasture.

MATERIALS AND METHODS

(i) Farm 'surveys'

Visits were confined to farms on which slurry was produced by cattle in their first winter (weanlings), since young cattle are likely to void trichostrongyle eggs in significant numbers.

Using sampling methods previously described (Downey et al., 1977; Moore, 1978), two types of material were investigated: (a) slurry stored under slatted-floor units and (b) slurry stored in slurry pits. Sampling was carried out in spring and autumn to correspond with the usual times at which each type is spread on land. For 'slatted-floor slurry' (SFS), sub-samples were taken randomly at several depths and locations in the tank, collecting 1 - 4.5 kg in all. Samples of slurry from pits or dungsteads (DS) could only be sampled from the periphery; sub-samples were taken at 4.5 m intervals around the perimeter to give a total of 2.5 kg per sample. On about half the farms DS was withdrawn at two depths.

On farms where SFS was sampled, individual samples of faeces were taken from a proportion of the animals producing the slurry.

(ii) Persistence of trichostrongyle eggs in slurry

Material for this study was produced by groups of weanlings known to be voiding eggs. They were maintained during winter (November to March/April) either on slats or in cubicles to provide SFS and DS respectively. Material produced in the first winter (1975/76) was randomly sampled on five occasions in the spring. Slurry produced in the next two winters was sampled at 14-day intervals from January to March and also during April (in 1978), while occasional samples were also obtained from May to August inclusive. Faeces from the cattle producing the slurry were sampled every month during the housing period.

(iii) Grazing experiments

Experiment 1. Trichostrongyle-contaminated slurry was applied early in the grazing seasons of 1976 and 1977 to an area of pasture designated plot A. An adjacent area of the same pasture, plot B, served as the control receiving no slurry. In both years, the plots were grazed for the whole season by separate groups of calves numbering initially 14 and 15 head in 1976 and 1977 respectively.

Experiment 2. Treatments consisted of (1) pasture fertilised with contaminated slurry and (2) pasture fertilised with non-contaminated slurry. Slurry was applied in March/April 1977 and in April 1978, there being two replicates of each treatment. Twelve calves grazed each treatment from early May until November in 1977 and 1978.

Experiment 3. An area of pasture was divided into two equally sized plots, one of which received an application of dungstead slurry on October 12 1977. Grazing with groups of six calves began on May 12 1978 and continued for the rest of the season.

Experiment 4. A further area of pasture was divided into two equally sized plots, one of which received an application of contaminated slurry in late April 1978, the other being the untreated control. The entire area was cut for silage on 21 June and eight calves grazed the aftermath of each plot from 5 July onwards.

The slurry used in these experiments was that which weanlings produced during persistency studies (see (ii) above).

The calves were weighed regularly throughout the grazing seasons.

(iv) Parasitology

The techniques used for examination of slurry samples and for examining individual samples of faeces of weanlings (see (i) and (ii) above) were described in earlier communications (Downey et al., 1977; Moore, 1978). To monitor infection levels in grazing Experiments 1 and 2, herbage larval counts (Parfitt, 1955) and egg counts in calves' faeces (Parfitt, 1958) were determined at weekly intervals, the results being expressed as larvae kg^{-1} of grass ($1\ kg^{-1}$) and eggs g^{-1} ($e\ g^{-1}$) respectively. Herbage larval counts and faecal egg counts were also determined in Experiment 3, while in Experiment 4 monitoring was limited to egg counts. In addition, necropsy worm counts were estimated on a limited number of 'tracer calves' and on calves contracting fatal parasitic infection.

RESULTS

(i) Farm 'surveys'

Although mean egg counts in slurry did not differ greatly between years, counts between farms varied much, although part of this variability may have been due to sampling error. Table 1 compares numbers of total and viable eggs indicating a heavy mortality or loss of fertility of eggs during storage, particularly in the case of dungsteads sampled in autumn. Nevertheless, viable eggs (predominantly *Ostertagia* spp.) were found in most of the samples collected in spring.

TABLE 1

AVERAGE NUMBERS OF TOTAL AND VIABLE* TRICHOSTRONGYLE EGGS (RANGES OF POSITIVE COUNTS IN PARENTHESIS) IN SAMPLES COLLECTED FROM SLURRY STORED IN FARMS

Time of sampling	No. of farms	Type of storage	Total eggs kg^{-1} ($x 10^3$)	Viable eggs* kg^{-1}	% with viable eggs
Spring 1976	7	Under slats	24 (1.7-71)	1670 (4-6150)	100
	3	Dung-stead	3 (2.8-3.4)	1340 (104-2435)	100
Autumn 1976	40	Dung-stead	21 (0.1-300)	0.4 (0.1-3.0	30
Spring 1977	34	Under slats	22 (0.3-75)	416 (1-12600)	77
Spring 1978	27	Under slats	15 (1.2-85)	1520 (1.2-16700)	96

* Viable eggs: estimated by incubating the slurry and enumerating 3rd stage larvae.

TABLE 2

FAECAL EGG COUNTS (e g^{-1}) - AVERAGE TRICHOSTRONGYLE EGG NUMBERS OF WEANLING CATTLE AND PERCENTAGE DISTRIBUTION OF EGG COUNTS

Year	No. of cattle	Average e g^{-1}	% Frequency, e g^{-1}		
			< 40	40 - 200	> 200
1976	181	112	49	39	12
1977	321	68	63	30	7
1978	137	49	73	23	4

TABLE 3

RESULTS FOR MEAN MONTHLY FAECAL EGG COUNTS (e g^{-1}) OF WEANLING CATTLE PRODUCING SLURRY DURING WINTERS 1975/76, 1976/77 AND 1977/78

Animal accommodation	No. of weanlings	Wintering period (Nov.-Mar)	e g^{-1} of faeces				
			Nov.	Dec.	Jan.	Feb.	Mar.
Slatted-floor, unroofed	25	1975/76	-	428	173	149	195
"	40	1976/77	205	208	175	152	-
Slatted-floor roofed	22*	1976/77	2	1	1	0	-
"	18	1977/78	585	443	291	546	222
"	18*	1977/78	4	0	2	0	-
Cubicles	12	1976/77	774	334	176	108	-
unroofed	12	1977/78	242	315	260	80	167

* Treated fortnightly with anthelmintic drugs to suppress egg output.

TABLE 4

PERSISTENCE OF TRICHOSTRONGYLE CONTAMINATION IN SLURRY DURING ACCUMULATION AND STORAGE: EGG CONCENTRATION AND LARVAL RECOVERIES FOLLOWING INCUBATION (NINE TYPICAL RESULTS TABULATED)

Winter storage period	Type of storage	Date of sampling	Eggs kg^{-1} (x 10^3)	Larvae kg^{-1}*	
				Ostertagia	*Cooperia*
1976/77	Under slats	March 23	38	1 361	84
	"	May 31	8	33	3
	"	August 5	0.6	0	0
1977/78	"	March 20	94	16 600	400
	"	April 24	52	432	30
	"	May 1	22	6	0
1977/78	Dung-stead	March 20	66	8 710	0
	"	May 31	25	281	0
	"	July 13	3	0	0

* Following incubation.

The results in Table 2 of the faecal egg counts of weanlings showed they were still voiding eggs into slurry in spring when the farm visits were made.

(ii) Persistence of trichostrongyle eggs in slurry

Table 3 gives the results of mean faecal egg counts of the weanling cattle in the three winters of the study which show that the faeces contributing to the contaminated slurry contained trichostrongyle eggs throughout the wintering period. A few eggs were occasionally found in samples from weanlings treated with anthelmintic drugs with the object of producing non-contaminated slurry for use in Experiment 2.

During accumulation and storage of the slurry, whether under slats or in dungsteads, the concentration of eggs remained fairly constant up to February/March when it decreased. In 1978 a further decline in egg numbers, accompanied by a fall in apparent egg viability was recorded in late April/early May. Apparent egg viability was at or near zero in samples examined in June, July and August. (Some typical results are shown in Table 4).

(iii) Grazing experiments

Experiment 1. Results 1976: Herbage was generally little infected between May and September due partly to severe drought, but the slurry-treated plot A had 191 larvae kg^{-1} (l kg^{-1}) on August 9, compared with 34 l kg^{-1} on the control plot B. Numbers were again low for most of September, but between September 28 and November 8 increased somewhat on both plots. In this period, the larval count was higher on plot A on all but one occasion, rising to a maximum value of 598 l kg^{-1} on November 8, with a corresponding count of 320 l kg^{-1} on plot B. Figure 2 records a similar trend in egg counts from calves. The calves on plot A showed consistently higher counts than those on plot B. The effect was statistically significant on 8 out of 21 occasions and highly significant ($P < 0.001$) when cumulative counts over the whole period were considered.

Results in 1977. On 25 April, the herbage larval count on the slurry treated pasture (plot A) was 313 l kg^{-1} compared with 33 l kg^{-1} on the control pasture (plot B) and the herbage on plot A had consistently more during May. In this period tracer calves on plot A acquired about seven times as many worms (*Ostertagia* and *Cooperia*) as those grazing on plot B. Herbage infection declined in early June but on 27 June the count on plot A was 172 l kg^{-1} compared with only 27 l kg^{-1} on plot B. Counts decreased again in July, though they remained more on plot A than on plot B, while on 8 August herbage on plot A had a count of 555 l kg^{-1} compared with 200 l kg^{-1} on plot B. Thereafter, the pattern of herbage infection was normal over the whole experimental area, obscuring the effects of slurry application.

All the calves on plot A developed rather severe parasitic gastro-enteritis (PGE) in early June 1977, whereas mild diarrhoea affecting only two calves occurred on plot B. On 5 August a calf on plot A died of PGE and symptoms among calves on this plot again became general towards the end of the month when calves on plot B weighed on average 19 kg more than calves on plot A ($P < 0.05$). Near the end of the season, calves on both plots exhibited clinical parasitism although the ones on plot B maintained their liveweight advantage.

Figure 3 shows that the pattern of clinical infection was reflected in the faecal egg counts which were significantly higher from the calves on plot A than from those on plot B on most occasions up to August 25.

Experiment 2. Larval numbers on herbage were generally rather low in both years. However, in the first year (1977) most recovered in each month from September to November were 103, 145 and 169 l kg^{-1} for the contaminated slurry (treatment 1) compared with 61, 52 and 25 l kg^{-1} for non-contaminated slurry (treatment 2). Similarly Figure 4 shows that faecal egg counts were always higher from calves on grass with contaminated slurry, this effect usually being statistically significant.

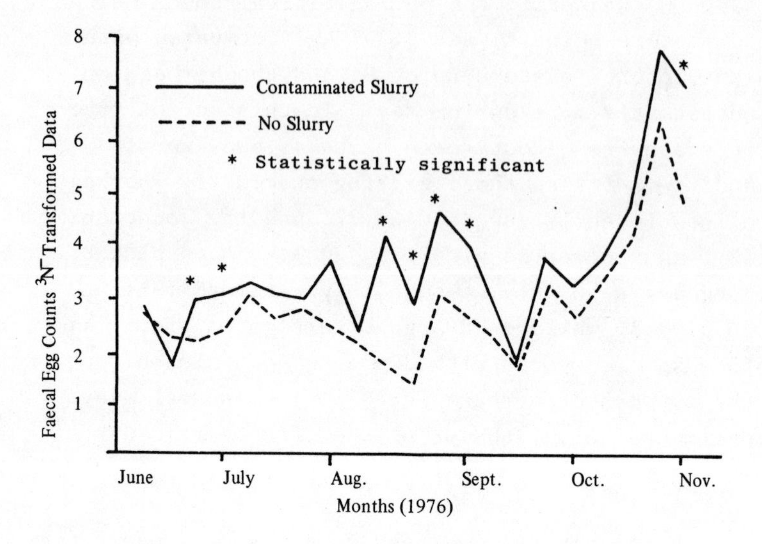

Fig. 2. Worm egg counts in faeces of calves Exp. 1. 1976.

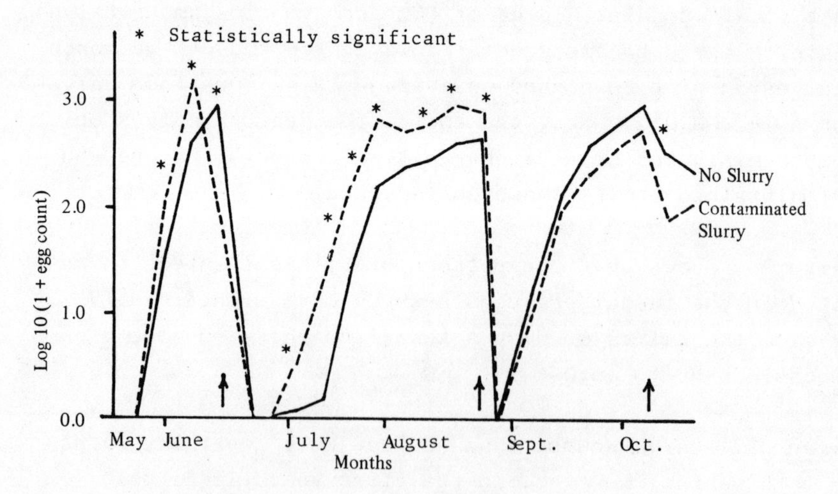

Fig. 3. Worm egg counts in faeces of calves Exp. 1. 1977.

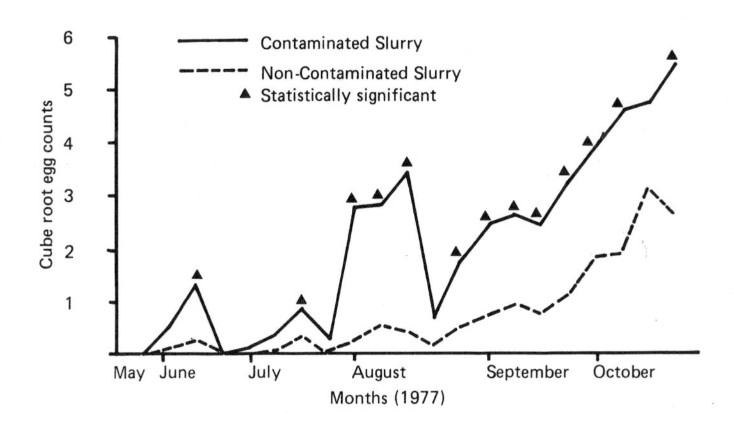

Fig. 4. Worm egg counts in faeces of calves Exp. 2. 1977.

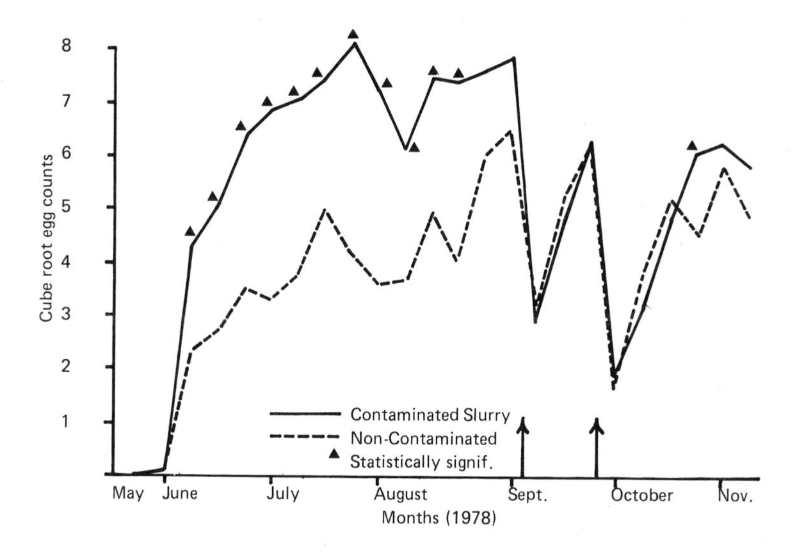

Fig. 5. Worm egg counts in faeces of calves Exp. 2. 1978.

In 1978, maximum herbage infection was recorded in early August when counts of 364 and 123 l kg^{-1} were recorded for grass with contaminated and uncontaminated slurry respectively. However, over most of the season few larvae were recovered from herbage and differences between treatments were not consistent. Under such circumstances, Figure 5 shows that faecal egg counts from calves were probably a better indicator of pasture infection: up to mid-August, calves on grass with contaminated slurry had significantly ($P < 0.01$) higher egg counts than calves on grass with uncontaminated slurry.

During August 1978, most of the calves on 'contaminated pasture', but none of those on 'uncontaminated pasture', showed symptoms of PGE. The symptoms eventually affected calves on both pastures. However, over the whole grazing season the calves on 'contaminated pasture' gained significantly less weight ($P < 0.01$) than their counterparts on 'uncontaminated pasture'.

Experiment 3. The application of dungstead slurry in the autumn did not increase herbage infection in the following season as determined by larval counts on herbage, egg counts of calves' faeces or liveweight gains of calves.

Experiment 4. There was no evidence from faecal egg counts or from worm burdens of 'tracer calves' that spring application of contaminated slurry before cutting grass for silage affected the level of infection on the aftermath.

DISCUSSION

Presumably the short-term storage of 'slatted-floor slurry' (SFS) during a period of low temperatures is partly responsible for the very high incidence of viable tricho-strongyle eggs in the material revealed in the farm surveys (Table 1). Another factor leading to such a high incidence in these investigations was the young age of animals producing the slurry, Although similar animals had produced the 'dungstead slurry' (DS), it appeared to be almost free of viable eggs, due evidently to more prolonged storage involving exposure to high summer temperatures. The effect of these storage factors was confirmed in the persistency studies in which, irrespective of storage method, numbers of viable eggs declined noticeably in April/May, becoming scarcely detectable from June onwards.

The application of SFS in spring, when eggs may be numerous and still viable, coincides with the start of grazing and of suitable conditions for larval development. This material has thus the potential to increase infection on pasture. Dungstead slurry has no comparable potential since it will contain few viable eggs and be applied in the autumn when conditions do not favour development of larvae.

Larval recoveries following incubation of the slurry varied much when investigating persistency. Such variation and that of egg concentration can, in part, be attributed to possible sampling errors resulting from uneven egg distribution. Such errors, which have been dealt with in earlier communications (Downey et al., 1977; Moore, 1978; Moore and Downey, 1978), coupled with variations resulting from failure always to achieve optimal laboratory conditions for larval development and extraction, illustrate the difficulty of assessing quantitatively the contamination potential of slurry. Possibly improved estimates of egg viability might be obtained by (a) observing embryonation in water; (b) hatching the eggs in water followed by larval incubation on agar; (c) extraction of l_3 by migration in place of Baermannisation.

Little information is available on trichostrongyle egg output by cattle in Ireland during the winter. The results presented in Table 3 show the amount of egg contamination that weanlings are capable of contributing to pasture, whether through application of their excrement as slurry or directly during grazing in winter/early spring.

There was very little residual herbage infection at the outset in Experiments 1 and 2. Subsequent results in these experiments suggest that, in these circumstances, the numbers of larvae arising from slurry are not sufficient in themselves to cause heavy pasture contamination. For heavy infection to arise, grazing of the treated area by susceptible calves is probably essential. Such calves may then increase infection to such an extent that the numbers of larvae over-wintering on the sward is sufficient to cause serious levels of parasitism in calves early in the following season. Alternatively, if sufficient residual larvae are present to combine with those emerging from the slurry, the calves may build up infection levels dangerous to themselves.

From the results of Experiments 3 and 4 neither DS applied in the previous autumn nor SFS applied in spring prior to cutting for conservation are likely to be important sources of infection on spring pasture or aftermath respectively.

The potential for parasitic contamination in SFS derived from weanlings, due to its comparatively high concentration of viable eggs at the time of application in spring (Tables 1 and 4), has been referred to. Experiments 1 and 2 show that this potential may well be realised in practice although animal performance will only be impaired under some conditions, as above.

Parasitic gastro-enteritis can be controlled by methods which could be applied whether or not slurry were spread on pasture (Michel and Lancaster, 1970; Leaver, 1970; Oostendorp and Harmsen, 1968). Unfortunately, these methods are not widely adopted. None of them is in any case universally applicable due to problems related to overall grassland management, farm lay-out, and other factors. Slurry produced by adult cattle probably does not give rise to the risks. Also, there is likely to be less risk with adult cattle grazing slurry-treated pasture. On any farm, however, convenience may dictate the use of slurry on paddocks close to the farm-yard, this often being the most suitable location for young calves. Moreover, producers who rear calves in summer are likely to retain them as weanlings during the winter thereby providing conditions suitable for the production of contaminated slurry and for the recycling of parasites.

Clearly, precautions are needed in the use of slurry as a manure so as to avoid an extra input of parasitic contamination to the animal environment. Failure to observe such precautions, particularly in relation to SFS produced by young cattle, could result in impairment of animal health and of production, as demonstrated in the present experiments. The findings in relation to production are in conformity with those of Nansen and Jorgensen in Denmark, cited by Persson (1978), who showed that calves grazing pasture fertilised in winter with liquid manure gained less weight than calves grazing on 'clean' pasture. These effects should be emphasised because production losses of this kind may go unnoticed since they will not necessarily be accompanied by overt disease symptoms.

Recommendations designed to minimise the hazard to animal hygiene posed by fertilising land with slurry were included in the proceedings of the CEC Workshop, Dublin (1978). If no practical alternative is available to spreading supposedly contaminated slurry on pasture, minimum storage times should be observed. Persson (1974b) showed that in Sweden the eggs and larvae of *O. ostertagi* and *C. oncophora* were destroyed in liquid manure after 1 - 2 months in summer (May to August) and after 3 - 5 months in autumn/winter. Under Irish conditions meaningful numbers of larvae could still be cultured from slurry in late May. Therefore, assuming the cattle leave the house in April, their excrement should preferably be stored for a further two months, that is, May and June, agreeing with results of the present investigations showing that few larvae, if any, could be cultured from slurry samples collected in June.

Grazing management precautions should also be taken: calves should not graze swards fertilised with slurry in winter or spring especially if young cattle have contributed to the slurry. Experimental results reported here indicate that infection would not be increased if grazing were delayed until after cutting the sward for silage. However, a definite recommendation cannot be based on this finding until further investigations have been made. Chemotherapy for cattle at the onset of winter housing with a broad-spectrum anthelmintic should be carried out as it will considerably decrease the number of eggs building up in the slurry.

These recommendations are only made in relation to trichostrongyles in cattle slurry, because insufficient knowledge is available at present about the survival of other parasites in slurry to enable recommendations to be made concerning storage times and grazing management. These other parasites include: *Fasciola hepatica, Ascaris suum, Dictyocaulus viviparus* and *Coccidia*, all of which are highly important economically.

ACKNOWLEDGMENTS

The authors wish to thank Mr. J. Sherington for statistical analysis; Messrs. M. O'Sullivan, A. Baldwin and J. Farrell for technical assistance; and Mr. T. Kendrick for the graphs.

REFERENCES

Anon., 1978. Proc. Workshop - 'Animal and human health hazards associated with the utilisation of animal effluents', Dublin, Ireland. Editor, W.R. Kelly. CEC, p. 300.

Bürger, H-J. and Stoye, M., 1978. Proc. Workshop - 'Animal and human health hazards associated with the utilisation of animal effluents', Dublin, Ireland. Editor, W.R. Kelly. CEC, p. 24.

Downey, N.E., 1973. Vet. Rec. 93, 505.

Downey, N.E. and Moore, J.F., 1977. Vet. Rec. 101, 478.

Downey, N.E., Moore, J.F. and Bradley, J., 1977. Proc. Seminar - 'Utilisation of manure by land spreading', Modena, Italy. Editor, J.H. Voorburg. CEC, p. 363.

Leaver, J.D., 1970. J. agric. Sci. Camb., 75, 265.

Michel, J.F., 1967. Proc. 4th Int. Meet. Wld. Soc. Buriatrics, Zurich.

Michel, J.F., 1969. J. Helminth., 43, 111.

Michel, J.F. and Lancaster, M.B., 1970. J. Helminth., 44, 107.

Moore, J.F., 1978. Ir. J. agric. Res., 17, 255.

Moore, J.F. and Downey, N.E., 1978. Proc. Workshop - 'Animal and human health hazards associated with the utilisation of animal effluents', Dublin, Ireland. Editor, W.R. Kelly. CEC, p. 97.

Oostendorp, D. and Harmsen, H.E., 1968. Neth. J. agric. Sci., 16, 177.

Parfitt, J.W., 1955. Lab. Pract. 4, 15.

Parfitt, J.W., 1958. Lab. Pract., 7, 353.

Persson, L., 1974a. Nord. Vet. Med., 26, 151.

Persson, L., 1974b. Zentbl. Vet. Med., B 21, 311.

Persson, L., 1974c. Zentbl. Vet. Med., B 21, 677.

Persson, L., 1974d. Zentbl. Vet. Med., B 21, 787.

Persson, L., 1978. Proc. Workshop - 'Animal and human health hazards associated with the utilisation of animal effluents', Dublin, Ireland. Editor, W.R. Kelly. CEC, p. 155.

DISCUSSION

H. Vetter *(West Germany)*

You recommend storing slurry for 60 days or more. If this is done in practice, surely two different storage tanks are needed? Is this a typical practice in your country? I would have thought that in most cases the slurry goes into the same tank for the whole winter, it is mixed, and then perhaps some is spread in January and the rest in the spring.

W.R. Kelly *(Ireland)*

You have raised one of the points that caused considerable concern and was discussed at length in our workshop meeting in Dublin. To answer your second question first, the answer is no; farms in Ireland do not have two storage tanks. We hope that when young stock in particular, are turned out from the spring onwards as soon as pasture becomes available, the amount of slurry put into the slurry tank from that source will at least be reduced. This perhaps is more relevant to parasitological problems than to potential bacterial ones. I have not got a convenient answer for you, but I agree that two slurry tanks would be required to meet the recommendations that we made in Dublin. At the moment it is a question of economics.

J.C. Hawkins *(UK)*

Storage is not the only solution to the problem because therapy is possible as well, and some control can be achieved by management of grazing and grasslands.

W.R. Kelly

Yes, I think that is true. The recommendations include the suggestions that, if possible, slurry should be applied to arable land (provided it is not to be used for vegetable crops for immediate consumption) and the amount applied should just be sufficient to meet the nutrient requirements of the growing plants. However, that might introduce some constraints, as far as disposal is concerned.

SUMMARY OF SESSION 5

W.R. Kelly

The recommendations from the Dublin workshop, to some of which I have already made reference earlier, appear relevant to our discussions. These recommended that slurry should be used for arable crops (excluding crops for fresh consumption) wherever possible. Second, slurry should be spread on grassland, that is used for conservation wherever possible. Before spreading slurry on grazing land, it should be stored for a minimum of 60 days and then 30 days should elapse before grazing with adult or non-susceptible animals. Third, utilisation of slurry should be related to plant nutrient requirements.

Until we have further information about the viability and potential hazard of micro-organisms, it is going to be extremely difficult to say whether a delay of 30 days after spreading is adequate or not. This is particularly true of *Leptospira* and there are other organisms to which attention might be given in this regard.

Some other aspects concern toxic elements. The question of copper was again raised and this element is still causing considerable concern, perhaps more in some countries than others. However, as has already been stated, there are large areas in Ireland where copper deficiency exists, so perhaps that country is less concerned with potential copper toxicity, except in limited areas.

Cadmium is causing concern but as far as I am aware no one has yet suggested explicitly that cadmium originating from slurry has been a disease hazard or even caused toxic effects in animals.

Arsenic and fluorine do not cause concern, and of the nutrient elements, the one that might give rise to problems is potassium, in relation to hypomagnesaemic tetany in cattle.

GENERAL DISCUSSION

During the discussion following the Chairman's summary and conclusions, the following points were emphasised:

1. The time of storage, particularly in summer, is important in decreasing the number of viable pathogens in slurry.
2. The maximum possible time should elapse between applying slurry and either cutting or grazing the herbage. Cutting followed by grazing the aftermath is the preferred sequence.

SESSION VI

REVIEW OF THE 'EFFLUENTS FROM INTENSIVE LIVESTOCK' PROGRAMME

Chairman: G. Wansink

1975 - 1978: RESEARCH ON EFFLUENTS FROM INTENSIVE LIVESTOCK IN THE EUROPEAN COMMUNITIES*

P. L'Hermite and J. Dehandtschutter
Commission of the European Communities,
Rue de la Loi - 200, 1040 Brussels,
Belgium.

1. BACKGROUND

In traditional farming, animal manures were used as the major source of nutrients for crop production and the two enterprises were closely integrated and mutually dependent. Animal manures were applied to land to grow crops which in turn were used as animal feed. Animals were housed in sheds, the floors of which were littered with straw which served as a bedding material in addition to absorbing urine. This mixture of straw and animal manure was allowed to accumulate and was spread on land in the spring and autumn. Under such circumstances, the farmer appreciated the manurial value of the manure and neighbours raised few objections to malodours.

The advent of chemical fertilisers on a large scale resulted in a situation in which crop and animal production were no longer interdependent. During the sixties, chemical fertilisers became the major source of plant nutrients with a consequent decrease in the demand for animal manures for their fertilising value. There was a marked increase in both meat consumption and livestock production and the latter was achieved against a background of a continuously declining agricultural labour force.

* Conclusions and Recommendations issued from the co-ordinated activities within the CEC 1975 - 1978 Research Programme on Effluents from Intensive Livestocks.

Council decision of 22 July 1975 No.75/460/EEC - Published in the Official Journal of the Commission of the European Communities, 30 July 1975, No.L.199.

All these publications of meetings were sponsored by the Commission of the European Communities, Directorate General for Agriculture, Co-ordination of Agricultural Research.

Major changes have taken place in livestock production techniques in recent years. One of the most important developments has been in the field of animal nutrition. Animals are now fed to gain maximum liveweight gain consistent with market requirements for carcass quality. Dietary formulation is based on scientific principles and feed compounders draw on a large variety of ingredients to supply optimum feed rations for different animals. It is not uncommon for feed to be transported long distances to animal production units, the location of which were chosen with emphasis on market outlets. This trend is particularly applicable to pigs, which are often located on very limited land areas. In addition, the fact that these enterprises can be carried out with purchased feed rations, has enabled farmers, traditionally located in areas of poor soil types in which grassland and tillage farming are difficult, to develop 'farmyard' type enterprises and recently on other soils. In addition, such farmyard enterprises are used to generate additional income on intensive grassland or tillage farms often of limited area.

Concurrent with the trend to stream agricultural enterprises into two distinct routes, viz, crop and livestock production, major changes in the housing of livestock were developed. Confined feeding of pigs is now standard practice and modern housing methods with the possible exception of broiler production require the use of no bedding material whatever. Present day housing techniques require animal manures to be handled in liquid rather than in solid form as heretofore. Having regard to the large volume of animal manure produced, its management and disposal require careful planning, to ensure there are no environmental problems.

The problem of animal manures is accentuated by the fact that most western European countries are densely populated. The population is in general engaged in non-agricultural employment and dislikes 'farmyard smells'.

Moreover, there is an increasing interest in open air activities, an expansion of urban centres into rural areas, and an increasing concern on the part of the general public for the quality of the environment.

The biological data when compared with that of municipal sewage indicates the enormous water pollution potential of animal manures.

The nutrient content of animal manures is considerable and, with a trend towards standardising feeding rations, there is a predictable nutrient content which, if utilised, could result in significant savings in chemical fertiliser costs to agriculture.

Since the energy crisis, the prices of fertilisers increased and having regard to the likelihood of a continuing increase in the cost of chemical fertilisers, it is considered that the utilisation of animal manures in crop production will become highly significant in terms of the economics of both livestock and crop production.

2. TERMS OF REFERENCE OF THE RESEARCH PROGRAMME

The research programme which resulted from these considerations was based on the following priority items:

a) Study of influence of physical, chemical and micro-biological characteristics of manure, in particular the amounts of fertilising elements it contains, organic matter and polluting agents on mainly:

- yield and quality of crop
- soil characteristics
- water quality

as a function of
- soil type
- amount of manure
- time of application,

in order to determine the maximum amounts of manure to be spread for optimum crop yields and quality, while creating minimum harmful effects on environment for any type of soil, climate, etc....

The results of these studies should contribute to the data input for the system analysis which will lead to improve the:

b) Establishment of a mathematical model for the whole system with all essential factors such as manure characteristics, soil type, different climatic factors and crops in the input. One of the aims of the model is the prediction of ecological effects, economic aspects, etc.... in order to enable the management of livestock production to be orderly in the context of planning and regional politics.

c) It was agreed that research on production and storage of pig manure would also be a first priority. The control of bad odour emitted directly from the farm as well as during manure spreading is a problem of eminent importance. The most efficient and economic measures to prevent odour during spreading can be taken during storage.

In addition to odour control, the research activities proposed in the following might contribute to projects aimed at a removal of pathogens and at a reduction of BOD and nitrogen content. Taking into account the different ways of implementation of the proposed action, the research project was subdivided into four actions:

- odour characterisation
- aeration in the storage facilities
- odour and pathogen control by chemical treatment in the storage facilities

- control of odour emission from the animal house.

3. RESEARCH ON ODOUR CHARACTERISATION AND CONTROL

An up-dating of the research done in the field of odours was made at the 10th - 13th May 1976 seminar on odour characterisation and odour control organised in Gent (Belgium) by Dr. A. Maton. It stressed the need to establish a close co-ordination between the research on odour characterisation and odour control and starting discussions with the idea of finding and standardising a simple technique for establishing an 'alarm test' for odour pollution based on the level of one or several specific components of the malodours.

From the resulting discussion, emphasis came on the need for relating each component from these malodours with the other components, in order to establish a kind of percentage scale between them, so that knowing the level of one component it would be possible to get an idea of the total composition of the malodour. Such results from chemical analysis, for example, percentages of fatty acids, nitrogen compounds, phenolic components, would lead to a comparison with the results of the panel tests in order to define between both methods a unique common parameter, odour intensity, and also lead to the study of a 'detector'. The first main problem is to exchange the methodology in order to co-ordinate the work within this field. This working group met for the first time at Wageningen* and the results of the discussion pointed out that the aim of the EEC was too ambitious and that the simplest technique available was too complicated to improve it directly on the spot. So this group should try to find an 'odour nuisance level' also as a method for identifying the level of malodours.

* Report of meeting not published

4. RESEARCH ON LANDSPREADING

The second seminar held in Modena under the chairmanship of Ir. J.H. Voorburg (20 - 24th September 1976) centred on the subject of landspreading of manure and modelling in order to assess the current practice in the Member States and to determine the amount of knowledge.

This subject was subdivided into the following topics:

- Crop production
- Manure and pollution of surface water
- Manure and pollution of soil and ground water

for which the conclusions were as follows:

1. More attention should be paid to the long term effect and to the impact on the environment of excessive amounts of manure. Therefore, balances should be produced of input/output and composition of the main nutrients and other relevant minerals.

2. To understand the results of experiments, an exact description of the manure applied is necessary. Effort should be made to develop a standardised formula to describe the chemical and physical properties of manure.

3. Research on manure application on grassland should take account of the influence of manure on the sward and on animal behaviour.

4. In regions and on farms with a surplus of manure, the amount of dilution water should be minimised in order to reduce transport costs.

5. More information is needed about the role of run-off in the pollution of surface waters.

6. Attention should be paid to the amount of zinc in concentrates and manure and its toxicity to animals and crops.

7. It is necessary for research to produce information to allow prediction of the long term effects of excessive amounts of manure, therefore, we should try to understand the mechanism of the composition and transport of minerals in the soil.

For the themes on:

- regional aspects
- veterinary aspects of landspreading of manure,

the conclusions can be read as:

1. Various systems of landspreading of slurry should be investigated with respect to their potential for spreading known pathogenic micro-organisms which might cause diseases in animals and man (e.g. as aerosols).

2. Investigations on persistence of pathogens on herbage and soil, in top-soil and throughout the soil profile and into the groundwater.

3. Influence of various slurry treatment methods (physical-chemical-biological) on the persistence of pathogens.

4. Determinations of the virulence of pathogens after varying periods in stored slurry and following its application to pasture and soil.

5. Determination of survival periods of tubercle bacilli, non-coated viruses and parasites in poultry manure and deep litter, and methods for their inactivation.

6. Investigations on the occurrence and persistence of *Ascaris suum* in pig slurry in relation to its transmission via landspreading to ruminants and man.

7. Investigations on sheep manure as a possible means of recycling helminths and other pathogens.

For the items on:

- the effects of separating slurry on its storage handling and spreading on land
- treatment in relation to land application of slurry
- models on landspreading of manure

the conclusions are:

1. That we should give attention to generalising results from experiments by the use of models which may be at the levels of research and of extension.

2. A long term policy for landspreading of animal manures should be related to the constraints of general regional policy, for example, recommendations on quality of surface waters.

3. Treatment and separation offer possibilities for modification of slurries which may be favourable financially in some situations.

An important part of this seminar was the session devoted to methods for sampling and analysis in use by the Institutes in the EEC, which are involved in research in landspreading of manure. A list of parameters produced has a first priority for harmonisation. The final choice of reference methods was delegated to a small working group of analytical experts. This group was also asked to pay special attention to the sampling technique and pre-treatment of the sample.

During the general discussions, suggestions were made for improving the international co-operation of research.

- The establishment of documentation and information facilities on the handling and utilisation of animal manure.
- Small specialised groups should meet annually in order to apply the conclusions and recommendations of this seminar.

5. HARMONISATION OF ANALYTICAL METHODS

This working group in the field of standardisation and harmonisation of reference methods started its activities on the basis of a first proposal made by Professor A. Cottenie et al., which was discussed at a meeting in Brussels in March 1977. After having received comments from the members of the group, a workshop was organised in Ghent, from 6 to 9 June, 1978.

This working group produced a document containing the methods finally adopted for analyses of manure, soils, plants and water. In presenting this document, the working group did not intend to standardise analytical methods and did not pretend to have selected the most adequate techniques overall. It was stated that the agreement on so-called reference methods aimed principally at improving the calibration of personal methods and at comparison between results obtained in different research centres.

This group also produced a table for the standardisation of the extension of the results in order to avoid any possible confusion and to prevent the necessity of permanent recalculation. It was agreed that this group should meet once a year in order to up-date the booklet of reference methods.

6. HEALTH HAZARDS ASSOCIATED WITH THE UTILISATION OF ANIMAL EFFLUENTS

In November 1977, a workshop, under the chairmanship of Professor W.R. Kelly, was held in Dublin on the subject of 'The Animal and Human Health Hazards Associated with the Utilisation of Animal Effluents'.

This topic was sub-divided into the following sections:

- identifying the priority contaminants
- sampling procedures
- analytical methodology
- animal effluents management and the survival and persistence of the priority contaminants.

Within the scope of these sections, consideration was confined to the main groups causing concern and included within the subject title of the workshop as follows:

- microbiological agents
- parasitological agents
- toxicological agents.

The responsibility placed upon the workshop was to evaluate the existing situation and the available knowledge within the subject of its merit.

The conclusions and the recommendations can be read as follows:

- Microbiological Agents:

I. Possible contaminants of animal wastes

The following list is considered to include those organisms which may be of particular concern for animal and/or human health when present in animal effluents and wastes:

Salmonella spp*

Leptospira spp

Treponema hyodysenteriae

Erysipelothrix insidiosa

Mycobacterium spp*, in particular *M. tuberculosis, M. bovis, M. avium* complex, *M. paratuberculosis* and the atypical Mycobacteria

Brucella spp*

Bacillus anthracis *

Escherichia coli including enteropathogenic strains and normal gut *E. coli* multiply resistant to antibiotics

Rickettsia spp

Chlamydia spp

Viruses

eg. Foot and mouth disease*

Swine fever*

Swine vesicular disease*

Swine influenza

Transmissible gastroenteritis (TGE) of pigs

Rotaviruses

Teschen disease*

Aujeszky disease*

Blue tongue virus*

Newcastle disease*

* Outbreaks involving these organisms are the subject of specific regulations which include control on the movement and utilisation of animal effluents/waste from infected farms/plants.

The importance of farm animal wastes as a source of diseases caused by these organisms should be evaluated by collaborative research.

2. Isolation methodology

1. *Salmonella* spp

The reference method for the isolation of *Salmonella* spp which was developed under the auspices of the EEC and adopted by ISO for meat and meat products (No. 3565) is recommended. Considering that little data is available on the application of this method to isolates from farm animal wastes, it is recommended that comparative studies be initiated.

2. *Escherichia coli*

The isolation of *E. coli* is primarily aimed at determining the risk of spreading R plasmids by farm animal wastes, including poultry waste (see Eur 5672e, Modena Meeting, Italy - 1976).

3. Standard methods for the isolation of e.g. *B. anthracis*, *Brucella* spp and *Mycobacterium* spp are available (WHO, ISO)*. These have not been evaluated for use in isolations from farm animal wastes but may be useful should the necessity to isolate arise. The method of isolation for *Leptospira* proposed by Diesch et al. (1975) may also be useful.

- Parasitological Agents:

I. Sampling

It is desirable to categorise animal effluents (i.e. slatted floor or dungstead cattle slurry, pig slurry, sheep slurry) and to formulate repeatable sampling procedures

* International Standard ISO - 3565
Meat and Meat Products - detection of salmonellae (Reference Method)
International Organisation for Standardisation
2 rue de Varembe, 1121 Geneva 20, Switzerland.

appropriate to each type and to the organisms that require to be studied.

II. Identifying contaminants

Comprehensive lists of the parasitological organisms that may occur in animal effluents are contained in the paper by Dr. H.J. Burger. The organisms that should be investigated are listed in order of research priority as follows:

a) For cattle

1. Strongyle nematodes
2. *Fasciola hepatica*
3. *Ascaris suum*
4. *Cysticercus bovis/Taenia saginata*
Sporozoa (e.g. *Sarcosporidia*)

b) For pigs

1. *Ascaris suum*
2. Strongyle nematodes
3. Sporozoa

c) For sheep

1. Strongyle nematodes
2. *Fasciola hepatica*
3. Sporozoa

d) For man

1. *Ascaris* spp
2. Sporozoa

III. Minimising health hazards

1. Whenever feasible, slurry should be applied to arable land.

2. Storage

A reduction of viable trichostrongylid eggs occurs in slurry during storage. Such eggs disappear after two months storage in summer and after three to four months storage in winter. Ideally, therefore, spring application of slurry to grassland is not advisable, it being preferable to delay application until July at the earliest. If this is not practicable, other recommendations, e.g. anthelmintic therapy (see 4 below), should be adopted.

In respect of *F. hepatica* in cattle* and sheep slurry and *A. suum* in pig slurry, further investigations into their persistence are needed in order to give storage recommendations.

3. Grazing management

Pasture to which 'slatted-floor' cattle slurry has been applied in spring should not be grazed by young susceptible calves. Investigations into the persistence of viable tricho-strongylid larvae on aftermath are required. Pending the results of these enquiries, aftermath following silage cutting of swards that received slurry in spring should not be grazed by other than mature cattle or sheep. With regard to *F. hepatica* in cattle and sheep slurry and *A. suum* in pig slurry, grazing recommendations await further studies.

4. Anthelmintic therapy

Cattle, particularly yearlings, should be treated with anthelmintic three weeks after being housed. Preferably, the anthelmintic used should be active against inhibited fourth stage trichostrongylid larvae.

5. Where possible, slurry should be injected into the soil rather than sprayed on the surface.

* *T. saginata* may occur in cattle slurry; information on its persistence is lacking.

6. Lastly, slurry may be treated (e.g. by aeration, chemical, etc.) in order to destroy pathogens.

- Toxicological Agents:

Copper is added to pig feed and thus leads to an imbalance between copper and N, P, K, in slurry, so that excess copper is supplied with normal dressings of N, P, K.

The group were concerned about the following dangers arising from the application of high levels of copper in pig slurries to soils or pastures:

a) The immediate hazard to sheep grazing a treated pasture

b) The long term build-up of copper levels which may endanger crop growth and decrease yield.

Recommendations

1. Sheep should not be grazed within the same season on pasture treated with pig slurry.

2. The EEC directive of 125 ppm added copper in pig feed should not be exceeded. Consideration should be given to a further reduction so that the copper-treated feed should only be fed for an eight week period.

3. Research should be undertaken to establish soil values in order to determine safe levels for plant growth and animal health. Soils could then be monitored and slurry applications adjusted over time.

4. Sheep livers should be monitored and values above 400 ppm should be investigated and prevented where possible. The use of antagonistic agents, e.g. molybdenum, should be considered.

5. Phosphorus dietary requirements of pigs should be re-examined with the objective of reducing the phosphorus content of pig feeds and therefore of pig slurries.

6. Maximum safe soil values for phosphorus and potassium should be determined.

7. The directive on feed additives and on undesirable substances should be designed to prevent the transmission of toxic substances, e.g. selenium, arsenic, cadmium, to slurries.

Management systems of slurries should be designed to conserve and recycle nutrients rather than disposing of effluents by dumping.

Uses of animal effluents other than by landspreading and crop nutrition, e.g. re-feeding, should be considered.

At the conclusion of this meeting, minimum guidelines were proposed by the expert groups.

1. Slurry should be utilised on tillage crops (excluding crops for fresh consumption), wherever possible.
2. If slurry is spread on grassland, then:
   a) Use on pasture for conservation, wherever possible
   b) If on grazing land -
      (i) storage of all slurry for a minimum of 60 days before spreading
      (ii) delay of 30 days before grazing
      (iii) graze with adult or non-susceptible animals.
3. Utilisation of slurry should be related to plant nutrient requirements.

7. REVIEW OF ENGINEERING PROBLEMS

A seminar was held in September 1978 in St. Catherine's College, Cambridge, under the Chairmanship of Mr. J.C. Hawkins, on the subject of 'Engineering Problems with Effluents from Livestock'. This seminar provided a good picture of recent and current work on the engineering problems dealing with animal manures. The objectives were as follows:

a) Economic and trouble-free storage of slurry,

b) Mechanisation and automation of the application of manure to land,

c) The efficient use of manures as fertilisers,

d) The control of pollution of air, soil and water,

e) The minimising of health hazards,

f) The re-use of manure as a feed,

g) The production of energy from manure,

h) The disposal of manure when there is a surplus.

Engineering research and development had contributed towards achieving these objectives in the following ways:

a) Storage

Improvements in the design of slurry stores and mixing equipment, together with the use of solid/liquid separation had made the emptying of existing types of store simpler and more efficient. Integrating storage into the building had reduced dilution and helped mixing.

b) Application

Accurate land application had been achieved by work on tanker and applicator design, and pipeline blockages eliminated by separation. Conversion of slurry to a solid, with or without adding straw had provided an alternative solution to application problems.

c) Efficient use

Buildings and systems to reduce dilution were important. Efficient mixing or separation was also necessary to provide consistent materials, which could be sampled with accuracy, and spread on land so that pre-determined quantities of plant nutrients were applied.

d) Pollution control

Aerobic and anaerobic treatment systems had been developed to control odour and give less polluting products for land application.

e) Health

The Dublin workshop has shown the importance of storage for pathogen control and that survival of the parasitological agents was the factor controlling the minimum safe storage time.

f) Re-feeding

Studies indicated that re-feeding after treatment might be technically and economically feasible: the optimum treatment had yet to be determined.

g) Energy

Experimental treatment plants, producing energy aerobically as low grade heat or anaerobically as methane gas, were in operation.

h) Disposal

Treatment systems producing solids and acceptable effluents or little or no liquid effluent had been developed.

Recommendations for the future

The results of research by microbiologists, chemists, agronomists, veterinarians and others working on livestock effluents can rarely reach the farmer except through developments in engineering. Therefore, future progress depends on close co-operation between these workers and engineers and suitable working groups should be formed where necessary.

Important areas for increased research effort on manures (including run-off from polluted areas) are given below, not necessarily in order of priority.

a) Pathogen control

Joint work on storage times and conditions is required from parasitologists, microbiologists and engineers.

b) Separation

More work is needed to find cheaper and more reliable methods of separating slurry on a farm scale, and the effects of separation on odour.

c) Aeration

Research should be increased to establish optimum on-farm aeration techniques to control odour in slurry at minimum cost and energy input.

d) Complete systems

Effort should be directed towards establishing efficient, safe and economic on-farm systems for dealing with slurry as well as towards the development of the individual items of equipment used in such systems. An important aspect of this would be comparisons and cost benefit analysis of aerobic and anaerobic systems.

e) Precipitation

Sedimentation of solids and sludge handling are facilitated by efficient flocculation. Increased effort is required to find cheaper techniques and flocculants suitable for farm use.

f) Odour measurement

Progress in engineering research requires a method of measuring odour intensity. Current EEC research on this should be encouraged.

g) Co-ordination

Closer contact should be established between workers on composting and on aerobic and anaerobic treatment.

h) Other areas of importance

Research effort should continue at the present level on anaerobic digestion, re-feeding, complete treatment systems for slurry disposal and the modelling of land application and the resulting changes in the soil.

8. MODELLING

A workshop was held in Chimay, in October 1978, under the Chairmanship of Professor H. Laudelout, (the Proceedings were edited by Dr. J.K.R. Gasser)on the subject of 'Modelling Nitrogen from Farm Wastes'.

This meeting proved that good progress was made in developing models that can explain the behaviour of nitrogen in the soil.

It was recommended that the work on models should be continued and distinguished between scientific and simplified models.

The scientific models are important to understand what is happening and to predict what will happen under various conditions. This work can be done by a small group of specialists and international co-operation (or co-ordination) is very useful. It is necessary that specialists who have to use the results of this model have good contact with the 'model-makers' in order:

- to prevent modelling becoming a goal in itself, e.g. there was some discussion on the question: 'How to model the effect of a crack in the soil';
- to select the factors which have to be taken into account e.g. are micro-elements important in a N-model for manure?;
- to develop simplified models.

The simplified model can be used in practice, e.g. to give advice about the amounts of manure to be spread or to predict the leaching of minerals in a given situation. They are only valid for the range of conditions used. They can be understood by extension workers and farmers. Usually the simplification does not result in inferior answers, because a part of the data to be put into the model has to be estimated, e.g.

- the expected rainfall
- the expected soil temperature
- the composition of the manure
- the amount that will be spread.

9. IMPACT OF SURFACE RUN-OFF ON THE UTILISATION OF EFFLUENTS FROM LIVESTOCK

A meeting to discuss this problem* was held in Brussels on 7 - 8 June 1979 whch was considered to have been very useful as:

- an inventory of the work done in the different countries of the EEC
- an opportunity to exchange experiences, ideas and techniques
- a start for more contacts and co-operation in research on run-off.

Definition of run-off

Horizontal transport of water over the land surface and through the topsoil to the surface water.

Conclusions

1. Run-off is an important phenomenon occurring in many places in the EEC depending on:

a) Slope of the soil (steepness and length), and the direction of tilling the soil.

b) Soil properties:
- infiltration rate in relation with structure, texture and moisture content.
- storage capacity for water
- groundwater table.

c) Landscape: hedges, field roads.

d) Cropping situation: arable, grassland or woods

e) The dry matter content of the slurry

* Report of meeting not published

f) Climatic and weather conditions:

- frost and snow
- rainfall (amount, intensity, distribution over the year).

2. From the agricultural point of view, run-off has many consequences varying from erosion to loss of nutrients. This meeting paid special attention to the aspects mentioned in conclusions 3 and 4.

3. In guidelines about the landspreading of organic manure or wastes, attention should be paid to:

- amount of manure, wastes and fertilisers
- time of spreading in relation to:
  - a) weather conditions
  - b) farm management
  - c) storage capacity
- way of spreading (e.g. injection).

4. A small working group should be formed to:

a) produce the model mentioned in point 2

b) make the guidelines from point 3

c) complete, as far as necessary, the standardisation of analytical methods

d) make a list of research going on and of the gaps in our knowledge.

10. GENERAL CONCLUSION

Following the energy crisis and the permanent increase in the cost of chemical fertilisers, we can always consider that the utilisation of animal manures in crop production will become more and more significant in economic terms; also, other refuse and waste from domestic and industrial activities, specifically agro-food industrial activities are expected to be used resulting in increasing risks for agriculture and the environment.

However, this increase will be influenced by the increasing demand for energy, more specifically biogas from manures, slurries, sewage sludges and town refuses, which constitutes a new area of research. Another point to be raised is the importance of the pressure of the national legislative bodies responsible for environmental problems. Both pressures are already influencing priorities of research and if, in New Zealand, it is the drastic need for biogas as replacement for fuel-oil which gives priority to research on the use of wastes and effluents; in Canada, it is a reinforcement of legislation on environmental protection which orientates the problems of treatment of sludge and slurry towards improving the anaerobic digestion process.

It is therefore evident that the terms of reference of the existing programme must be reviewed for the future, taking into consideration both the gaps and failures which appeared during the last three years and new objectives arising from pressures which technically and legally will come more from environmental legislative pressures than from biogas demand for energy.

In addition to the scientific terms of reference of the programme, we must also think about the efficiency of the work done, more specifically the efficiency of the co-ordination activities where some of you took a great part freely and without any contract with the Commission. There is obviously a need for improving the management of the co-ordination activities and for reinforcing the contacts between scientists and the formation of groups involved on specific subjects. Such an approach needs to consider the relative balance between the money allocated for contracts and the money allocated for the continuation and improvement of co-ordinated activities, the latter having proved to be more profitable and efficient in all the research programmes undertaken either under the auspices of the Coordination of Agricultural Research, or under the auspices of other parts of the Commission. This would imply a stricter selection of projects proposed in the future and a decrease in the number of those which would be selected for very specific priority items. We hope that this will help to improve development and enlargement of the co-ordination activities which have proved to be the most beneficial effect on the results obtained within the 1975 - 1979 research programme on Effluents from Intensive Livestocks. Although many problems still have to be solved with regard to the urgency of the problems on practical farming, more attention should be paid to the transfer of knowledge through the proper channels.

REFERENCES

EUR 5746, 1976. Odour characterisation and odour control: Gent, May 1976, edited by Dr. A. Maton.

EUR 5672, 1976. Utilisation of manure by landspreading: Modena, September 1976, edited by Ir. J.H. Voorburg.

EUR 6368, 1978. Reference methods: Ghent, June 1978, edited by Professor A. Cottenie.

EUR 6009, 1977. Animal and Human Health Hazards Associated with the Utilisation of Animal Effluents: Dublin, November 1977, edited by Professor W.R. Kelly.

EUR 6249, 1978. Engineering Problems with Effluents from Livestock: Cambridge, September 1978, edited by Mr. J.C. Hawkins.

EUR 6361, 1978. Modelling Nitrogen from Farm Wastes: Chimay, October 1978, edited by Dr. J.K.R. Gasser.

DISCUSSION

G. Wansink *(The Netherlands)*

Ladies and gentlemen, do you have any remarks on the report prepared by Dr. L'Hermite and Mr. Dehandtschutter?

Following the discussion of a number of points of wording and emphasis, which were subsequently included in an amended paper by Dr. L'Hermite and Mr. Dehandtschutter, the following more general points were raised:

H. Tunney *(Ireland)*

Much of the work that has been done has been summarised in the report by Dr. L'Hermite and Mr Dehandtschutter. As a point of general information, I would like to know, from your experience, how this programme compares with other programmes within the Commission?

P. L'Hermite *(Commission of the European Communities)*

Comparison of this programme with others is difficult. It has been a good and efficient programme because starting with less material and fewer existing studies than other programmes, much has been achieved. For example, with beef production, we and the experts knew exactly what to do. Also for the veterinary programmes the subject was more specific and therefore it was easier to obtain results than in the programme on effluents from livestock. From the start, results were more difficult to obtain and this aspect makes it a good research area within the action for co-ordination of our research, that is without the financing of projects or of institutes, compared with concerted action. Nevertheless I do not think that the Commission has stressed enough the co-ordination aspects and we should have created more subject groups for specific workshops which would have created better efficiency in relation to the other programmes. At present there is no difference in efficiency from other programmes.

J.K. Grundey *(UK)*

I noticed you were somewhat concerned about the need to improve the management of the co-ordination, which was a very good statement of fact. Perhaps I, as a visitor, had greater expectations than you, and one wonders if others are satisfied with the co-ordination of the programme outside the meeting. What happens between meetings? There is evidence that topics have been pursued without much apparent co-ordination, so that this is an area for a great deal of improvement in the future. Let us hope that the hallmark of the future will be greater co-ordination than in the past. Mr. Lecomte (Belgium) raised the question of disseminating the information. Some years ago, in England, we had difficulties in obtaining credit which particularly affected agriculture. One point made was that, if farmers would use all the information that was available, they would not have a financial problem at all. Therefore I would support the idea of spending more effort on the spreading of ideas and the information obtained from these programmes.

CLOSING REMARKS

G. Wansink

Ladies and gentlemen, that concludes the agenda except for 'Any Other Business'. The first item is our next meeting and I find it very difficult to say when we will be able to hold a similar meeting. This year we have suffered from procedural and administrative difficulties, because the present meeting was originally planned to be held in the month of May, as being a suitable time of year to discuss the previous results in relation to the development of a programme for 1980 and later. We had organised 1979 as an interim year, but we are already so close to 1980 that we are practically under the same pressure as we were at the end of 1978. However, these decisions were made at a high level, so we must accept them. Therefore, I cannot give you any definite dates. However, we would wish to consider a date in May in 1980, but I cannot be sure that is possible.

At the end of this meeting my pleasant duty as Chairman is to thank all who helped to make it so successful. I would like to start with the ladies, Mrs. Molly Robins and Miss Gillian Cookes, of Janssen Services, who probably worked the hardest during the meeting, and their work in preparing the Proceedings for publication means that the image of the Committee is well presented and our knowledge is transferred.

The second group that I must mention is that of Professor Vetter. He was not only a 'landspreading maniac' in the technical sessions but we have also discovered that he has directed his energy towards organisation, although ably assisted by Mr. Steffens and Mr. Kowalewsky. We thank them both.

Third on my list is the Commission, who made the meeting possible by paying our expenses. That is very important and is part of the co-ordinating activities which are a start for real co-ordination between workers; you must be able to get together as people. We helped the Commission by using English as the working language and I think we had no problem so long as the English did not speak!

The wisest decision I took was ,last year when I delegated the Chairmanship of the sessions to some people who knew the areas of work. That has been a very fruitful approach. You have been guided in the preparation of the agenda and during the meeting by the five presidents; I am very grateful for their contributions.

Above all, I must mention the participants. Co-ordination is a matter of giving and taking. Everyone has done his or her best in reporting the work that might be helpful to this programme. On the other hand we have all had the opportunity to improve our own work, orientation and safety to the environment by the contributions. Perhaps we were a little overloaded, and at the next meeting, we should consider the opportunity of a visit to provide practical information relevant to the papers being presented.

To avoid making my closing remarks too lengthy I will now close the meeting. Thank you all for your participation and thank you especially for the friendly atmosphere on this occasion: we look forward to the next time.

LIST OF PARTICIPANTS

Dr. A. AUMAITRE	INRA, Station de Recherches sur L'Elévage des Porcs, CNRZ, 78350 Jouy-en-Josas, France.
Dr. C. BESNARD	Centre technique du Genie Rural des Eaux et Forêts, Ministère de l'Agriculture, Division Qualité des Eaux, Pêche et Pisciculture, 14 Avenue de Saint Mandé, 75012 Paris, France.
Mr. J.C. BROGAN	The Agricultural Institute, Johnstown Castle Research Centre, Wexford, Co. Wexford, Ireland.
Dr. G. CATROUX	Centre de Recherches de Dijon, BV-1540, 21034 Dijon-Cedex, France.
Dr. D.P. COLLINS	The Agricultural Institute, Grange, Dunsany, Co. Meath, Ireland.
Dr. J.P. CURRY	Department of Agricultural Biology, University College, Dublin, Ireland.
Dr. A. DAM KOFOED	Forsogsstationen ved Askov, 6600 Vejen, Denmark.
Dr. R. DE BORGER	Institut de Recherches chimiques, 5, Museumlaan, 1980 Tervuren, Belgium.
Dr. F.A.M. DE HAAN	Landbouwscheikunde, De Dreyen 3, Wageningen, The Netherlands.

Mr. J. DEHANDTSCHUTTER | Commission of the European Communities, DG-VI, 200 rue de la Loi, 1049 Brussels, Belgium.

Dr. A. DORLING | Warren Spring Laboratory, PO Box 20, Stevenage, Hertfordshire, SG1 2BX, United Kingdom.

Dr. N.E. DOWNEY | The Agricultural Institute, Dunsinea Research Centre, Castleknock, Co. Dublin, Ireland.

Dr. G. DROEVEN | Station de Chimie de Physique agricoles, 115 Chaussée de Havre, 5800 Gembloux, Belgium.

Dr. C. DUTHION | Centre de Recherches de Dijon, BV-1540, 21034 Dijon-Cedex, France.

Dr. J.K.R. GASSER | Agricultural Research Council, 160 Great Portland Street, London W1N 6DT, United Kingdom.

Mr. J.K. GRUNDEY | Ministry of Agriculture, Fisheries and Food, Government Offices, Block A, Coley Park, Reading RG1 6DT, United Kingdom.

Dr. J. HARTUNG | Institut für Tierhygiene der Tierärztlichen Hochschule Hannover, Bunteweg 17 P, 3000 Hannover 71, West Germany.

Mr. J.C. HAWKINS | National Institute of Agricultural Engineering, Wrest Park, Silsoe, Bedford MK45 4HS, United Kingdom.

Dr. A. HEDUIT | Centre technique du Genie Rural des Eaux et Forêts, Ministère de l'Agriculture, Division Qualité des Eaux, Pêche et Pisciculture, 14 Avenue de Saint Mandé, 75012 Paris, France.

Dr. P.D. HERLIHY | The Agricultural Institute, 19 Sandymount Avenue, Dublin 4, Ireland.

Prof. Dr. H.G. HILLIGER | Institut für Tierhygiene der Tierärztlichen Hochschule Hannover, Bunteweg 17 P, 3000 Hannover 71, West Germany.

Dr. T. HULD | Statens Byggeforskningsinstitut, PO Box 119, 2970 Horsholm, Denmark.

Ir. A.A. JONGEBREUR | Institute for Agricultural Engineering, Wageningen, The Netherlands.

Professor W.R. KELLY | Department of Veterinary Medicine and Pharmacology, Faculty of Veterinary Medicine, Veterinary College, Ballsbridge, Dublin 4, Ireland.

Dr. H.H. KOWALEWSKY | Landwirtschaftliche Untersuchungs-und Forschungsanstalt, Mars-la-Tour Strasse 4, 2900 Oldenburg, West Germany.

Mr. J. KUYL | Commission of the European Communities, DG VI, 200 rue de la Loi, 1049 Brussels, Belgium.

Professor F. LANZA | Istituto Sperimentale Agronomico, Via C. Ulpiani 5, 70100 Bari, Italy.

Professor H. LAUDELOUT — Departement Science du Sol,
Place Croix du Sud 2,
Universitē de Louvain,
1348 Louvain-la-Neuve,
Belgium.

Dr. L. LECOMTE — Centre de Recherches Agronomiques de Gembloux,
22 Avenue de la Facultē d'Agronomie,
5800 Gembloux,
Belgium.

Dr. Th. M. LEXMOND — Landbouwscheikunde,
De Dreyen 3,
Wageningen,
The Netherlands.

Dr. P. L'HERMITE — Commission of the European Communities,
DG XII,
200 rue de la Loi,
1049 Brussels,
Belgium.

Dr. A. MATON — Ministerie van Landbouw,
Rijksstation voor Landbouwtechniek,
Van Gansberghelaan 115,
9220 Merelbeke,
Belgium.

Dr. D. McGRATH — The Agricultural Institute,
Johnstown Castle Research Centre,
Wexford,
Co. Wexford,
Ireland.

Dr. O. NEMMING — Forsogsstationen ved Askov,
6600 Vejen,
Denmark.

Professor J.R. O'CALLAGHAN — Department of Agricultural Engineering,
The University,
Newcastle-upon-Tyne NE1 7RU,
United Kingdom.

Mr. H. OTT — Commission of the European Communities,
DG XII
200 rue de la Loi,
1049 Brussels,
Belgium.

Dr. B.F. PAIN — National Institute for Research in Dairying,
Shinfield,
Reading RG2 9AT,
United Kingdom.

Dr. D.B.R. POOLE	The Agricultural Research Institute, Dunsinea Research Centre, Castleknock, Co. Dublin, Ireland.
Dr. L. RIXHON	Station de Phytotechnie, Centre de Recherches Agronomiques de Gembloux, 22 Avenue de la Faculté d'Agronomie, 5800 Gembloux, Belgium.
Dr. J.L. ROUSTAN	INRA, Station de Recherches sur L'Elévage des Porcs, CNRZ, 78350 Jouy-en-Josas, France.
Dr. W. RÜPRICH	Institut für Agrartechnik, Universität Hohenheim, Postfach 106, 7000 Stuttgart 70, West Germany.
Dr. J. SCHAEFER	CIVO, Central Institute for Nutrition and Food Research TNO, Utrechtseweg 48, Zeist, The Netherlands.
Dr. N. SCHAMP,	Rijksuniversiteit Gent, Coupure Links 533, 9000 Gent, Belgium.
Professor P. SEQUI	Laboratorio del CNR per la Chimica del terreno, Via Corridoni 78, 56100 Pisa, Italy.
Mrs. Marie SHERWOOD	The Agricultural Institute, Johnstown Castle Research Centre, Wexford, Co. Wexford, Ireland.
Dr. K.W. SMILDE	Institute for Soil Fertility, Oosterweg 92, Haren (Gr.), The Netherlands.
Dr. P. SPALLACCI	Istituto Sperimentale Agronomico, Sezione di Modena, Viale Caduti in Guerra 134, Modena, Italy.

Mr. G. STEFFENS	Landwirtschaftliche Untersuchungs-und Forschungsanstalt, Mars-la-Tour Strasse 4, 2900 Oldenburg, West Germany.
Dr. H. TUNNEY	The Agricultural Institute, Johnstown Castle Research Centre, Wexford, Co. Wexford, Ireland.
Dr. F. VAN DE MAELE	Rijksuniversiteit Gent, Rue Coupure 533, 9000 Gent, Belgium.
Professor Dr. H. VETTER	Landwirtschaftliche Untersuchungs-und Forschungsanstalt, Mars-la-Tour Strasse 4, 2900 Oldenburg, West Germany.
Dr. R. VETTER	Institut für Agrartechnik, Universität Hohenheim, Postfach 106, 7000 Stuttgart 70, West Germany.
Ir. J.H. VOORBURG	Rijks Agrarische Afvalwaterdienst, Kemperbergerweg 67, Arnhem, The Netherlands.
Mr. G. WANSINK	NRLO-TNO, Adelheidstraat 84, Postbus 297, 's-Gravenhage, The Netherlands.

RECORDING PERSONNEL

Mrs. Molly ROBINS	Janssen Services, 14 The Quay, Lower Thames Street, London EC3R 6BU, United Kingdom.
Miss Gillian COOKES	Janssen Services, 14 The Quay, Lower Thames Street, London EC3R 6BU, United Kingdom.